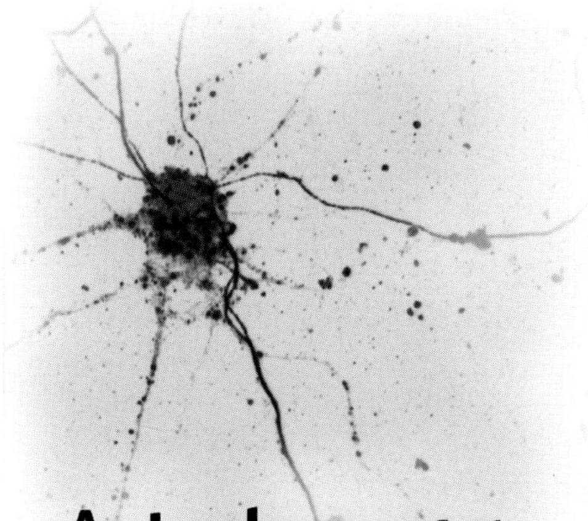

Autophagy of the Nervous System

Cellular Self-Digestion in Neurons and Neurological Diseases

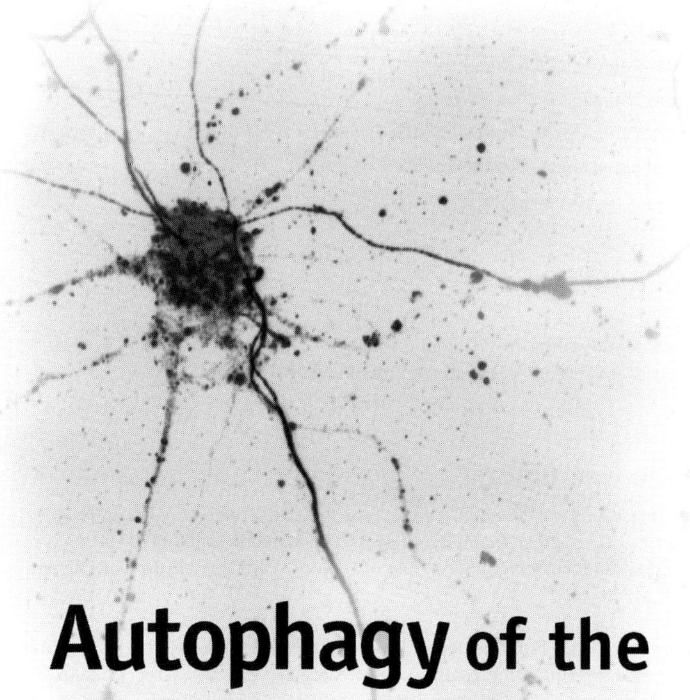

Autophagy of the Nervous System

Cellular Self-Digestion in Neurons and Neurological Diseases

Editors

Zhenyu Yue
Mount Sinai School of Medicine, USA

Charleen T Chu
University of Pittsburgh, USA

World Scientific

NEW JERSEY • LONDON • SINGAPORE • BEIJING • SHANGHAI • HONG KONG • TAIPEI • CHENNAI

Published by

World Scientific Publishing Co. Pte. Ltd.
5 Toh Tuck Link, Singapore 596224
USA office: 27 Warren Street, Suite 401-402, Hackensack, NJ 07601
UK office: 57 Shelton Street, Covent Garden, London WC2H 9HE

British Library Cataloguing-in-Publication Data
A catalogue record for this book is available from the British Library.

ABOUT THE COVER IMAGES:

Front cover: Mouse cortical neuronal culture was fixed and stained with antibodies to visualize autophagy proteins LC3 in green fluorescence and WIPI2 in red fluorescence. The nucleus was stained in blue. The original confocal image is a courtesy of Nicole C. McKnight in the Yue Laboratory at Mount Sinai School of Medicine.

Back cover: A postnatal rat Purkinje neuron subjected to trophic factor withdrawal exhibits an early autophagosome and several autophagolysosomes. Original magnification: 22,000x. Transmission electron micrograph courtesy of Charleen T. Chu, University of Pittsburgh.

AUTOPHAGY OF THE NERVOUS SYSTEM
Cellular Self-Digestion in Neurons and Neurological Diseases

ISBN 978-981-4350-44-0

Typeset by Stallion Press
Email: enquiries@stallionpress.com

Printed by FuIsland Offset Printing (S) Pte Ltd Singapore

CONTENTS

CONTRIBUTORS

Boya, Patricia
Department of Cell Proliferation and Development
Centro de Investigaciones Biológicas, CSIC
Ramiro de Maetzu 9, E-28040 Madrid, Spain
patricia.boya@csic.es

Cecconi, Francesco
Dulbecco Telethon Institute at the Department of Biology
University of Tor Vergata, Rome, 00133, Italy
francesco.cecconi@uniroma2.it

Cherra III, Salvatore J.
Department of Pathology
University of Pittsburgh School of Medicine
Pittsburgh, PA 15261, USA

**Current affiliation:*
Division of Biological Sciences
University of California San Diego, USA
scherra@ucsd.edu

Chu, Charleen T.
Department of Pathology
University of Pittsburgh School of Medicine
200 Lothrop St., Rm. S417 BST
Pittsburgh, PA 15261, USA
ctc4@pitt.edu

Di Bartolomeo, Sabrina
Dulbecco Telethon Institute at the Department of Biology
University of Tor Vergata, Rome, 00133, Italy
sabrina.dibartolomeo@uniroma2.it

Duff, Karen E.
Department of Pathology and Cell Biology
Taub Institute, Columbia University Medical Center
Department of Integrative Neuroscience, New York State Psychiatric Institute
P&S 12–461, 630 W. 168th St., New York, NY 10032, USA
ked2115@columbia.edu

Friedman, Lauren G.
Department of Neurology and Neuroscience
Friedman Brain Institute
Mount Sinai School of Medicine, New York, NY 10029, USA
lauren.friedman@mssm.edu

Gal, Jozsef
Department of Molecular and Cellular Biochemistry
College of Medicine, University of Kentucky
741 South Limestone Street, Lexington, KY 40536, USA
jgal2@email.uky.edu

Green-Thompson, Zeyn W.
Department of Medical Genetics
University of Cambridge
Cambridge Institute for Medical Research
Wellcome Trust/MRC Building
Hills Road, Cambridge CB2 0XY, UK

Hernandez, Daniela
Center for Neurobiology and Behavior
Columbia University, New York, NY 10032, USA
dfhernandezr@gmail.com

Ichimura, Yoshinobu
Protein Metabolism Project
Tokyo Metropolitan Institute of Medical Science
Setagaya-ku, Tokyo 156–8506, Japan
ichimura-ys@igakuken.or.jp

Johansen, Terje
Molecular Cancer Research Group
Institute of Medical Biology
University of Tromsø, 9037 Tromsø, Norway
terje.johansen@uit.no

Koike, Masato
Department of Cell Biology and Neurosciences
Juntendo University Graduate School of Medicine
2–1–1 Hongo, Bunkyo-ku, Tokyo 113–8421, Japan

Koike, Tatsuro
Department of Biosciences and Applied Chemistry
School of Life Science
Hosei University, Tokyo 184–8584, Japan
koike@sci.hokudai.ac.jp

Komatsu, Masaaki
Protein Metabolism Project
Tokyo Metropolitan Institute of Medical Science
Setagaya-ku, Tokyo 156–8506, Japan
komatsu-ms@igakuken.or.jp

Lim, Junghyun
Department of Neurology
Friedman Brain Institute
Mount Sinai School of Medicine, New York, NY 10029, USA
junghyun.lim@mssm.edu

McKnight, Nicole C.
Department of Neurology
Friedman Brain Institute
Mount Sinai School of Medicine, New York, NY 10029, USA
nicole.mcknight@mssm.edu

Mizushima, Noboru
Department of Physiology and Cell Biology
Tokyo Medical and Dental University, Tokyo 113–8519, Japan
nmizu.phy2@tmd.ac.jp

Nixon, Ralph A.
Center for Dementia Research
Nathan Kline Institute
140 Old Orangeburg Road
Orangeburg, NY 10962, USA

and

Departments of Psychiatry and Cell Biology
New York University Langone Medical Center
550 First Avenue, New York, NY 10016, USA
nixon@nki.rfmh.org

Pankiv, Serhiy
Department of Biochemistry
Institute of Basic Medical Sciences
University of Oslo, 0317 Oslo, Norway

Peng, Hsiao-Yu
Division of Basic Neurosciences
Biomedical Research Foundation of the Academy of Athens
Athens, Greece

Rubinsztein, David C.
Department of Medical Genetics
University of Cambridge
Cambridge Institute for Medical Research
WellcomeTrust/MRC Building
Hills Road, Cambridge CB2 0XY, UK
dcr1000@hermes.cam.ac.uk

Shacka, John J.
Department of Pathology, Neuropathology Division
University of Alabama at Birmingham and Birmingham VA Medical Center
Birmingham, AL 35294, USA
shacka@uab.edu

Sheng, Zu-Hang
Synaptic Function Section
National Institute of Neurological Disorders and Stroke
National Institutes of Health
35 Convent Drive, Building 35, Room 2B-215
Bethesda, MD 20892–3706, USA
shengz@ninds.nih.gov

Simonsen, Anne
Department of Biochemistry
Institute of Basic Medical Sciences
University of Oslo
PB 1112 Blindern, 0317 Oslo, Norway
anne.simonsen@medisin.uio.no

Stefanis, Leonidas
Division of Basic Neurosciences
Biomedical Research Foundation of the Academy of Athens
Athens, Greece

and

Second Department of Neurology
University of Athens Medical School
Athens, Greece
lstefanis@bioacademy.gr

Sulzer, David
Departments of Neurology, Psychiatry, Pharmacology
Columbia University, New York, NY 10032, USA
ds43@columbia.edu

Uchiyama, Yasuo
Department of Cell Biology and Neurosciences
Juntendo University Graduate School of Medicine
2–1–1 Hongo, Bunkyo-ku, Tokyo 113–8421, Japan
y-uchi@juntendo.ac.jp

Winslow, Ashley R.
Department of Medical Genetics
University of Cambridge
Cambridge Institute for Medical Research
Wellcome Trust/MRC Building
Hills Road, Cambridge CB2 0XY, UK
arw61@cam.ac.uk

Wolfe, Devin M.
Center for Dementia Research
Nathan Kline Institute
140 Old Orangeburg Road
Orangeburg, NY 10962, USA
dwolfe@nki.rfmh.org

Xilouri, Maria
Division of Basic Neurosciences
Biomedical Research Foundation of the Academy of Athens
Athens, Greece

Yamamoto, Ai
Department of Neurology and Department of Pathology and Cell Biology
Columbia University, New York, NY 10032, USA
ay46@columbia.edu

Yang, Yi
Qiushi Academy for Advanced Studies
Department of Biomedical Engineering
Key Laboratory of Biomedical Engineering of Ministry of Education
Zhejiang University, Hangzhou, 310027, P. R. China
yangyizju@hotmail.com

Yu, W. Haung
Department of Pathology and Cell Biology
Columbia University Medical Center
P&S 12–440, 630 W. 168th St., New York, NY 10032, USA
why2102@columbia.edu

Yue, Zhenyu
Department of Neurology and Neuroscience
Friedman Brain Institute
Mount Sinai School of Medicine, New York, NY 10029, USA
zhenyu.yue@mssm.edu

Zheng, Xiaoxiang
Qiushi Academy for Advanced Studies
Department of Biomedical Engineering
Key Laboratory of Biomedical Engineering of Ministry of Education
Zhejiang University, Hangzhou, 310027, P. R. China
zxx@mail.bme.zju.edu.cn

Zhu, Haining
Department of Molecular and Cellular Biochemistry
College of Medicine, University of Kentucky
741 South Limestone Street, Lexington, KY 40536, USA
haining@uky.edu

PREFACE

Discovered a half century ago as a self-eating process in mammalian cells, autophagy was studied in the context of tissue development, hormonal regulation and chemotherapy through the 1970s before entering a relatively dormant phase. The identification of the first autophagy gene in the 1990s marks the beginning of a molecular era of autophagy research that triggered an ongoing explosion of interest in autophagy with relation to its basic regulation, role in disease mechanisms and potential therapeutic applications. We now know that autophagy represents a basally active cellular quality control mechanism that is exquisitely regulated in response to a diverse array of cellular signals in health and disease.

Autophagy has three subtypes, which employ distinct mechanisms to sequester and deliver specific cargo for digestion in the lysosome. The regulatory mechanisms underlying digestion by autophagy, however, varies in different cell types and tissues, and the outcome of autophagy is dependent on the larger context of various patho-physiological conditions. With the rapidly increasing molecular knowledge about autophagy mechanisms, we hereby compile expert reviews focused on the topic of neuronal autophagy and the role of autophagy in neurodegeneration and other neurological disorders.

The chapters within this volume elaborate upon how the conserved autophagic process has evolved to include accelerators and brakes in the mammalian nervous system, and how autophagy participates actively in the regulation of neuronal survival and death in the context of neurological diseases. A section is devoted to critical discussion regarding the implication of autophagy dysregulation in brain and retinal development and in major human brain diseases: Alzheimer's, Parkinson's, Huntington's, amyotrophic lateral sclerosis, ischemia/hypoxia and lysosomal storage diseases. Emerging views of autophagy as a promising therapeutic target and new strategies for developing "autophagy-modulatory drugs" to treat neurological diseases are highlighted. The final section is devoted to emerging frontiers in neuronal autophagy research involving synaptic structure/function and axo-dendritic mitochondrial content.

We expect that this book will be helpful for a range of basic, translational, clinical and industrial research scientists. We are grateful to our colleagues who have contributed their expert viewpoints to this book despite busy schedules, and hope that this volume will function as a useful springboard for exploring the seminal contributions of many other groups in the primary literature, who could not be included due to space constraints.

Zhenyu Yue, New York, NY
Charleen T. Chu, Pittsburgh, PA
January 2012

SECTION 1

NEURONAL AUTOPHAGY: CELLULAR PROCESS AND REGULATION

CHAPTER 1

THE CELLULAR PROCESS OF AUTOPHAGY AND CONTROL OF AUTOPHAGY IN NEURONS

Nicole C. McKnight, Noboru Mizushima and Zhenyu Yue

ABSTRACT

Autophagy is a cell self-digestive, lysosomal degradation pathway. The three subtypes of autophagy, macroautophagy, microautophagy and chaperone-mediated autophagy, differ in the way that materials are delivered to the lysosomes for degradation. While recent studies reveal that the cellular process of macroautophagy (involving critical steps such as formation, trafficking, fusion and degradation of autophagosomes) is evolutionarily conserved from yeast to mammals, the regulation and functional adaptation of autophagy in various cells and tissues is poorly understood. This chapter provides an overview of what is known about autophagy machinery and highlights the recent findings of autophagy regulation in neurons. Elucidation of neuronal autophagy function will ultimately aid in drug target identification and perhaps lead to a rational therapeutic strategy to combat neurodegenerative diseases.

INTRODUCTION

Neurons are highly differentiated and their complexity arises from their specialized, polarized domains that rely on active membrane transport to connect the cell body to axons and dendrites, which are often long and far away. The neuron has long served as a model for studying autophagy[1] and the presence of double-membrane autophagosomes was described as early as the mid-1960s.[2–6] Upon physical or chemical injury (e.g. axotomy or excitotoxic insult), neurons developed a large number of autophagosomes that sequestered a portion of cytoplasm and were seen at cell bodies (soma) as well as axons. The phenomenon is associated with neuronal subcompartments devoid of organelles and filled with various types of vesicles,[1] contributing to the histologic appearance of "chromatolysis." Subsequent ultrastructural analysis assisted with electron microscopy (EM) showed accumulation of autophagosomes in neurons of postmortem brains with neurodegenerative diseases, leading to the early hypothesis that autophagy is acutely turned on under pathological conditions perhaps as an attempt to repair/restructure via degradation, but sometimes with destructive or suicidal outcomes. The historical view of autophagy, however, has evolved due to recent genetic and molecular evidence that neurons rely on basal autophagy to ward off intracellular aggregate accumulation and resulting neurotoxicity. In addition, functional autophagy is required under neurodegenerative disease conditions triggered by the expression of disease-associated mutant proteins as these are eliminated by the autophagy. The neuroprotective function of autophagy is conceptually in line with the conservation of autophagy as a survival mechanism. But the idea that autophagy only functions to rescue may not fit with the previous observations in neurons that suppression of autophagy (genetically or pharmacologically) sometimes alleviates the neurotoxicity[7] (for review see Yue *et al.*[7]). Therefore, these parallel hypotheses co-exist, though seemingly conflicting, and may reflect the multi-faceted characteristics of autophagy under different physiological functions.

Our ultimate goal is to resolve these important issues and understand autophagy in detail by thorough investigation of the autophagic process, its regulation and its relevance to human diseases. Several outstanding questions pertinent to this goal remain to be answered: what is the exact cellular process of basal autophagy and how is it controlled in neurons that are highly differentiated, polarized, postmitotic cell types? How does the basal autophagy machinery adapt to the neural pathways that are responsible for rapidly processing and transmitting information by electrical and chemical signals across different subcompartments? How do neurons recognize disease protein aggregates to signal and mobilize the autophagic machinery to selectively degrade the aggregates? Moreover, how does each human disease that exhibits autophagy-associated neuropathology cause altered autophagy? Do these alterations contribute to the disease process and is there any neuronal cell type-specific autophagy regulation? In turn, can targeting autophagy reverse or modify a disease process and as a therapeutic strategy? In fact, emerging evidence suggests that autophagy is a potential drug target. To this end, the research effort has now been increasingly devoted to the search for drugs that enhance autophagy in the brain in order to increase the elimination of disease proteins that are prone to aggregation, though we do not know all the ramifications of such an increase and whether this is even possible in human brains. A recent surge of research into the mechanics of autophagy has, however, greatly increased our knowledge of the process in non-electrical cells. The knowledge will facilitate our understanding of the conservation and distinct features of neuronal autophagy. Indeed, autophagy occurs in almost every cell of our body and its molecular machinery is highly conserved between organisms and tissues.

AUTOPHAGY PATHWAYS

Autophagy and the ubiquitin-proteasome system (UPS) are the two major degradative pathways in mammalian cells. The latter is responsible for degrading damaged proteins in the proteasome and is specific in its selection of proteins and protein aggregates that are tagged with ubiquitin. Autophagy uses the lysosome for degradation and has three subtypes: chaperone-mediated autophagy (CMA), microautophagy, and macroautophagy. CMA is induced after prolonged nutrient deprivation or oxidative stress[8] and is selective in its recycling of a subset of cytosolic proteins (see Chapter 4 for details). Microautophagy involves inward invagination of the lysosomal membrane to take up small pieces of cytoplasmic material. Macroautophagy is the main form and is a catabolic, mostly unselective pathway that channels bulk portions of the cytoplasm to the lysosome to generate amino acids and macromolecules; it is generally responsible for degrading long-lived proteins and large cellular structures. By balancing these degradative pathways with new protein synthesis, cells are able to maintain equilibrium of protein pools, which are essential for the survival, development, and proper function of the cells.

Macroautophagy, hereafter referred to as autophagy, is a term originating from the Greek words, "phagy," meaning to eat and "auto," for self. It is a highly conserved intracellular membrane trafficking pathway that is exhibited in all eukaryotes from yeast to humans. It responds to nutrient deprivation as its most evolutionarily conserved function, but is also triggered by cellular stress and the accumulation of protein aggregates and damaged organelles, especially in higher organisms. Autophagy generates the means for survival by degrading cytosolic proteins and whole organelles and recycling these back to the cytosol as amino acids and macromolecules. As a result, autophagy maintains cellular homeostasis by clearing the cell of misfolded or long-lived proteins and damaged parts and has also developed to rid the cell of invading microorganisms.

Autophagy is characterized by the formation of double-membrane vesicles, called autophagosomes that travel to the lysosome where they fuse and the contents of which are degraded by hydrolases. The autophagosome was originally observed using EM in the late 1950s and was first characterized by Christian de Duve in 1966 when he described degradative vesicles that were seen to contain both cytoplasmic components and mitochondria.[9] He saw the vesicles merge with lysosomes and called the process of "non-specific bulk segregation and digestion" autophagy.

Since a massive torrent of studies on the process of autophagy beginning in the early 1990s, cell biologists have dissected the progression of the autophagosome from its initiation or nucleation, its expansion, its maturation including fusion with components of the endocytic pathway and its eventual degradation by the lysosome. Also, great strides have been made to determine how the process is triggered and how the autophagy machinery is initiated. Many autophagy-specific proteins, called Atg proteins, have been discovered in yeast through humans. We now know that the process requires these components as well as other cytosolic proteins and intracellular organelles such as the endoplasmic reticulum (ER), Golgi apparatus, and perhaps mitochondria, in addition to the endosome-lysosome system.

During autophagy, cytoplasmic components are sequestered in a growing phagophore, also known as an isolation membrane (IM), which is generated at the preautophagosomal structure (PAS) in yeast (Figure 1). This double-membrane structure expands around portions of the cytoplasm containing proteins, aggregated proteins and organelles and then closes, forming the autophagic vacuole (AV) or autophagosome. Most mature autophagosomes are approximately 0.5 to 1 μm in diameter; they can fuse with endocytic pathway vesicles including early endosomes, multi-vesicular bodies (MVBs), and late endosomes thus forming amphisomes. Eventually, the autophagosome fuses with the lysosome, becoming an autolysosome, and the acidic hydrolases within break down the captured components to macromolecules and amino acids that are released back to the cytosol by transporters and permeases delivered with the lysosomal membrane.[10] After degradation by autophagy, the recycled products can be used in biosynthetic pathways and to generate new proteins.

In yeast, the cytoplasm-to-vacuole (CVT) pathway, which shares many of the core components with autophagy, and both pexophagy and mitophagy (the specific autophagic degradation of peroxisomes and mitochondria, respectively) are considered selective forms of yeast autophagy. Until recently mammalian autophagy was thought to be a non-selective degradation process, capturing random parts of the cell, but now it is believed to also specifically engulf protein aggregates, organelles and bacteria as these cargos have been seen to selectively and exclusively incorporate into autophagosomes. The organelle-specific pathways of pexophagy, mitophagy and even ER-phagy have been observed in human cells though how these organelles are recognized and targeted by the autophagosome is largely unknown. In addition, the source of the membrane(s) is a subject of intense research and great debate. Although autophagy allows for cellular survival in times of stress such as hypoxia, infection, and starvation, much of the prenucleation process and the steps leading to its initiation remain to be elucidated.

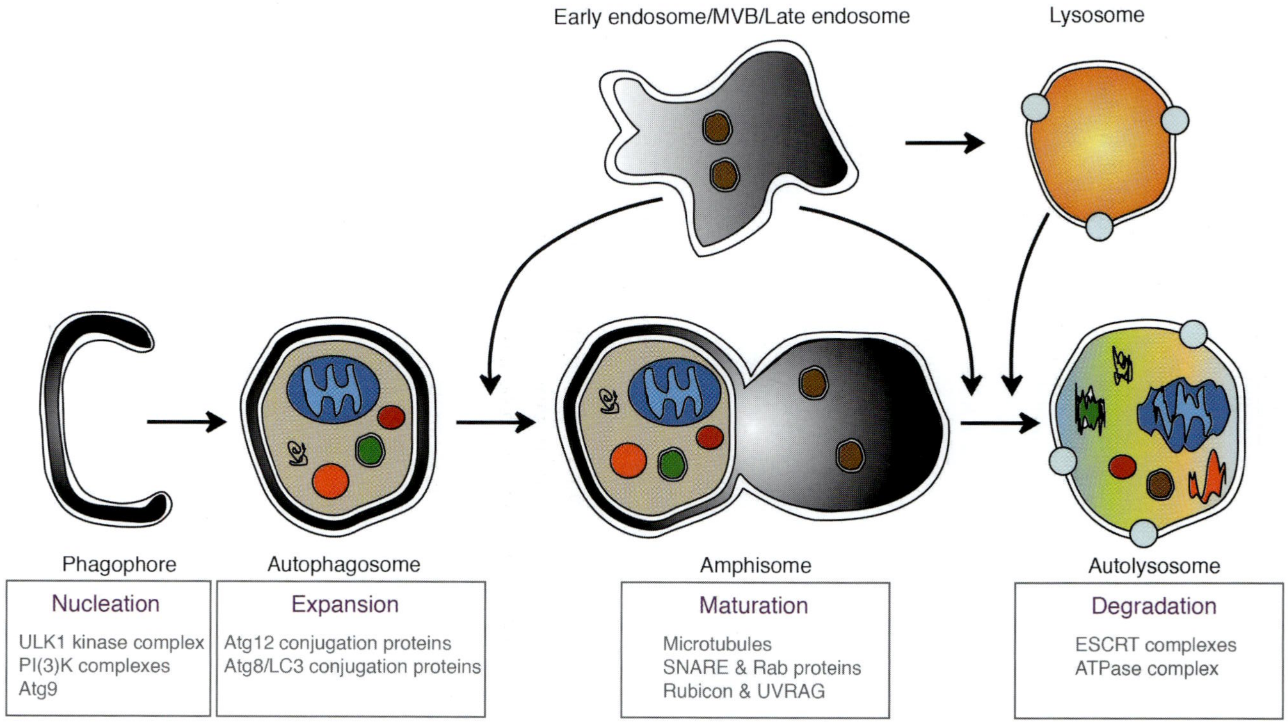

Figure 1. **Mammalian autophagy**. During mammalian autophagy, first the phagophore (or isolation membrane) is formed or nucleated. The phagophore engulfs portions of the cytoplasm and can take up mitochondria (blue), misfolded proteins and various vesicles in a step called expansion. During maturation, the autophagosome can fuse with endocytic vesicles to become an amphisome. The mature autophagosome or amphisome fuses with the lysosome and becomes degradative.

AUTOPHAGY CORE MACHINERY

The molecular machinery responsible for autophagy initiation and autophagosome nucleation and expansion was discovered first in *Saccharomyces cerevisiae* through genetic screens that identified essential autophagy genes in response to amino acid starvation.[11,12] Currently, there are 35 known Atg proteins involved in both autophagy and the *CVT* pathway in yeast and most, if not all, human counterparts of the autophagy-specific proteins have been discovered.[13] In mammals, there are multiple core and auxiliary proteins that contribute to and control autophagy. The human Atg proteins can be divided into five groups listed in the order that they are thought to act: the ULK1 kinase complex, the Atg9 cycling complex, the Vps34/class III PI3-kinase complexes, the lipid-binding Atg18 homologs, and the ubiquitin-like proteins Atg12 and Atg8/LC3 (and its related proteins) and their conjugation systems. Also, a set of proteins is responsible for transferring degraded material back into the cytoplasm. The five functional groups act at an early stage in the autophagy pathway and are recruited to the phagophore. The recent explosion of autophagy research has led to a somewhat comprehensive view of how they operate to form the autophagosome.

ULK1/ULK2 Kinase Complex (Atg1)

The first yeast autophagy gene identified was *ATG1*.[14] It transcribes a serine/threonine kinase that initiates autophagy downstream of the amino acid sensor and the negative regulator of autophagy, TOR kinase. Atg1 initiates both starvation-induced autophagy and the CVT pathway in yeast and does so through interactions with at least seven other Atg proteins. Autophagy is initiated through the starvation-dependent dephosphorylation of the interacting partner Atg13 that causes it and the third member of the complex, Atg17, to bind Atg1 tightly, activating its kinase activity.[15] This occurs as a result of the direct inhibition of TOR either by amino acid deprivation or treatment with the TOR inhibitor rapamycin. But it is Atg17, not Atg1 or Atg13 that is at the top of the hierarchy of Atg proteins, as determined by systematic analysis of yeast mutants.[16] Atg1 in *Drosophila* and *C. elegans* (in which it is called UNC-51) together with Atg13 and Atg17 are also critical to the control of autophagy in these organisms, but Atg1 is also involved in apoptosis and development through multiple non-Atg binding partners.[17] In addition, Atg1/UNC-51 is required for neuronal vesicular trafficking and development possibly through its indirect interaction with microtubule motors.[18] As in yeast, *Drosophila* Atg13 binds more strongly to Atg1 during autophagy but opposite to yeast, induction of autophagy upon nutrient deprivation in flies is characterized by increased phosphorylation of Atg13 and a loss of phosphorylation of Atg1 by TOR.[19] Also, TOR remains complexed with Atg1 in both nutrient-rich and starvation conditions.

Two of the mammalian homologues of Atg1, called unc-51-like kinase 1 and 2 (ULK1 and 2) were first discovered to be involved in autophagy in 2007.[20] The discovery of the Atg17 ortholog came shortly after and though it shares little sequence homology to the yeast protein, FIP200, also known as RB1CC1, shares a functional role as an autophagy scaffolding protein.[21] FIP200 has been shown to bind ULK1 in both nutrient-rich and amino acid-free conditions,[22] but ULK1 phosphorylates FIP200 in starvation and they then travel together to the site of autophagosome formation.[23] The Atg13 homologue was discovered through sequence analysis and was shown to not only bind both ULK1 and ULK2 but is also a substrate for both kinases.[24] Atg13 itself binds FIP200[23,25] and it is proposed that Atg13 serves to stabilize the interaction of ULK1 and FIP200. Again, ULK1 kinase activity is highest when it is complexed together with Atg13 (or FIP200) and Atg13 is seen to go to the phagophore upon starvation.[22,24,25]

Exciting new research has shown that the mammalian TOR kinase complex, mTORC1 directly binds to the ULK1-Atg13-FIP200 complex and that Raptor and ULK1 bind in nutrient-rich conditions but dissociate upon starvation.[25] In addition, multiple groups have shown that mTOR directly phosphorylates ULK1/2 and also Atg13.[22,23,25] Atg101, a fourth Atg13-binding protein that is not conserved in yeast was seen to interact with the Atg13-ULK1-FIP200 complex in an Atg13-dependent manner and is required for and possibly promotes autophagy by stabilizing Atg13.[26,27] Corroborating the importance of ULK1 and its interactions with its adaptor proteins and mTOR itself for the initiation of autophagy, a recent study of the hierarchy of mammalian autophagy proteins places ULK1 and FIP200 as the most upstream at the site of autophagosome formation.[28] It has also been recently shown that ULK1 is phosphorylated by AMPK,[29–31] the energy-sensing kinase, adding new information for the mechanism of autophagy initiation after nutrient deprivation.

Atg9 Recycling Complex

The only membrane-spanning autophagy core complex protein is Atg9; it has six transmembrane domains. Although there is no known function designated to mammalian Atg9 (mAtg9), its association with membrane and extrapolation from yeast studies suggests that mAtg9 may deliver membrane to forming autophagosomes. In yeast, Atg9 is seen to cycle between the PAS and a peripheral pool that has recently been shown to comprise of clusters of vesicles and tubules often found close to mitochondria.[32] The Q/t-SNAREs Sso1/2 and Sec9 are required for formation of the Atg9-positive tubular structures.[33] Atg1 is required for this cycling but Atg9's movement is not dependent on Atg1 kinase activity.[34] The localization of Atg9 to the PAS depends on Atg17 and Atg9 retrieval from the PAS, which occurs prior to autophagosome formation, requires Atg1, Atg2, and Atg18.[16]

Mammalian Atg9 is required for autophagy and is an essential gene for survival; Atg9 knockout mice cannot survive past P0 much like Atg3, Atg5 and Atg7 knockout mice.[35] Similar to yeast Atg9, mAtg9 cycles between two pools, residing in a juxta-nuclear Golgi pool in full medium but dispersing to a peripheral, perhaps endosomal, pool upon amino acid withdrawal.[36] As in yeast, this translocation requires ULK1 and also Atg13.[24] Regulation of autophagy was recently linked to the p38 MAP kinase pathway through Atg9, as it was demonstrated that the cycling of Atg9 requires the Atg9- and p38-interacting protein p38IP and through these interactions, p38α acts as a negative regulator of autophagy.[37]

PI3-Kinase Complexes (Vps34-Beclin 1)

The action of the autophagy-regulating class III PI3-kinase (PIK3C3), also known as Vps34, and its complex occurs early in the process and is required for nucleation of the phagophore. It was discovered previously that treatment with 3-methyladenine (3-MA), an inhibitor of the PI3-kinases, inhibits autophagy.[38] Since then, it has emerged that the production of Phosphatidylinositol(3)P, also known as PtdIns(3)P or PI(3)P, is essential for nucleation and formation of the autophagosome. Vps34 catalyzes the phosphorylation of phosphoinositides (PIs) at position 3 on their inositol ring to produce PI(3)P.[39]

The lipid kinase vacuolar protein sorting (Vps) 34 and its complex, including the Vps15 regulatory subunit and the accessory autophagy proteins Atg14 and Atg6 were first described as requirements for autophagy in yeast.[40] The mammalian homolog of Vps15 is p150 and is also called PIK3R4 for phosphoinositied-3-kinase, regulatory subunit 4; this protein anchors the Vps34 complex to membranes. The mammalian homologue of Atg6 is Beclin 1 or BECN, a coiled-coil protein, and together with the newly identified human ortholog Atg14L (Atg14-like), they form the mammalian Vps34 Complex I.

Human Atg14L, also called BARKOR,[41,42] was shown to be required for autophagosome formation and more specifically for the recruitment of Atg16L and LC3, essential proteins of the phagophore (discussed below) to the site of autophagosome formation.[41,43,44] Unique as a member of the Vps34 kinase complex, Atg14L interacts with both Beclin 1 and Vps34 and its coiled-coil domain is key for the binding. Atg14L is recruited to phagophore structures, but Atg14L is found on these structures without Beclin 1 and Vps34 and the kinase activity of Vps34 is not required, suggesting Atg14L dictates when and where autophagosome nucleation occurs, possibly by recruiting Vps34 to this site and/or stimulating its activity.[44] Overexpression of Atg14L increases autophagosome formation and it is required for the autophagy-mediated elimination of *Salmonella*.[42]

Another coiled-coil protein UVRAG (UV radiation resistance associated gene) binds to Beclin 1[45] and UVRAG does not associate directly with Vps34 Complex I containing Atg14L, but forms a distinct complex containing p150, Vps34 and Beclin 1.[41,43,46] UVRAG seems to compete with Atg14 for Beclin 1 binding leading to two mutually exclusive complexes.[42] There are conflicting views as to the requirement of UVRAG for autophagy[41,47] and the role of the UVRAG complex in autophagy is still unknown. However, its function in the endocytic pathway was recently shown.[47] The protein Bif-1 (also called SH3GLB1, SH3-domain GRB2-like endophilin B1, or simply endophilin B1) may provide some insight; Bif-1 is required for autophagy.[48] Like many others, it has a coiled-coil domain and interacts with Beclin 1 through direct binding to UVRAG. Bif-1 has other roles in membrane traffic and is required for fission of Golgi carriers. It has been suggested that Bif-1 may be required for autophagosome membrane expansion and curvature through its N-BAR membrane curvature domain and its binding to the tethering factor ARF-GAP.[39] Recently Bif-1 was shown not to act at the level of autophagy initiation, but to regulate degradative endocytic traffic and in turn, autophagosome maturation.[49]

Among the recent eruption of research surrounding the Vps34 kinase complex is the data describing yet-another Beclin 1-binding protein, RUBICON (an acronym for run domain Beclin-1 interacting and cysteine-rich containing protein, still annotated as KIAA0226).[43,44] Rubicon resides with partners Beclin 1, UVRAG, and Vps34 to form the Rubicon complex. Opposite to Atg14L, RUBICON knockdown increases autophagosomes by inhibiting fusion of the autophagosome with the lysosome and also the maturation of endosomal vesicles to lysosomes.[44] Tagged RUBICON is found on endosomes and lysosomes and overexpression of RUBICON inhibits autophagosome turnover confirming its role as a negative regulator of fusion; overexpression of the protein also causes aberrant expansion of late endosomes and lysosomes.[43]

A sixth Beclin 1-binding protein, AMBRA1 may also contribute to the regulation of Vps34 activity.[50] The WD-repeat-containing protein is not only a positive regulator of Beclin 1, but is also required for starvation-induced autophagy. AMBRA1 appears to be an ULK1 substate and mediates autophagy regulation by the dynein motor complex,[51] though the exact molecular function of AMBRA1 in autophagy initiation is not known.

Though p150 is a shared core component of the Vps34 complex, Beclin 1 is the protein that mediates the interactions of the auxiliary proteins that distinguish the multiple known complexes. Through its differential binding, Beclin 1 may act as a switch between the complexes in order to coordinate the regulation of autophagy in highly evolved species. As a master regulator, Beclin 1 may balance autophagosome nucleation and maturation by toggling between Complex1 and the UVRAG complex and also the RUBICON complex.[46] Beclin 1 also plays a role in apoptosis through its binding to the anti-apoptotic protein Bcl-2.[52] Bcl-2 negatively regulates Beclin 1 by keeping it away from its Vps34 kinase complex partners and thereby preventing autophagy. Though it should be said that under certain pathological conditions, autophagy can be induced independently of Beclin 1 to promote cell death.[53,54] Beclin 1 may not only be important for the balance of autophagosome formation and maturation, but is positioned between autophagy and apoptosis, making it an important modulator of cell death versus survival.

Atg18/WIPI PI(3)P-Binding Protein and Atg2

In yeast, retrieval of Atg9 from the PAS, which occurs prior to autophagosome formation requires Atg1, Atg2, and Atg18, which recruits Atg2 and Vps34.[34] In fact, the Atg18-Atg2 complex is essential for nucleation.[16]

Atg18 interacts with Atg9 in yeast and recently the mammalian homologues of Atg18, WIPI (WD-repeat protein Interacting with PhosphoInosides) family proteins, have been characterized.[55,56] Atg18 and the WIPI family proteins contain a non-canonical PI(3)P binding region. Among the WIPI family proteins, WIPI2 is the major ubiquitous isoform, is required for autophagy, and forms puncta upon starvation that partially localize with Atg16L1, GFP-LC3 and ULK1, placing it very early in the hierarchy of autophagy proteins.[56] As it does not bind to endosomes, WIPI2 is thought to be a specific effector of the autophagy-specific cellular pool of PI(3)P. Yeast Atg18, like Atg9, complexes with Atg2 but neither of these interactions has been shown with their mammalian counterparts. The mammalian homologue of Atg2 has been annotated but to this date, has not been characterized.

Ubiquitin-like Conjugation Systems

Two ubiquitin-like conjugation reactions produce the Atg5-Atg12-Atg16 complex and lipidated Atg8 and are essential for the formation and elongation of the autophagosome. The proteins required for these two conjugation reactions are critical for life; Atg3, Atg5, Atg7 and Atg16L1 are essential genes and are important for maintenance of the amino acid pool during the postnatal starvation period.[57] Many detailed studies have resulted in a thorough understanding of how these critical autophagy proteins contribute to the expansion of the double-membrane vesicle.[58]

Atg12 conjugation reaction

In mammalian cells, Atg12, an ubiquitin-like (UBL) protein is conjugated to Atg5 through a multi-step conjugation reaction that requires the enzymatic activities of E1-like and E2-like proteins. First Atg12 is activated by the E1-like enzyme Atg7 and then transferred to the E2-like enzyme Atg10 which then catalyzes the covalent conjugation of Atg5 to Atg12.[59] Subsequently, the Atg5-Atg12 complex is bound to Atg16L1 and this trimeric complex homodimerizes to become a large complex.[60,61] Mammalian Atg5 is an important marker for early autophagosome formation and it localizes to the phagophore.[62] Mammalian Atg16 (Atg16L1) differs from yeast in that it contains a C-terminal WD40 repeat domain for multi-protein complex assembly and, as mentioned above, is recruited to the phagophore.

Atg8/LC3 conjugation reaction

Atg8, the yeast UBL protein that is lipidated and integrated into the autophagosomal membrane, has multiple homologues in mammals grouped into LC3-like and GABARAP-like subgroups. LC3 (microtubule-associated protein 1 light chain 3) is the best characterized Atg8 homologue and was the first to be shown to localize to the autophagosome using EM.[63] To begin the LC3 lipidation reaction, LC3 is primed by the cysteine endopeptidase Atg4 family proteins, exposing a C-terminal glycine,[64] referred to as the LC3-I form. Next, Atg7, which serves as the E1-like enzyme for both conjugation reactions, activates LC3 and then the E2-like enzyme Atg3 facilitates the conjugation of phosphatidylethanolamine (PE) to LC3 to become LC3-II (or LC3-PE).[65] In the same way, GABARAP and GABARAPL2 (also known as GATE16) can be post-translationally processed and bound to PE.[66] Once LC3 is lipidated and becomes part of the autophagosome, it remains membrane-bound and can be followed from phagophore expansion until fusion with and degradation by the lysosome. As a result, LC3 is a bone fide marker of the autophagosome[67] and we can measure LC3's flux through the pathway visually or by Western blot analysis in order to monitor autophagy.[68]

After closure of the autophagosome, Atg4 cleaves LC3 from PE thereby removing LC3 from the mature autophagosome,[69] a step that is critical for its fusion with endosomes and lysosomes. Experiments with an Atg4 mutant have shown that priming LC3 is critical for the early Atg7-mediated step of the conjugation pathway but the Atg4 mutant also led to defects in the closure of the expanded autophagosome, suggesting that the lipidated, or LC3-II form of the protein is needed for the membrane curvature required to close the vesicle.[61]

Current evidence points to an interplay between the ubiquitin-like conjugation pathways, beyond sharing their E1-like enzyme Atg7. It is generally believed that the Atg5-Atg12 reaction occurs upstream of the LC3 lipidation pathway and Atg16L1 is required for the recruitment of LC3II to the site of autophagosome formation.[61] In addition, some evidence shows that the Atg5-Atg12 complex acts as an E3-like enzyme, conferring specificity to the conjugation of LC3 to PE[70] and conversely, overexpressed Atg3 enhances Atg5-Atg12 conjugation.[65]

SOURCE OF THE AUTOPHAGOSOME MEMBRANE

Although many of the essential proteins and machinery required for autophagosome nucleation and expansion are known (as discussed above), the source of the membrane and the mechanism for phagophore formation are mostly unknown. The origin of the membrane has been in question since the discovery of the double membraned vacuoles and is recently the subject of intense debate and concentrated research. Autophagy scientists strive to determine whether the membrane forms *de novo* from localized lipid synthesis or is derived from a pre-existing organelle or organelles. In yeast autophagosomes always form at the PAS and next to the vacuole. But in mammalian cells, autophagosomes seem to be able to form anywhere in the cytoplasm including the axo-dendritic extensions, and a distinct place of origin, like the PAS, does not exist. Phagophores may form from a PAS-like structure, or the phagophore itself could be the mammalian equivalent of the PAS, but this is not known. Perhaps the difficulty arises from a lack of protein markers, both for the autophagosome and of other membrane sources found in the autophagosomal membrane.[71] Using current methods, it is difficult to differentiate between the autophagosome membrane and the membrane of the organelles it engulfs and the two primary marker proteins, Atg5 and LC3 associate with the phagophore only after it begins to form so earlier, non-transient markers are needed.[72] Multiple recent studies have increased our knowledge of the origin of the autophagosome and point to the ER, mitochondria and even the plasma membrane as possible sources of the membrane.[73–76]

AUTOPHAGY SIGNALING/REGULATION OF AUTOPHAGY

In yeast, autophagy is directly activated by the loss of negative regulation by TOR through amino acid starvation or TOR inhibition with rapamycin.[77] In mammalian cells, the mTOR kinase complex mTORC1 also acts as the key negative regulator of autophagy as well a positive regulator of cell growth. Amino acids stimulate mTOR and promote the phosphorylation of ribosomal S6 protein (S6) by p70S6 kinase and a decrease in phosphorylated S6 reflects an induction of autophagy by starvation or rapamycin.[78] Recently, Rag GTPases and three proteins in a complex called the "Ragulator" have been shown to signal mTOR in response to amino acid starvation and also translocate mTOR to the lysosome membrane.[79] mTOR loss stimulates autophagy by directly affecting ULK1, but adding complexity, ULK1 and ULK2 have been shown to feedback to the mTOR-p70S6 kinase pathway (for review, see Ref. 17). In fact, in *Drosophila* autophagy inhibits cell growth and Atg1 itself can directly inhibit Tor signaling suggesting positive feedback.[80]

Nutrients like amino acids can directly signal mTOR through the Rag-Ragulator complex but a lack of amino acids can also activate autophagy independently from mTOR. For instance, amino acids inhibit the Raf-1-MEK1/2-ERK1/2 signaling cascade that in turn inhibits autophagy.[81] Other mTOR-independent autophagy activation pathways have been reported including that induced by lithium chloride, which appears to act through modulation of myo-inositol-1,4,5 triphosphate (IP_3) levels.[82] Although conflicting data does exist, one group reports that resveratrol also drives autophagy independent of mTOR by activating sirtuins that deacetylate Atg's, which activates autophagy.[83] PKA was also shown to directly modulate autophagy proteins; it suppresses autophagy by directly phosphorylating both LC3 and ULK1 kinase complex proteins.[84]

As previously discussed, 3-MA and also wortmannin inhibit autophagy through the inhibition of the class III PI3-K product PI(3)P; but 3-MA also inhibits class I PI3-K which is a negative regulator of autophagy through the Akt-TSC1/2-Rheb-mTOR signaling pathway.[85] Growth factor (insulin) signaling inhibits autophagy through stimulation of IRS1 and IRS2, which activate the class I PI3-kinase. Alternately, overexpression of PTEN stimulates autophagy by inhibiting the inhibitory Akt pathway.[86] Interestingly, a recent study has shown an unexpected role of class IA PI3K catalytic p110-β in positively regulating autophagy, adding to the complexity of autophagy control by the PI3K family.[87]

The anti-apoptotic protein Bcl-2 inhibits autophagy by sequestering Beclin 1 away from Vps34 and it was recently demonstrated that starvation removes Bcl-2, through direct phosphorylation by JNK1, from the class III PI3-kinase complex 1.[88]

In addition, energy inhibits LKB1, a positive regulator of autophagy that activates AMPK, and this pathway is activated in stress conditions;[89] AMPK itself senses a reduced ATP:AMP ratio and is reported to directly phosphorylate ULK1.[29–31] Autophagy is also influenced by p53, a master regulator of cell cycle progression and apoptosis, in opposing ways. Basal levels of the cytoplasmic form of the protein inhibit autophagy and its degradation leads to the induction of autophagy;[90] but in times of cellular stress, p53 can activate both autophagy and apoptosis.[91]

THE ROLE OF LC3-BINDING PROTEINS IN AUTOPHAGY

p62, also known as sequestosome 1 (SQSTM1) is a multi-functional adaptor protein that binds polyubiquitinated proteins and aggregates by oligomerization.[92] Though p62 is not required for starvation-induced autophagy, a body of evidence shows that it is a specific substrate for the process and its autophagy-induced degradation can be used as a read-out for autophagy function.[93–95] In addition to binding ubiquitin (Ub), it also binds to LC3, perhaps recruiting the ubiquitinated proteins to the autophagosome.[96,97] Loss of autophagy resulted in accumulation of ubiquitinated proteins and the formation of ubiquitinated inclusion relies on the presence of p62, suggesting a role for p62 in assembly of ubiquitinated proteins into aggregates.[98,99] p62 has recently been demonstrated to translocate into the nucleus where it co-localizes with promyelocytic leukemia (PML) bodies. The nucleoplasm to cytoplasm shuffling of p62, which is possibly controlled by phosphorylation events, may serve as a regulatory mechanism for its role in autophagy.[100] Though the previous view was that the UPS and autophagy act in a mutually exclusive way, recent work has shown that the two pathways are linked through their common modulator, p62/SQSTM1.[101]

Another LC3- and Ub-binding protein, neighbor of BRCA1 gene 1 (NBR1), also serves as autophagic receptor for ubiquitinated targets.[102] Like p62, NBR1 contains an LC3-interacting region (LIR) and this motif is required for NBR1's degradation by autophagy. It seems that NBR1 cooperates with p62 to direct autophagic degradation of ubiquitinated targets and both can serve as markers of pathological inclusions. The clearance of these inclusions is an important homeostatic function that autophagy has gained in higher organisms, although the exact mechanism of how autophagy selectively degrades these substrates is not known.

AUTOPHAGY IN NEURONS

Autophagy is critical for the maintenance of homeostasis for intracellular proteins and cellular organelles to ensure their proper functions as demonstrated in recent studies with autophagy-deficient mouse models. Suppression of basal autophagy through the genetic deletion of *ATG5* or *ATG7* specifically in neural tissue causes neurodegeneration in mice.[103,104] The mice experience motor and reflex deficits but perhaps most striking pathology is the ubiquitin-positive inclusion bodies that accumulate in the neurons. Atg7 itself is essential for the maintenance of axonal homeostasis; without it, intracellular proteins and aberrant membrane structures accumulate to toxic levels and axons degenerate.[105] Neurons are especially sensitive to the accumulation of damaged or aggregated cytosolic proteins or membranes because they are extremely polarized, large cells; constitutive basal autophagy thus serves an important protective role in neurons.[7]

Multiple studies have suggested that the levels of lipidated LC3 and the numbers of autophagosomes detected in neurons are relatively low in normal as well as nutrient withdrawal conditions.[106,107] One hypothesis is that autophagosomes are formed normally, but there is an extremely high rate of turnover of autophagosomes due to efficient autophagy-lysosomal degradation in neurons.[107,108] An alternate hypothesis, however, is that autophagosome synthesis is tightly controlled in neurons; the constantly low rate of formation of typical autophagosomes may be sufficient to maintain the homeostatic balance. Perhaps neurons are prevented from undergoing extensive autophagy induction in response to starvation because this could lead to loss of mass and cells, as neurons cannot be regenerated. When an organism encounters a period of nutrient deprivation the brain is typically shielded by peripheral production of nutrients through autophagy, insuring that they can maintain essential normal function regardless of such conditions. Future studies of the unique regulation of autophagy in neurons should examine these possibilities.

Compartmentalized Autophagy in Neurons

The degree of functional subcompartmentalization (soma, axons and dendrites) of the neuron distinguishes it from other cell types (Figure 2). Accordingly, autophagy may be adapted to different neuronal functions in different subcompartments. It is known that an intact microtubule network is required for the formation, maturation and degradation of autophagosomes in mitotic cells[109] and that autophagosomes traffic towards the centrosome, where most lysosomes reside.[110] Similarly in neurons, the retrograde trafficking of autophagosomes from axons or neurites towards the cell body has been observed.[111] Recently it was shown that though autophagosomes and autolysosomes travel in both anterograde and retrograde directions (using kinesin and dynein, respectively), there is bias of traffic towards the nucleus of PC12 neurites during starvation.[112] In Alzheimer's

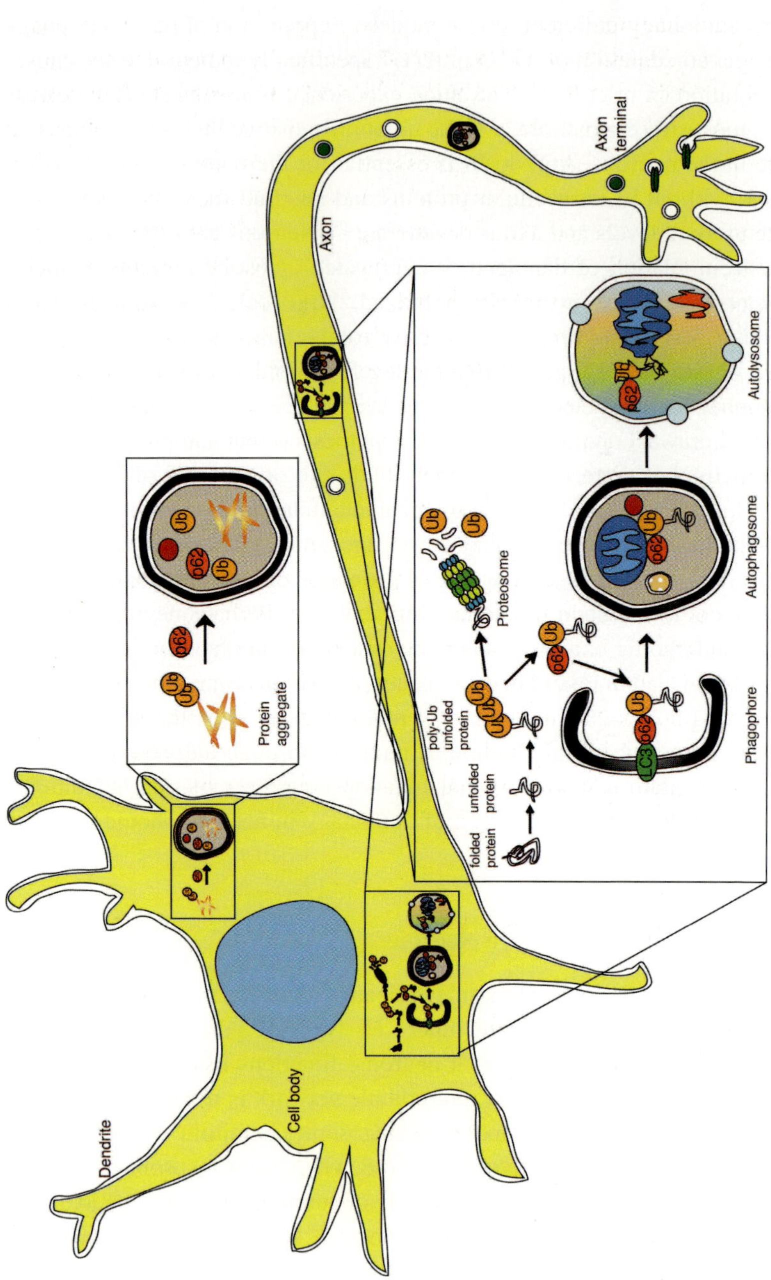

Figure 2. Autophagy in neurons. Membrane trafficking directs vesicles including endosomes and autophagosomes from the axons or the dendrites to the cell body. Ubiquitinated (Ub) proteins are either degraded by the proteasome or carried by the LC3- and Ub-binding protein p62 to the phagophore. Growth factors and neurotrophins bind to receptors at the axon terminal and are endocytosed. Adapted from Tooze and Schiavo.[116]

brains, undigested autophagosomes in dystrophic neurites are unable to travel to the lysosomes for degradation and accumulate, a disease hallmark further exacerbated by impaired lysosomal proteolysis.[113] Motor proteins such as dynein were shown to participate in autophagosome trafficking in axons[111,113] and another protein that may be related to autophagosome trafficking in the axon is MAP1B, a microtubule- and LC3-binding protein.[93] There is increased autophagosome production in the axon after injury,[114] suggesting that autophagosome can be synthesized locally in axon terminals. Perhaps autophagy is also important during axonal outgrowth and development. MAP1B is implicated in the autophagy that participates in the axon remodeling that takes place after excitotoxic stress or neurodegeneration.[93] Axonal or neuritic autophagy is further discussed in Chapters 16 and 17.

Autophagy Signaling in Neurons

Autophagy is a constitutive, basal process in the central nervous system (CNS). Autophagy is activated without starvation signals in the brain,[115] suggesting that it is not amino acid sensitive (and also suggesting tissue-specific functions for autophagy in organisms). Several lines of evidence suggest that neurotrophin receptor signaling and responses are connected to the regulation of autophagy.[116] Reduced levels of ULK1/2 causes impaired endocytosis of nerve growth factor (NGF), excessive axon arborization, and retarded axon elongation, suggesting that the autophagy machinery plays a role in controlling NGF/TrkA levels and signaling.[117] The investigations as to whether autophagy signaling in neurons requires mTOR have yielded somewhat mixed results. Some evidence points to an mTOR-independent mechanism[118–120] and in fact, a small molecule screen for inducers of autophagy in neuroblastoma cells found compounds that induced autophagy through distinct mechanism from rapamycin.[121] Recently, an siRNA screen searching for inducers of basal autophagy in neuroblastoma cells suggested that the Vps34 kinase complex but not mTORC1 regulates autophagy.[122] However, other evidence has been provided that mTOR inhibition by rapamycin does indeed induce autophagic flux in neurons.[123–125] Furthermore, an interesting study searching for autophagy regulators in cell culture medium showed that insulin withdrawal triggers the autophagy process[126] (and the authors suggest these effects are mTOR-dependent). Taken together, these data suggest that there are parallel signaling pathways that regulate neuronal autophagy with distinct mechanisms.

Despite the observations noted above regarding autophagy signaling in neurons, it remains unknown whether or not similar results occur in the intact brain *in vivo*.

It has been shown that neuronal autophagy can be induced by hypoxic-ischemia, overexcited glutamate receptors (caused by excitotoxic stimuli such as NMDA and kainic acid), methamphetamine, proteasome inhibition, and secondary to alterations in lysosomal enzyme activity or lipid storage (for review see Yue *et al.*[7]). There is evidence that Akt signaling negatively regulates autophagy in various cell types, but very recently the Akt/Rheb/mTor signaling pathways were shown to regulate acute retrograde axon degeneration which is achieved not by programmed cell death but by autophagy.[127] New research adds other signaling pathways including IGF-1,[128] Cdk5/endophilin B1(Bif-1),[129] JNK/FoxO,[130] and mutant LRRK2[84] to the list of autophagy inducers in pathophysiological conditions.

Autophagic Clearance of Protein Aggregates in Neurons

The studies of *ATG5* and *ATG7* conditional knockout brains reveal a critical role for autophagy in the suppression of spontaneous protein aggregation.[103,104] They also suggest that basal autophagy in neurons is a protein quality control pathway for continuous turnover of misfolded and/or unfolded proteins. Homeostatic levels of p62 control cytoplasmic inclusion body formation in autophagy-deficient mice[99] and proper turnover of p62 by autophagy is critical to prevent spontaneous aggregate formation, which is directly mediated by p62 itself.[99] Furthermore, if p62 accumulates, through inhibition of autophagy, proteasomal function is hindered.[101] Autophagy is likely an alternative degradation pathway for ubiquitinated proteins and supports the UPS to clear potential aggregating proteins when the UPS is overloaded[131] (Figure 2). There are at least two main hypotheses that explain how autophagy protects against neurodegeneration: (1) the autophagy machinery specifically recognizes and eliminates protein aggregates and inclusions, or (2) inclusion bodies that form after impaired protein turnover are degraded by autophagy through an unspecific, secondary mechanism. Supporting the latter, it is suggested that the generation of inclusions and aggresomes of mutant proteins is actually a cytoprotective mechanism.[132,133] Yet most recent studies, especially those involving p62 and other ubiquitin- and LC3-binding proteins suggest that these function as adaptors to specifically identify damaged proteins and direct the degradation of aggregates, almost all of which are tagged with ubiquitin.[98,134]

Aggrephagy, or the specific autophagocytosis of protein aggregates or inclusion, is an example of a selective form of autophagy.[135] Alfy, a FYVE domain-containing protein associated with autophagy, interacts with PI(3)P, p62, and Atg5, and recently has emerged as a conductor of aggrephagy.[134] Therefore, autophagy can be selective, and its selectivity is mediated by adaptor-like proteins such as p62 and Alfy. Protein inclusions in both liver and brain of Atg7 KO mice disappear in Atg7/p62 double KO mice.[99] Whereas the liver damage was dramatically improved, neurological dysfunction was not improved even though inclusion bodies were not generated. This study suggests that prevention/elimination of inclusions is not a major role of autophagy in the brain, and that autophagy may have more general roles in protein/organelle turnover in the brain. Histone deacetylase 6 (HDAC6), which is associated with microtubules and regulates a variety of cellular functions, may also participate in autophagic degradation of polyubiquitinated proteins[136,137] and promotes the clearance of polyQ containing protein aggregates.[136]

Regulation of Neuronal Autophagy by Endosomal and ESCRT Machinery

ESCRT complexes I–III are required for multi-vesicular body (MVB) formation and growth factor signaling in neurons.[116] The fusion of autophagosomes with MVBs creates amphisomes, which likely represent an important route for the degradation of the contents of both original organelles. It was demonstrated that subunits of these complexes, including CHMP2B, plus some regulatory proteins are required for autophagy, and specifically the maturation and delivery of autophagosomes to the lysosome.[138,139] Interestingly, mutations in CHMP2B are implicated in both frontotemporal dementia (FTD)[140] and amyotropic lateral sclerosis (ALS).[141] Though intact ESCRT complexes are required for dendritic development and maintenance, the pathology of these mutations is due to dysfunctional autophagosome degradation.[142]

It seems that neurons are especially vulnerable to aberrant ESCRT function. The emerging evidence strongly suggests the convergence of the endocytic/MVB and autophagic pathways in neurons.[143] The merge may provide an additional layer of control or an alternative pathway for autophagosome turnover in higher eukaryotes; thus, the ESCRT-mediated route may be a critical one in neurons. ESCRT-deficient neurons phenocopy the *ATG5* and *ATG7* deleted neurons, suggesting that ESCRTs/MVBs may be indispensible for basal autophagy. Moreover, functional MVBs are required for the clearance of the disease-related protein TDP-43 and polyQ-containing proteins.[139] In fact, the accumulation of ubiquitin- and p62-labeled inclusions and non-digested autophagosomes in ESCRT-deficient neurons suggests that ESCRT is not only important for homeostasis, but also for selective autophagy in the brain. Finally, membrane dynamics of ESCRT-mediated multi-vesicular body formation is similar to microautophagy, and in fact, a recent paper showed that endosomal microautophagy degrades cytosolic proteins in a somewhat selective manner.[144]

Neuroprotective versus Neurotoxic Role of Autophagy

While most evidence points to autophagy as a protective process in neurons, there is some intriguing evidence to suggest that autophagy is responsible for cell death in the brain. Increased autophagosome number has been observed during hippocampal neuron death and in fact, Atg7 is critical for neuronal cell death after hypoxic/ischemic brain injury.[145] In another model of neuronal cell loss, FTD, inhibition of autophagy actually delays the neurodegeneration progression.[146] Also, loss of the critical autophagy protein Atg7 increases resistance to retrograde degeneration in axons after toxin injury.[127] A consensus in the literature is emerging that the axon, which in part incorporates autophagic cell death, dies differently than the soma, which uses apoptotic cell death.[147] The paradox of autophagy also extends to cancer: (1) autophagy has tumor suppressor properties and helps to keep the cell alive, (2) yet, it is a form of apoptosis-independent cell death. Also, growing tumors can hijack the process and live off of autophagic recycling until angiogenesis occurs. Perhaps basal autophagy in the brain is essential for neuronal homeostasis, but in the case of acute neural injuries that induce massive expansion of autophagy-lysosomal degradation, it becomes harmful and responsible for degeneration, especially in axodendritic projections. The context is therefore very important when considering the application of drugs to increase autophagy in the brain.

AUTOPHAGY IN NEURODEGENERATIVE DISEASES

Basal autophagy serves a neuroprotective function by maintaining neural homeostasis under healthy conditions; but perhaps the more impressive role of autophagy is its ability to degrade and clear disease-related mutant proteins. A body of evidence has increasingly shown that autophagy is primarily responsible for the turnover of aggregate-prone proteins or macromolecular complexes. The examples include polyQ-containing proteins that cause neurodegenerative diseases such as Huntington's disease (huntingtin),[148] spinocerebellar ataxia (SCA1),[149] and spinobulbar muscular atrophy (androgen receptor).[137] In addition, disease-linked mutants of alpha-synuclein in Parkinson's disease, mutant SOD1 in ALS, and tau in Alzheimer's disease and other tauopathies were also shown to be autophagy substrates.[106,150–152] An intact, efficient autophagy system is undoubtedly important and beneficial for the neurons to prevent the build-up of these disease proteins.

Autophagy in mammals is highly evolved and has become a tightly regulated process; but when autophagy is disrupted or misregulated multiple diseases can occur.[153] The evidence for the disturbance of various autophagy pathways by disease proteins has been presented. For example, alpha-synuclein (mutant or wildtype) is linked to impaired autophagy[150] and CMA[154] separately; Parkin and PINK1 mutants were shown to inhibit autophagy of mitochondria (mitophagy);[155] PolyQ-Htt may take part in compromising the efficiency of autophagy.[156] The most remarkable evidence was provided by Nixon's group in their study of Alzheimer's disease (AD)-associated mutations and their effects on multiple stages of autophagy. In particular, the impairment of acidification and substrate proteolysis within autolysosomes is a potential mechanism underlying autophagy failure in AD.[157] In the most common form of familial onset AD, mutant presenilin 1 blocks autophagy by impeding lysosomal proteolysis.

CONCLUSION

Despite the recent surge of publications delving into the molecular mechanisms of autophagic function, there are many remaining questions. Of critical importance is to determine how and where the phagophore (isolation membrane) and thus the autophagosome forms in mammalian cells. What other regulatory proteins are required to act at this site and how might this differ in various tissues and cell types, particularly neurons? In addition, it is imperative to understand how endocytic, MVB/ESCRT, and lysosomal acidification pathways participate in and are required for autophagy completion in neurons. Autophagy is critical for maintaining proper cellular function, especially in differentiated, specialized neurons that are especially vulnerable to aggregated protein accumulation. In turn, disrupted autophagy is implicated in many neurodegenerative diseases such as Alzheimer's, Huntington's, and Parkinson's (discussed in detail in chapters to follow). Striving to uncover the mysteries of the process, autophagy scientists will find new drug targets that may provide the framework for therapies for multiple conditions including aging, cancer and neurodegeneration.

ACKNOWLEDGMENTS

This work was supported by United States National Institutes of Health–National Institute of Neurological Disorders and Stroke Grants R01NS060123, NS060809, RNS055683A; the Michael J. Fox Foundation; the Bachmann-Strauss Dystonia and Parkinson Foundation (to Z. Yue). We thank members of Dr. Yue's lab for critical reading and comments.

REFERENCES

1. J. S. Dixon, "Phagocytic" lysosomes in chromatolytic neurones, *Nature* **215**: 657–658 (1967).
2. R. Wettstein and J. R. Sotelo, Electron microscope study on the regenerative process of peripheral nerves of mice, *Z Zellforsch Mikrosk Anat* **59**: 708–730 (1963).
3. W. Schlote, [The structure of stratified bodies in the axoplasm of dissected fibers of the optic nerve situated distally from the lesion], *J Ultrastruct Res* **16**: 548–568 (1966).
4. E. Holtzman, A. B. Novikoff and H. Villaverde, Lysosomes and GERL in normal and chromatolytic neurons of the rat ganglion nodosum, *Cell Biol* **33**: 419–435 (1967).
5. P. W. Lampert and S. S. Schochet, Jr., Demyelination and remyelination in lead neuropathy. Electron microscopic studies, *J Neuropathol Exp Neurol* **27**: 527–545 (1968).
6. S. Blumcke, H. R. Niedorf, J. Rode and E. Gehlen, [Degeneration of peripheral aminergic nerves], *Verh Dtsch Ges Pathol* **52**: 228–233 (1968).
7. Z. Yue, L. Friedman, M. Komatsu and K. Tanaka, The cellular pathways of neuronal autophagy and their implication in neurodegenerative diseases, *Biochim Biophys Acta* **1793**: 1496–1507 (2009).
8. J. F. Dice, Chaperone-mediated autophagy, *Autophagy* **3**: 295–299 (2007).
9. C. De Duve and R. Wattiaux, Functions of lysosomes, *Annu Rev Physiol* **28**: 435–492 (1966).
10. Z. Xie and D. J. Klionsky, Autophagosome formation: core machinery and adaptations, *Nat Cell Biol* **9**: 1102–1109 (2007).
11. K. Takeshige, M. Baba, S. Tsuboi, T. Noda and Y. Ohsumi, Autophagy in yeast demonstrated with proteinase-deficient mutants and conditions for its induction, *J Cell Biol* **119**: 301–311 (1992).
12. M. Tsukada and Y. Ohsumi, Isolation and characterization of autophagy-defective mutants of *Saccharomyces cerevisiae*, *FEBS Lett* **333**: 169–174 (1993).
13. D. J. Klionsky, J. M. Cregg, W. A. Dunn, Jr., S. D. Emr, Y. Sakai, I. V. Sandoval, A. Sibirny, S. Subramani, M. Thumm, M. Veenhuis and Y. Ohsumi, A unified nomenclature for yeast autophagy-related genes, *Dev Cell* **5**: 539–545 (2003).
14. A. Matsuura, M. Tsukada, Y. Wada and Y. Ohsumi, Apg1p, a novel protein kinase required for the autophagic process in *Saccharomyces cerevisiae*, *Gene* **192**: 245–250 (1997).
15. Y. Kabeya, Y. Kamada, M. Baba, H. Takikawa, M. Sasaki and Y. Ohsumi, Atg17 functions in cooperation with Atg1 and Atg13 in yeast autophagy, *Mol Biol Cell* **16**: 2544–2553 (2005).
16. K. Suzuki, Y. Kubota, T. Sekito and Y. Ohsumi, Hierarchy of Atg proteins in pre-autophagosomal structure organization, *Genes Cells* **12**: 209–218 (2007).
17. E. Y. Chan and S. A. Tooze, Evolution of Atg1 function and regulation, *Autophagy* **5**: 758–765 (2009).

18. H. Toda, H. Mochizuki, R. Flores, 3rd, R. Josowitz, T. B. Krasieva, V. J. Lamorte, E. Suzuki, J. G. Gindhart, K. Furukubo-Tokunaga and T. Tomoda, UNC-51/ATG1 kinase regulates axonal transport by mediating motor-cargo assembly, *Genes Dev* **22**: 3292–3307 (2008).
19. Y. Y. Chang and T. P. Neufeld, An Atg1/Atg13 complex with multiple roles in TOR-mediated autophagy regulation, *Mol Biol Cell* **20**: 2004–2014 (2009).
20. E. Y. Chan, S. Kir and S. A. Tooze, siRNA screening of the kinome identifies ULK1 as a multidomain modulator of autophagy, *J Biol Chem* **282**: 25464–25474 (2007).
21. T. Hara, A. Takamura, C. Kishi, S. Iemura, T. Natsume, J. L. Guan and N. Mizushima, FIP200, a ULK-interacting protein, is required for autophagosome formation in mammalian cells, *J Cell Biol* **181**: 497–510 (2008).
22. I. G. Ganley, H. Lam du, J. Wang, X. Ding, S. Chen and X. Jiang, ULK1.ATG13. FIP200 complex mediates mTOR signaling and is essential for autophagy, *J Biol Chem* **284**: 12297–12305 (2009).
23. C. H. Jung, C. B. Jun, S. H. Ro, Y. M. Kim, N. M. Otto, J. Cao, M. Kundu and D. H. Kim, ULK-Atg13-FIP200 complexes mediate mTOR signaling to the autophagy machinery, *Mol Biol Cell* **20**: 1992–2003 (2009).
24. E. Y. Chan, A. Longatti, N. C. McKnight and S. A. Tooze, Kinase-inactivated ULK proteins inhibit autophagy via their conserved C-terminal domains using an Atg13-independent mechanism, *Mol Cell Biol* **29**: 157–171 (2009).
25. N. Hosokawa, T. Hara, T. Kaizuka, C. Kishi, A. Takamura, Y. Miura, S. Iemura, T. Natsume, K. Takehana, N. Yamada, J. L. Guan, N. Oshiro and N. Mizushima, Nutrient-dependent mTORC1 association with the ULK1-Atg13-FIP200 complex required for autophagy, *Mol Biol Cell* **20**: 1981–1991 (2009).
26. C. A. Mercer, A. Kaliappan and P. B. Dennis, A novel, human Atg13 binding protein, Atg101, interacts with ULK1 and is essential for macroautophagy, *Autophagy* **5**: 649–662 (2009).
27. N. Hosokawa, T. Sasaki, S. Iemura, T. Natsume, T. Hara and N. Mizushima, Atg101, a novel mammalian autophagy protein interacting with Atg13, *Autophagy* **5**: 973–979 (2009).
28. E. Itakura and N. Mizushima, Characterization of autophagosome formation site by a hierarchical analysis of mammalian Atg proteins, *Autophagy* **6**: 764–776 (2010).
29. D. F. Egan, D. B. Shackelford, M. M. Mihaylova, S. Gelino, R. A. Kohnz, W. Mair, D. S. Vasquez, A. Joshi, D. M. Gwinn, R. Taylor, J. M. Asara, J. Fitzpatrick, A. Dillin, B. Viollet, M. Kundu, M. Hansen and R. J. Shaw, Phosphorylation of ULK1 (hATG1) by AMP-activated protein kinase connects energy sensing to mitophagy, *Science* **331**: 456–461.
30. J. Kim, M. Kundu, B. Viollet and K. L. Guan, AMPK and mTOR regulate autophagy through direct phosphorylation of Ulk1, *Nat Cell Biol* **13**: 132–141.
31. L. Shang, S. Chen, F. Du, S. Li, L. Zhao and X. Wang, Nutrient starvation elicits an acute autophagic response mediated by Ulk1 dephosphorylation and its subsequent dissociation from AMPK, *Proc Natl Acad Sci USA* **108**: 4788–4793 (2011).

32. M. Mari, J. Griffith, E. Rieter, L. Krishnappa, D. J. Klionsky and F. Reggiori, An Atg9-containing compartment that functions in the early steps of autophagosome biogenesis, *J Cell Biol* **190**: 1005–1022 (2010).
33. U. Nair, A. Jotwani, J. Geng, N. Gammoh, D. Richerson, W. L. Yen, J. Griffith, S. Nag, K. Wang, T. Moss, M. Baba, J. A. McNew, X. Jiang, F. Reggiori, T. J. Melia and D. J. Klionsky, SNARE proteins are required for macroautophagy, *Cell* **146**: 290–302 (2011).
34. F. Reggiori, K. A. Tucker, P. E. Stromhaug and D. J. Klionsky, The Atg1-Atg13 complex regulates Atg9 and Atg23 retrieval transport from the pre-autophagosomal structure, *Dev Cell* **6**: 79–90 (2004).
35. T. Saitoh, N. Fujita, T. Hayashi, K. Takahara, T. Satoh, H. Lee, K. Matsunaga, S. Kageyama, H. Omori, T. Noda, N. Yamamoto, T. Kawai, K. Ishii, O. Takeuchi, T. Yoshimori and S. Akira, Atg9a controls dsDNA-driven dynamic translocation of STING and the innate immune response, *Proc Natl Acad Sci USA* **106**: 20842–20846 (2009).
36. A. R. Young, E. Y. Chan, X. W. Hu, R. Kochl, S. G. Crawshaw, S. High, D. W. Hailey, J. Lippincott-Schwartz and S. A. Tooze, Starvation and ULK1-dependent cycling of mammalian Atg9 between the TGN and endosomes, *J Cell Sci* **119**: 3888–3900 (2006).
37. J. L. Webber and S. A. Tooze, Coordinated regulation of autophagy by p38alpha MAPK through mAtg9 and p38IP, *EMBO J* **29**: 27–40 (2010).
38. P. O. Seglen and P. B. Gordon, 3-methyladenine: specific inhibitor of autophagic/lysosomal protein degradation in isolated rat hepatocytes, *Proc Natl Acad Sci USA* **79**: 1889–1892 (1982).
39. A. Simonsen and S. A. Tooze, Coordination of membrane events during autophagy by multiple class III PI3-kinase complexes, *J Cell Biol* **186**: 773–782 (2009).
40. A. Kihara, T. Noda, N. Ishihara and Y. Ohsumi, Two distinct Vps34 phosphatidylinositol 3-kinase complexes function in autophagy and carboxypeptidase Y sorting in *Saccharomyces cerevisiae*, *J Cell Biol* **152**: 519–530 (2001).
41. E. Itakura, C. Kishi, K. Inoue and N. Mizushima, Beclin 1 forms two distinct phosphatidylinositol 3-kinase complexes with mammalian Atg14 and UVRAG, *Mol Biol Cell* **19**: 5360–5372 (2008).
42. Q. Sun, W. Fan, K. Chen, X. Ding, S. Chen and Q. Zhong, Identification of Barkor as a mammalian autophagy-specific factor for Beclin 1 and class III phosphatidylinositol 3-kinase, *Proc Natl Acad Sci USA* **105**: 19211–19216 (2008).
43. Y. Zhong, Q. J. Wang, X. Li, Y. Yan, J. M. Backer, B. T. Chait, N. Heintz and Z. Yue, Distinct regulation of autophagic activity by Atg14L and Rubicon associated with Beclin 1-phosphatidylinositol-3-kinase complex, *Nat Cell Biol* **11**: 468–476 (2009).
44. K. Matsunaga, T. Saitoh, K. Tabata, H. Omori, T. Satoh, N. Kurotori, I. Maejima, K. Shirahama-Noda, T. Ichimura, T. Isobe, S. Akira, T. Noda and T. Yoshimori, Two Beclin 1-binding proteins, Atg14L and Rubicon, reciprocally regulate autophagy at different stages, *Nat Cell Biol* **11**: 385–396 (2009).

45. C. Liang, P. Feng, B. Ku, I. Dotan, D. Canaani, B. H. Oh and J. U. Jung, Autophagic and tumour suppressor activity of a novel Beclin1-binding protein UVRAG, *Nat Cell Biol* **8**: 688–699 (2006).
46. S. F. Funderburk, Q. J. Wang and Z. Yue, The Beclin 1-VPS34 complex — at the crossroads of autophagy and beyond, *Trends Cell Biol* **20**: 355–362 (2010).
47. C. Liang, J. S. Lee, K. S. Inn, M. U. Gack, Q. Li, E. A. Roberts, I. Vergne, V. Deretic, P. Feng, C. Akazawa and J. U. Jung, Beclin1-binding UVRAG targets the class C Vps complex to coordinate autophagosome maturation and endocytic trafficking, *Nat Cell Biol* **10**: 776–787 (2008).
48. Y. Takahashi, D. Coppola, N. Matsushita, H. D. Cualing, M. Sun, Y. Sato, C. Liang, J. U. Jung, J. Q. Cheng, J. J. Mule, W. J. Pledger and H. G. Wang, Bif-1 interacts with Beclin 1 through UVRAG and regulates autophagy and tumorigenesis, *Nat Cell Biol* **9**: 1142–1151 (2007).
49. S. B. Thoresen, N. M. Pedersen, K. Liestol and H. Stenmark, A phosphatidylinositol 3-kinase class III sub-complex containing VPS15, VPS34, Beclin 1, UVRAG and BIF-1 regulates cytokinesis and degradative endocytic traffic, *Exp Cell Res* **316**: 3368–3378 (2010).
50. G. M. Fimia, A. Stoykova, A. Romagnoli, L. Giunta, S. Di Bartolomeo, R. Nardacci, M. Corazzari, C. Fuoco, A. Ucar, P. Schwartz, P. Gruss, M. Piacentini, K. Chowdhury and F. Cecconi, Ambra1 regulates autophagy and development of the nervous system, *Nature* **447**: 1121–1125 (2007).
51. S. Di Bartolomeo, M. Corazzari, F. Nazio, S. Oliverio, G. Lisi, M. Antonioli, V. Pagliarini, S. Matteoni, C. Fuoco, L. Giunta, M. D'Amelio, R. Nardacci, A. Romagnoli, M. Piacentini, F. Cecconi and G. M. Fimia, The dynamic interaction of AMBRA1 with the dynein motor complex regulates mammalian autophagy, *J Cell Biol* **191**: 155–168 (2010).
52. S. Pattingre, A. Tassa, X. Qu, R. Garuti, X. H. Liang, N. Mizushima, M. Packer, M. D. Schneider and B. Levine, Bcl-2 antiapoptotic proteins inhibit Beclin 1-dependent autophagy, *Cell* **122**: 927–939 (2005).
53. J. H. Zhu, C. Horbinski, F. Guo, S. Watkins, Y. Uchiyama and C. T. Chu, Regulation of autophagy by extracellular signal-regulated protein kinases during 1-methyl-4-phenylpyridinium-induced cell death, *Am J Pathol* **170**: 75–86 (2007).
54. F. Scarlatti, R. Maffei, I. Beau, P. Codogno and R. Ghidoni, Role of non-canonical Beclin 1-independent autophagy in cell death induced by resveratrol in human breast cancer cells, *Cell Death Differ* **15**: 1318–1329 (2008).
55. T. Proikas-Cezanne, S. Waddell, A. Gaugel, T. Frickey, A. Lupas and A. Nordheim, WIPI-1alpha (WIPI49), a member of the novel 7-bladed WIPI protein family, is aberrantly expressed in human cancer and is linked to starvation-induced autophagy, *Oncogene* **23**: 9314–9325 (2004).
56. H. E. Polson, J. de Lartigue, D. J. Rigden, M. Reedijk, S. Urbe, M. J. Clague and S. A. Tooze, Mammalian Atg18 (WIPI2) localizes to omegasome-anchored phagophores and positively regulates LC3 lipidation, *Autophagy* **6**: 506–522 (2010).

57. Z. Yang and D. J. Klionsky, Eaten alive: a history of macroautophagy, *Nat Cell Biol* **12**: 814–822 (2010).
58. D. J. Klionsky, The molecular machinery of autophagy: unanswered questions, *J Cell Sci* **118**: 7–18 (2005).
59. N. Mizushima, H. Sugita, T. Yoshimori and Y. Ohsumi, A new protein conjugation system in human. The counterpart of the yeast Apg12p conjugation system essential for autophagy, *J Biol Chem* **273**: 33889–33892 (1998).
60. N. Mizushima, A. Kuma, Y. Kobayashi, A. Yamamoto, M. Matsubae, T. Takao, T. Natsume, Y. Ohsumi and T. Yoshimori, Mouse Apg16L, a novel WD-repeat protein, targets to the autophagic isolation membrane with the Apg12-Apg5 conjugate, *J Cell Sci* **116**: 1679–1688 (2003).
61. N. Fujita, M. Hayashi-Nishino, H. Fukumoto, H. Omori, A. Yamamoto, T. Noda and T. Yoshimori, An Atg4B mutant hampers the lipidation of LC3 paralogues and causes defects in autophagosome closure, *Mol Biol Cell* **19**: 4651–4659 (2008).
62. N. Mizushima, A. Yamamoto, M. Hatano, Y. Kobayashi, Y. Kabeya, K. Suzuki, T. Tokuhisa, Y. Ohsumi and T. Yoshimori, Dissection of autophagosome formation using Apg5-deficient mouse embryonic stem cells, *J Cell Biol* **152**: 657–668 (2001).
63. Y. Kabeya, N. Mizushima, T. Ueno, A. Yamamoto, T. Kirisako, T. Noda, E. Kominami, Y. Ohsumi and T. Yoshimori, LC3, a mammalian homologue of yeast Apg8p, is localized in autophagosome membranes after processing, *EMBO J* **19**: 5720–5728 (2000).
64. J. Hemelaar, V. S. Lelyveld, B. M. Kessler and H. L. Ploegh, A single protease, Apg4B, is specific for the autophagy-related ubiquitin-like proteins GATE-16, MAP1-LC3, GABARAP, and Apg8L, *J Biol Chem* **278**: 51841–51850 (2003).
65. I. Tanida, E. Tanida-Miyake, M. Komatsu, T. Ueno and E. Kominami, Human Apg3p/Aut1p homologue is an authentic E2 enzyme for multiple substrates, GATE-16, GABARAP, and MAP-LC3, and facilitates the conjugation of hApg12p to hApg5p, *J Biol Chem* **277**: 13739–13744 (2002).
66. Y. Kabeya, N. Mizushima, A. Yamamoto, S. Oshitani-Okamoto, Y. Ohsumi and T. Yoshimori, LC3, GABARAP and GATE16 localize to autophagosomal membrane depending on form-II formation, *J Cell Sci* **117**: 2805–2812 (2004).
67. D. C. Rubinsztein, A. M. Cuervo, B. Ravikumar, S. Sarkar, V. Korolchuk, S. Kaushik and D. J. Klionsky, In search of an "autophagomometer", *Autophagy* **5**: 585–589 (2009).
68. D. J. Klionsky, H. Abeliovich, P. Agostinis, D. K. Agrawal, *et al.*, Guidelines for the use and interpretation of assays for monitoring autophagy in higher eukaryotes, *Autophagy* **4**: 151–175 (2008).
69. I. Tanida, Y. S. Sou, J. Ezaki, N. Minematsu-Ikeguchi, T. Ueno and E. Kominami, HsAtg4B/HsApg4B/autophagin-1 cleaves the carboxyl termini of three human Atg8 homologues and delipidates microtubule-associated protein light chain 3- and GABAA receptor-associated protein-phospholipid conjugates, *J Biol Chem* **279**: 36268–36276 (2004).

70. T. Hanada, N. N. Noda, Y. Satomi, Y. Ichimura, Y. Fujioka, T. Takao, F. Inagaki and Y. Ohsumi, The Atg12-Atg5 conjugate has a novel E3-like activity for protein lipidation in autophagy, *J Biol Chem* **282**: 37298–37302 (2007).
71. A. Overbye, M. Fengsrud and P. O. Seglen, Proteomic analysis of membrane-associated proteins from rat liver autophagosomes, *Autophagy* **3**: 300–322 (2007).
72. S. A. Tooze and T. Yoshimori, The origin of the autophagosomal membrane, *Nat Cell Biol* **12**: 831–835 (2010).
73. M. Hayashi-Nishino, N. Fujita, T. Noda, A. Yamaguchi, T. Yoshimori and A. Yamamoto, A subdomain of the endoplasmic reticulum forms a cradle for autophagosome formation, *Nat Cell Biol* **11**: 1433–1437 (2009).
74. P. Yla-Anttila, H. Vihinen, E. Jokitalo and E. L. Eskelinen, 3D tomography reveals connections between the phagophore and endoplasmic reticulum, *Autophagy* **5**: 1180–1185 (2009).
75. D. W. Hailey, A. S. Rambold, P. Satpute-Krishnan, K. Mitra, R. Sougrat, P. K. Kim and J. Lippincott-Schwartz, Mitochondria supply membranes for autophagosome biogenesis during starvation, *Cell* **141**: 656–667 (2010).
76. B. Ravikumar, K. Moreau, L. Jahreiss, C. Puri and D. C. Rubinsztein, Plasma membrane contributes to the formation of pre-autophagosomal structures, *Nat Cell Biol* **12**: 747–757 (2010).
77. T. Noda and Y. Ohsumi, Tor, a phosphatidylinositol kinase homologue, controls autophagy in yeast, *J Biol Chem* **273**: 3963–3966 (1998).
78. E. F. Blommaart, J. J. Luiken, P. J. Blommaart, G. M. van Woerkom and A. J. Meijer, Phosphorylation of ribosomal protein S6 is inhibitory for autophagy in isolated rat hepatocytes, *J Biol Chem* **270**: 2320–2326 (1995).
79. Y. Sancak, L. Bar-Peled, R. Zoncu, A. L. Markhard, S. Nada and D. M. Sabatini, Ragulator-Rag complex targets mTORC1 to the lysosomal surface and is necessary for its activation by amino acids, *Cell* **141**: 290–303 (2010).
80. R. C. Scott, G. Juhasz and T. P. Neufeld, Direct induction of autophagy by Atg1 inhibits cell growth and induces apoptotic cell death, *Curr Biol* **17**: 1–11 (2007).
81. E. Ogier-Denis, S. Pattingre, J. El Benna and P. Codogno, Erk1/2-dependent phosphorylation of Galpha-interacting protein stimulates its GTPase accelerating activity and autophagy in human colon cancer cells, *J Biol Chem* **275**: 39090–39095 (2000).
82. S. Sarkar, R. A. Floto, Z. Berger, S. Imarisio, A. Cordenier, M. Pasco, L. J. Cook and D. C. Rubinsztein, Lithium induces autophagy by inhibiting inositol monophosphatase, *J Cell Biol* **170**: 1101–1111 (2005).
83. E. Morselli, M. C. Maiuri, M. Markaki, E. Megalou, A. Pasparaki, K. Palikaras, A. Criollo, L. Galluzzi, S. A. Malik, I. Vitale, M. Michaud, F. Madeo, N. Tavernarakis and G. Kroemer, The life span-prolonging effect of sirtuin-1 is mediated by autophagy, *Autophagy* **6**: 186–188 (2010).
84. S. J. Cherra, 3rd, S. M. Kulich, G. Uechi, M. Balasubramani, J. Mountzouris, B. W. Day and C. T. Chu, Regulation of the autophagy protein LC3 by phosphorylation, *J Cell Biol* **190**: 533–539 (2010).

85. A. Petiot, E. Ogier-Denis, E. F. Blommaart, A. J. Meijer and P. Codogno, Distinct classes of phosphatidylinositol 3′-kinases are involved in signaling pathways that control macroautophagy in HT-29 cells, *J Biol Chem* **275**: 992–998 (2000).
86. S. Arico, A. Petiot, C. Bauvy, P. F. Dubbelhuis, A. J. Meijer, P. Codogno and E. Ogier-Denis, The tumor suppressor PTEN positively regulates macroautophagy by inhibiting the phosphatidylinositol 3-kinase/protein kinase B pathway, *J Biol Chem* **276**: 35243–35246 (2001).
87. Z. Dou, M. Chattopadhyay, J. A. Pan, J. L. Guerriero, Y. P. Jiang, L. M. Ballou, Z. Yue, R. Z. Lin and W. X. Zong, The class IA phosphatidylinositol 3-kinase p110-beta subunit is a positive regulator of autophagy, *J Cell Biol* **191**: 827–843 (2010).
88. Y. Wei, S. Pattingre, S. Sinha, M. Bassik and B. Levine, JNK1-mediated phosphorylation of Bcl-2 regulates starvation-induced autophagy, *Mol Cell* **30**: 678–688 (2008).
89. J. Liang, S. H. Shao, Z. X. Xu, B. Hennessy, Z. Ding, M. Larrea, S. Kondo, D. J. Dumont, J. U. Gutterman, C. L. Walker, J. M. Slingerland and G. B. Mills, The energy sensing LKB1-AMPK pathway regulates p27(kip1) phosphorylation mediating the decision to enter autophagy or apoptosis, *Nat Cell Biol* **9**: 218–224 (2007).
90. E. Tasdemir, M. Chiara Maiuri, E. Morselli, A. Criollo, M. D'Amelio, M. Djavaheri-Mergny, F. Cecconi, N. Tavernarakis and G. Kroemer, A dual role of p53 in the control of autophagy, *Autophagy* **4**: 810–814 (2008).
91. K. H. Vousden and K. M. Ryan, p53 and metabolism, *Nat Rev Cancer* **9**: 691–700 (2009).
92. M. W. Wooten, T. Geetha, J. R. Babu, M. L. Seibenhener, J. Peng, N. Cox, M. T. Diaz-Meco and J. Moscat, Essential role of sequestosome 1/p62 in regulating accumulation of Lys63-ubiquitinated proteins, *J Biol Chem* **283**: 6783–6789 (2008).
93. Q. J. Wang, Y. Ding, D. S. Kohtz, N. Mizushima, I. M. Cristea, M. P. Rout, B. T. Chait, Y. Zhong, N. Heintz and Z. Yue, Induction of autophagy in axonal dystrophy and degeneration, *J Neurosci* **26**: 8057–8068 (2006).
94. Z. Yue, Regulation of neuronal autophagy in axon: implication of autophagy in axonal function and dysfunction/degeneration, *Autophagy* **3**: 139–141 (2007).
95. S. Pankiv, T. H. Clausen, T. Lamark, A. Brech, J. A. Bruun, H. Outzen, A. Overvatn, G. Bjorkoy and T. Johansen, p62/SQSTM1 binds directly to Atg8/LC3 to facilitate degradation of ubiquitinated protein aggregates by autophagy, *J Biol Chem* **282**: 24131–24145 (2007).
96. E. Shvets, E. Fass, R. Scherz-Shouval and Z. Elazar, The N-terminus and Phe52 residue of LC3 recruit p62/SQSTM1 into autophagosomes, *J Cell Sci* **121**: 2685–2695 (2008).
97. Y. Ichimura, T. Kumanomidou, Y. S. Sou, T. Mizushima, J. Ezaki, T. Ueno, E. Kominami, T. Yamane, K. Tanaka and M. Komatsu, Structural basis for sorting mechanism of p62 in selective autophagy, *J Biol Chem* **283**: 22847–22857 (2008).
98. G. Bjorkoy, T. Lamark, A. Brech, H. Outzen, M. Perander, A. Overvatn, H. Stenmark and T. Johansen, p62/SQSTM1 forms protein aggregates degraded by

autophagy and has a protective effect on huntingtin-induced cell death, *J Cell Biol* **171**: 603–614 (2005).

99. M. Komatsu, S. Waguri, M. Koike, Y. S. Sou, T. Ueno, T. Hara, N. Mizushima, J. Iwata, J. Ezaki, S. Murata, J. Hamazaki, Y. Nishito, S. Iemura, T. Natsume, T. Yanagawa, J. Uwayama, E. Warabi, H. Yoshida, T. Ishii, A. Kobayashi, M. Yamamoto, Z. Yue, Y. Uchiyama, E. Kominami and K. Tanaka, Homeostatic levels of p62 control cytoplasmic inclusion body formation in autophagy-deficient mice, *Cell* **131**: 1149–1163 (2007).
100. S. Pankiv, T. Lamark, J. A. Bruun, A. Overvatn, G. Bjorkoy and T. Johansen, Nucleocytoplasmic shuttling of p62/SQSTM1 and its role in recruitment of nuclear polyubiquitinated proteins to promyelocytic leukemia bodies, *J Biol Chem* **285**: 5941–5953 (2010).
101. V. I. Korolchuk, A. Mansilla, F. M. Menzies and D. C. Rubinsztein, Autophagy inhibition compromises degradation of ubiquitin-proteasome pathway substrates, *Mol Cell* **33**: 517–527 (2009).
102. V. Kirkin, T. Lamark, Y. S. Sou, G. Bjorkoy, J. L. Nunn, J. A. Bruun, E. Shvets, D. G. McEwan, T. H. Clausen, P. Wild, I. Bilusic, J. P. Theurillat, A. Overvatn, T. Ishii, Z. Elazar, M. Komatsu, I. Dikic and T. Johansen, A role for NBR1 in autophagosomal degradation of ubiquitinated substrates, *Mol Cell* **33**: 505–516 (2009).
103. M. Komatsu, S. Waguri, T. Chiba, S. Murata, J. Iwata, I. Tanida, T. Ueno, M. Koike, Y. Uchiyama, E. Kominami and K. Tanaka, Loss of autophagy in the central nervous system causes neurodegeneration in mice, *Nature* **441**: 880–884 (2006).
104. T. Hara, K. Nakamura, M. Matsui, A. Yamamoto, Y. Nakahara, R. Suzuki-Migishima, M. Yokoyama, K. Mishima, I. Saito, H. Okano and N. Mizushima, Suppression of basal autophagy in neural cells causes neurodegenerative disease in mice, *Nature* **441**: 885–889 (2006).
105. M. Komatsu, Q. J. Wang, G. R. Holstein, V. L. Friedrich, Jr., J. Iwata, E. Kominami, B. T. Chait, K. Tanaka and Z. Yue, Essential role for autophagy protein Atg7 in the maintenance of axonal homeostasis and the prevention of axonal degeneration, *Proc Natl Acad Sci USA* **104**: 14489–14494 (2007).
106. W. H. Yu, A. M. Cuervo, A. Kumar, C. M. Peterhoff, S. D. Schmidt, J. H. Lee, P. S. Mohan, M. Mercken, M. R. Farmery, L. O. Tjernberg, Y. Jiang, K. Duff, Y. Uchiyama, J. Naslund, P. M. Mathews, A. M. Cataldo and R. A. Nixon, Macroautophagy — a novel beta-amyloid peptide-generating pathway activated in Alzheimer's disease, *J Cell Biol* **171**: 87–98 (2005).
107. R. A. Nixon, J. Wegiel, A. Kumar, W. H. Yu, C. Peterhoff, A. Cataldo and A. M. Cuervo, Extensive involvement of autophagy in Alzheimer disease: an immuno-electron microscopy study, *J Neuropathol Exp Neurol* **64**: 113–122 (2005).
108. N. Mizushima, Methods for monitoring autophagy, *Int J Biochem Cell Biol* **36**: 2491–2502 (2004).
109. R. Kochl, X. W. Hu, E. Y. Chan and S. A. Tooze, Microtubules facilitate autophagosome formation and fusion of autophagosomes with endosomes, *Traffic* **7**: 129–145 (2006).

110. S. Kimura, T. Noda and T. Yoshimori, Dynein-dependent movement of autophagosomes mediates efficient encounters with lysosomes, *Cell Struct Funct* **33**: 109–122 (2008).

111. K. Katsumata, J. Nishiyama, T. Inoue, N. Mizushima, J. Takeda and M. Yuzaki, Dynein- and activity-dependent retrograde transport of autophagosomes in neuronal axons, *Autophagy* **6**: 378–385 (2010).

112. Y. Yang, L. Q. Feng and X. X. Zheng, Microtubule and kinesin/dynein-dependent, bi-directional transport of autolysosomes in neurites of PC12 cells, *Int J Biochem Cell Biol* **43**: 1147–1156 (2011).

113. S. Lee, Y. Sato and R. A. Nixon, Lysosomal proteolysis inhibition selectively disrupts axonal transport of degradative organelles and causes an Alzheimer's-like axonal dystrophy, *J Neurosci* **31**: 7817–7830 (2011).

114. M. R. Matthews and V. H. Nelson, Detachment of structurally intact nerve endings from chromatolytic neurones of rat superior cervical ganglion during the depression of synaptic transmission induced by post-ganglionic axotomy, *J Physiol* **245**: 91–135 (1975).

115. N. Mizushima, A. Yamamoto, M. Matsui, T. Yoshimori and Y. Ohsumi, *In vivo* analysis of autophagy in response to nutrient starvation using transgenic mice expressing a fluorescent autophagosome marker, *Mol Biol Cell* **15**: 1101–1111 (2004).

116. S. A. Tooze and G. Schiavo, Liaisons dangereuses: autophagy, neuronal survival and neurodegeneration, *Curr Opin Neurobiol* **18**: 504–515 (2008).

117. X. Zhou, J. R. Babu, S. da Silva, Q. Shu, I. A. Graef, T. Oliver, T. Tomoda, T. Tani, M. W. Wooten and F. Wang, Unc-51-like kinase 1/2-mediated endocytic processes regulate filopodia extension and branching of sensory axons, *Proc Natl Acad US* **104**: 5842–5847 (2007).

118. A. Yamamoto, M. L. Cremona and J. E. Rothman, Autophagy-mediated clearance of huntingtin aggregates triggered by the insulin-signaling pathway, *J Cell Biol* **172**: 719–731 (2006).

119. S. Sarkar, B. Ravikumar, R. A. Floto and D. C. Rubinsztein, Rapamycin and mTOR-independent autophagy inducers ameliorate toxicity of polyglutamine-expanded huntingtin and related proteinopathies, *Cell Death Differ* **16**: 46–56 (2009).

120. A. S. Tsvetkov, J. Miller, M. Arrasate, J. S. Wong, M. A. Pleiss and S. Finkbeiner, A small-molecule scaffold induces autophagy in primary neurons and protects against toxicity in a Huntington disease model, *Proc Natl Acad Sci USA* **107**:16982–16987 (2010).

121. L. Zhang, J. Yu, H. Pan, P. Hu, Y. Hao, W. Cai, H. Zhu, A. D. Yu, X. Xie, D. Ma and J. Yuan, Small molecule regulators of autophagy identified by an image-based high-throughput screen, *Proc Natl Acad Sci USA* **104**: 19023–19028 (2007).

122. M. M. Lipinski, G. Hoffman, A. Ng, W. Zhou, B. F. Py, E. Hsu, X. Liu, J. Eisenberg, J. Liu, J. Blenis, R. J. Xavier and J. Yuan, A genome-wide siRNA screen reveals multiple mTORC1 independent signaling pathways regulating autophagy under normal nutritional conditions, *Dev Cell* **18**: 1041–1052 (2010).

123. B. Boland, A. Kumar, S. Lee, F. M. Platt, J. Wegiel, W. H. Yu and R. A. Nixon, Autophagy induction and autophagosome clearance in neurons: relationship to autophagic pathology in Alzheimer's disease, *J Neurosci* **28**: 6926–6937 (2008).
124. C. Rose, F. M. Menzies, M. Renna, A. Acevedo-Arozena, S. Corrochano, O. Sadiq, S. D. Brown and D. C. Rubinsztein, Rilmenidine attenuates toxicity of polyglutamine expansions in a mouse model of Huntington's disease, *Hum Mol Genet* **19**: 2144–2153 (2010).
125. L. Crews, B. Spencer, P. Desplats, C. Patrick, A. Paulino, E. Rockenstein, L. Hansen, A. Adame, D. Galasko and E. Masliah, Selective molecular alterations in the autophagy pathway in patients with Lewy body disease and in models of alpha-synucleinopathy, *PloS One* **5**: e9313 (2010).
126. J. E. Young, R. A. Martinez and A. R. La Spada, Nutrient deprivation induces neuronal autophagy and implicates reduced insulin signaling in neuroprotective autophagy activation, *J Biol Chem* **284**: 2363–2373 (2009).
127. H. C. Cheng, S. R. Kim, T. F. Oo, T. Kareva, O. Yarygina, M. Rzhetskaya, C. Wang, M. During, Z. Talloczy, K. Tanaka, M. Komatsu, K. Kobayashi, H. Okano, N. Kholodilov and R. E. Burke, Akt suppresses retrograde degeneration of dopaminergic axons by inhibition of macroautophagy, *J Neurosci* **31**: 2125–2135 (2011).
128. M. Bains, V. Zaegel, J. Mize-Berge and K. A. Heidenreich, IGF-I stimulates Rab7-RILP interaction during neuronal autophagy, *Neurosci Lett* **488**: 112–117 (2011).
129. A. S. Wong, R. H. Lee, A. Y. Cheung, P. K. Yeung, S. K. Chung, Z. H. Cheung and N. Y. Ip, Cdk5-mediated phosphorylation of endophilin B1 is required for induced autophagy in models of Parkinson's disease, *Nat Cell Biol* **13**: 568–579 (2011).
130. P. Xu, M. Das, J. Reilly and R. J. Davis, JNK regulates FoxO-dependent autophagy in neurons, *Genes Dev* **25**: 310–322 (2011).
131. V. I. Korolchuk, F. M. Menzies and D. C. Rubinsztein, Mechanisms of cross-talk between the ubiquitin-proteasome and autophagy-lysosome systems, *FEBS Lett* **584**: 1393–1398 (2010).
132. M. Arrasate, S. Mitra, E. S. Schweitzer, M. R. Segal and S. Finkbeiner, Inclusion body formation reduces levels of mutant huntingtin and the risk of neuronal death, *Nature* **431**: 805–810 (2004).
133. M. Tanaka, Y. M. Kim, G. Lee, E. Junn, T. Iwatsubo and M. M. Mouradian, Aggresomes formed by alpha-synuclein and synphilin-1 are cytoprotective, *J Biol Chem* **279**: 4625–4631 (2004).
134. M. Filimonenko, P. Isakson, K. D. Finley, M. Anderson, H. Jeong, T. J. Melia, B. J. Bartlett, K. M. Myers, H. C. Birkeland, T. Lamark, D. Krainc, A. Brech, H. Stenmark, A. Simonsen and A. Yamamoto, The selective macroautophagic degradation of aggregated proteins requires the PI3P-binding protein Alfy, *Mol Cell* **38**: 265–279 (2010).
135. A. Yamamoto and A. Simonsen, The elimination of accumulated and aggregated proteins: a role for aggrephagy in neurodegeneration, *Neurobiol Dis* **43**: 17–28 (2011).

136. A. Iwata, B. E. Riley, J. A. Johnston and R. R. Kopito, HDAC6 and microtubules are required for autophagic degradation of aggregated huntingtin, *J Biol Chem* **280**: 40282–40292 (2005).
137. U. B. Pandey, Z. Nie, Y. Batlevi, B. A. McCray, G. P. Ritson, N. B. Nedelsky, S. L. Schwartz, N. A. DiProspero, M. A. Knight, O. Schuldiner, R. Padmanabhan, M. Hild, D. L. Berry, D. Garza, C. C. Hubbert, T. P. Yao, E. H. Baehrecke and J. P. Taylor, HDAC6 rescues neurodegeneration and provides an essential link between autophagy and the UPS, *Nature* **447**: 859–863 (2007).
138. T. E. Rusten, T. Vaccari, K. Lindmo, L. M. Rodahl, I. P. Nezis, C. Sem-Jacobsen, F. Wendler, J. P. Vincent, A. Brech, D. Bilder and H. Stenmark, ESCRTs and Fab1 regulate distinct steps of autophagy, *Curr Biol* **17**: 1817–1825 (2007).
139. M. Filimonenko, S. Stuffers, C. Raiborg, A. Yamamoto, L. Malerod, E. M. Fisher, A. Isaacs, A. Brech, H. Stenmark and A. Simonsen, Functional multivesicular bodies are required for autophagic clearance of protein aggregates associated with neurodegenerative disease, *J Cell Biol* **179**: 485–500 (2007).
140. G. Skibinski, N. J. Parkinson, J. M. Brown, L. Chakrabarti, S. L. Lloyd, H. Hummerich, J. E. Nielsen, J. R. Hodges, M. G. Spillantini, T. Thusgaard, S. Brandner, A. Brun, M. N. Rossor, A. Gade, P. Johannsen, S. A. Sorensen, S. Gydesen, E. M. Fisher and J. Collinge, Mutations in the endosomal ESCRTIII-complex subunit CHMP2B in frontotemporal dementia, *Nat Genet* **37**: 806–808 (2005).
141. N. Parkinson, P. G. Ince, M. O. Smith, R. Highley, G. Skibinski, P. M. Andersen, K. E. Morrison, H. S. Pall, O. Hardiman, J. Collinge, P. J. Shaw and E. M. Fisher, ALS phenotypes with mutations in CHMP2B (charged multivesicular body protein 2B), *Neurology* **67**: 1074–1077 (2006).
142. J. A. Lee, A. Beigneux, S. T. Ahmad, S. G. Young and F. B. Gao, ESCRT-III dysfunction causes autophagosome accumulation and neurodegeneration, *Curr Biol* **17**: 1561–1567 (2007).
143. K. E. Larsen and D. Sulzer, Autophagy in neurons: a review, *Histol Histopathol* **17**: 897–908 (2002).
144. R. Sahu, S. Kaushik, C. C. Clement, E. S. Cannizzo, B. Scharf, A. Follenzi, I. Potolicchio, E. Nieves, A. M. Cuervo and L. Santambrogio, Microautophagy of cytosolic proteins by late endosomes, *Dev Cell* **20**: 131–139 (2011).
145. M. Koike, M. Shibata, M. Tadakoshi, K. Gotoh, M. Komatsu, S. Waguri, N. Kawahara, K. Kuida, S. Nagata, E. Kominami, K. Tanaka and Y. Uchiyama, Inhibition of autophagy prevents hippocampal pyramidal neuron death after hypoxic-ischemic injury, *Am J Pathol* **172**: 454–469 (2008).
146. J. A. Lee and F. B. Gao, Inhibition of autophagy induction delays neuronal cell loss caused by dysfunctional ESCRT-III in frontotemporal dementia, *J Neurosci* **29**: 8506–8511 (2009).
147. M. C. Raff, A. V. Whitmore and J. T. Finn, Axonal self-destruction and neurodegeneration, *Science* **296**: 868–871 (2002).

148. B. Ravikumar, R. Duden and D. C. Rubinsztein, Aggregate-prone proteins with polyglutamine and polyalanine expansions are degraded by autophagy, *Hum Mol Genet* **11**: 1107–1117 (2002).
149. H. Y. Zoghbi and H. T. Orr, Pathogenic mechanisms of a polyglutamine-mediated neurodegenerative disease, spinocerebellar ataxia type 1, *J Biol Chem* **284**: 7425–7429 (2009).
150. J. L. Webb, B. Ravikumar, J. Atkins, J. N. Skepper and D. C. Rubinsztein, Alpha-synuclein is degraded by both autophagy and the proteasome, *J Biol Chem* **278**: 25009–25013 (2003).
151. T. Kabuta, Y. Suzuki and K. Wada, Degradation of amyotrophic lateral sclerosis-linked mutant Cu,Zn-superoxide dismutase proteins by macroautophagy and the proteasome, *J Biol Chem* **281**: 30524–30533 (2006).
152. Z. Berger, B. Ravikumar, F. M. Menzies, L. G. Oroz, B. R. Underwood, M. N. Pangalos, I. Schmitt, U. Wullner, B. O. Evert, C. J. O'Kane and D. C. Rubinsztein, Rapamycin alleviates toxicity of different aggregate-prone proteins, *Hum Mol Genet* **15**: 433–442 (2006).
153. N. Mizushima, B. Levine, A. M. Cuervo and D. J. Klionsky, Autophagy fights disease through cellular self-digestion, *Nature* **451**: 1069–1075 (2008).
154. A. M. Cuervo, L. Stefanis, R. Fredenburg, P. T. Lansbury and D. Sulzer, Impaired degradation of mutant alpha-synuclein by chaperone-mediated autophagy, *Science* **305**: 1292–1295 (2004).
155. R. J. Youle and D. P. Narendra, Mechanisms of mitophagy, *Nat Rev Mol Cell Biol* **12**: 9–14 (2011).
156. M. Martinez-Vicente, Z. Talloczy, E. Wong, G. Tang, H. Koga, S. Kaushik, R. de Vries, E. Arias, S. Harris, D. Sulzer and A. M. Cuervo, Cargo recognition failure is responsible for inefficient autophagy in Huntington's disease, *Nat Neurosci* **13**: 567–576 (2010).
157. J. H. Lee, W. H. Yu, A. Kumar, S. Lee, P. S. Mohan, C. M. Peterhoff, D. M. Wolfe, M. Martinez-Vicente, A. C. Massey, G. Sovak, Y. Uchiyama, D. Westaway, A. M. Cuervo and R. A. Nixon, Lysosomal proteolysis and autophagy require presenilin 1 and are disrupted by Alzheimer-related PS1 mutations, *Cell* **141**: 1146–1158 (2010).

CHAPTER 2

AUTOPHAGOSOME MATURATION, ENDOCYTOSIS AND NEURODEGENERATIVE DISEASE

Ai Yamamoto and Anne Simonsen

ABSTRACT

Although autophagy and endocytosis has been largely considered discrete cellular events and studied separately, in macroautophagy, these two pathways are known to converge during autophagosome maturation. It thus may be unsurprising that genetic disruptions in endocytosis lead to cytoplasmic protein accumulation in adult brain and neurodegeneration. In this chapter, we will discuss how endocytic compartments contribute towards the formation of a macroautophagic intermediate structure known as the amphisome. We will also survey the current literature on how disruption of these pathways can negatively impact the adult brain.

INTRODUCTION

The balance between protein and lipid biosynthesis and their eventual degradation is a critical component of cellular health. The lysosome, discovered in 1955 by de Duve and colleagues,[1] is responsible for the degradation of much of the protein and lipid in cells. The importance of proper lysosomal function is particularly evident in neurons, as its dysfunction has been attributed to a wide range of neurological disorders from the lysosomal storage diseases to Parkinson's disease and Alzheimer's disease. Even when lysosome-dysfunction has not been directly implicated in pathogenesis, there are a growing number of studies that suggest that enhancing lysosome-mediated degradation may counter the degenerative process. Taken together, understanding lysosome-mediated degradation can shed new light onto the molecular underpinnings of different neurodegenerative diseases.

The lysosomes receive cargoes destined for degradation through the convergence of two cellular pathways, endocytosis and autophagy. The main difference between these two pathways lies in the cargo that is being trafficked: While endocytosis involves the uptake and transport of extracellular (heterophagic) and plasma membrane bound materials, autophagy involves the transport of intracellular cytoplasmic materials. Initial studies in lower organisms indicated that endocytosis and autophagy are largely mutually exclusive, and thus these pathways are often studied separately. Nonetheless for mammalian cells, biochemical and morphological studies have revealed that endocytosis is an important component for mammalian autophagy. This is particularly the case for macroautophagy, for which the autophagosome can fuse to endocytic compartments to deliver its cargo to the lysosome. The molecular machinery involved in this process and the functional significance of autophagosome maturation in neurodegeneration will be discussed here.

MACROAUTOPHAGY

A Brief Introduction

Autophagy is described as the degradation of cytoplasmic material by the lysosome, and can be separated into three main forms: microautophagy, chaperone-mediated autophagy (CMA), and macroautophagy. In microautophagy, cytoplasmic components are directly sequestered into the lysosome by an invagination of the lysosomal membrane and subsequent budding of vesicles into the lysosomal lumen.[2] This process has been studied particularly in yeast, and has not been mechanistically described in mammalian cells. In CMA, proteins containing a KFERQ motif is delivered to the lysosome by a direct, chaperone-mediated import through the lysosomal membrane.[3] In contrast to microautophagy, CMA is a process originally identified and most studied in mammalian cells and has not been described in yeast. Recently, a dysfunction of CMA has been described in Parkinson's disease and is the topic of discussion in Chapter 4.

The third form of autophagy, macroautophagy, is a highly conserved process observed in all eukaryotes. It involves the sequestration of its cargo into double-membrane vesicles called autophagosomes which then deliver their contents to the lysosome for degradation.[2] Macroautophagy was originally identified as a bulk, non-specific degradation process to produce amino acids during periods of nutrient deprivation; however recent studies have extended this role, and now macroautophagy is implicated in development, aging, elimination of microorganisms, cell death, cancer, neurodegeneration and antigen presentation.[4]

Furthermore, there is growing evidence that cells use macroautophagy to drive the selective elimination of complex proteins[5], lipid droplets[6] and organelles.[7]

The molecular machinery involved in autophagosome formation was identified using yeast genetic screens and named the autophagy-related (Atg) genes.[8,9] Analyses of Atg proteins in yeast and mammals have provided a growing understanding of the complex autophagic molecular machinery and its role in various physiological and pathological conditions.[10] The molecular machinery involved in autophagosome formation is the topic of Chapter 1 and has already been extensively reviewed,[11–13] and thus we will focus on the points at which macroautophagy and endocytosis meet.

Autophagosome Maturation: A Convergence with Endocytosis

For the purpose of simplicity, macroautophagy can be divided into two stages: autophagosome formation and maturation. Autophagosome formation involves autophagosome genesis to its ultimate closure, while maturation involves the evolution of the autophagosome until its ultimate destruction by fusion with the lysosome. Although the autophagosome only fuses directly with the lysosome in yeast, in mammalian cells, autophagosomes can also fuse sequentially with different endocytic vesicles and to mature into a structure called an amphisome.[14] The amphisome then fuses with lysosomes to form the final structure known as an autolysosome. To describe mammalian autophagosome maturation fully, we will first briefly summarize the endocytic compartments involved.

Endocytosis

Endocytosis is the cellular mechanism that uses portions of the plasma membrane to engulf and internalize extracellular components, membrane proteins and other molecules localized to the plasma membrane. Upon internalization, endocytosis then transports and targets the internalized cargo to specific organelles, including the lysosome (Figure 1). Endocytosis is a multidirectional process involved in many physiological events such as the uptake of extracellular nutrients, regulation of cell-surface receptor expression, cellular cholesterol homeostasis, antigen presentation and others.[15] Additionally in neurons, endocytosis is also crucial for axon growth, dendritic morphology and synaptic vesicle recycling.[16]

Once the plasma membrane pinches off to internalize extracellular and plasma membrane proteins, the resulting vesicle quickly fuses to the early endosome, also known as the sorting endosome (Figure 1). At the sorting endosome, the internalized proteins are selected either to be recycled or degraded.

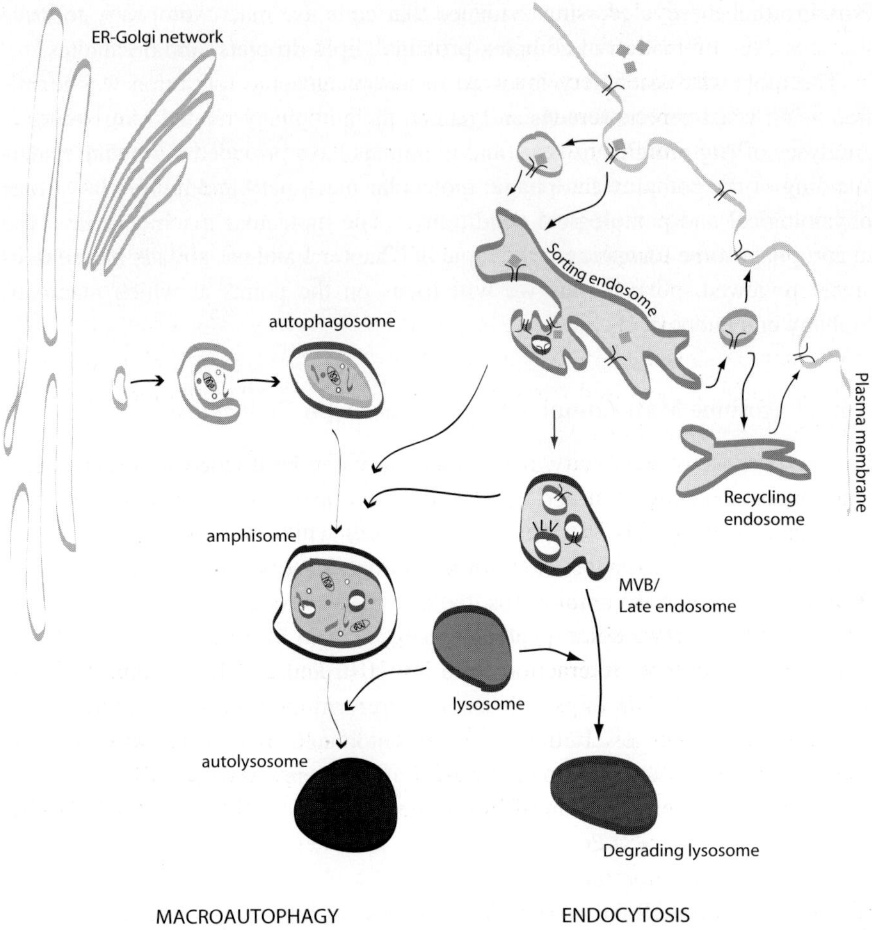

Figure 1. Autophagosome maturation involves fusion with endocytic compartments. Schematic view of how endocytosis and macroautophagy converge onto the lysosomal pathway. Macroautophagy involves sequestration of cytoplasmic cargo by an expanding membrane (called the phagophore or isolation membrane), forming double-membrane autophagosomes which upon sequential fusion with different endocytic vesicles form the hybrid organelle called the amphisome. Critical to autophagosome formation is localization of the Atg protein-based machinery to the expanding phagophore. Although the source of the phagophore membrane is still a matter of debate,[30,94–97] studies using live cell imaging and electron microscopy have shown that an ER-associated structure containing DFCP1 (double-FYVE containing protein 1) and PI3P seems to forms a cradle for autophagosome formation in cells subject to nutrient deprivation.[95–97] Termed the omegasome, this structure is the point at which other autophagy-specific proteins such as Atg14L[98] and WIPI2 (mammalian Atg18)[99] are recruited. Although this data strongly suggests that autophagosome-formation may initiate at the ER, the contribution of other membrane sources such as TGN,[100,101] mitochondria (not shown),[94] plasma membrane (not shown)[30] and also endocytic compartments cannot be excluded, as they may

Proteins to be recycled, such as the transferrin receptor and other nutrient receptors, are often trafficked back to the cell surface after ligand-dissociation in recycling endosomes. Proteins to be degraded, such as downregulated epidermal growth factor receptors (EGFRs), are trafficked to the lysosomes for their ultimate destruction. Prior to trafficking to the lysosome, however, the degradation cargo is first ubiquitinated and then sorted into specific membrane domains within the vacuolar region of the sorting endosome through the action of a series of protein complexes called ESCRTs (endosomal sorting complexes required for transport).[17]

The key role of the ESCRTs is the creation of an endocytic structure known as the multi-vesicular body (MVB). The MVB is a distinct late endosome structure in which endocytosed cargo are amassed prior to degradation via fusion with the lysosome (Figure 1). The role of ESCRTs in MVB formation has been extensively reviewed;[17] briefly it requires a stepwise process that requires cargo and membrane recognition and vesicle formation. First, the ubiquitinated cargo is recognized by ESCRT-0 (Hrs, STAM), which is recruited to the endosomal membrane through interaction with the phosphorylated lipid, phosphatidylinositol-3-monophosphate (PI3P). Hrs then recruits the ESCRT-I complex, comprised of Tsg101/Vacuolar protein sorting (Vps) 28/Vps37/Mvb12, to the endosome through its interaction with Tsg101. The ESCRT-I subunit Vps28 then recruits ESCRT-II (Vps22/Vps25/Vps36), which also has ubiquitin- and lipid-binding properties. Initiated by Vps25, ESCRT-II then recruits the proteins of the ESCRT-III complex (Vps2(CHMP2)/Vps20Vps24/Vps32(Snf7)), which mediates the formation of intralumenal vesicles (ILVs). ESCRT-III also recruits enzymes which de-ubiquitinate the cargo and the AAA-ATPase Vps4, or its mammalian homolog SKD1, mediating ESCRT disassembly and ILV abscission, ultimately resulting in formation of the MVB. The MVB then matures further into a late endosome and fuses with lysosomes to deliver the cargo of ILVs for degradation.

Figure 1. ***(Continued)*** be critical to elongate and expand the phagophore to form the autophagosome. Continued studies using high resolution microscopy and genetics will make much needed progress in this field. The amphisome then fuses with lysosomes to form the final structure known as an autolysosome, where the sequestered cargo becomes degraded. In yeast, the autophagosomes can fuse directly with the vacuole. Endocytosis involves internalization of extracellular and plasma membrane proteins. The internalized cargo is first delivered to the early endosome, also known as the sorting endosome, where the internalized proteins are selected either to be recycled or degraded. Integral membrane proteins to be degraded in the lysosomes first become ubiquitinated and then sorted into intraluminal vesicles (ILV) of the multi-vesicular body (MVB) by the action of the ESCRT complexes. The MVB then undergoes maturation before fusion with the lysosome.

The maturation of an autophagosome to an autolysosome: Fusion with endocytic compartments

Although there is clear evidence that, like in yeast, mammalian autophagosomes can fuse directly with lysosomes, biochemical and morphological evidence since the late 1980s clearly suggests that autophagosomes undergo sequential fusion with early endosomes, MVBs and late endosomes prior to fusion with the lysosome.[14,18–22] These findings are further supported by recent genetic studies in mammalian cells and *Drosophila melanogaster* that find that the coat protein complex I (COPI), a protein found at early endosomes, as well as different subunits of the ESCRTs and the ATPase Vps4/SKD1 are required for autophagosome maturation.[23–27]

A general principle of vesicle fusion implies tethering, docking and fusion. Both tethering complexes and endocytic fusion proteins impact autophagosome maturation. Vesicle fusion along the endocytic pathway relies on the function of the Rab family of small GTPases as well as SNARE (soluble N-ethylmaleimide-sensitive fusion protein (NSF) attachment receptors) proteins. Several Rab proteins have been implicated in autophagosome formation and maturation. Rab5 and Rab33, which localize to early endosomes and the Golgi, were both found to be important for the early stages of autophagosome formation by regulating Atg5–12 conjugation and binding to Atg16L, respectively.[28,29] This autophagic function of Rab5 might be related to a recent study showing that blocking clathrin-mediated endocytosis leads to impaired autophagosome formation.[30] The ER-localized Rab32 seems to be important for basal autophagy and not for starvation-induced autophagy and its depletion was found to result in accumulation of protein aggregates.[31] Genetic knockdown of the late endosomal Rab7 have been shown to lead to amphisome accumulation in mammalian cells, most likely by preventing fusion of amphisome with late endosomes and lysosomes.[32,33] Other Rab members such as Rab11 and Rab24 also have been implicated in autophagy-associated fusion processes.[21,34] The endocytic v-SNARE Vesicle-associated membrane protein (VAMP)-3 is involved in the fusion of autophagosomes to MVBs in reticulocyte maturation,[35] while VAMP-7 and VAMP-8, and the t-SNARE Vti1b is required for co-localization of the autophagosomal membrane associated protein MAP1LC3 (LC3) with the lysosomal membrane protein 1 (LAMP1) both during canonical and antimicrobial macroautophagy.[36] Interestingly, Vti1b seems to be derived from autophagic compartments, whereas VAMP8 originates from lysosomes, indicating that this SNARE pair mediates fusion of both antimicrobial and canonical autophagosomes with lysosomes.[36] The role for VAMP-7 and Vti1b is much in line with yeast for which the NSF homologue Sec18p and the SNARE proteins Vam3p and Vti1p are required for fusion of the autophagosome to the vacuole.[37,38] Interestingly, depletion of other endocytic t-SNAREs such as Syntaxin-7 and Syntaxin-8 had little effect on autophagosome maturation.[36]

The endosomal class C Vps (Vps-C) tethering complexes HOPS (homotypic vacuole fusion and protein sorting) and CORVET (class C core vacuole/endosome tethering) are known to couple Rab activation and SNARE assembly during endosome fusion[39] and may thus regulate the fusion events along the late endocytic pathway.[40] Although the function of these large multiprotein complexes in macroautophagy is only recently being explored, their importance in late endosome and lysosomal integrity make it unsurprising that there is early evidence implicating them in autophagosome maturation. The most studied of the Vps-C complexes, the HOPS complex, is comprised of the core complex Vps11/Vps16/Vps18 and Vps33 in addition to the selective proteins Vps39/Vps41, and in yeast is known to regulate tethering and fusion of endosomes to the yeast vacuole. Studies in *Drosophila* have revealed that the Vps18 homologue, Deep orange (Dor), mediates fusion of MVBs to lysosomes, and also has revealed that it is required for lysosomal delivery and autophagosome maturation.[41–43] HOPS is also an established guanine nucleotide exchange factor (GEF) for the late endosomal Rab7,[44] but also can interact with the early endosomal Rab5.[45] Although the CORVET complex is known to interact with the yeast Rab GTPase Vps21 (Rab5)[46] little is known about this complex in mammalian systems and even less in autophagosome maturation.

Vps34 complexes

The class III phosphatidyl inositol 3 kinase (PI3K) Vps34 is responsible for the generation of the lipid phosphatidyl-inositol-3-monophosphate (PI3P). PI3P and Vps34 activity is essential for the formation and maturation of the autophagosome, and accordingly 3-methyladenine (3-MA) or controlled application of wortmannin, which antagonize Vps34 activity,[47,48] often is used to inhibit macroautophagy. Interestingly, PI3P and Vps34 also play a critical role in the maturation and trafficking of endosomes to Golgi,[41,49] suggesting a necessary cross-talk between macroautophagy and endocytosis.

Vps34 can be found in two complexes, Vps34/class III PI3K complex I and complex II. Vps34/class III PI3K complex I is comprised of the namesake class III PI3K Vps34, together with the Vps15-like serine/threonine kinase p150, the mammalian Atg6 protein Beclin 1,[13] and Atg14L (also known as Barkor).[50,51] It is the core complex that is required for autophagosome nucleation and can be further regulated by several Beclin 1-associated proteins including UVRAG, Ambra-1, Bif-1 and Bcl-2.

Endocytosis, on the other hand relies on Vps34/class III PI3K complex II. In yeast, PI3K complex II contains Vps34/Vps15/Atg6(Vps30) and contains Vps38 rather than Atg14.[52,53] Although a mammalian counterpart of Vps 38 has not been identified, recent evidence suggests this role for ultraviolet irradiation resistant-associated gene (UVRAG)[50,51,54,55] Vps34/p150/Beclin 1/UVRAG is important for efficient EGFR degradation, most likely through its interaction with late endocytic structures, and can regulate autophagosome maturation through its interaction with Vps-c.[56] Further regulation can also be achieved by an interaction between Vps34/class III PI3K complex II and Rubicon (RUN domain and cysteine-rich domain containing, Beclin1-interacting protein),[54,55] which acts to negatively regulate the maturation of late endosomal/lysosomal structures and autophagosomes. RUN domains are often found in several proteins involved in regulation of Ras-like GTPase activity (e.g. Rab proteins), and thus it is tempting to speculate that Rubicon may be a Rab effector protein that regulates the activity of endocytic Rab proteins. All-in-all, the tethering complexes may be regulated by Vps34/Beclin1 complexes and Rab proteins, and together define organelle biogenesis in the endocytic, as well as in the macroautophagic pathway.

Fusion with the lysosome

The final stage of autophagosome maturation, the fusion of the mature amphisome with the lysosome, depends on low lysosomal pH, lysosomal membrane proteins as well as lysosomal enzymes. Drugs affecting lysosomal pH, such as Bafilomycin A1, and inhibitors of lysosomal enzymes have been described to inhibit autophagosome maturation. In agreement with the effect of lysosomal inhibitors, mice deficient for the lysosomal protease cathepsin D show an accumulation of autophagic vacuoles.[57] Moreover, hepatocytes lacking the lysosomal membrane protein LAMP-2 had increased amounts of both early and late autophagic vacuoles.[58,59] It has been difficult, however, to distinguish amphisomes from autolysosomes by other means than electron microscopy, a method not well suited for high-throughput quantification. The recently developed tandem-fluorochrome-tagged LC3 (mCherry-GFP- or mRFP-GFP-LC3) might be used to separate autophagosomes and amphisomes (yellow) from autolysosomes (red) due to the acid-sensitivity of GFP,[60,61] although also amphisomes are somewhat acidic.[62] Finally, autophagosomes, must traffic to the site of fusion, a process shown to depend on microtubules and dynein motor proteins.[63–66] This process is particularly important in neurons where autophagosomes must undergo efficient retrograde transport from the distal processes to the lysosomes located in the soma. The protein Snapin was recently found to mediate retrograde transport of late endosome by linking them to dynein. Neurons null for SNAPin showed impaired clearance of autolysosomes, combined with reduced neuron viability and neurodegeneration.[67,68] Recently, the Rab7-, PI3P- and LC3-binding protein FYCO1 was found to promote microtubule plus end-directed transport of autophagosomes,[69] thereby connecting transport of autophagic vacuoles to the fusion with lysosomes.

AUTOPHAGOSOME MATURATION AND NEURODEGENERATION

Deficits in lysosomal function and macroautophagy are invariably linked to neuronal dysfunction. Although the Purkinje cells of the cerebellum has been shown to be especially sensitive to excessive or inhibition of macroautophagy,[70–72] the inability of autophagosomes to efficiently mature and degrade its cargo is emerging to be problematic for differing neuronal subtypes. In Alzheimer's disease for example, axonal accumulation of autophagosomes is found throughout the neocortex.[73] Recently, an inability of the lysosome to efficiently acidify and hence drive the formation of autolysosomes and degradation has been implicated in this autophagosome accumulation.[74]

The most direct suggestion that links endocytosis, autophagosome maturation and neurodegeneration is supported by the identification of an autosomal dominant mutation in the ESCRT-III subunit CHMP2B in patients with chromosome-3 linked frontotemporal dementia (FTD3).[75,76] FTD3 is one of several forms of frontotemporal dementia, a progressive neurodegenerative condition considered to be the most common form of senile dementia that affects individuals before the age of 65 years.[77] Unlike Alzheimer's disease, FTD-derived dementia can also co-morbidize with motor neuron disease, and may share similarities with amyotrophic lateral sclerosis (ALS).[78] Indeed, similar CHMP2B mutations have been identified in a subset of ALS patients.[76,79] CHMP2B is a 213-amino acid protein that is the ortholog of yeast Vps2, a component of the ESCRT-III complex. The mutation, a G-to-C transition in the acceptor splice site of exon 6, leads to the production of two abnormal products: one with a 210 bp insertion of intron 5, leading to the production of a protein product with a 36-amino acid C-terminus truncation ($CHMPB2^{intron5}$); and the other with a short deletion from a cryptic splice site from the 5′ end of exon 6, leading to the production of a protein whose C-terminus is replaced with an abnormal 29-amino acid sequence ($CHMPB2^{\Delta 10}$).[75] siRNA-mediated knockdown of CHMP2B in primary neurons led to no overt neurotoxicity, likely due to the presence of a Vps2A isoform in mammalian cells, suggesting that the truncation mutations do not lead to simple loss of function. In contrast, overexpression of CHMP2B mutations in stable cell lines lead to an accumulation of vesicle structures, and deficient endocytic trafficking and MVB maturation.[75] Similar abnormal endocytic structures were seen in CHMP2B mutation-positive patient brains and in patient fibroblasts.[80] Studies of the CHMP2B mutants in cortical and hippocampal neurons led to dendritic retraction, altered spine morphology and an accumulation of autophagosomes.[25,81] Further evidence for a neuroprotective role of the ESCRTs come from a study showing that the Tsg101-interacting protein SIMPLE is mutated in Charcot-Marie-Tooth type 1C, a hereditary disease causing demyelinating peripheral neuropathy in humans.[82] Moreover, a null mutation of Mahogunin, an E3 ligase involved in ubiquitination of Tsg101, was shown to cause spongiform neurodegeneration in mice, a hallmark of prion–like disease.[83] These findings are consistent with these mutations leading to a disruption of the endolysosomal system. In line with this, targeted disruption of subunits of ESCRT-I (Tsg101), ESCRT-II (Vps22) or ESCRT-III (Vps24) in neurons and non-neuronal cells, or *Drosophila* larvae deficient for ESCRT-I, -II, or –III showed a decreased number of lysosomal structures, diminished dendritic branching and a corresponding accumulation of autophagosomes.[25,84,85] Exactly how these deletion mutations disrupt endocytosis is under continued investigation and should lead to new insights into the importance of ESCRTs on neuronal function.

A pathogenic hallmark of FTD and other neurodegenerative disorder is the accumulation of ubiquitinated proteins, and patient brains with CHMP2B mutations demonstrate TDP-43- and tau-negative, ubiquitin-positive inclusions.[86] Although macroautophagy was originally characterized as a non-specific pathway involving random sequestration of cytoplasmic components, recently there has been growing evidence that cargo can be selectively and exclusively incorporated into autophagosomes. One of these cargoes, ubiquitinated protein aggregates, is of particular interest, in light of the large number of neurological disorders which share this common pathological feature.[5] Termed aggrephagy, this macroautophagic process is required for the elimination of accumulating proteins in both non-neuronal and neuronal cells. Neurons may be particularly dependent upon aggrephagy since they do not have the capacity to dilute their cytosol through cell division. Moreover, as mentioned above, the size and geometry of neurons may also make neurons particularly vulnerable to protein trafficking deficits. It is interesting to note that a protein required for aggrephagy, the PI3P-binding protein Alfy, as well as the macroautophagy machinery has some of the highest level of expression in the brain.[87,88]

Table 1. A summary of the current knowledge of proteins involved in autophagosome maturation.

Protein	Organism	Type of experiment	Autophagy-related phenotype
ESCRT- O			
Hrs	*C.e.* *H.s* (HeLa), *M.m.* (MEFs) *M.m.*	LOF mutant shRNA, KO Neuron-specific KO	Accumulation of AP[102] Accumulation of AP[103] Neuronal death; accumulation of Ub-protein aggregates[103]
ESCRT- I			
Tsg101	*H.s.* (HeLa)	siRNA	Accumulation of AP, Amph and Ub-protein aggregates[24]
Tsg1010, Vps28	*D.m.*	LOF mutant	Neuronal death; accumulation of AP, lack of Amph and AL[84]
ESCRT- II			
Vps22	*H.s.* (HeLa)	siRNA	Accumulation of AP, Amph and Ub-protein aggregates[24]
Vps25	*D.m.*	LOF mutant	Neuronal death; accumulation of AP, lack of Amph and AL[84]
ESCRT- III			
Vps32	*M.m.* (neurons), *D.m.*	siRNA LOF mutant	Accumulation of AP; decreased number of AL; neuronal death[25,84]
Vps24	*H.s.* (HeLa)	siRNA	Accumulation of AP, Amph, Ub-protein aggregates and Huntingtin-inclusions[24]
Vps2B/ CHMP2B	*H.s.* (HeLa) *M.m.* (neurons)	DN mutant associated with FTLD-U and ALS	Accumulation of AP and Ub-protein aggregates; neuronal death[24,25]
ESCRT accessory			
Vps4/ SKD1	*S.c.* *H.s.* (HeLa) *D.m.*	GOF mutant DN mutant DN mutant	Increased autophagy[104] Accumulation of AP and AL[23] Accumulation of AP[84]
Rabs			
Rab7	*H.s. (HeLa)*	OE, DN mutant	Accumulation of AP and Amph[33]
Rab11	*C.g. (CHO)* *H.s. (K562)*	OE, DN mutant	Accumulation of AP[32]

(*Continued*)

Table 1. (*Continued*)

Protein	Organism	Type of experiment	Autophagy-related phenotype
SNAREs			
VAMP3	*H.s. (K562)*	OE	Accumulation of AP[35]
VAMP7	*H.s. (K562)*	OE	Accumulation of AP[35]
VAMP8	*H.s. (HeLa)*	siRNA	Accumulation of AP and
Vti1b	*H.s.*	siRNA	GAS-containing AP[36]
Others			
COP-I	*H.s. (293T)*	siRNA	Accumulation of AP and Ub-protein aggregates[27]
UVRAG	*H.s. (HeLa)*	OE	Increased autophagy[56]
Rubicon	*H.s. (A549)*	shRNA	Enhanced AP maturation; AL[54]
	M.m. (3T3)	OE	accumulation of AP[105]
Vps Class C			
Vps16	*H.s. (HeLa)*	siRNA	Accumulation of Amph[56]
Vps18	*S.c.*	ts LOF	Accumulation of AP[106]
	D.m.	LOF	Accumulation of AP[41]
FYCO-I	*H.s. (HeLa)*	siRNA	Accumulation of AP in perinuclear clusters[69]

AP, autophagosome; Amph, amphisome; AL, autolysosome; *C.e., Caenorhabditis elegans*; *C.g., Cricetulus griseus*; *D.m., Drosophila melanogaster*; DN, dominant-negative; GOF, gain of function; *H.s., Homo sapiens*; LOF, loss of function; KO, knockout; MEFs, mouse embryonic fibroblasts; *M.m., Mus musculus*; *S.c., Saccharomyces cerevisiae*; shRNA, short hairpin RNA; siRNA, small interfering RNA; Ub, ubiquitin. The table is modified from Ref. 15.

In light of the dependence of cells on macroautophagy to eliminate aggregate-prone proteins, it is not surprising that deficits in autophagosome maturation often lead to the appearance of ubiquitinated inclusions. Although this does not comment on the toxicity of the inclusions themselves, their appearance is a clear indicator that a selective arm of protein degradation is impeded. The suggestion of macroautophagic disruption can be further strengthened if the accumulation of ubiquitin-positive membrane-free protein aggregates is accompanied by concomitant accumulation of p62/SQSTM1, a protein that helps to sequester and target inclusions to the autophagosome by recognizing ubiquitin-positive substrates. Although p62 levels are often inversely correlated to the rate of macroautophagic degradation, it is important to note that this correlation can sometimes be lost,[89] and thus multiple measures should be made in addition to p62 to ensure that degradation is indeed impeded. For example, cells having depleted various ESCRT subunits or overexpressing the CHMP2B^{Intron5} mutant, corresponding to the mutation found in the FTD patients, showed accumulation of not only p62, but ubiquitin and LC3 as well, indicating impeded autophagic degradation in these cells.[24] In line with these data, ubiquitin- and p62-positive inclusions have been found in CHMP2B mutant patients,[76,86] indicating that impaired autophagic degradation is responsible for the CHMP2B-linked neurodegenerative phenotype.

Another protein in which mutations have been linked to FTD is p97/VCP (valosin-containing protein), a member of the AAA-ATPase protein family. Mutations in VCP are associated with inclusion body myopathy associated with Paget's disease of the bone and fronto temporal dementia (IBMPFD), a progressive autosomal dominant disorder.[90] VCP was recently suggested to participate in autophagosome maturation, as autophagic structures accumulated in patient and transgenic animal tissue having IBMPFD-associated VCP mutations.[91,92] Histone deacetylase 6 (HDAC6), known to target protein aggregates and mitochondria, was also found to be specifically involved in maturation of autophagosomes formed under basal quality control conditions. HDAC6 may recruit the actin-remodeling machinery, which facilitates F-actin assembly leading to autophagosome-lysosome fusion and substrate degradation.[93] Thus, several proteins linked to neurodegenerative disease have been found to be involved in autophagosome maturation, suggesting that proper autophagosome maturation may be important to maintain proper neuronal function.

CONCLUDING REMARKS

The importance of protein trafficking in neuronal health has been well-established for many years, but only recently have studies begun to examine endocytosis and autophagy side-by-side. Since deficits in either pathway lead to arguably similar pathogenic events in the developing and adult brain, it also becomes critical that we pursue more of these studies in the correct context, as many questions still remained unanswered, especially within the CNS. This may be especially relevant since the intertwined nature of these pathways may explain why many neurodegenerative diseases not only share protein accumulation as a hallmark, but vesicle trafficking alterations as well.

REFERENCES

1. D. De Duve, B. C. Pressman, R. Gianetto, R. Wattiaux and F. Appelmans, Tissue fractionation studies. 6. Intracellular distribution patterns of enzymes in rat-liver tissue, *Biochem J* **60**: 604–617 (1955).
2. Z. Xie and D. J. Klionsky, Autophagosome formation: core machinery and adaptations, *Nat Cell Biol* **9**: 1102–1109 (2007).

3. M. Kon and A. M. Cuervo, Chaperone-mediated autophagy in health and disease, *FEBS Lett* **584**: 1399–1404 (2010).
4. N. Mizushima and B. Levine, Autophagy in mammalian development and differentiation, *Nat Cell Biol* **12**: 823–830 (2010).
5. A. Yamamoto and A. Simonsen, The elimination of accumulated and aggregated proteins: a role for aggrephagy in neurodegeneration, *Neurobiol Dis* **43**: 17–28 (2011).
6. R. Singh, S. Kaushik, Y. Wang, Y. Xiang, I. Novak, M. Komatsu, K. Tanaka, A. M. Cuervo and M. J. Czaja, Autophagy regulates lipid metabolism, *Nature* **458**: 1131–1135 (2009).
7. C. Kraft, M. Peter and K. Hofmann, Selective autophagy: ubiquitin-mediated recognition and beyond, *Nat Cell Biol* **12**: 836–841 (2010).
8. M. Tsukada and Y. Ohsumi, Isolation and characterization of autophagy-defective mutants of *Saccharomyces cerevisiae*, *FEBS Lett* **333**: 169–174 (1993).
9. D. J. Klionsky, J. M. Cregg, W. A. Dunn, Jr., S. D. Emr, Y. Sakai, I. V. Sandoval, A. Sibirny, S. Subramani, M. Thumm, M. Veenhuis and Y. Ohsumi, A unified nomenclature for yeast autophagy-related genes, *Dev Cell* **5**: 539–545 (2003).
10. Z. Yang and D. J. Klionsky, Eaten alive: a history of macroautophagy, *Nat Cell Biol* **12**: 814–822 (2010).
11. C. He and D. J. Klionsky, Regulation mechanisms and signaling pathways of autophagy, *Annu Rev Genet* **43**: 67–93 (2009).
12. A. Longatti and S. A. Tooze, Vesicular trafficking and autophagosome formation, *Cell Death Differ* **16**: 956–965 (2009).
13. A. Simonsen and S. A. Tooze, Coordination of membrane events during autophagy by multiple class III PI3-kinase complexes, *J Cell Biol* **186**: 773–782 (2009).
14. P. B. Gordon and P. O. Seglen, Prelysosomal convergence of autophagic and endocytic pathways, *Biochem Biophys Res Commun* **151**: 40–47 (1988).
15. T. E. Rusten and H. Stenmark, How do ESCRT proteins control autophagy?, *J Cell Sci* **122**: 2179–2183 (2009).
16. S. A. Tooze and G. Schiavo, Liaisons dangereuses: autophagy, neuronal survival and neurodegeneration, *Curr Opin Neurobiol* **18**: 504–515 (2008).
17. C. Raiborg and H. Stenmark, The ESCRT machinery in endosomal sorting of ubiquitylated membrane proteins, *Nature* **458**: 445–452 (2009).
18. W. A. Dunn, Jr., Studies on the mechanisms of autophagy: maturation of the autophagic vacuole, *J Cell Biol* **110**: 1935–1945 (1990).
19. J. Lucocq and D. Walker, Evidence for fusion between multilamellar endosomes and autophagosomes in HeLa cells, *Eur J Cell Biol* **72**: 307–313 (1997).
20. E. L. Eskelinen, Maturation of autophagic vacuoles in mammalian cells, *Autophagy* **1**: 1–10 (2005).
21. C. M. Fader, D. Sanchez, M. Furlan and M. I. Colombo, Induction of autophagy promotes fusion of multivesicular bodies with autophagic vacuoles in k562 cells, *Traffic* **9**: 230–250 (2008).

22. J. Morvan, R. Kochl, R. Watson, L. M. Collinson, H. B. Jefferies and S. A. Tooze, *In vitro* reconstitution of fusion between immature autophagosomes and endosomes, *Autophagy* **5**: 676–689 (2009).
23. A. Nara, N. Mizushima, A. Yamamoto, Y. Kabeya, Y. Ohsumi and T. Yoshimori, SKD1 AAA ATPase-dependent endosomal transport is involved in autolysosome formation, *Cell Struct Funct* **27**: 29–37 (2002).
24. M. Filimonenko, S. Stuffers, C. Raiborg, A. Yamamoto, L. Malerod, E. M. Fisher, A. Isaacs, A. Brech, H. Stenmark and A. Simonsen, Functional multivesicular bodies are required for autophagic clearance of protein aggregates associated with neurodegenerative disease, *J Cell Biol* **179**: 485–500 (2007).
25. J. A. Lee, A. Beigneux, S. T. Ahmad, S. G. Young and F. B. Gao, ESCRT-III dysfunction causes autophagosome accumulation and neurodegeneration, *Curr Biol* **17**: 1561–1567 (2007).
26. T. E. Rusten, M. Filimonenko, L. M. Rodahl, H. Stenmark and A. Simonsen, ESCRTing autophagic clearance of aggregating proteins, *Autophagy* **4**: 233–236 (2007).
27. M. Razi, E. Y. Chan and S. A. Tooze, Early endosomes and endosomal coatomer are required for autophagy, *J Cell Biol* **185**: 305–321 (2009).
28. B. Ravikumar, S. Imarisio, S. Sarkar, C. J. O'Kane and D. C. Rubinsztein, Rab5 modulates aggregation and toxicity of mutant huntingtin through macroautophagy in cell and fly models of Huntington disease, *J Cell Sci* **121**: 1649–1660 (2008).
29. T. Itoh, N. Fujita, E. Kanno, A. Yamamoto, T. Yoshimori and M. Fukuda, Golgi-resident small GTPase Rab33B interacts with Atg16L and modulates autophagosome formation, *Mol Biol Cell* **19**: 2916–2925 (2008).
30. B. Ravikumar, K. Moreau, L. Jahreiss, C. Puri and D. C. Rubinsztein, Plasma membrane contributes to the formation of pre-autophagosomal structures, *Nat Cell Biol* **12**: 747–757 (2010).
31. Y. Hirota and Y. Tanaka, A small GTPase, human Rab32, is required for the formation of autophagic vacuoles under basal conditions, *Cell Mol Life Sci* **66**: 2913–2932 (2009).
32. M. G. Gutierrez, D. B. Munafo, W. Beron and M. I. Colombo, Rab7 is required for the normal progression of the autophagic pathway in mammalian cells, *J Cell Sci* **117**: 2687–2697 (2004).
33. S. Jager, C. Bucci, I. Tanida, T. Ueno, E. Kominami, P. Saftig and E. L. Eskelinen, Role for Rab7 in maturation of late autophagic vacuoles, *J Cell Sci* **117**: 4837–4848 (2004).
34. D. B. Munafo and M. I. Colombo, Induction of autophagy causes dramatic changes in the subcellular distribution of GFP-Rab24, *Traffic* **3**: 472–482 (2002).
35. C. M. Fader and M. I. Colombo, Autophagy and multivesicular bodies: two closely related partners, *Cell Death Differ* **16**: 70–78 (2009).
36. N. Furuta, N. Fujita, T. Noda, T. Yoshimori and A. Amano, Combinational soluble N-ethylmaleimide-sensitive factor attachment protein receptor proteins VAMP8 and Vti1b mediate fusion of antimicrobial and canonical autophagosomes with lysosomes, *Mol Biol Cell* **21**: 1001–1010 (2010).

37. T. Darsow, S. E. Rieder and S. D. Emr, A multispecificity syntaxin homologue, Vam3p, essential for autophagic and biosynthetic protein transport to the vacuole, *J Cell Biol* **138**: 517–529 (1997).
38. N. Ishihara, M. Hamasaki, S. Yokota, K. Suzuki, Y. Kamada, A. Kihara, T. Yoshimori, T. Noda and Y. Ohsumi, Autophagosome requires specific early Sec proteins for its formation and NSF/SNARE for vacuolar fusion, *Mol Biol Cell* **12**: 3690–3702 (2001).
39. D. F. Seals, G. Eitzen, N. Margolis, W. T. Wickner and A. Price, A Ypt/Rab effector complex containing the Sec1 homolog Vps33p is required for homotypic vacuole fusion, *Proc Natl Acad Sci USA* **97**: 9402–9407 (2000).
40. D. P. Nickerson, C. L. Brett and A. J. Merz, Vps-C complexes: gatekeepers of endolysosomal traffic, *Curr Opin Cell Biol* **21**: 543–551 (2009).
41. K. Lindmo, A. Simonsen, A. Brech, K. Finley, T. E. Rusten and H. Stenmark, A dual function for Deep orange in programmed autophagy in the *Drosophila melanogaster* fat body, *Exp Cell Res* **312**: 2018–2027 (2006).
42. E. A. Sevrioukov, J. P. He, N. Moghrabi, A. Sunio and H. Kramer, A role for the deep orange and carnation eye color genes in lysosomal delivery in Drosophila, *Mol Cell* **4**: 479–486 (1999).
43. S. Pulipparacharuvil, M. A. Akbar, S. Ray, E. A. Sevrioukov, A. S. Haberman, J. Rohrer and H. Kramer, Drosophila Vps16A is required for trafficking to lysosomes and biogenesis of pigment granules, *J Cell Sci* **118**: 3663–3673 (2005).
44. A. E. Wurmser, T. K. Sato and S. D. Emr, New component of the vacuolar class C-Vps complex couples nucleotide exchange on the Ypt7 GTPase to SNARE-dependent docking and fusion, *J Cell Biol* **151**: 551–562 (2000).
45. J. Rink, E. Ghigo, Y. Kalaidzidis and M. Zerial, Rab conversion as a mechanism of progression from early to late endosomes, *Cell* **122**: 735–749 (2005).
46. K. Peplowska, D. F. Markgraf, C. W. Ostrowicz, G. Bange and C. Ungermann, The CORVET tethering complex interacts with the yeast Rab5 homolog Vps21 and is involved in endo-lysosomal biogenesis, *Dev Cell* **12**: 739–750 (2007).
47. K. B. Hendil, A. M. Lauridsen and P. O. Seglen, Both endocytic and endogenous protein degradation in fibroblasts is stimulated by serum/amino acid deprivation and inhibited by 3-methyladenine, *Biochem J* **272**: 577–581 (1990).
48. E. F. Blommaart, U. Krause, J. P. Schellens, H. Vreeling-Sindelarova and A. J. Meijer, The phosphatidylinositol 3-kinase inhibitors wortmannin and LY294002 inhibit autophagy in isolated rat hepatocytes, *Eur J Biochem* **243**: 240–246 (1997).
49. S. S. Skanland, S. Walchli, A. Utskarpen, A. Wandinger-Ness and K. Sandvig, Phosphoinositide-regulated retrograde transport of ricin: crosstalk between hVps34 and sorting nexins, *Traffic* **8**: 297–309 (2007).
50. E. Itakura, C. Kishi, K. Inoue and N. Mizushima, Beclin 1 forms two distinct phosphatidylinositol 3-kinase complexes with mammalian Atg14 and UVRAG, *Mol Biol Cell* **19**: 5360–5372 (2008).

51. Q. Sun, W. Fan, K. Chen, X. Ding, S. Chen and Q. Zhong, Identification of Barkor as a mammalian autophagy-specific factor for Beclin 1 and class III phosphatidylinositol 3-kinase, *Proc Natl Acad Sci USA* **105**: 19211–19216 (2008).
52. M. N. Seaman, E. G. Marcusson, J. L. Cereghino and S. D. Emr, Endosome to Golgi retrieval of the vacuolar protein sorting receptor, Vps10p, requires the function of the VPS29, VPS30, and VPS35 gene products, *J Cell Biol* **137**: 79–92 (1997).
53. P. Burda, S. M. Padilla, S. Sarkar and S. D. Emr, Retromer function in endosome-to-Golgi retrograde transport is regulated by the yeast Vps34 PtdIns 3-kinase, *J Cell Sci* **115**: 3889–3900 (2002).
54. K. Matsunaga, T. Saitoh, K. Tabata, H. Omori, T. Satoh, N. Kurotori, I. Maejima, K. Shirahama-Noda, T. Ichimura, T. Isobe, S. Akira, T. Noda and T. Yoshimori, Two Beclin 1-binding proteins, Atg14L and Rubicon, reciprocally regulate autophagy at different stages, *Nat Cell Biol* **11**: 385–396 (2009).
55. Y. Zhong, Q. J. Wang, X. Li, Y. Yan, J. M. Backer, B. T. Chait, N. Heintz and Z. Yue, Distinct regulation of autophagic activity by Atg14L and Rubicon associated with Beclin 1-phosphatidylinositol-3-kinase complex, *Nat Cell Biol* **11**: 468–476 (2009).
56. C. Liang, J. S. Lee, K. S. Inn, M. U. Gack, Q. Li, E. A. Roberts, I. Vergne, V. Deretic, P. Feng, C. Akazawa and J. U. Jung, Beclin1-binding UVRAG targets the class C Vps complex to coordinate autophagosome maturation and endocytic trafficking, *Nat Cell Biol* **10**: 776–787 (2008).
57. M. Koike, H. Nakanishi, P. Saftig, J. Ezaki, K. Isahara, Y. Ohsawa, W. Schulz-Schaeffer, T. Watanabe, S. Waguri, S. Kametaka, M. Shibata, K. Yamamoto, E. Kominami, C. Peters, K. von Figura and Y. Uchiyama, Cathepsin D deficiency induces lysosomal storage with ceroid lipofuscin in mouse CNS neurons, *J Neurosci* **20**: 6898–6906 (2000).
58. Y. Tanaka, G. Guhde, A. Suter, E. L. Eskelinen, D. Hartmann, R. Lullmann-Rauch, P. M. Janssen, J. Blanz, K. von Figura and P. Saftig, Accumulation of autophagic vacuoles and cardiomyopathy in LAMP-2-deficient mice, *Nature* **406**: 902–906 (2000).
59. E. L. Eskelinen, C. K. Schmidt, S. Neu, M. Willenborg, G. Fuertes, N. Salvador, Y. Tanaka, R. Lullmann-Rauch, D. Hartmann, J. Heeren, K. von Figura, E. Knecht and P. Saftig, Disturbed cholesterol traffic but normal proteolytic function in LAMP-1/LAMP-2 double-deficient fibroblasts, *Mol Biol Cell* **15**: 3132–3145 (2004).
60. S. Pankiv, T. H. Clausen, T. Lamark, A. Brech, J. A. Bruun, H. Outzen, A. Overvatn, G. Bjorkoy and T. Johansen, p62/SQSTM1 binds directly to Atg8/LC3 to facilitate degradation of ubiquitinated protein aggregates by autophagy, *J Biol Chem* **282**: 24131–24145 (2007).
61. S. Kimura, T. Noda and T. Yoshimori, Dissection of the autophagosome maturation process by a novel reporter protein, tandem fluorescent-tagged LC3, *Autophagy* **3**: 452–460 (2007).

62. P. E. Stromhaug and P. O. Seglen, Evidence for acidity of prelysosomal autophagic/endocytic vacuoles (amphisomes), *Biochem J* **291**(Pt 1): 115–121 (1993).
63. R. Kochl, X. W. Hu, E. Y. Chan and S. A. Tooze, Microtubules facilitate autophagosome formation and fusion of autophagosomes with endosomes, *Traffic* **7**: 129–145 (2006).
64. S. Kimura, T. Noda and T. Yoshimori, Dynein-dependent movement of autophagosomes mediates efficient encounters with lysosomes, *Cell Struct Funct* **33**: 109–122 (2008).
65. E. Fass, E. Shvets, I. Degani, K. Hirschberg and Z. Elazar, Microtubules support production of starvation-induced autophagosomes but not their targeting and fusion with lysosomes, *J Biol Chem* **281**: 36303–36316 (2006).
66. L. Jahreiss, F. M. Menzies and D. C. Rubinsztein, The itinerary of autophagosomes: from peripheral formation to kiss-and-run fusion with lysosomes, *Traffic* **9**: 574–587 (2008).
67. Q. Cai, L. Lu, J. H. Tian, Y. B. Zhu, H. Qiao and Z. H. Sheng, Snapin-regulated late endosomal transport is critical for efficient autophagy-lysosomal function in neurons, *Neuron* **68**: 73–86 (2010).
68. M. Yuzaki, Snapin snaps into the dynein complex for late endosome-lysosome trafficking and autophagy, *Neuron* **68**: 4–6 (2010).
69. S. Pankiv, E. A. Alemu, A. Brech, J. A. Bruun, T. Lamark, A. Overvatn, G. Bjorkoy and T. Johansen, FYCO1 is a Rab7 effector that binds to LC3 and PI3P to mediate microtubule plus end-directed vesicle transport, *J Cell Biol* **188**: 253–269 (2010).
70. Z. Yue, A. Horton, M. Bravin, P. L. DeJager, F. Selimi and N. Heintz, A novel protein complex linking the delta 2 glutamate receptor and autophagy: implications for neurodegeneration in lurcher mice, *Neuron* **35**: 921–933 (2002).
71. T. Hara, K. Nakamura, M. Matsui, A. Yamamoto, Y. Nakahara, R. Suzuki-Migishima, M. Yokoyama, K. Mishima, I. Saito, H. Okano and N. Mizushima, Suppression of basal autophagy in neural cells causes neurodegenerative disease in mice, *Nature* **441**: 885–889 (2006).
72. M. Komatsu, S. Waguri, T. Chiba, S. Murata, J. Iwata, I. Tanida, T. Ueno, M. Koike, Y. Uchiyama, E. Kominami and K. Tanaka, Loss of autophagy in the central nervous system causes neurodegeneration in mice, *Nature* **441**: 880–884 (2006).
73. B. Boland, A. Kumar, S. Lee, F. M. Platt, J. Wegiel, W. H. Yu and R. A. Nixon, Autophagy induction and autophagosome clearance in neurons: relationship to autophagic pathology in Alzheimer's disease, *J Neurosci* **28**: 6926–6937 (2008).
74. J. H. Lee, W. H. Yu, A. Kumar, S. Lee, P. S. Mohan, C. M. Peterhoff, D. M. Wolfe, M. Martinez-Vicente, A. C. Massey, G. Sovak, Y. Uchiyama, D. Westaway, A. M. Cuervo and R. A. Nixon, Lysosomal proteolysis and autophagy require presenilin 1 and are disrupted by Alzheimer-related PS1 mutations, *Cell* **141**: 1146–1158 (2010).
75. G. Skibinski, N. J. Parkinson, J. M. Brown, L. Chakrabarti, S. L. Lloyd, H. Hummerich, J. E. Nielsen, J. R. Hodges, M. G. Spillantini, T. Thusgaard, S. Brandner, A. Brun, M. N. Rossor, A. Gade, P. Johannsen, S. A. Sorensen,

S. Gydesen, E. M. Fisher and J. Collinge, Mutations in the endosomal ESCRTIII-complex subunit CHMP2B in frontotemporal dementia, *Nat Genet* **37**: 806–808 (2005).
76. N. Parkinson, P. G. Ince, M. O. Smith, R. Highley, G. Skibinski, P. M. Andersen, K. E. Morrison, H. S. Pall, O. Hardiman, J. Collinge, P. J. Shaw and E. M. Fisher, ALS phenotypes with mutations in CHMP2B (charged multivesicular body protein 2B), *Neurology* **67**: 1074–1077 (2006).
77. J. A. Lee and F. B. Gao, ESCRT, autophagy, and frontotemporal dementia, *BMB Rep* **41**: 827–832 (2008).
78. I. R. Mackenzie, M. Neumann, E. H. Bigio, N. J. Cairns, I. Alafuzoff, J. Kril, G. G. Kovacs, B. Ghetti, G. Halliday, I. E. Holm, P. G. Ince, W. Kamphorst, T. Revesz, A. J. Rozemuller, S. Kumar-Singh, H. Akiyama, A. Baborie, S. Spina, D. W. Dickson, J. Q. Trojanowski and D. M. Mann, Nomenclature and nosology for neuropathologic subtypes of frontotemporal lobar degeneration: an update, *Acta Neuropathol* **119**: 1–4 (2010).
79. L. E. Cox, L. Ferraiuolo, E. F. Goodall, P. R. Heath, A. Higginbottom, H. Mortiboys, H. C. Hollinger, J. A. Hartley, A. Brockington, C. E. Burness, K. E. Morrison, S. B. Wharton, A. J. Grierson, P. G. Ince, J. Kirby and P. J. Shaw, Mutations in CHMP2B in lower motor neuron predominant amyotrophic lateral sclerosis (ALS), *PloS One* **5**: e9872 (2010).
80. H. Urwin, A. Authier, J. E. Nielsen, D. Metcalf, C. Powell, K. Froud, D. S. Malcolm, I. Holm, P. Johannsen, J. Brown, E. M. Fisher, J. van der Zee, M. Bruyland, C. Van Broeckhoven, J. Collinge, S. Brandner, C. Futter and A. M. Isaacs, Disruption of endocytic trafficking in frontotemporal dementia with CHMP2B mutations, *Hum Mol Genet* **19**: 2228–2238 (2010).
81. A. Belly, G. Bodon, B. Blot, A. Bouron, R. Sadoul and Y. Goldberg, CHMP2B mutants linked to frontotemporal dementia impair maturation of dendritic spines, *J Cell Sci* **123**: 2943–2954 (2010).
82. A. J. Shirk, S. K. Anderson, S. H. Hashemi, P. F. Chance and C. L. Bennett, SIMPLE interacts with NEDD4 and TSG101: evidence for a role in lysosomal sorting and implications for Charcot-Marie-Tooth disease, *J Neurosci Res* **82**: 43–50 (2005).
83. B. Y. Kim, J. A. Olzmann, G. S. Barsh, L. S. Chin and L. Li, Spongiform neurodegeneration-associated E3 ligase Mahogunin ubiquitylates TSG101 and regulates endosomal trafficking, *Mol Biol Cell* **18**: 1129–1142 (2007).
84. T. E. Rusten, T. Vaccari, K. Lindmo, L. M. Rodahl, I. P. Nezis, C. Sem-Jacobsen, F. Wendler, J. P. Vincent, A. Brech, D. Bilder and H. Stenmark, ESCRTs and Fab1 regulate distinct steps of autophagy, *Curr Biol* **17**: 1817–1825 (2007).
85. N. T. Sweeney, J. E. Brenman, Y. N. Jan and F. B. Gao, The coiled-coil protein shrub controls neuronal morphogenesis in Drosophila, *Curr Biol* **16**: 1006–1011 (2006).
86. I. E. Holm, E. Englund, I. R. Mackenzie, P. Johannsen and A. M. Isaacs, A reassessment of the neuropathology of frontotemporal dementia linked to chromosome 3, *J Neuropathol Exp Neurol* **66**: 884–891 (2007).

87. A. Simonsen, H. C. Birkeland, D. J. Gillooly, N. Mizushima, A. Kuma, T. Yoshimori, T. Slagsvold, A. Brech and H. Stenmark, Alfy, a novel FYVE-domain-containing protein associated with protein granules and autophagic membranes, *J Cell Sci* **117**: 4239–4251 (2004).
88. M. Filimonenko, P. Isakson, K. D. Finley, M. Anderson, H. Jeong, T. J. Melia, B. J. Bartlett, K. M. Myers, H. C. Birkeland, T. Lamark, D. Krainc, A. Brech, H. Stenmark, A. Simonsen and A. Yamamoto, The selective macroautophagic degradation of aggregated proteins requires the PI3P-binding protein Alfy, *Mol Cell* **38**: 265–279 (2010).
89. P. Szyniarowski, E. Corcelle-Termeau, T. Farkas, M. Hoyer-Hansen, J. Nylandsted, T. Kallunki and M. Jaattela, A comprehensive siRNA screen for kinases that suppress macroautophagy in optimal growth conditions, *Autophagy* **7**: 892–903 (2011).
90. G. D. Watts, J. Wymer, M. J. Kovach, S. G. Mehta, S. Mumm, D. Darvish, A. Pestronk, M. P. Whyte and V. E. Kimonis, Inclusion body myopathy associated with Paget disease of bone and frontotemporal dementia is caused by mutant valosin-containing protein, *Nat Genet* **36**: 377–381 (2004).
91. J. S. Ju, R. A. Fuentealba, S. E. Miller, E. Jackson, D. Piwnica-Worms, R. H. Baloh and C. C. Weihl, Valosin-containing protein (VCP) is required for autophagy and is disrupted in VCP disease, *J Cell Biol* **187**: 875–888 (2009).
92. E. Tresse, F. A. Salomons, J. Vesa, L. C. Bott, V. Kimonis, T. P. Yao, N. P. Dantuma and J. P. Taylor, VCP/p97 is essential for maturation of ubiquitin-containing autophagosomes and this function is impaired by mutations that cause IBMPFD, *Autophagy* **6**: 217–227 (2010).
93. J. Y. Lee, H. Koga, Y. Kawaguchi, W. Tang, E. Wong, Y. S. Gao, U. B. Pandey, S. Kaushik, E. Tresse, J. Lu, J. P. Taylor, A. M. Cuervo and T. P. Yao, HDAC6 controls autophagosome maturation essential for ubiquitin-selective quality-control autophagy, *EMBO J* **29**: 969–980 (2010).
94. D. W. Hailey, A. S. Rambold, P. Satpute-Krishnan, K. Mitra, R. Sougrat, P. K. Kim and J. Lippincott-Schwartz, Mitochondria supply membranes for autophagosome biogenesis during starvation, *Cell* **141**: 656–667 (2010).
95. E. L. Axe, S. A. Walker, M. Manifava, P. Chandra, H. L. Roderick, A. Habermann, G. Griffiths and N. T. Ktistakis, Autophagosome formation from membrane compartments enriched in phosphatidylinositol 3-phosphate and dynamically connected to the endoplasmic reticulum, *J Cell Biol* **182**: 685–701 (2008).
96. M. Hayashi-Nishino, N. Fujita, T. Noda, A. Yamaguchi, T. Yoshimori and A. Yamamoto, A subdomain of the endoplasmic reticulum forms a cradle for autophagosome formation, *Nat Cell Biol* **11**: 1433–1437 (2009).
97. P. Yla-Anttila, H. Vihinen, E. Jokitalo and E. L. Eskelinen, 3D tomography reveals connections between the phagophore and endoplasmic reticulum, *Autophagy* **5**: 1180–1185 (2009).
98. K. Matsunaga, E. Morita, T. Saitoh, S. Akira, N. T. Ktistakis, T. Izumi, T. Noda and T. Yoshimori, Autophagy requires endoplasmic reticulum targeting of the PI3-kinase complex via Atg14L, *J Cell Biol* **190**: 511–521 (2010).

99. H. E. Polson, J. de Lartigue, D. J. Rigden, M. Reedijk, S. Urbe, M. J. Clague and S. A. Tooze, Mammalian Atg18 (WIPI2) localizes to omegasome-anchored phagophores and positively regulates LC3 lipidation, *Autophagy* **6**: 506–522 (2010).
100. J. L. Webber and S. A. Tooze, New insights into the function of Atg9, *FEBS Lett* **584**: 1319–1326 (2010).
101. M. Mari, J. Griffith, E. Rieter, L. Krishnappa, D. J. Klionsky and F. Reggiori, An Atg9-containing compartment that functions in the early steps of autophagosome biogenesis, *J Cell Biol* **190**: 1005–1022 (2010).
102. N. Roudier, C. Lefebvre and R. Legouis, CeVPS-27 is an endosomal protein required for the molting and the endocytic trafficking of the low-density lipoprotein receptor-related protein 1 in *Caenorhabditis elegans*, *Traffic* **6**: 695–705 (2005).
103. K. Tamai, M. Toyoshima, N. Tanaka, N. Yamamoto, Y. Owada, H. Kiyonari, K. Murata, Y. Ueno, M. Ono, T. Shimosegawa, N. Yaegashi, M. Watanabe and K. Sugamura, Loss of hrs in the central nervous system causes accumulation of ubiquitinated proteins and neurodegeneration, *Am J Pathol* **173**: 1806–1817 (2008).
104. K. Shirahama, T. Noda and Y. Ohsumi, Mutational analysis of Csc1/Vps4p: involvement of endosome in regulation of autophagy in yeast, *Cell Struct Funct* **22**: 501–509 (1997).
105. Y. Zhong, Q. J. Wang, X. Li, Y. Yan, J. M. Backer, B. T. Chait, N. Heintz and Z. Yue, Distinct regulation of autophagic activity by Atg14L and Rubicon associated with Beclin 1-phosphatidylinositol-3-kinase complex, *Nat Cell Biol* **11**: 468–476 (2009).
106. S. E. Rieder and S. D. Emr, A novel RING finger protein complex essential for a late step in protein transport to the yeast vacuole, *Mol Biol Cell* **8**: 2307–2327 (1997).

99. [illegible] D. J. [illegible], M. [illegible], S. J. [illegible], M. J. Clague and S. [illegible] [illegible] (2010) [illegible] and [illegible] [illegible] (2010).
100. [illegible] and [illegible] [illegible] (2010).
101. M. Matt, J. [illegible], F. [illegible], L. [illegible], T. [illegible] and R. [illegible] [illegible] in the early steps of [illegible] [illegible] 1005–1016 (2009).
102. [illegible] (VPS) [illegible] is an essential protein required for the sorting and [illegible] trafficking of the [illegible] receptor-related protein [illegible] (2005).
103. [illegible] K. [illegible] and K. [illegible] [illegible] in the central nervous system causes accumulation of ubiquitinated proteins and neurodegeneration, *Am J Pathol* 172, [illegible] (2008).
104. [illegible] (2008).
105. [illegible] with [illegible] (2009).
106. [illegible] for [illegible]

CHAPTER 3

CROSS-TALK BETWEEN THE UBIQUITIN-PROTEASOME SYSTEM AND MACROAUTOPHAGY

Serhiy Pankiv and Terje Johansen

ABSTRACT

Regulation of intracellular signaling events, maintenance of homeostasis and adaptation to changing environmental factors are just a few examples of sets of processes that rely heavily on protein recycling. Eukaryotes have two main proteolytic systems that degrade intracellular proteins: the ubiquitin-proteasomal system and the autophagy-lysosomal system. They have been considered to be two completely independent protein-recycling systems, but it is now becoming increasingly clear that both systems interact extensively with each other. Hence, in this chapter we will focus on current findings on the cross-talk between proteasomal and autophagic protein degradation pathways, both at the regulatory and substrate selection levels. The mechanisms for mutual regulation of the UPS and autophagy, as well as the interplay between these two systems in recognition and processing of physiological proteinaceous substrates and neurodegenerative diseases-causing intermediates, will be discussed in some detail.

INTRODUCTION

Proteins are the most abundant and functionally versatile cellular macromolecules. New proteins are constantly synthesized and old, damaged, non-functional or redundant proteins become degraded inside the cell to help maintain cellular homeostasis and to provide the adaptation to changing environmental conditions. In this chapter we will discuss two intracellular proteolytic systems that are currently believed to be the main executors of the degradation step in protein turnover, the ubiquitin-proteasome system (UPS) and the autophagy-lysosomal systems. Since chaperone-mediated autophagy will be discussed in Chapter 4, we will focus here mainly on macroautophagy, and will refer to it simply as autophagy.

The idea that there are specialized intracellular mechanisms for degradation of endogenous proteins is relatively new. At the early 20th century tissue proteins were considered to be stable and their metabolism in adult organisms limited only to the minimal replacement of proteins lost due to "wear and tear.[1]" The high stability of intracellular proteins was questioned in 1934 by Henry Borsook and Geofrey Keighley who, based on the results of their studies on sulphur and nitrogen balance, suggested that most of the nitrogen and sulphur excreted in the urine does not come from ingested food, but rather originate from metabolized endogenous proteins.[2] The work of Rudolph Schoenheimer and colleagues on incorporation of ingested amino acids labeled with the radioactive isotope of nitrogen (^{15}N) in intracellular proteins confirmed this suggestion and demonstrated experimentally the highly dynamic nature of endogenous proteins in live organism.[3]

While studying the activity of acidic phosphatase, cathepsin, β-glucoronidase and several other hydrolytic enzymes in different subcellular fractions, Christian de Duve identified in 1955 the subcellular fraction containing novel organelles which he named lysosomes (from Greek "lysis" — destruction and "soma"— body).[4] Electron microscopy visualized lysosomes as small cytosolic lipid membrane-surrounded vesicles that stained positively for acidic phosphatase.[5] In 1957 the first report was published where electron microscopy was used to demonstrate the presence of mitochondria in the lumen of lysosome-like vesicles.[6] This suggested that cytosolic content and even a large organelle, like the mitochondrion, can be degraded inside the lysosomes. Subsequently, Christian de Duve introduced the concept and term "autophagy" (from Greek self-eating) in 1963.[7] By the mid-1960s autophagic vacuoles had been described in normal cells and tissues and in cells subjected to glucagon treatment or metabolic stress, like starvation or hypoxia.[8] Based on the electron microscopy data, the concept of cellular autophagy was created where part of the cytoplasm becomes surrounded by a cistern that fuses on itself, creating a double-membrane vesicle, the autophagosome. The luminal content of the autophagosome is degraded by acidic lysosomal hydrolases after fusion of the autophagosome with the lysosome.[8] For more than two decades, lysosomal protein degradation remained the only known mechanism of intracellular protein recycling. By the mid-1970s data had accumulated pointing to the existence of an alternative intracellular mechanism for protein degradation. Inactivation of lysosomal hydrolases with mild bases, like ammonia and chloroquine or with protease inhibitors only partially suppressed the turnover of proteins. It was also difficult to explain the different half-lives of different proteins using only the lysosomal model of protein degradation.[9] In 1978 Aaron Ciechanover, Avram Hershko and colleagues identified the small 76-amino acid long polypeptide ubiquitin as the essential component of an alternative, ATP-dependent proteolytic system.[10] Ubiquitin was shown to be covalently linked to and work as a degradation signal for proteins targeted for recycling by this proteolytic system.[11] In 1986–1987 the groups of Martin Rechsteiner and Alfred Goldberg identified the second component of this ATP-dependent proteolytic system, the large multi-subunit chambered protease, called the proteasome, that could recognize and degrade the ubiquitinated substrates.[12,13]

Upon discovery, the ubiquitin-proteasome system (UPS) and autophagy-lysosome systems were viewed as two independent intracellular degradation mechanisms with clearly different approaches for substrate selection and processing. This concept is gradually changing to one involving extensive cross-talk between these vital systems for cellular protein quality control and homeostasis.[14–17]

THE UBIQUITIN-PROTEASOME SYSTEM (UPS)

The term UPS is commonly used to describe the proteasome and components of the enzymatic cascade that catalyzes the ubiquitination of substrates for this large chambered protease. The proteasome is a large multi-subunit protease, consisting of the catalytic core particle and the regulatory particle.[18,19] The catalytic particle, commonly called 20S proteasome has a barrel-like shape and is formed by four heptameric rings stacked on top of each other: two central rings with β-subunits (β1–β7) and two external rings with α-subunits (α1–α7). Together they form a cylindrical particle with a central channel with a diameter of only ~1.5 nm. The catalytic centers of three proteolytically active proteasomal subunits β1, β2 and β5 face the inside of this channel, making them inaccessible for folded proteins. The access of substrates is further controlled by "gates" created on both sides of the 20S proteasome by N-terminal domains of α-subunits and by several alternative proteasomal regulatory particles that associate with one or both ends of the 20S proteasome.[18–20] The most common and best described 26S proteasome contains two 19S regulatory subunits that mediate the recognition of substrates, their unfolding and transfer into the catalytic chamber of the proteasome. The 19S regulatory particle can be subdivided into two subcomplexes: the base and the lid. The base is composed of six AAA-type ATPases (Rpt1–Rpt6) forming the hexameric ring, and four non-ATPase subunits (Rpn1, Rpn2, Rpn10 and Rpn13). The hexameric ring unfolds proteasomal substrates and, together with the Rpn1–Rpn2 complex, opens the gate into the central proteolytic channel of the 20S catalytic particle. Rpn10 and Rpn13 subunits, on the other hand, recognize and recruit proteasomal substrates by binding to the K48-linked polyubiquitin degradation tag.[21] The lid complex of the 19S regulatory particle is composed of nine Rpn subunits (Rpn 3, Rpn 5–9, Rpn11–12 and Rpn15). The organization of the lid complex and its function is currently poorly characterized. However, one of its subunits, Rpn11, has de-ubiquitination activity and mediates the recycling of ubiquitin for further cycles of ubiquitination.[18,19,21]

As mentioned earlier, substrates for the 26S proteasome are usually labeled with the 76-amino acid polypeptide tag ubiquitin.[22] The attachment of ubiquitin to its target is covalent and formed by an amide bond between the carboxyl group of the C-terminal glycine of the ubiquitin and an amino-group of the substrate, usually the ε-amino group of lysine. The ubiquitination reaction is catalyzed by a cascade reaction of three groups of enzymes: E1 activation, E2 conjugation and E3 ligation enzymes.[23] In the initial activation step the E1 enzyme catalyzes the adenylation of the C-terminal glycine of ubiquitin, using ATP. The carboxyl group of the activated C-terminal glycine of ubiquitin-AMP then reacts with the sulfhydryl group of the catalytic cysteine in E1, resulting in an intermediate ubiquitin-E1 thioester conjugate. Ubiquitin-loaded E1 can recruit E2-conjugation enzyme, using its ubiquitin-like (Ubl) domain, and transfers ubiquitin to the catalytic sulfhydryl group of E2 in a trans-thiolation reaction. In the final ligation step, ubiquitin-loaded E2 either directly binds to the target and transfers the activated C-terminal carboxyl group of ubiquitin to the amino group of the target, or is assisted by a scaffold E3 ligase that binds both E2 and the target and bring them to the reaction distance (RING-domain E3 enzymes). Alternatively, E2 transfers ubiquitin to the sulfhydryl group of an E3 ligase, which in turn binds to the target and catalyzes the ligation of ubiquitin to the substrate (HECT-domain E3 enzymes).[22,24] Ubiquitin itself has seven internal lysines (K6, K11, K27, K29, K33, K48 and K63) that can be ubiquitinated, forming polyubiquitin chains.[25] To be efficiently degraded by the proteasome a substrate needs to be tagged with chains of at least four ubiquitin molecules. K48-linked polyubiquitin chains function as the canonical proteasomal degradation tag, but K11-linkages are also used and certain substrates carrying K63-linked polyubiquitin may also be degraded by the proteasome.[17] The human genome encodes two ubiquitin-specific E1 activation enzymes, around 30 E2-conjugation enzymes and more than 1000 E3 ligases. This provides a great versatility in recognition of different substrates as well as enabling diversity in the types of ubiquitin chain linkages added to substrates.[26–28]

AUTOPHAGY

Autophagy is a lysosomal degradation pathway for cytoplasmic or nucleoplasmic material. There are at least three distinct forms of autophagy, which differ in the mode of delivery of substrates to the lysosome: macroautophagy, microautophagy and chaperone-mediated autophagy. In macroautophagy, originally described by de Duve, a flat membrane cistern called the isolation membrane or phagophore wraps around part of the cytoplasm including organelles, like mitochondria, and sequesters the material inside the double-membrane bound autophagosome. Fusion of autophagosomes with lysosomes delivers autophagic lumenal content for degradation by acidic lysosomal hydrolases. A breakthrough in the elucidation of the molecular machinery acting in autophagy was provided by yeast genetics at the turn of the century.[29,30] Today we know of 35 Autophagy-related (Atg) genes in yeast many of which have homologues in mammals.[29,31] The evolutionary conserved protein complexes involved in forming the autophagosomes consist of two ubiquitin (Ub)-like protein conjugation systems containing the ubiquitin-like modifiers Atg8 and Atg12. Atg8 is conjugated to phosphatidylethanolamine (PE) and inserted into the lipid bilayers of the double membrane of the forming autophagosome. In addition, four more protein complexes constitute the rest of the molecular machinery required for autophagosome formation. These are the serine-threonine protein kinase ULK1 complex (unc-51-like kinase 1, a homologue of yeast Atg1), the class III phosphatidylinositol (PI) 3-kinase complex, the WIPI2-Atg2 complex and the transmembrane protein Atg9. In yeast there is a single Atg8 protein while mammals have seven Atg8 proteins grouped into two subfamilies. The MAP1 light chain 3 subfamily with LC3A, -B and -C and the GABARAP (gamma-aminobutyrate receptor associated protein) subfamily with GABARAP and GABARAP-like proteins GABARAPL1–3.[32,33] The phosphatidylethanolamine-conjugated form II of Atg8 proteins serves as an autophagic marker protein.[34,35] In mammals LC3B is regarded as the main Atg8 homologue required for starvation induced autophagy, but also GABARAP, GABARAP-L2 and L1 are lipid-conjugated and act in autophagy.[34,36]

During microautophagy a portion of the cytosol or nucleus becomes sequestered inside the direct invaginations of the limiting lysosomal membrane. Vesicles containing cytoplasmic material that pinch off from these invaginations inside the lysosomal lumen becomes degraded by acidic hydrolases.[37] Topologically, microautophagy resembles the formation of intralumenal vesicles of the multivesicular body (late endosome), but the molecular mechanisms involved are still poorly characterized. Unlike macro- or microautophagy, the chaperone-mediated autophagy is a single molecule delivery mechanism that does not involve the formation of new vesicle compartments. Proteinaceous substrates containing KFERQ-like motif in their sequence form complexes with Hsc70 chaperone and becomes directly transferred across the limiting lysosomal membrane by the LAMP2A-translocation complex.[38,39] We urge readers to read Chapter 4 on the topic of chaperone-mediated autophagy. In the current chapter we will focus on macroautophagy and cross-talk with the UPS both at the regulatory and at substrate levels.

UPS AND AUTOPHAGY: CROSS-TALK AT THE REGULATORY LEVEL

The UPS and autophagy are highly controlled and regulated proteolytic systems that also can cross-talk to regulate the activity of each other (see Figure 1).[14–17] Impairment of the UPS induces autophagy,[40–42] likely as a compensatory mechanism in order to cope with increased amounts of undigested proteasomal substrates. In the cytosol, such substrates become sequestered in specialized protein inclusions such as aggresomes[43] and p62 bodies (also called aggresome-like inducible structures[44–46]) that can be degraded by autophagy.[46–51] Interestingly, for aggregates made by polyglutamine-expansion mutants of huntingtin, sequestration and inactivation of mTOR, a phosphatidylinositol kinase-related kinase that acts as a key inhibitor of autophagy, has been suggested as one of the mechanisms for activation of autophagy.[52] The accumulation of misfolded proteins in the endoplasmic reticulum, due to failure of the proteasome-dependent ERAD (endoplasmic reticulum (ER)-associated degradation) system also activates autophagy as one of the manifestations of the ER-stress response.[53,54] The signaling from at least two ER-stress sensors, PERK (PKR-like ER kinase) and IRE1 (inositol requiring enzyme 1) is involved in this process. Activated by ER stress the PERK transmembrane receptor phosphorylates translation initiation factor eIF2α leading to inhibition of general translation, selective stimulation of translation of the transcription factor ATF4 and upregulation of the autophagy-related genes Atg5, Atg7, Atg12 and LC3.[55–57] Alternatively, activated IRE1 recruits the E3-ligase TRAF2 and stimulates the ASK1-JNK signal transduction pathway. JNK1 has been proposed to activate autophagy through Bcl-2 phosphorylation preventing Bcl-2 from binding to and inactivating Beclin 1.[40,58] Beclin 1 is a regulatory subunit of the class III PI 3-kinase complex required for autophagosome formation. ER stress can also trigger the release of Ca^{2+} ions into the cytosol. This activates calcium/calmodulin-dependent kinase β which in turn activates AMP-

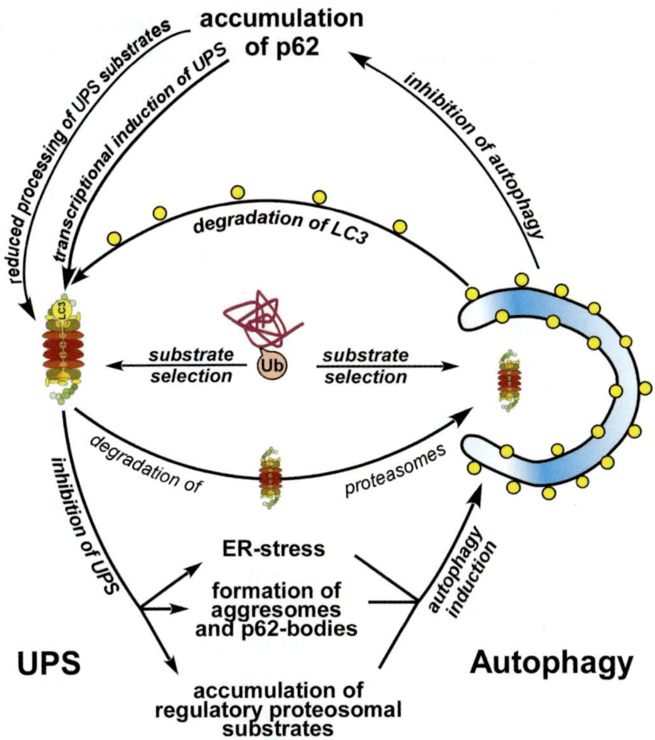

Figure 1. Cross-talk between the ubiquitin-proteasome system (UPS) and autophagy. Both the UPS and autophagy mutually regulate the activity of each other. Inhibition of autophagy leads to stabilization of the UPS-substrates in p62-containing protein inclusions and compensatory transcriptional induction of the UPS components. Impairment of the UPS in turn activates autophagy via several alternative mechanisms like ER-stress, sequestration and inactivation of negative or stabilization of positive mediators of autophagy. UPS and autophagy also cooperate with each other on the substrate selection level by utilizing ubiquitin as a substrate-recognition tag.

activated protein kinase (AMPK) leading to inactivation of mTOR and stimulation of autophagy.[59]

Autophagy can be induced not only as a general cellular response to the accumulation of misfolded proteasomal substrates in the cytosol and ER, but also as a consequence of stabilization of specific cell signaling proteins that are normally kept at low levels by proteasomal degradation. One such signaling protein is the tumor suppressor protein p53. The level of p53 is tightly controlled by Mdm2-mediated ubiquitination and proteasomal degradation. In cells where the UPS is inhibited, p53 is stabilized and can induce autophagy via activation of AMPK and subsequent inhibition of mTOR kinase.[60,61] Other proteins that have been suggested to be regulated by proteasomal degradation are the LC3 family members, the essential components of autophagic membranes. Both LC3A and LC3B can be processed and inactivated by the 20S proteasome in a ubiquitin-independent manner and their levels increase in cells with inhibited proteasomal activity.[62]

The transcription factor FoxO3 is involved in the regulation of both the UPS and autophagy.[63,64] Overexpression of constitutively active FoxO3 causes atrophy of muscles and cultured myotubes by inducing the transcription of a set of atrophy-related genes including critical ubiquitin ligases. The muscle-specific ubiquitin ligase Atrogin-1/MAFbx is induced to increase proteasomal proteolysis while autophagy related genes, such as LC3B and Bnip3, are also activated by FoxO3.[63,64] Hence, FoxO3 may perhaps not mediate a direct cross-talk but rather a coordinate response to boost both proteolytic systems.

While the impairment of the UPS clearly leads to the compensatory activation of autophagy, inactivation of autophagy itself seems to have a dual effect on the UPS. The deficiency in the autophagic activity manifests itself by accumulation of specific autophagic substrates inside the cell, including the autophagic cargo receptor p62.[47,65] p62/SQSTM1 is a 440-amino acid adaptor protein that can homo-polymerize via its N-terminal PB1 domain and interact with ubiquitinated proteins using the C-terminal ubiquitin-associated (UBA) domain and with LC3 via a LC3-interacting region (LIR).[47,51,65,66] It also binds to KEAP1, a component of Cullin-3-type ubiquitin ligase, using its KEAP1-interacting region. Binding of KEAP1 to accumulated p62 sequesters KEAP1 from its substrate protein Nrf2 and stabilizes this transcription factor.[67–71] This in turn leads to transcriptional upregulation of Nrf2-response genes, including UPS-related genes such as E3 ubiquitin ligases Smurf1 and FBXO32, E2 conjugation enzyme UBE2N and de-ubiquitination enzymes Usp19, Usp20, BAP1.[72] In some cells and tissue types, like skin fibroblasts or liver, Nrf2 can also upregulate the expression of proteasomal subunits.[73,74]

Proteasomes have also been reported to be degraded by autophagy.[75] Thus, impairment of autophagy can lead to transcriptional upregulation of UPS-related genes and stabilize proteasomes (Figure 1). Surprisingly this does not result in increased degradation of proteins via UPS. On the contrary, the levels of several endogenous UPS substrates, like p53, β-catenin or the artificial UPS substrate Ub^{G76V}-GFP increase in autophagy-deficient cells, suggesting the impairment of substrate flux through the UPS.[76] Sequestration of ubiquitinated UPS substrates together with upregulated p62 was in this case suggested as the main mechanism preventing delivery of UPS substrates to proteasomes, since knockdown of p62 reduced the level of the previously mentioned substrates to normal level.[76]

To fully understand the functional significance of these interesting findings several questions remain to be answered.

— While inhibition of autophagy decreases the flux of p53, β-catenin and Ub^{G76V}-GFP due to their sequestration with p62, it is currently unclear whether all UPS substrates behave the same. Alternatively, decrease in proteasomal flux of some of the UPS substrates can lead to a reprogramming of proteasomes for degradation of other specific ubiquitinated substrates, or non-ubiquitinated proteins.
— Are ubiquitinated proteasomal substrates sequestered into p62-containing structures functionally active or not? In the latter case the accumulation of inactive UPS substrates in p62-containing protein bodies is functionally similar to their degradation by proteasomes. Reprogramming of the cell to sequester UPS substrates instead of their degradation may gradually deplete the intracellular depot of amino acids. This will in turn lead to inactivation of mTOR and function as an additional stimulus to upregulate autophagy.

The reason for the different compensatory approaches by the two intracellular proteolytic systems to the impairment of their counterpart, i.e. autophagy tries to work for two and degrade both its own and UPS-pathway substrates, while UPS tries to stimulate and jump-start impaired autophagy, most likely lies in intrinsic differences between these two systems. Autophagy can potentially be reprogrammed to degrade UPS-substrates by p62-mediated sequestration of these substrates into p62 bodies (inclusions containing p62 and ubiqutinated proteins), followed by engulfment of these p62 bodies by the autophagic isolation membrane. However, the small size of the proteolytic chamber makes it impossible for proteasomes to degrade large and bulky autophagic substrates, such as entire organelles or protein aggregates that are resistant to unfolding.

UPS AND AUTOPHAGY: CROSS-TALK AT THE SUBSTRATE LEVEL

Excepting protein aggregates, all proteinaceous substrates processed by intracellular proteolytic systems can be subdivided into three broad categories: (1) misfolded proteins, (2) short-lived proteins and (3) long-lived cytosolic proteins. Complete proteolysis of misfolded proteins eliminates potentially dangerous protein intermediates that interfere with proper folding of other proteins and can generate toxic substances, like reactive oxygen species. Proteolysis of properly folded, short-lived proteins has a regulatory role, widely used in the regulation of numerous signal transduction pathways. The unselective degradation of long-lived cytosolic proteins at normal growth conditions provides a constant slow exchange of the intracellular proteome. Under conditions of limited nutrients supply this degradation provides the cell with amino acids that can be used for synthesis of new proteins essential for cell survival.

Until recently the turnover of the first two groups of substrates was attributed solely to the activity of the UPS, while the function of autophagy was thought to be limited to the random slow recycling of long-lived proteins, the third group of substrates. Evidence that autophagy actively participates in the degradation of all three groups of substrates has accumulated during the last decade.

Most cellular proteins have to acquire a certain three-dimensional structure, called the native fold to perform their function. Improper folding of newly synthesized proteins or misfolding of functional proteins due to genetic mutations or influence of extracellular proteotoxic stress lead to surface exposure of hydrophobic residues that are normally buried inside the native structure. Such hydrophobic protein surfaces can bind to the hydrophobic surfaces of other partially-folded or misfolded proteins, triggering protein aggregation. Alternatively, they attract chaperones that prevent unspecific protein aggregation and assist in protein refolding. Chaperones that fail to properly refold their substrates recruit ubiquitin E3 ligases, like CHIP (carboxyl terminus of Hsc70-interacting protein),[77] Parkin,[78] Cullin 5[79] or E6-AP,[80] which trigger ubiquitination of misfolded proteins. The ubiquitin tag functions as a binding epitope for the group of proteins called ubiquitin receptors which are proteins containing ubiquitin-binding domains or motifs. In the UPS two subunits of the proteasomal 19S cap, Rpn10 and Rpn13, have ubiquitin-binding domains/motifs and function as integral proteasomal ubiquitin receptors. In addition, several soluble UPS Ub-receptors of the Ubl-UBA family, like Rad23, DDI1 and Ubiquilin-1 that bind to ubiquitinated substrates with ubiquitin-associated domains and to the proteasomal Rpn10/Rpn13 subunits with their Ubl-domain have been described.[81] The autophagy-lysosomal system use similar soluble ubiquitin receptors, namely p62/SQSTM1, NBR1, NDP52 and several others that recognize ubiquitinated substrates with ubiquitin-binding domains and recruit sequestering autophagosomal isolation membranes using their LC3-interacting regions (LIRs).[48–51,66] LIRs are short peptide sequence motifs that bind to the ATG8 family of autophagosomal membrane-resident proteins.[50,51,66,82] Both p62 and Ubiquilin-1 have been implicated as ubiquitin receptors channeling substrates both to autophagy and the UPS.[47,51,66,83–85]

Histone deacetylase 6 (HDAC6) is a cytoplasmic tubulin deacetylase which associates with microtubules and the retrograde molecular motor dynein and is involved in aggresome formation at the microtubule organizing center (MTOC).[41,86] HDAC6 binds polyubiquitin via a C-terminal BUZ domain and is reported to show preference for K63-linked ubiquitin chains.[87] HDAC6 is involved in transporting ubiquitinated misfolded proteins and protein aggregates along microtubules to form aggresomes.[14,41,86] The HDAC6-interacting chaperone, p97/VCP, dissociates the HDAC6-ubiquitin complexes and may modulate protein aggregation versus degradation by the UPS.[88] HDAC6 is required not only for the formation of aggresomes but also for degradation of the aggregated proteins by autophagy.[41,42] Strikingly, in a Drosophila model the neurodegeneration elicited by inhibition of the UPS, either by a temperature-sensitive mutation of a proteasomal subunit or by overexpression of a polyQ expansion protein, could be ameliorated by overexpressing HDAC6.[42] The positive effect of HDAC6 overexpression was dependent on autophagy, strongly emphasizing cross-talk between the UPS and autophagy. Recent studies suggest that HDAC6 may not be required for autophagy activation as such but helps to assemble an F-actin network that stimulates autophagosome-lysosome fusion.[89] Together with p62, HDAC6 is also involved in selective autophagy of damaged mitochodria (mitophagy).[90] This is highly relevent to Parkinson's disease since disease-causing mutants of the E3 ligase Parkin impair mitophagy of damaged mitochondria. Activation of Parkin leads to ubiqutination of mitochondria and recruitment of p62 and HDAC6 and subsequent aggregation of the damaged mitochondria at the MTOC followed by their degradation by autophagy.[90] The involvement of p62 in mitophagy of damaged mitochondria has been studied by several groups and there is some disagreement on the exact role played by p62 after it has been recruited to ubiquitinated mitochondria.[91–93]

Recruitment of either the proteasome or the autophagic isolation membrane initiates the degradation of misfolded proteins. The organization of misfolded proteins and their aggregation propensity determines the choice of proteolytic system used. Whereas the soluble monomeric misfolded proteins are preferentially degraded by the proteasome, their aggregated counterparts often cannot be processed by the proteasome because proteasomes need to unfold their substrates before cleavage. Large autophagic vesicles on the other hand, with the help of p62, NBR1 or other autophagic ubiquitin receptors can specifically sequester protein aggregates and target them for lysosomal degradation without their unfolding.[51] However, the selection of soluble substrates is not as efficient due to the large volume to surface ratio of the autophagosome. The precise mechanisms that may switch misfolded proteins from proteasomal to autophagosomal substrates is currently unclear. The editing of ubiquitin chains from proteasome-specific K48-linked polyUb to autophagy-specific K63-linked polyUb was suggested as one possible mechanism.[94] Using the mass spectrometry-based AQUA method to quantify different ubiquitin chain linkages Riley *et al.* found no selective enrichment of particular ubiquitin chain linkages in autophagy-deficient livers and brains of transgenic mice.[72] This finding argues against ubiquitin chain linkages being involved in cargo selection for autophagy. Alternatively, the mere aggregation of proteins followed with recognition by an autophagic cargo receptor alone may target them for autophagic degradation as seen for example from the degradation of the maternally derived P granules via the autophagic cargo receptor SEPA-1 in *C. elegans* and a mutant form of human SOD1 interacting directly with p62.[95,96] Both these cases occur in a ubiquitin-independent manner. A recent paper also proposes that protein oligomerization drives autophagic substrate selection and that the accumulation of ubiquitinated proteins seen when autophagy is blocked may be an indirect effect caused by Nrf2 activation.[72] It is clear that oligomerization or polymerization is a required element in efficient selective autophagy mediated by cargo receptors such as p62, SEPA-1 and the yeast cytoplasm to vacuole targeting pathway receptor Atg19.[31] However, more studies are required to clarify the role of ubiquitin or other labels or signals for degradation of substrates by selective autophagy.

Studies of the Hsp/Hsc70 co-chaperones of the Bag family shed some light on switching mechanisms between UPS and autophagic degradation. A shift from Bag1 to Bag3 expression during aging correlated with a shift from proteasomal degradation to autophagic degradation of misfolded proteins.[97] Bag3 induces formation of p62 bodies,[97] and the small heat shock protein HspB8 binds to Bag3 and helps clear misfolded or damaged proteins by selective autophagy.[98–100] Interestingly, HspB8 expression is induced by inhibition of the proteasome.[100]

Persistent imbalance between the production and elimination of certain misfolded protein intermediates results in protein conformation diseases. Together with systemic and localized amyloidosis, neurodegenerative diseases represent the largest group of protein conformation diseases.[101] The nature of proteins that undergo aggregation is different in different neurodegenerative diseases, and range in size from the 40–42 amino acid-long amyloid-β peptides in patients with Alzheimer's disease to the more than 3144 amino acid-long huntingtin protein in patients with Huntington's disease. The common feature of these pathological states is the accumulation of intracellular amyloid-like protein inclusions in the cytosol or nucleus of affected neurons.[101,102] Both increased production of misfolded proteins due to hereditary mutations interfering with proper folding of the affected protein and age-related decrease in the activity of the protein quality systems contribute to disease development.[103–105] Selective neuronal inactivation of either the UPS or autophagy in mouse results in the development of functional and pathological symptoms of neurodegeneration, supporting this hypothesis.[106–109]

Almost without exception ubiquitin is found in the protein inclusions observed in neurodegenerative disorders and in the model systems used to study these diseases.[110–112] Similarly, the selective autophagy substrate and cargo receptor p62 is almost invariantly found in the ubiquitinated inclusions.[110,113–116] A number of studies during the past decade show that aggregation prone proteins causing neurodegenerative diseases can be degraded by both the UPS and by autophagy. This is the case for alpha-synuclein involved in Parkinson's disease.[117] A mutant of SOD1 linked to amyotrophic lateral sclerosis is also a substrate of both proteolytic systems.[118] Polyglutamine (polyQ) expanded mutants of huntingtin have perhaps been the most studied aggregate prone proteins with respect to degradation by both the UPS and autophagy.[119,120] These mutant proteins can be degraded by proteasomes, but the efficiency decreases with increasing polyQ size.[121] There is ample evidence that polyQ mutant huntingtin accumulates in aggregates both because the proteasome cannot degrade aggregated proteins and because the proteasome is directly inhibited by the polyQ proteins.[122–128] It has also been reported that there may be a global impairment of the UPS by nuclear or cytoplasmic protein aggregates that precedes inclusion body formation.[129] Whereas microscopically-visible amyloid-like intracellular inclusions is the common denominator of these neurodegenerative disorders, it is currently believed that the soluble oligomers of misfolded proteins are the mediators that exert the toxic effect inside the cell.[130]

UPS and autophagic proteolytic system are likely to work on different stages in controlling the formation of these toxic substances. Proteasomes degrade the monomeric precursors, preventing the formation of toxic oligomers. The accumulation of unfoldable protein substrates that cannot be degraded by and can sterically inhibit proteasomes likely explains the commonly observed decrease in the activity of the UPS in the model systems of neurodegenerative diseases.[122–129] Macroautophagy, as a bulk protein degradation system on the other hand degrades mainly oligomeric and aggregated forms of misfolded proteins, decreasing the amount of already formed toxic protein intermediates. Interestingly, the autophagic ubiquitin receptor p62 helps decrease toxicity of misfolded proteins using several mechanisms. Due to its polymerization activity, p62 collects soluble misfolded proteins into less toxic compact inclusions, which we denote p62 bodies, in addition to facilitating the recognition of these structures by autophagic isolation membranes.[47,51,131] The spontaneous formation of amyloid protein aggregates is driven by hydrophobic interactions and hydrogen bonding between misfolded proteins, while the assembly of p62 bodies is orchestrated by ionic bonds between the positively and negatively charged surfaces of the PB1 domain of p62.[51,132] Additionally, interactions between the UBA domain of polymerized p62 and ubiquitinated proteins are involved in building the p62 bodies.[47] Unlike hydrophobic interactions, ionic bonds are highly sensitive to changes in pH. Therefore the disassembly and degradation of p62 bodies inside the acidic lysosomes is likely to be more efficient than the degradation of amyloid aggregates.

In addition to degradation of toxic misfolded intermediates, intracellular proteolytic systems also control the stability and half-life of properly folded proteins. Selective degradation of functional proteins is a common mechanism for down-regulation of their function. It is widely used in regulation of almost every aspect of cellular metabolism, including the regulation of proliferation, cell cycle progression, transcription and apoptosis. The role of the UPS in selective degradation of functional regulatory proteins is well established and will not be discussed in this chapter. Interestingly, autophagy was recently identified as an alternative mechanism for degradation of some of the regulatory proteins. The scaffold protein Dishevelled acting in the Wnt-signaling pathway,[133] the KEAP1 substrate adapter for the cullin-3 ubiquitin ligase complex involved in proteasomal degradation of the antioxidative response transcription factor Nrf2,[67–71] the apoptosis effector protease caspase 8[134] and the Drosophila inhibitor of apoptosis dBruce are examples of such proteins.[135]

The evolutionarily conserved Wnt-signal transduction pathway is a key mediator of embryonic development, affecting cell proliferation, cell polarity and cell fate. Its components are also expressed in adult brain where they regulate synaptic plasticity and adult neurogenesis. The canonical Wnt signaling regulate the stability of transcriptional coactivator β-catenin. In resting cells the level of β-catenin is low due to it constant ubiquitination and proteosomal degradation. The binding of the secreted Wnt ligand to its receptor Frizzled initiate the activation and plasma membrane recruitment of the scaffold phosphoprotein Dishevelled, which in turn sequesters the β-catenin destruction complex formed by glycogen synthase kinase 3, adenomatous polyposis coli and axin, stabilize β-catenin and initiate β-catenin-specific transcriptional response. As a key linker between the plasma membrane receptor and downstream cytosolic effectors, Dishevelled plays a crucial role in regulation of Wnt signal transduction. Its level is regulated by ubiquitination mediated by the pVHL-containing ubiquitin E3 ligase complex, followed by proteasomal or autophagic degradation. The autophagic sequestration of Dishevelled is facilitated by p62, which binds to ubiquitinated Dishevelled and recruits the LC3-containing autophagic membranes[133]. Interestingly, the activity of Wnt-signaling is commonly inhibited in brains affected with neurodegeneration, possibly reflecting the high autophagic activity of affected neurons.

Caspases are the main effectors of apoptosis, one form of programmed cell death. These conserved cystein proteases become activated due to oligomerization or controlled proteolysis in response to extra- or intracellular stimuli and initiate the cleavage of various cytosolic and nuclear substrates, causing characteristic morphological features of apoptotic cell death. Caspase 8 is the primary caspase that initiates the apoptotic reaction in response to extracellular stimuli. Surprisingly, the group of Hannah Rabinowich has found that autophagy can rescue cells from the apoptotic cell death by sequestering and degrading an activated form of this caspase.[134] p62 can be speculated as a possible autophagic receptor for caspase 8 since it has previously been shown to promote the aggregation of ubiquitinated forms of this caspase.[136] A second example of autophagic regulation of life and death signaling is provided by Nezis *et al.* who showed that autophagic degradation of a *Drosophila* inhibitor of apoptosis called dBruce acts as a novel mechanism of triggering cell death.[135] During development of the egg chamber the nurse cells that have supported the developing oocyte undergo programmed cell death. Autophagy is required for this cell death to occur and the selective autophagic degradation of dBruce is required for DNA fragmentation and cell death of the nurse cells.

Cells have numerous compensatory response mechanisms to different stress stimuli. Nrf2 is one of the transcription factors that stimulate the expression of stress-response genes. At optimal growth conditions Nrf2 is complexed with KEAP1, the substrate-recognition subunit of Cullin-3 ubiquitin ligase, and becomes ubiquitinated and degraded by the proteasome. Exposure to oxidative agents or electrophiles leads to oxidation of critical cystein residues in KEAP1, its dissociation from Nrf2 and transcriptional activation of Nrf2-response genes.[137–139] Recently, several groups have found that p62, which accumulates in the cytosol in response to proteotoxic stress or inhibition of autophagy, can bind to KEAP1 stabilizing Nrf2 so that Nrf2-response genes become activated.[67–71] It seems clear that ectopically expressed KEAP1 is degraded by selective autophagy in a p62-dependent manner,[69] but more studies of endogenous KEAP1 are required to determine if KEAP1 is a bona fide substrate for selective autophagy. It may be sufficient that p62 sequesters KEAP1 into p62 bodies in a temporary fashion to allow activation of Nrf2 for the required time for the cell to cope with oxidative damage. It is also suggested that KEAP1 may function to facilitate the p62-mediated autophagic degradation of ubiquitin aggregates as such.[68] Clearly, p62 is an autophagy substrate itself,[47] and its degradation by autophagy is critical to avoid aberrant activation of Nrf2.[140]

CONCLUSIONS

There is cross-talk between the UPS and autophagy both at the regulatory level and at the substrate selection level. At the gene regulatory level transcription factors induced by ER stress or failure of proteasome dependent ERAD, like ATF4, can activate autophagy genes. The p53 tumor suppressor protein is stabilized when UPS is inhibited and will then activate autophagy. The transcription factor FoxO3 activates genes both involved in UPS and autophagy and Nrf2 induces the expression of UPS components upon accumulation of p62 caused by autophagy inhibition. The elucidation of mechanisms of selective autophagy involving cargo receptors such as p62 has changed the view of autophagy as a process only employed for unspecific, bulk degradation to a process also accommodating selectivity. p62 is a selective autophagy substrate.[47,51,66] It is also an adaptor protein acting in several signaling pathways involving activation of NFκB.[141] Until recently, the only substrates known to be shared by the UPS and autophagy were misfolded proteins.[14–17] p62 is mainly degraded by autophagy, but we can now add a regulatory protein in a signaling pathway as a substrate for both the UPS and selective autophagy. The Wnt signaling pathway adapter, Dishevelled, is degraded by both systems and is interacting directly with LC3B via a LIR motif in addition to being assisted by p62 for efficient autophagic degradation.[133] Also, activated caspase 8 can be degraded by autophagy, and in *Drosophila* dBruce, an inhibitor of apoptosis, can be degraded by autophagy to allow controlled tissue-specific apoptosis during development. It will be interesting to see if more examples of regulatory proteins degraded by both the UPS and selective autophagy will be identified in future studies.

A glimpse into possible switching mechanisms between the UPS and autophagy is provided by recent studies of the Bag family of co-chaperones. A Bag3-containing chaperone complex acts together with p62 in selective autophagy of some substrates. Presently, we know too little about the interplay between chaperones, co-chaperones, different ubiquitin-binding proteins, including cargo receptors for selective autophagy and ubiquitin E3 ligases and how substrates are selected for degradation by the UPS or autophagy or both. How the cross-talk is occurring at the level of signaling pathways and at the gene regulatory level has just begun to be studied. Given the importance of this topic in understanding the molecular pathology of neurodegenerative diseases more emphasis should be put on studies of cross-talk between the UPS and autophagy in the future.

REFERENCES

1. O. Folin, A Theory of Protein Metabolism, *Am J Physiol* **13**: 117–138 (1905).
2. H. Borsook and G. Keighley, A theory of protein metabolism in man, *Proc Natl Acad Sci USA* **20**: 179–183 (1934).
3. R. Schoenheimer, S. Ratner and D. Rittenberg, Studies in protein metabolism, *J Biol Chem* **127**: 333–344 (1939).
4. C. De Duve, B. C. Pressman, R. Gianetto, R. Wattiaux and F. Appelmans, Tissue fractionation studies 6. Intracellular distribution patterns of enzymes in rat-liver tissue, *Biochem J* **60**: 604–617 (1955).
5. A. B. Novikoff, H. Beaufay and C. De Duve, Electron microscopy of lysosomerich fractions from rat liver, *J Biophys Biochem Cytol* **2**: 179–184 (1956).
6. S. L. Clark, Cellular differentiation in the kidneys of newborn mice studied with the electron microscope, *J Biophys Biochem Cytol* **3**: 349–362 (1957).
7. A. V. S. De Reuck, M. P. Cameron and C. De Duve, *Lysosomes; [proceedings]*, (Little Brown, Boston, 1963), p. xiii, 446 pp.
8. C. De Duve and R. Wattiaux, Functions of lysosomes, *Annu Rev Physiol* **28**: 435–492 (1966).
9. A. L. Goldberg and A. C. St John, Intracellular protein degradation in mammalian and bacterial cells: part 2, *Annu Rev Biochem* **45**: 747–803 (1976).
10. A. Ciechanover, Y. Hod and A. Hershko, A heat-stable polypeptide component of an ATP-dependent proteolytic system from reticulocytes, *Biochem Biophys Res Commun* **81**: 1100–1105 (1978).
11. A. Ciechanover, H. Heller, S. Elias, A. L. Haas and A. Hershko, ATP-dependent conjugation of reticulocyte proteins with the polypeptide required for protein degradation, *Proc Natl Acad Sci USA* **77**: 1365–1368 (1980).
12. R. Hough, G. Pratt and M. Rechsteiner, Ubiquitin-lysozyme conjugates. Identification and characterization of an ATP-dependent protease from rabbit reticulocyte lysates, *J Biol Chem* **261**: 2400–2408 (1986).
13. L. Waxman, J. M. Fagan and A. L. Goldberg, Demonstration of two distinct high molecular weight proteases in rabbit reticulocytes, one of which degrades ubiquitin conjugates, *J Biol Chem* **262**: 2451–2457 (1987).
14. W. X. Ding and X. M. Yin, Sorting, recognition and activation of the misfolded protein degradation pathways through macroautophagy and the proteasome, *Autophagy* **4**: 141–150 (2008).
15. V. I. Korolchuk, F. M. Menzies and D. C. Rubinsztein, Mechanisms of cross-talk between the ubiquitin-proteasome and autophagy-lysosome systems, *FEBS Lett* **584**: 1393–1398 (2010).
16. T. Lamark and T. Johansen, Autophagy: links with the proteasome, *Curr Opin Cell Biol* **22**: 192–198 (2010).
17. E. Wong and A. M. Cuervo, Integration of clearance mechanisms: the proteasome and autophagy, *Cold Spring Harb Perspect Biol* **2**: 0006734 (2010).

18. A. J. Marques, R. Palanimurugan, A. C. Matias, P. C. Ramos and R. J. Dohmen, Catalytic mechanism and assembly of the proteasome, *Chem Rev* **109**: 1509–1536 (2009).
19. K. Tanaka, The proteasome: overview of structure and functions, *Proc Jpn Acad Ser B Phys Biol Sci* **85**: 12–36 (2009).
20. F. Forster, K. Lasker, S. Nickell, A. Sali and W. Baumeister, Toward an integrated structural model of the 26S proteasome, *Mol Cell Proteomics* **9**: 1666–1677 (2010).
21. D. Finley, Recognition and processing of ubiquitin-protein conjugates by the proteasome, *Annu Rev Biochem* **78**: 477–513 (2009).
22. A. Hershko and A. Ciechanover, The ubiquitin system, *Annu Rev Biochem* **67**: 425–479 (1998).
23. A. Hershko, H. Heller, S. Elias and A. Ciechanover, Components of ubiquitin-protein ligase system. Resolution, affinity purification, and role in protein breakdown, *J Biol Chem* **258**: 8206–8214 (1983).
24. O. Kerscher, R. Felberbaum and M. Hochstrasser, Modification of proteins by ubiquitin and ubiquitin-like proteins, *Annu Rev Cell Dev Biol* **22**: 159–180 (2006).
25. D. Komander, The emerging complexity of protein ubiquitination, *Biochem Soc Trans* **37**: 937–953 (2009).
26. C. M. Pickart, Mechanisms underlying ubiquitination, *Annu Rev Biochem* **70**: 503–533 (2001).
27. B. A. Schulman and J. W. Harper, Ubiquitin-like protein activation by E1 enzymes: the apex for downstream signalling pathways, *Nat Rev Mol Cell Biol* **10**: 319–331 (2009).
28. S. J. van Wijk and H. T. Timmers, The family of ubiquitin-conjugating enzymes (E2s): deciding between life and death of proteins, *FASEB J* **24**: 981–993 (2010).
29. H. Nakatogawa, K. Suzuki, Y. Kamada and Y. Ohsumi, Dynamics and diversity in autophagy mechanisms: lessons from yeast, *Nat Rev Mol Cell Biol* **10**: 458–467 (2009).
30. Z. Xie and D. J. Klionsky, Autophagosome formation: core machinery and adaptations, *Nat Cell Biol* **9**: 1102–1109 (2007).
31. M. A. Lynch-Day and D. J. Klionsky, The Cvt pathway as a model for selective autophagy, *FEBS Lett* **584**: 1359–1366 (2010).
32. H. He, Y. Dang, F. Dai, Z. Guo, J. Wu, X. She, Y. Pei, Y. Chen, W. Ling, C. Wu, S. Zhao, J. O. Liu and L. Yu, Post-translational modifications of three members of the human MAP1LC3 family and detection of a novel type of modification for MAP1LC3B, *J Biol Chem* **278**: 29278–29287 (2003).
33. Y. Xin, L. Yu, Z. Chen, L. Zheng, Q. Fu, J. Jiang, P. Zhang, R. Gong and S. Zhao, Cloning, expression patterns, and chromosome localization of three human and two mouse homologues of GABA(A) receptor-associated protein, *Genomics* **74**: 408–413 (2001).
34. Y. Kabeya, N. Mizushima, A. Yamamoto, S. Oshitani-Okamoto, Y. Ohsumi and T. Yoshimori, LC3, GABARAP and GATE16 localize to autophagosomal membrane depending on form-II formation, *J Cell Sci* **117**: 2805–2812 (2004).

35. Y. S. Sou, I. Tanida, M. Komatsu, T. Ueno and E. Kominami, Phosphatidylserine in addition to phosphatidylethanolamine is an *in vitro* target of the mammalian Atg8 modifiers, LC3, GABARAP, and GATE-16, *J Biol Chem* **281**: 3017–3024 (2006).
36. F. Z. Chakrama, S. Seguin-Py, J. N. Le Grand, A. Fraichard, R. Delage-Mourroux, G. Despouy, V. Perez, M. Jouvenot and M. Boyer-Guittaut, GABARAPL1 (GEC1) associates with autophagic vesicles, *Autophagy* **6**: 495–505 (2010).
37. J. B. Kunz, H. Schwarz and A. Mayer, Determination of four sequential stages during microautophagy *in vitro*, *J Biol Chem* **279**: 9987–9996 (2004).
38. J. F. Dice, Chaperone-mediated autophagy, *Autophagy* **3**: 295–299 (2007).
39. M. Kon and A. M. Cuervo, Chaperone-mediated autophagy in health and disease, *FEBS Lett* **584**: 1399–1404 (2010).
40. W. X. Ding, H. M. Ni, W. Gao, T. Yoshimori, D. B. Stolz, D. Ron and X. M. Yin, Linking of autophagy to ubiquitin-proteasome system is important for the regulation of endoplasmic reticulum stress and cell viability, *Am J Pathol* **171**: 513–524 (2007).
41. A. Iwata, B. E. Riley, J. A. Johnston and R. R. Kopito, HDAC6 and microtubules are required for autophagic degradation of aggregated huntingtin, *J Biol Chem* **280**: 40282–40292 (2005).
42. U. B. Pandey, Z. Nie, Y. Batlevi, B. A. McCray, G. P. Ritson, N. B. Nedelsky, S. L. Schwartz, N. A. DiProspero, M. A. Knight, O. Schuldiner, R. Padmanabhan, M. Hild, D. L. Berry, D. Garza, C. C. Hubbert, T. P. Yao, E. H. Baehrecke and J. P. Taylor, HDAC6 rescues neurodegeneration and provides an essential link between autophagy and the UPS, *Nature* **447**: 859–863 (2007).
43. J. A. Johnston, C. L. Ward and R. R. Kopito, Aggresomes: a cellular response to misfolded proteins, *J Cell Biol* **143**: 1883–1898 (1998).
44. H. Lelouard, V. Ferrand, D. Marguet, J. Bania, V. Camosseto, A. David, E. Gatti and P. Pierre, Dendritic cell aggresome-like induced structures are dedicated areas for ubiquitination and storage of newly synthesized defective proteins, *J Cell Biol* **164**: 667–675 (2004).
45. H. Lelouard, E. Gatti, F. Cappello, O. Gresser, V. Camosseto and P. Pierre, Transient aggregation of ubiquitinated proteins during dendritic cell maturation, *Nature* **417**: 177–182 (2002).
46. J. Szeto, N. A. Kaniuk, V. Canadien, R. Nisman, N. Mizushima, T. Yoshimori, D. P. Bazett-Jones and J. H. Brumell, ALIS are stress-induced protein storage compartments for substrates of the proteasome and autophagy, *Autophagy* **2**: 189–199 (2006).
47. G. Bjørkøy, T. Lamark, A. Brech, H. Outzen, M. Perander, A. Øvervatn, H. Stenmark and T. Johansen, p62/SQSTM1 forms protein aggregates degraded by autophagy and has a protective effect on huntingtin-induced cell death, *J Cell Biol* **171**: 603–614 (2005).
48. V. Kirkin, T. Lamark, Y. S. Sou, G. Bjorkoy, J. L. Nunn, J. A. Bruun, E. Shvets, D. G. McEwan, T. H. Clausen, P. Wild, I. Bilusic, J. P. Theurillat, A. Øvervatn, T. Ishii, Z. Elazar, M. Komatsu, I. Dikic and T. Johansen, A role for NBR1 in autophagosomal degradation of ubiquitinated substrates, *Mol Cell* **33**: 505–516 (2009).

49. V. Kirkin, D. G. McEwan, I. Novak and I. Dikic, A role for ubiquitin in selective autophagy, *Mol Cell* **34**: 259–269 (2009).
50. T. Lamark, V. Kirkin, I. Dikic and T. Johansen, NBR1 and p62 as cargo receptors for selective autophagy of ubiquitinated targets, *Cell Cycle* **8**: 1986–1990 (2009).
51. S. Pankiv, T. H. Clausen, T. Lamark, A. Brech, J. A. Bruun, H. Outzen, A. Øvervatn, G. Bjørkøy and T. Johansen, p62/SQSTM1 binds directly to Atg8/LC3 to facilitate degradation of ubiquitinated protein aggregates by autophagy, *J Biol Chem* **282**: 24131–24145 (2007).
52. B. Ravikumar, C. Vacher, Z. Berger, J. E. Davies, S. Luo, L. G. Oroz, F. Scaravilli, D. F. Easton, R. Duden, C. J. O'Kane and D. C. Rubinsztein, Inhibition of mTOR induces autophagy and reduces toxicity of polyglutamine expansions in fly and mouse models of Huntington disease, *Nat Genet* **36**: 585–595 (2004).
53. T. Yorimitsu, U. Nair, Z. Yang and D. J. Klionsky, Endoplasmic reticulum stress triggers autophagy, *J Biol Chem* **281**: 30299–30304 (2006).
54. M. Ogata, S. Hino, A. Saito, K. Morikawa, S. Kondo, S. Kanemoto, T. Murakami, M. Taniguchi, I. Tanii, K. Yoshinaga, S. Shiosaka, J. A. Hammarback, F. Urano and K. Imaizumi, Autophagy is activated for cell survival after endoplasmic reticulum stress, *Mol Cell Biol* **26**: 9220–9231 (2006).
55. Y. Kouroku, E. Fujita, I. Tanida, T. Ueno, A. Isoai, H. Kumagai, S. Ogawa, R. J. Kaufman, E. Kominami and T. Momoi, ER stress (PERK/eIF2alpha phosphorylation) mediates the polyglutamine-induced LC3 conversion, an essential step for autophagy formation, *Cell Death Differ* **14**: 230–239 (2007).
56. M. Milani, T. Rzymski, H. R. Mellor, L. Pike, A. Bottini, D. Generali and A. L. Harris, The role of ATF4 stabilization and autophagy in resistance of breast cancer cells treated with Bortezomib, *Cancer Res* **69**: 4415–4423 (2009).
57. K. Zhu, K. Dunner, Jr. and D. J. McConkey, Proteasome inhibitors activate autophagy as a cytoprotective response in human prostate cancer cells, *Oncogene* **29**: 451–462 (2010).
58. Y. Wei, S. Pattingre, S. Sinha, M. Bassik and B. Levine, JNK1-mediated phosphorylation of Bcl-2 regulates starvation-induced autophagy, *Mol Cell* **30**: 678–688 (2008).
59. M. Hoyer-Hansen and M. Jaattela, Connecting endoplasmic reticulum stress to autophagy by unfolded protein response and calcium, *Cell Death Differ* **14**: 1576–1582 (2007).
60. Y. Du, D. Yang, L. Li, G. Luo, T. Li, X. Fan, Q. Wang, X. Zhang, Y. Wang and W. Le, An insight into the mechanistic role of p53-mediated autophagy induction in response to proteasomal inhibition-induced neurotoxicity, *Autophagy* **5**: 663–675 (2009).
61. K. H. Vousden and K. M. Ryan, p53 and metabolism, *Nat Rev Cancer* **9**: 691–700 (2009).
62. Z. Gao, N. Gammoh, P. M. Wong, H. Erdjument-Bromage, P. Tempst and X. Jiang, Processing of autophagic protein LC3 by the 20S proteasome, *Autophagy* **6**: 126–137 (2010).

63. C. Mammucari, G. Milan, V. Romanello, E. Masiero, R. Rudolf, P. Del Piccolo, S. J. Burden, R. Di Lisi, C. Sandri, J. Zhao, A. L. Goldberg, S. Schiaffino and M. Sandri, FoxO3 controls autophagy in skeletal muscle *in vivo*, *Cell Metab* **6**: 458–471 (2007).
64. J. Zhao, J. J. Brault, A. Schild, P. Cao, M. Sandri, S. Schiaffino, S. H. Lecker and A. L. Goldberg, FoxO3 coordinately activates protein degradation by the autophagic/lysosomal and proteasomal pathways in atrophying muscle cells, *Cell Metab* **6**: 472–483 (2007).
65. M. Komatsu, S. Waguri, M. Koike, Y. S. Sou, T. Ueno, T. Hara, N. Mizushima, J. Iwata, J. Ezaki, S. Murata, J. Hamazaki, Y. Nishito, S. Iemura, T. Natsume, T. Yanagawa, J. Uwayama, E. Warabi, H. Yoshida, T. Ishii, A. Kobayashi, M. Yamamoto, Z. Yue, Y. Uchiyama, E. Kominami and K. Tanaka, Homeostatic levels of p62 control cytoplasmic inclusion body formation in autophagy-deficient mice, *Cell* **131**: 1149–1163 (2007).
66. Y. Ichimura, T. Kumanomidou, Y. S. Sou, T. Mizushima, J. Ezaki, T. Ueno, E. Kominami, T. Yamane, K. Tanaka and M. Komatsu, Structural basis for sorting mechanism of p62 in selective autophagy, *J Biol Chem* **283**: 22847–22857 (2008).
67. I. M. Copple, A. Lister, A. D. Obeng, N. R. Kitteringham, R. E. Jenkins, R. Layfield, B. J. Foster, C. E. Goldring and B. K. Park, Physical and functional interaction of sequestosome 1 with Keap1 regulates the Keap1-Nrf2 cell defense pathway, *J Biol Chem* **285**: 16782–16788 (2010).
68. W. Fan, Z. Tang, D. Chen, D. Moughon, X. Ding, S. Chen, M. Zhu and Q. Zhong, Keap1 facilitates p62-mediated ubiquitin aggregate clearance via autophagy, *Autophagy* **6**: 614–621 (2010).
69. A. Jain, T. Lamark, E. Sjøttem, K. Bowitz Larsen, J. A. Awuh, A. Øvervatn, M. McMahon, J. D. Hayes and T. Johansen, p62/SQSTM1 is a target gene for transcription factor NRF2 and creates a positive feedback loop by inducing antioxidant response element-driven gene transcription, *J Biol Chem* **285**: 22576–22591 (2010).
70. M. Komatsu, H. Kurokawa, S. Waguri, K. Taguchi, A. Kobayashi, Y. Ichimura, Y. S. Sou, I. Ueno, A. Sakamoto, K. I. Tong, M. Kim, Y. Nishito, S. I. Iemura, T. Natsume, T. Ueno, E. Kominami, H. Motohashi, K. Tanaka and M. Yamamoto, The selective autophagy substrate p62 activates the stress responsive transcription factor Nrf2 through inactivation of Keap1, *Nat Cell Biol* **12**: 213–223 (2010).
71. A. Lau, X. J. Wang, F. Zhao, N. F. Villeneuve, T. Wu, T. Jiang, Z. Sun, E. White and D. D. Zhang, A noncanonical mechanism of Nrf2 activation by autophagy deficiency: direct interaction between Keap1 and p62, *Mol Cell Biol* **30**: 3275–3285 (2010).
72. B. E. Riley, S. E. Kaiser, T. A. Shaler, A. C. Ng, T. Hara, M. S. Hipp, K. Lage, R. J. Xavier, K. Y. Ryu, K. Taguchi, M. Yamamoto, K. Tanaka, N. Mizushima, M. Komatsu and R. R. Kopito, Ubiquitin accumulation in autophagy-deficient mice is dependent on the Nrf2-mediated stress response pathway: a potential role for protein aggregation in autophagic substrate selection, *J Cell Biol* **191**: 537–552 (2010).

73. D. C. Kraft, C. C. Deocaris, R. Wadhwa and S. I. Rattan, Preincubation with the proteasome inhibitor MG-132 enhances proteasome activity via the Nrf2 transcription factor in aging human skin fibroblasts, *Ann N Y Acad Sci* **1067**: 420–424 (2006).
74. M. K. Kwak, N. Wakabayashi, J. L. Greenlaw, M. Yamamoto and T. W. Kensler, Antioxidants enhance mammalian proteasome expression through the Keap1-Nrf2 signaling pathway, *Mol Cell Biol* **23**: 8786–8794 (2003).
75. A. M. Cuervo, A. Palmer, A. J. Rivett and E. Knecht, Degradation of proteasomes by lysosomes in rat liver, *Eur J Biochem* **227**: 792–800 (1995).
76. V. I. Korolchuk, A. Mansilla, F. M. Menzies and D. C. Rubinsztein, Autophagy inhibition compromises degradation of ubiquitin-proteasome pathway substrates, *Mol Cell* **33**: 517–527 (2009).
77. S. Murata, Y. Minami, M. Minami, T. Chiba and K. Tanaka, CHIP is a chaperone-dependent E3 ligase that ubiquitylates unfolded protein, *EMBO Rep* **2**: 1133–1138 (2001).
78. Y. C. Tsai, P. S. Fishman, N. V. Thakor and G. A. Oyler, Parkin facilitates the elimination of expanded polyglutamine proteins and leads to preservation of proteasome function, *J Biol Chem* **278**: 22044–22055 (2003).
79. E. S. Ehrlich, T. Wang, K. Luo, Z. Xiao, A. M. Niewiadomska, T. Martinez, W. Xu, L. Neckers and X. F. Yu, Regulation of Hsp90 client proteins by a Cullin5-RING E3 ubiquitin ligase, *Proc Natl Acad Sci USA* **106**: 20330–20335 (2009).
80. A. Mishra, S. K. Godavarthi, M. Maheshwari, A. Goswami and N. R. Jana, The ubiquitin ligase E6-AP is induced and recruited to aggresomes in response to proteasome inhibition and may be involved in the ubiquitination of Hsp70-bound misfolded proteins, *J Biol Chem* **284**: 10537–10545 (2009).
81. C. Grabbe and I. Dikic, Functional roles of ubiquitin-like domain (ULD) and ubiquitin-binding domain (UBD) containing proteins, *Chem Rev* **109**: 1481–1494 (2009).
82. N. N. Noda, Y. Ohsumi and F. Inagaki, Atg8-family interacting motif crucial for selective autophagy, *FEBS Lett* **584**: 1379–1385 (2010).
83. M. L. Seibenhener, J. R. Babu, T. Geetha, H. C. Wong, N. R. Krishna and M. W. Wooten, Sequestosome 1/p62 is a polyubiquitin chain binding protein involved in ubiquitin proteasome degradation, *Mol Cell Biol* **24**: 8055–8068 (2004).
84. C. Rothenberg, D. Srinivasan, L. Mah, S. Kaushik, C. M. Peterhoff, J. Ugolino, S. Fang, A. M. Cuervo, R. A. Nixon and M. J. Monteiro, Ubiquilin functions in autophagy and is degraded by chaperone-mediated autophagy, *Hum Mol Genet* **19**: 3219–3232 (2010).
85. H. S. Ko, T. Uehara, K. Tsuruma and Y. Nomura, Ubiquilin interacts with ubiquitylated proteins and proteasome through its ubiquitin-associated and ubiquitin-like domains, *FEBS Lett* **566**: 110–114 (2004).
86. Y. Kawaguchi, J. J. Kovacs, A. McLaurin, J. M. Vance, A. Ito and T. P. Yao, The deacetylase HDAC6 regulates aggresome formation and cell viability in response to misfolded protein stress, *Cell* **115**: 727–738 (2003).

ERRATUM FOR PAGE 92 AND PAGE 39
OF *AUTOPHAGY OF THE NERVOUS SYSTEM: CELLULAR SELF-DIGESTION IN NEURONS AND NEUROLOGICAL DISEASES*

(ISBN: 978-981-4350-44-0)

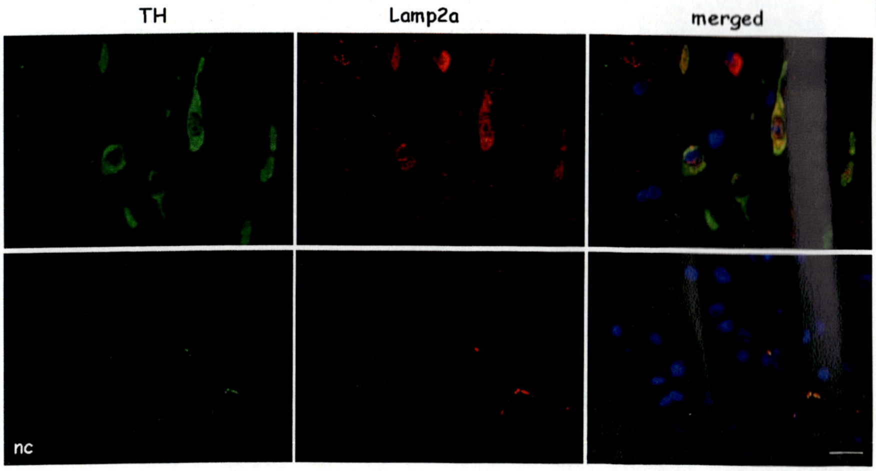

Figure 2. Lamp2a immunohistochemistry in the dopaminergic neurons of the rat substantia nigra pars compacta. Ten micron cryostat-cut midbrain sections were immunostained for TH (1:1000, Chemicon, msAb) and Lamp2a (1:400, Zymed, rbAb). Merged image, with the characteristic punctuate perinuclear staining of endogenous Lamp2a in TH-positive neurons is shown on the right. Scale bar, 20 μm. Midbrain section in which the primary antibodies were omitted is shown in the bottom panel (nc, negative control).

Amended color version of Figure 2 (Chapter 4, page 92)

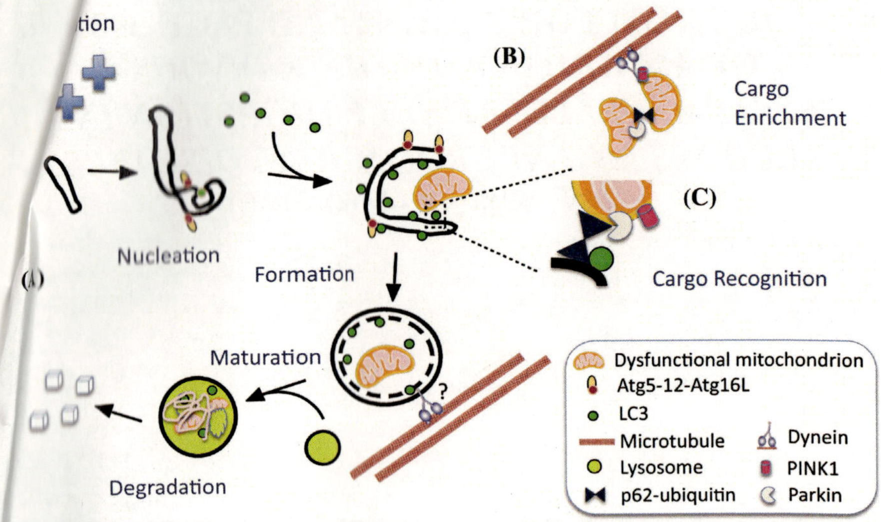

Figure 6. Schematic diagram illustrating major steps in mitochondrial autophagy. (A) Cellular conditions trigger-signaling pathways that promote and suppress autophagy. If integration of these signals results in autophagy induction, localized increases in endoplasmic reticulum phosphatidylinositol 3-phosphate content mediate the nucleation of autophagy protein deposition, to include the Atg5-12-16L complex, which serves to promote covalent conjugation of LC3 to phosphatidylethanolamine. LC3 lipidation is necessary for growth and complete formation of the autophagosome. Selective sequestration of cargo occurs during LC3-dependent extension of autophagic membrane, potentially through two distinct mechanisms that are not mutually exclusive (**B, C**). The transiently double membrane autophagosome undergoes maturation by fusing with acidic compartments such as endosomes and lysosomes, a process that is often microtubule dependent. The cargo is degraded inside lysosomes, with release of cellular building blocks. **(B)** In the first model of cargo selection, enrichment for damaged mitochondria is mediated by selective transport to perinuclear microtubule organizing centers. Concentrating damaged cargo to be cleared in aggresomes, a process dependent upon p62 binding to ubiquitinated cargo, would enhance the relative efficiency of autophagic clearance even in the absence of more specific means of cargo enrichment. PINK1 interactions with Miro, or VDAC1 interactions with dynein may be involved in this process, as full length PINK1 is stabilized and VDAC1 is ubiquitinated by parkin on chemically depolarized mitochondria. **(C)** There is also data to indicate that autophagosomes can form and be transported within neurites of primary neurons and differentiated neuroblastoma cells, although the adapters remain to be identified. Selective mitophagy in this context would depend upon direct interactions involving the altered mitochondrial surface, or adaptor proteins that recognize damaged mitochondria, and LC3 itself. In a manner analogous to recognition of protein aggregates, PINK1 recruitment of parkin could result in protein ubiquitination at the surface of mitochondria, with recruitment of p62, an adapter protein that binds both ubiquitin and LC3. This would function in analogy to the Atg32-Atg11 system in yeast. Both of these models have precedent in non-neuronal and non-mitochondrial contexts for autophagic cargo selection, and further studies in primary neurons are needed to clarify mechanisms that function to ensure mitochondrial quality control in such a highly polarized, mitochondrially-dependent cell type.

Amended color version of Figure 6 (Chapter 16, page 393)

87. J. A. Olzmann, L. Li, M. V. Chudaev, J. Chen, F. A. Perez, R. D. Palmiter and L. S. Chin, Parkin-mediated K63-linked polyubiquitination targets misfolded DJ-1 to aggresomes via binding to HDAC6, *J Cell Biol* **178**: 1025–1038 (2007).
88. C. Boyault, B. Gilquin, Y. Zhang, V. Rybin, E. Garman, W. Meyer-Klaucke, P. Matthias, C. W. Muller and S. Khochbin, HDAC6-p97/VCP controlled polyubiquitin chain turnover, *EMBO J* **25**: 3357–3366 (2006).
89. J. Y. Lee, H. Koga, Y. Kawaguchi, W. Tang, E. Wong, Y. S. Gao, U. B. Pandey, S. Kaushik, E. Tresse, J. Lu, J. P. Taylor, A. M. Cuervo and T. P. Yao, HDAC6 controls autophagosome maturation essential for ubiquitin-selective quality-control autophagy, *EMBO J* **29**: 969–980 (2010).
90. J. Y. Lee, Y. Nagano, J. P. Taylor, K. L. Lim and T. P. Yao, Disease-causing mutations in Parkin impair mitochondrial ubiquitination, aggregation, and HDAC6-dependent mitophagy, *J Cell Biol* **189**: 671–679 (2010).
91. S. Geisler, K. M. Holmstrom, D. Skujat, F. C. Fiesel, O. C. Rothfuss, P. J. Kahle and W. Springer, PINK1/Parkin-mediated mitophagy is dependent on VDAC1 and p62/SQSTM1, *Nat Cell Biol* **12**: 119–131 (2010).
92. D. P. Narendra, L. A. Kane, D. N. Hauser, I. M. Fearnley and R. J. Youle, p62/SQSTM1 is required for Parkin-induced mitochondrial clustering but not mitophagy; VDAC1 is dispensable for both, *Autophagy* **6**: 1090–1106 (2010).
93. K. Okatsu, K. Saisho, M. Shimanuki, K. Nakada, H. Shitara, Y. S. Sou, M. Kimura, S. Sato, N. Hattori, M. Komatsu, K. Tanaka and N. Matsuda, p62/SQSTM1 cooperates with Parkin for perinuclear clustering of depolarized mitochondria, *Genes Cells* **15**: 887–900 (2010).
94. J. M. Tan, E. S. Wong, D. S. Kirkpatrick, O. Pletnikova, H. S. Ko, S. P. Tay, M. W. Ho, J. Troncoso, S. P. Gygi, M. K. Lee, V. L. Dawson, T. M. Dawson and K. L. Lim, Lysine 63-linked ubiquitination promotes the formation and autophagic clearance of protein inclusions associated with neurodegenerative diseases, *Hum Mol Genet* **17**: 431–439 (2008).
95. J. Gal, A. L. Strom, D. M. Kwinter, R. Kilty, J. Zhang, P. Shi, W. Fu, M. W. Wooten and H. Zhu, Sequestosome 1/p62 links familial ALS mutant SOD1 to LC3 via an ubiquitin-independent mechanism, *J Neurochem* **111**: 1062–1073 (2009).
96. Y. Zhang, L. Yan, Z. Zhou, P. Yang, E. Tian, K. Zhang, Y. Zhao, Z. Li, B. Song, J. Han, L. Miao and H. Zhang, SEPA-1 mediates the specific recognition and degradation of P granule components by autophagy in *C. elegans*, *Cell* **136**: 308–321 (2009).
97. M. Gamerdinger, P. Hajieva, A. M. Kaya, U. Wolfrum, F. U. Hartl and C. Behl, Protein quality control during aging involves recruitment of the macroautophagy pathway by BAG3, *EMBO J* **28**: 889–901 (2009).
98. V. Arndt, N. Dick, R. Tawo, M. Dreiseidler, D. Wenzel, M. Hesse, D. O. Furst, P. Saftig, R. Saint, B. K. Fleischmann, M. Hoch and J. Hohfeld, Chaperone-assisted selective autophagy is essential for muscle maintenance, *Curr Biol* **20**: 143–148 (2010).

99. S. Carra, S. J. Seguin, H. Lambert and J. Landry, HspB8 chaperone activity toward poly(Q)-containing proteins depends on its association with Bag3, a stimulator of macroautophagy, *J Biol Chem* **283**: 1437–1444 (2008).
100. V. Crippa, D. Sau, P. Rusmini, A. Boncoraglio, E. Onesto, E. Bolzoni, M. Galbiati, E. Fontana, M. Marino, S. Carra, C. Bendotti, S. De Biasi and A. Poletti, The small heat shock protein B8 (HspB8) promotes autophagic removal of misfolded proteins involved in amyotrophic lateral sclerosis (ALS), *Hum Mol Genet* **19**: 3440–3456 (2010).
101. F. Chiti and C. M. Dobson, Protein misfolding, functional amyloid, and human disease, *Annu Rev Biochem* **75**: 333–366 (2006).
102. E. Herczenik and M. F. Gebbink, Molecular and cellular aspects of protein misfolding and disease, *FASEB J* **22**: 2115–2133 (2008).
103. G. Carrard, A. L. Bulteau, I. Petropoulos and B. Friguet, Impairment of proteasome structure and function in aging, *Int J Biochem Cell Biol* **34**: 1461–1474 (2002).
104. E. Cohen, J. Bieschke, R. M. Perciavalle, J. W. Kelly and A. Dillin, Opposing activities protect against age-onset proteotoxicity, *Science* **313**: 1604–1610 (2006).
105. A. M. Cuervo and J. F. Dice, Age-related decline in chaperone-mediated autophagy, *J Biol Chem* **275**: 31505–31513 (2000).
106. T. Hara, K. Nakamura, M. Matsui, A. Yamamoto, Y. Nakahara, R. Suzuki-Migishima, M. Yokoyama, K. Mishima, I. Saito, H. Okano and N. Mizushima, Suppression of basal autophagy in neural cells causes neurodegenerative disease in mice, *Nature* **441**: 885–889 (2006).
107. M. Komatsu, S. Waguri, T. Chiba, S. Murata, J. Iwata, I. Tanida, T. Ueno, M. Koike, Y. Uchiyama, E. Kominami and K. Tanaka, Loss of autophagy in the central nervous system causes neurodegeneration in mice, *Nature* **441**: 880–884 (2006).
108. L. Bedford, D. Hay, A. Devoy, S. Paine, D. G. Powe, R. Seth, T. Gray, I. Topham, K. Fone, N. Rezvani, M. Mee, T. Soane, R. Layfield, P. W. Sheppard, T. Ebendal, D. Usoskin, J. Lowe and R. J. Mayer, Depletion of 26S proteasomes in mouse brain neurons causes neurodegeneration and Lewy-like inclusions resembling human pale bodies, *J Neurosci* **28**: 8189–8198 (2008).
109. E. Martin-Aparicio, A. Yamamoto, F. Hernandez, R. Hen, J. Avila and J. J. Lucas, Proteasomal-dependent aggregate reversal and absence of cell death in a conditional mouse model of Huntington's disease, *J Neurosci* **21**: 8772–8781 (2001).
110. E. Kuusisto, A. Salminen and I. Alafuzoff, Ubiquitin-binding protein p62 is present in neuronal and glial inclusions in human tauopathies and synucleinopathies, *Neuroreport* **12**: 2085–2090 (2001).
111. J. Lowe, A. Blanchard, K. Morrell, G. Lennox, L. Reynolds, M. Billett, M. Landon and R. J. Mayer, Ubiquitin is a common factor in intermediate filament inclusion bodies of diverse type in man, including those of Parkinson's disease, Pick's disease, and Alzheimer's disease, as well as Rosenthal fibres in cerebellar astrocytomas, cytoplasmic bodies in muscle, and mallory bodies in alcoholic liver disease, *J Pathol* **155**: 9–15 (1988).

112. H. Mori, J. Kondo and Y. Ihara, Ubiquitin is a component of paired helical filaments in Alzheimer's disease, *Science* **235**: 1641–1644 (1987).
113. E. Kuusisto, T. Kauppinen and I. Alafuzoff, Use of p62/SQSTM1 antibodies for neuropathological diagnosis, *Neuropathol Appl Neurobiol* **34**: 169–180 (2008).
114. E. Kuusisto, L. Parkkinen and I. Alafuzoff, Morphogenesis of Lewy bodies: dissimilar incorporation of alpha-synuclein, ubiquitin, and p62, *J Neuropathol Exp Neurol* **62**: 1241–1253 (2003).
115. E. Kuusisto, A. Salminen and I. Alafuzoff, Early accumulation of p62 in neurofibrillary tangles in Alzheimer's disease: possible role in tangle formation, *Neuropathol Appl Neurobiol* **28**: 228–237 (2002).
116. K. Zatloukal, C. Stumptner, A. Fuchsbichler, H. Heid, M. Schnoelzer, L. Kenner, R. Kleinert, M. Prinz, A. Aguzzi and H. Denk, p62 is a common component of cytoplasmic inclusions in protein aggregation diseases, *Am J Pathol* **160**: 255–263 (2002).
117. J. L. Webb, B. Ravikumar, J. Atkins, J. N. Skepper and D. C. Rubinsztein, Alpha-synuclein is degraded by both autophagy and the proteasome, *J Biol Chem* **278**: 25009–25013 (2003).
118. T. Kabuta, Y. Suzuki and K. Wada, Degradation of amyotrophic lateral sclerosis-linked mutant Cu,Zn-superoxide dismutase proteins by macroautophagy and the proteasome, *J Biol Chem* **281**: 30524–30533 (2006).
119. Z. Berger, B. Ravikumar, F. M. Menzies, L. G. Oroz, B. R. Underwood, M. N. Pangalos, I. Schmitt, U. Wullner, B. O. Evert, C. J. O'Kane and D. C. Rubinsztein, Rapamycin alleviates toxicity of different aggregate-prone proteins, *Hum Mol Genet* **15**: 433–442 (2006).
120. B. Ravikumar, R. Duden and D. C. Rubinsztein, Aggregate-prone proteins with polyglutamine and polyalanine expansions are degraded by autophagy, *Hum Mol Genet* **11**: 1107–1117 (2002).
121. N. R. Jana, E. A. Zemskov, G. Wang and N. Nukina, Altered proteasomal function due to the expression of polyglutamine-expanded truncated N-terminal huntingtin induces apoptosis by caspase activation through mitochondrial cytochrome c release, *Hum Mol Genet* **10**: 1049–1059 (2001).
122. N. F. Bence, R. M. Sampat and R. R. Kopito, Impairment of the ubiquitin-proteasome system by protein aggregation, *Science* **292**: 1552–1555 (2001).
123. Y. Chai, S. L. Koppenhafer, S. J. Shoesmith, M. K. Perez and H. L. Paulson, Evidence for proteasome involvement in polyglutamine disease: localization to nuclear inclusions in SCA3/MJD and suppression of polyglutamine aggregation *in vitro*, *Hum Mol Genet* **8**: 673–682 (1999).
124. C. I. Holmberg, K. E. Staniszewski, K. N. Mensah, A. Matouschek and R. I. Morimoto, Inefficient degradation of truncated polyglutamine proteins by the proteasome, *EMBO J* **23**: 4307–4318 (2004).
125. J. N. Keller, K. B. Hanni and W. R. Markesbery, Impaired proteasome function in Alzheimer's disease, *J Neurochem* **75**: 436–439 (2000).

126. L. Stefanis, K. E. Larsen, H. J. Rideout, D. Sulzer and L. A. Greene, Expression of A53T mutant but not wild-type alpha-synuclein in PC12 cells induces alterations of the ubiquitin-dependent degradation system, loss of dopamine release, and autophagic cell death, *J Neurosci* **21**: 9549–9560 (2001).
127. P. Venkatraman, R. Wetzel, M. Tanaka, N. Nukina and A. L. Goldberg, Eukaryotic proteasomes cannot digest polyglutamine sequences and release them during degradation of polyglutamine-containing proteins, *Mol Cell* **14**: 95–104 (2004).
128. S. Waelter, A. Boeddrich, R. Lurz, E. Scherzinger, G. Lueder, H. Lehrach and E. E. Wanker, Accumulation of mutant huntingtin fragments in aggresome-like inclusion bodies as a result of insufficient protein degradation, *Mol Biol Cell* **12**: 1393–1407 (2001).
129. E. J. Bennett, N. F. Bence, R. Jayakumar and R. R. Kopito, Global impairment of the ubiquitin-proteasome system by nuclear or cytoplasmic protein aggregates precedes inclusion body formation, *Mol Cell* **17**: 351–365 (2005).
130. M. Arrasate, S. Mitra, E. S. Schweitzer, M. R. Segal and S. Finkbeiner, Inclusion body formation reduces levels of mutant huntingtin and the risk of neuronal death, *Nature* **431**: 805–810 (2004).
131. T. H. Clausen, T. Lamark, P. Isakson, K. Finley, K. B. Larsen, A. Brech, A. Øvervatn, H. Stenmark, G. Bjørkøy, A. Simonsen and T. Johansen, p62/SQSTM1 and ALFY interact to facilitate the formation of p62 bodies/ALIS and their degradation by autophagy, *Autophagy* **6**: 330–344 (2010).
132. T. Lamark, M. Perander, H. Outzen, K. Kristiansen, A. Øvervatn, E. Michaelsen, G. Bjørkøy and T. Johansen, Interaction codes within the family of mammalian Phox and Bem1p domain-containing proteins, *J Biol Chem* **278**: 34568–34581 (2003).
133. C. Gao, W. Cao, L. Bao, W. Zuo, G. Xie, T. Cai, W. Fu, J. Zhang, W. Wu, X. Zhang and Y. G. Chen, Autophagy negatively regulates Wnt signalling by promoting Dishevelled degradation, *Nat Cell Biol* **12**: 781–790 (2010).
134. W. Hou, J. Han, C. Lu, L. A. Goldstein and H. Rabinowich, Autophagic degradation of active caspase-8: a crosstalk mechanism between autophagy and apoptosis, *Autophagy* **6**: 891–900 (2010).
135. I. P. Nezis, B. V. Shravage, A. P. Sagona, T. Lamark, G. Bjørkøy, T. Johansen, T. E. Rusten, A. Brech, E. H. Baehrecke and H. Stenmark, Autophagic degradation of dBruce controls DNA fragmentation in nurse cells during late *Drosophila melanogaster* oogenesis, *J Cell Biol* **190**: 523–531 (2010).
136. Z. Jin, Y. Li, R. Pitti, D. Lawrence, V. C. Pham, J. R. Lill and A. Ashkenazi, Cullin3-based polyubiquitination and p62-dependent aggregation of caspase-8 mediate extrinsic apoptosis signaling, *Cell* **137**: 721–735 (2009).
137. T. Nguyen, P. Nioi and C. B. Pickett, The Nrf2-antioxidant response element signaling pathway and its activation by oxidative stress, *J Biol Chem* **284**: 13291–13295 (2009).

138. J. D. Hayes and M. McMahon, NRF2 and KEAP1 mutations: permanent activation of an adaptive response in cancer, *Trends Biochem Sci* **34**: 176–188 (2009).
139. T. W. Kensler, N. Wakabayashi and S. Biswal, Cell survival responses to environmental stresses via the Keap1-Nrf2-ARE pathway, *Annu Rev Pharmacol Toxicol* **47**: 89–116 (2007).
140. Y. Inami, S. Waguri, A. Sakamoto, T. Kouno, K. Nakada, O. Hino, S. Watanabe, J. Ando, M. Iwadate, M. Yamamoto, M. S. Lee, K. Tanaka and M. Komatsu, Persistent activation of Nrf2 through p62 in hepatocellular carcinoma cells, *J Cell Biol* **193**: 275–284 (2011).
141. J. Moscat, M. T. Diaz-Meco and M. W. Wooten, Signal integration and diversification through the p62 scaffold protein, *Trends Biochem Sci* **32**: 95–100 (2007).

CHAPTER 4

CHAPERONE-MEDIATED AUTOPHAGY (CMA) IN NEURONS

Maria Xilouri, Hsiao-Yu Peng and Leonidas Stefanis

ABSTRACT

Chaperone-mediated autophagy (CMA) is a selective mechanism, described mainly in mammalian cells, for the lysosomal degradation of specific soluble proteins. The main difference between CMA and the other types of autophagy (macro-, micro-autophagy), is the fact that it does not involve vesicle formation; the proteins to be degraded reach the lysosomal lumen by directly crossing the lysosomal membrane. The two intrinsic characteristics of CMA are the selective targeting and the direct translocation of substrate proteins into the lysosomal lumen. CMA was initially identified as a stress-induced pathway described mainly in the liver, but soon it became obvious that basic levels of CMA activity are detectable in most tissues, including neurons. Macroautophagy and CMA work in a coordinated manner, even though the molecular mechanisms that modulate this crosstalk are not fully understood. Importantly, CMA activity declines with age and such decline may contribute to tissue dysfunction and, possibly, neurodegeneration. Little is known about CMA in the nervous system. In the current review, we discuss recent findings regarding the physiologic role of CMA in the nervous system and the potential link of CMA to various different neurodegenerative diseases, with an emphasis on Parkinson's disease.

INTRODUCTION

The process of Chaperone-Mediated Autophagy (CMA) is one of the three major autophagic pathways used in mammalian cells and organisms. All autophagic pathways involve the degradation of autophagocytosed components within lysosomes, and the reutilization of the produced building blocks; however, CMA is the only non-vesicular autophagic pathway. Furthermore, though the once held notion that macroautophagy is involved only in non-specific bulk protein degradation has not withstood the test of time, it is still appropriate to stress the high qualitative and temporal selectivity of CMA relative to other autophagic pathways. Only cytosolic substrates with the loose pentapeptide motif KFERQ or a biochemically related sequence are recognized and degraded through this pathway.[1] Furthermore, due to the fact that the pathway depends on discrete protein interactions with the main elements of CMA, Hsc70 and Lysosome-associated membrane protein type 2a (Lamp 2a), the substrates are degraded one by one in a piecemeal fashion.[2,3] CMA is a process that has been characterized mainly in mammalian systems, and it does not appear to be functional in lower organisms, such as invertebrates.[4]

Dysfunctional intracellular protein degradation has been implicated as an important component for the initiation and propagation of neurodegenerative diseases. Although the Ubiquitin Proteasome System (UPS) and macroautophagy have received the main focus of attention in this regard, there is emerging evidence that CMA dysfunction may also be an important element in neurodegeneration. The main tissue where CMA has been studied has traditionally been the liver, where starvation induces robust CMA activation. Cells utilized have mainly been fibroblasts or other non-neuronal cells. As a consequence, there is little known about the physiologic role of CMA in the nervous system.

MECHANISM OF CMA

It is not common in biological sciences that one or two researchers have had such a vast impact on a particular field. This is, however, the case in the field of CMA research. This pathway was described for the first time and characterized in depth by the laboratory of Fred Dice in a series of remarkable publications. In most of these publications, the involvement of the then post-doc Ana Maria Cuervo was essential. The torch has been carried on by Ana Maria Cuervo's laboratory, leading to further refinements in the understanding of the mechanisms involved, and to hitherto unappreciated links to various pathological conditions, including neurodegenerative disorders.

The process of CMA starts with the recognition of cytosolic proteins bearing the loose pentapeptide motif KFERQ by the heat-shock protein of 70 kDa (Hsc70), a cytosolic member of the Hsp70 chaperone family.[2] Hsc70 cooperates with a group of Hsp70 co-chaperones (such as Hsp40, Hsp90, Hip, Hop, and Bag-1) in this action.[5] The KFERQ motif is present in about 30% of cytosolic proteins, but few of them have been tested for their actual degradation by this pathway.[6] Prototypical substrates include glyceraldehyde-3-phosphate dehydrogenase (GAPDH), the inhibitor of NFkB (IkB), and others. It is possible that other proteins may undergo post-translational modifications that enable them to acquire the loose pentapeptide motif that targets them to CMA.

Following binding to the Hsc70 complex, substrates are translocated to the level of the lysosomal membrane. There lies awaiting the receptor and rate-limiting step in the CMA pathway, Lamp2a,[7] which is one of three isoforms of the gene Lamp2, encoded through alternative splicing. The other two isoforms, Lamp2b and Lamp2c, do not appear to be linked to CMA. All isoforms are lysosomal transmembrane proteins, and what differentiates Lamp2a from the others are regions within the transmembrane domain and cytosolic tail. The substrate binds to Lamp2a and is subsequently threaded into the lysosomal lumen, in an ATP-dependent fashion. This process is assisted by a resident lysosomal chaperone, lys Hsc70. The substrate is then finally degraded into amino acids.

For this process to work optimally, lysosomal membrane levels and conformation of Lamp2a are critical. The monomeric conformation of Lamp2a is required for substrate binding, which then drives Lamp2a multimerization in a 700 kDa complex, which enables translocation of the substrates into the lysosomal lumen. Following such translocation, lys Hsc70 enables disassembly of the Lamp2a complex, freeing Lamp2a monomers for further substrates.[8] Another fate of Lamp2a may be its translocation into the lysosomal lumen, where it does not participate in CMA and can be degraded by cathepsin A. Lysosomal (lys) Hsp90 stabilizes the transition phases of Lamp2a from monomeric to multimeric forms and vice versa, and is also critical for smooth CMA function. A recent study from the Cuervo group has added a further layer of complexity to this process. Bandyopadhyay *et al.* have identified GFAP and EF1alpha as regulators of the disassembly of the transmembrane translocation complex;[9] through these regulatory proteins, GTP mediates an inhibitory effect on CMA (Figure 1).

The rate of CMA depends largely on the levels of lysosomal membrane-associated Lamp2a, and on the presence within the lysosomal lumen of lys Hsc70.[10–12] In fact, although all lysosomes contain membrane-associated Lamp2a, only a subset of lysosomes contain hsc70 in their lumen, and they are the only ones competent for CMA.[3] CMA is operative at basal conditions in most mammalian cell models studied so far, but its main role comes into play when it responds to stressors, such as trophic deprivation or oxidative stress. Under such conditions, CMA is induced through mechanisms such as increased Lamp2a transcription, decreased Lamp2a clearance within the lysosomal lumen, or increased abundance of luminal lys Hsc70. In the classical paradigm of trophic deprivation, macroautophagy is activated early, and CMA later on, likely because in situations of prolonged starvation it is advantageous for the cell to have some selectivity in the degradation process.

As mentioned, the liver has been a tissue where a lot of the original CMA research was performed. It was found that with aging, CMA processing was

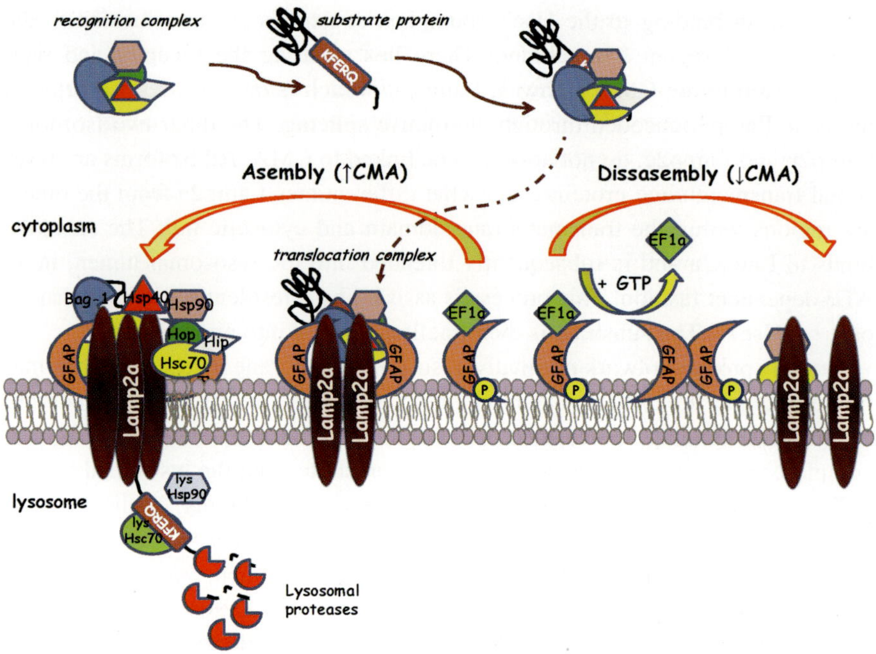

Figure 1. CMA pathway machinery. Two main protein complexes, a group of cytosolic chaperones/co-chaperones (*cargo recognition complex*) and a group of lysosomal proteins located at both sides of the lysosomal membrane (*cargo translocation complex*), are involved in the binding, unfolding and translocation of substrate proteins across the lysosomal membrane. The cytosolic Hsc70, recognizes the CMA-targeting motif in the substrate protein, and along with chaperones/co-chaperones delivers it to the surface of the lysosomes, where it binds to Lamp2a monomers. The lysosomal forms of Hsc70 and Hsp90 are required for stabilization, assembly and disassembly of the Lamp2a-enriched multimeric complexes. In addition, GFAP and EF1α proteins modulate the stability of the CMA translocation complex, where GFAP binds to Lamp2a at the complex and stabilizes it, whilst in the presence of GTP, Lamp2a-bound GFAP is exchanged with EF1α. GFAP is retrieved out of the Lamp2a multimeric complex by interaction with a phosphorylated form of GFAP at the lysosomal membrane (modified from Ref. 9).

significantly attenuated in this tissue.[13] This was due to increased instability of Lamp2a at the level of the lysosomal membrane, possibly in part due to alterations in the composition of the lipid bilayer.[14] In fact, directed overexpression of Lamp2a in the liver in a transgenic mouse not only improved lysosomal function, but also led to general improvements of liver function, suggesting that the aging-associated CMA impairment had functional consequences, and that these could be reversible with Lamp2a overexpression.[15]

Macroautophagy and CMA are interconnected, as experimental blockage of one of them results in compensatory upregulation of the other, revealing a close cross-talk between these two autophagic pathways. Blockage of CMA via Lamp2a silencing in cultured cells led to constitutive upregulation of macroautophagy,[16,17] but rendered cells more sensitive to oxidative stress and exposure to UV light.[17] Similarly, mouse embryonic fibroblasts null for *Atg5* had higher constitutive CMA activity,[18] leading to higher resistance to oxidative stress but more vulnerability to other stressors.[19] A potential molecular link of CMA to macroautophagy could be provided by ubiquilin. This molecule is degraded by both CMA and macroautophagy and promotes macroautophagy.[20,21] Thus, it would be expected, in the face of CMA inhibition, to increase and enhance macroautophagy. This hypothesis, however, was not specifically tested, and whether ubiquilin could also function as a CMA enhancer, to complete the loop of association between the two processes, is unknown.

The subject of the monitoring of CMA activity in cells and tissues is covered extensively in an excellent review article.[22] The basic assays include the following. (1) Measurement of long-lived protein degradation, where CMA-dependent activity is calculated as the difference between total lysosomal-dependent degradation (inhibited by general lysosomal inhibitors) and macroautophagy-dependent degradation (inhibited by 3-methyl-adenine (3-MA) or other inhibitors of class III PI3 kinases). This method is applicable only to cell culture, and has the drawback that it does not take into account the potential contribution of microautophagy or beclin 1/PI3 kinase independent autophagy.[23] (2) Measurement, by Western immunoblot, of levels of key lysosome-associated components of the CMA pathway, and more specifically of Lamp2a and Hsc70. (3) Immunocytochemical, immunohistochemical, or immunogold-EM (Electron Microscopy) evidence of the presence of Lamp-2a and Hsc-70 on lysosomes. (4) Measurement of steady state levels or half-lives of known CMA substrates. (5) The gold standard for CMA, the assessment of the *in vitro* binding, uptake and degradation of purified proteins by isolated lysosomes.

CMA IN THE NERVOUS SYSTEM

Doubts have been raised about the importance or even existence of a CMA-dependent pathway for protein degradation in the nervous system, due to the fact that proteins in the brain with KFERQ motifs did not appear to be altered following starvation,[24] and that transcripts for the specific Lamp2a isoform appeared to be very low in the brain.[25] The starvation paradigm appears to be analogous to the situation for macroautophagy, since starvation also fails to promote membrane-associated LC3-II formation in GFP-LC3 mice.[26] This of course does not negate the importance of macroautophagy in the nervous system, which is now widely

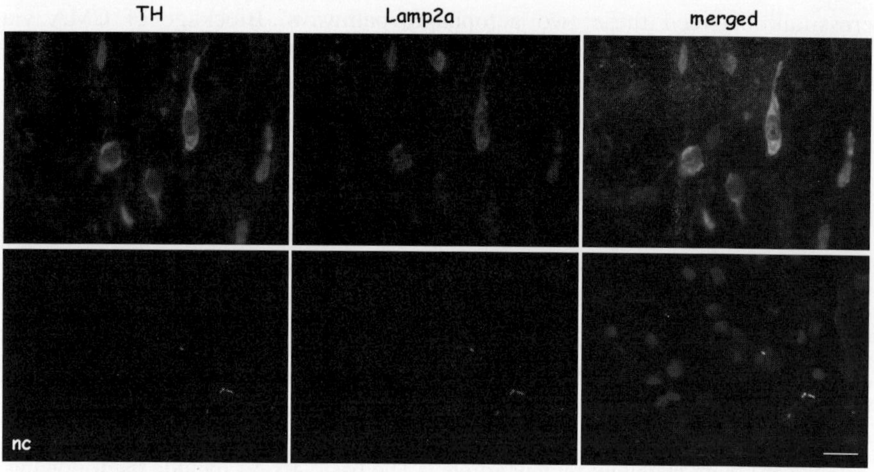

Figure 2. Lamp2a immunohistochemistry in the dopaminergic neurons of the rat substantia nigra pars compacta. Ten microns cryostat-cut midbrain sections were immunostained for TH (1:1000, Chemicon, msAb) and Lamp2a (1:400, Zymed, rbAb). Merged image, with the characteristic punctuate perinuclear staining of endogenous Lamp2a in TH-positive neurons is shown on the right. Scale bar, 20 μm. Midbrain section in which the primary antibodies were omitted is shown in the bottom panel (nc, negative control).

appreciated. In fact, we have found Lamp2a levels in the brain, at least in rodents, to be quite robust at the protein level, and to be developmentally regulated.[16] We did not find any significant decrease of Lamp2a levels in brain tissue with aging, but it should be noted that we only assessed total, and not lysosomal membrane-specific Lamp2a levels.[16] Consequently, whether indeed lysosomal membrane-associated levels of Lamp2a decline with aging in the brain, as in other tissues, is still unknown.

We have recently, due to our specific interest in the dopaminergic system, examined whether Lamp2a is expressed in dopaminergic neurons of the rat substantia nigra. With double immunofluorescence imaging, we have found significant co-expression of punctuate Lamp2a and TH immunostaining within nigral neurons (Figure 2). Similar results have recently been reported by Mak *et al.*[27] It appears, therefore, that not only is Lamp2a expressed in the nervous system, but also it is specifically expressed within neuronal populations that are vulnerable in neurodegenerative conditions. This notion is reinforced by studies that are beginning to examine the presence of the main CMA components, Hsc70 and Lamp2a, in human brain. Using immunohistochemistry or Western immunoblotting, Hsc70 and Lamp2a have both been detected in human neuropathological material in areas such as the amygdala and the substantia nigra.[28,29] It should be noted, however, that the specificity of available Lamp2 antibodies against the human Lamp2a isoform is debatable.[22]

Furthermore, it appears that the CMA pathway is indeed functional in neuronal cells, as its inhibition, through RNAi-mediated downregulation of Lamp2a, leads to alterations in lysosomal degradation of long-lived proteins and CMA substrates in such cellular systems.[16,28] Interestingly, in our own work, we have found that Lamp2a downregulation has a more pronounced effect in primary neurons, in terms of leading to more profound accumulation of CMA substrates, compared to neuronal cell lines, where differences are only observed in the turnover of substrates, but not in their steady state levels.[16]

As in other cellular systems, on many occasions macroautophagy is activated as a compensatory response to CMA inhibition[16]; however this is not always the case.[30] It will be interesting to decipher exactly how this compensatory activation occurs, and why it does not always operate. Of note, steady state levels of the prototypical CMA substrate GAPDH were not altered following Lamp2a downregulation in cultured cortical neurons, even though CMA-dependent degradation of long-lived proteins was compromised.[16] This suggests that CMA-dependent degradation of particular substrates may be tissue and cell type-specific. Alternatively, compensatory degradation mechanisms or issues of half-life may be at play.

The recent study of Mak *et al.*[27] also provides *in vivo* evidence that CMA can be modulated in the CNS and lead to functional consequences, as Lamp2a can be induced under stress conditions and lead to enhancement of CMA-dependent degradation in lysosomes. Martinez-Vicente *et al.* noted that Lamp2a downregulation in primary ventral midbrain cultures was neurotoxic, suggesting that basal CMA functioning is essential for neuronal survival.[31] We have not noted such effects in cultured cortical neurons, but differences in the duration and magnitude of Lamp2a downregulation (more prolonged and potent in the Martinez-Vicente *et al.* study) or the particular neuronal population studied may account for this discrepancy.

Overall, such studies leave little doubt about the importance of CMA as a degradation mechanism in the nervous system. However, the physiological importance of CMA in the nervous system has not been verified in an animal model, in part due to the difficulty of achieving specific targeting of Lamp2a relative to the other Lamp2 splice variants.

CMA IN NEURODEGENERATION: THE CASE FOR PARKINSON'S DISEASE

The first link between CMA and neurodegeneration, or, for that matter, the nervous system more generally, was provided through the study of Cuervo *et al.*

which investigated the relationship of CMA to the protein alpha-synuclein.[32] Genetic, neuropathological and biochemical evidence indicates that alterations in alpha-synuclein levels or conformations are critical in Parkinson's Disease (PD) pathogenesis.[33–38] Alpha-synuclein is the main constituent of Lewy bodies (LBs) and Lewy neurites (LNs) that characterize the disease.[39,40] The protein has a tendency to assume a beta-sheet, aggregated conformation under certain conditions, and is thought to constitute the building block of LBs and LNs. Point mutations or multiplications of the alpha-synuclein (*SCNA*) locus lead to autosomal dominant PD in rare families, and genome-wide association studies (GWAS) further indicate that variations within this locus confer a risk of developing sporadic PD.[33–38] These data suggest that control of alpha-synuclein levels is critical in PD pathogenesis, and that mechanisms of alpha-synuclein degradation deserve attention in this regard.

Cuervo *et al.* investigated whether alpha-synuclein could be a CMA substrate.[32] Indeed, the sequence VKKDQ fulfilled criteria for the loose KFERQ motif. Furthermore, in an *in vitro* assay where recombinant alpha-synuclein was bound, taken up and degraded within liver-derived isolated lysosomes in a CMA-dependent fashion. This process was competitively inhibited by classical CMA substrates Ribonuclease A or GAPDH, but not by ovalbumin, which is not a CMA substrate. Interestingly, mutant forms of alpha-synuclein, present in familial PD, bound to lysosomal membranes, but were not taken up and degraded within lysosomes. These mutant forms bound more tightly to Lamp2a at the lysosomal membrane compared to the Wild Type (WT) protein, and were not subsequently released into the lumen. Tight binding of the mutant forms was confirmed in a neuronal cell context, in PC12 cells.[32] The exact mechanism through which such dysfunction is conferred by the A30P and A53T forms of alpha-synuclein remains to be determined.

These findings indicated that CMA could be a mechanism for WT alpha-synuclein degradation, but it was unclear whether this actually occurred in neuronal cells and in particular primary neurons that are more closely related to PD. To address this, we used two parallel approaches: Application of RNA is targeted against Lamp2a, and expression of artificially mutated forms of human alpha-synuclein, which, due to the loss of the pentapeptide recognition motif, should not be degraded by CMA (these forms were called ΔDQs, although in reality there were no deletions, but rather substitutions of ${}_{95}DQ_{99}$ with AA). We predicted that, if alpha-synuclein was turned over to a significant extent by CMA in neuronal cells, then downregulation of Lamp2a should lead to more prolonged half-life and higher steady state levels, and the ΔDQ form should be degraded more slowly that the WT form. We performed such studies in neuronal cell lines, such

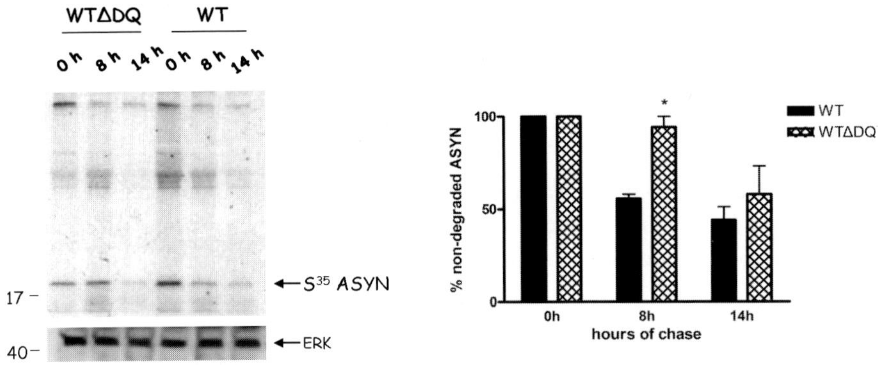

Figure 3. Adenoviral overexpressed WTΔDQ alpha-synuclein (ASYN) exhibits slower turnover compared to WT ASYN in rat cortical neurons. Five-day-old cortical cultures were transduced with adenoviruses (MOI 150) expressing human WT or WTΔDQ ASYN. Two days after transduction, cells were labeled with ^{35}S cysteine/methionine for 24 hours and pulse-chased for 0, 8 and 14 hours. ASYN was immunoprecipitated from hot lysates with C20 Ab and its levels were assessed by autoradiography. ***Left:*** A representative pulse-chase experiment is shown. The band corresponding to ASYN is depicted by the arrow. ***Right:*** Quantification of the turnover of WTΔDQ and WT ASYN. All data are presented as the relative OD values of each time point relative to time point 0. The graph represents the mean + SE of 2 independent experiments. (*$p < 0.05$, student's t-test comparing WTΔDQ to WT ASYN).

as PC12 and SH-SY5Y human neuroblastoma cells, and in primary rodent ventral midbrain and cortical neuron cultures. The ΔDQ form of human alpha-synuclein indeed had a longer half-life compared to the WT in PC12 and SH-SY5Y cells.[16] In subsequent experiments, we have verified that this is also the case in cultured rat cortical neurons, where WT or WTΔDQ alpha-synuclein were expressed through adenoviral transduction (Figure 3).

Using the siRNA approach, we found that 50% downregulation of Lamp2a had significant effects on the turnover of overexpressed alpha-synuclein in PC12 cells, although steady state levels of the protein did not change. The results were even clearer in primary neurons, where lentivirus-mediated-downregulation of Lamp2a led to robust increases in steady state levels of endogenous rodent alpha-synuclein. Alpha-synuclein increase was observed throughout the cell soma and the neuritic extensions.[16] Importantly, levels of beta-synuclein which lacks the pentapeptide targeting motif, were not affected by Lamp2a downregulation, stressing the specificity of CMA degradation for alpha-synuclein.[41] Differences in the potency of the effect on alpha-synuclein between cell lines and primary neurons may be due to the activation of compensatory mechanisms in the cell lines, such as macroautophagy induction, which enable total lysosomal-dependent degradation to remain unchanged, despite CMA blockade.[16] Although there were some cell-dependent differences, macroautophagy overall was also involved in alpha-synuclein degradation.[16,41]

These results were recently confirmed by Alvarez-Erviti *et al.*,[28] who found prolongation of WT, but not A53T, alpha-synuclein half-life with Lamp2a RNAi in SH-SY5Y cells. Interestingly, in their work, macroautophagy-dependent degradation of alpha-synuclein occurred only when CMA was compromised, as in the case of mutant alpha-synuclein expression.

In our experiments we also examined whether CMA dysfunction would lead to accumulation of aberrant forms of alpha-synuclein. We found that High Molecular Weight (HMW) soluble species, likely representing oligomeric forms, accumulated in cells overexpressing WT alpha-synuclein following exposure to Lamp2a siRNA. Likewise, in primary cortical and ventral midbrain cultures, detergent-insoluble monomeric conformations also increased with lentivirus-mediated transduction of Lamp2a shRNA. At the immunocytochemical levels, no obvious alpha-synuclein aggregates or inclusions were detected. Whether more sustained and potent CMA dysfunction would lead to the formation of bona fide alpha-synuclein inclusions remains to be demonstrated. In any case, based on such findings we suggested that CMA dysfunction may represent a contributing factor to LB formation.[16,41]

Whether alpha-synuclein is a CMA substrate *in vivo* has not been proven beyond doubt, but there are strong indications that this is the case. We found that endogenous or transgenically overexpressed alpha-synuclein and Lamp2a co-immunoprecipated from cortical neuron cultures and various regions of the rodent brain.[16] Furthermore, Lamp2a and alpha-synuclein displayed similar temporal pattern of expression in the developing rat brain. Importantly, Mak *et al.* have recently reported, in an elegant study, that, following exposure to paraquat or to transgenically expressed human alpha-synuclein, lysosomes respond with increased turnover of alpha-synuclein in lysosomes in a CMA-dependent fashion.[27] This is mediated, at least in part, by a robust induction of lysosomal membrane-associated Lamp2a.

Recent studies have also begun to investigate alterations of the main CMA components, Hsc70 and Lamp2a, in human neuropathological material derived from PD patients. Chu *et al.* found decreased Hsc70 levels in PD nigra by immunohistochemistry, however other lysosomal components were also decreased, suggesting a more generalized lysosomal impairment, not specific to CMA.[29] Alvarez-Erviti *et al.* used Western immunoblotting and detected significant decreases in both Hsc70 and Lamp2a in nigra and amygdala of PD brains compared to controls, whereas Lamp1 levels were unchanged.[28] Isolation of lysosomal membranes from the amygdala confirmed that the decrease in PD brains also applied to lysosome-associated Lamp2a. There were no alterations in these CMA components in AD brains. Despite some concerns about the specificity of the Lamp2a antibody used for this specific isoform, mentioned above, these results are potentially important, as they suggest that there may be a specific impairment of CMA that is widespread in PD brains, beyond areas with the confounding factor of neuronal cell loss and gliosis.

Another important facet of the connection of alpha-synuclein with CMA was also raised by the Cuervo *et al.* manuscript.[32] The mutant forms A30P and A53T, identified in familial PD, not only were not degraded by CMA, as mentioned above, but also impeded access of other CMA substrates to the process and, as a consequence, inhibited their CMA-dependent degradation. In PC12 cells, the mutants, but not the WT form, inhibited total lysosomal-dependent degradation, despite an increase of macroautophagy-dependent degradation, suggesting that these effects were also occurring in the context of neuronal cell lines.[32]

A further important contribution to the field was the discovery that in the *in vitro* assay with isolated lysosomes, even WT alpha-synuclein, post-translationally modified through formation of adducts with dopamine or its metabolites, could behave like the mutants, and impede CMA-dependent degradation of other CMA substrates. In fact, dopamine application to ventral midbrain cultures led to CMA dysfunction that was dependent on the presence of alpha-synuclein.[31] This raises the possibility that even in cases with sporadic PD, where there are no mutations in alpha-synuclein, excess WT alpha-synuclein in dopaminergic neurons could cause CMA dysfunction.

In our laboratory, we have explored in depth the significance of alpha-synuclein-dependent CMA dysfunction. We have used multiple neuronal cell culture systems, in order to get a sense of the general applicability of our findings. In proliferating PC12 and SH-SY5Y cells, in neuronally differentiated SH-SY5Y cells and in primary cortical neurons, we have found in every case that expression of mutant A53T alpha-synuclein causes inhibition of CMA-dependent long-lived protein degradation.[30] Expression of double mutant ΔDQ-A53T alpha-synuclein did not lead to such effects, ensuring that they were dependent on CMA targeting. This suggests that such inhibition is a direct effect, as predicted by the *in vitro* assays, and that it is universal. CMA inhibition was only associated with death when there was compensatory activation of macroautophagy or generalized severe lysosomal dysfunction; death was significantly less when the double mutant ΔDQ-A53T was expressed, showing that CMA inhibition is a crucial death-promoting effect of A53T alpha-synuclein. Furthermore, molecular or pharmacological inhibition of macroautophagy partially prevented neuronal death, suggesting that activation of macroautophagy in these settings was aberrant and deleterious for neuronal survival.[30] Interestingly, in neuronally differentiated SH-SY5Y cells even WT alpha-synuclein caused inhibition of CMA-dependent long-lived protein degradation, which did not occur with the ΔDQ mutant, and cell death, that was again significantly attenuated when the ΔDQ variant was expressed instead. This raised the possibility that dopamine adducts on WT alpha-synuclein may have conferred a CMA inhibitory activity, as proposed by Martinez-Vicente *et al.*[31] Consistent with this idea, application of alpha-methyl-p-tyrosine, an inhibitor of dopamine synthesis, attenuated both lysosomal dysfunction and death.[16]

Apart from aberrant activation of macroautophagy, suggested by our studies as a cause of CMA dysfunction-induced neuronal death, effects on specific CMA substrates have also been implicated. In particular, overexpression of either WT or A53T alpha-synuclein led to CMA dysfunction, which in turn led to mislocalization of MEF2D, a pro-survival transcription factor, to the cytosol, and loss of its activity. Interestingly, high levels of MEF2D mislocalized to the cytosol were detected in A53T transgenic mouse brain and in neuropathological material from PD nigra, suggesting that this effect may be important *in vivo*, and indirectly arguing for the occurrence of CMA dysfunction in PD.[42]

In conjunction, such data argue that CMA dysfunction may be a pathologically important and direct effect of aberrant alpha-synuclein. Taking also into account the previously mentioned data indicating that CMA is a major pathway for alpha-synuclein degradation in neuronal cells, we suggest that promotion of CMA function may represent a fruitful strategy for novel therapeutics against alpha-synuclein-mediated neurodegeneration in PD, as it would not only enhance alpha-synuclein clearance, but also prevent its detrimental effects on lysosomal function, in effect "killing two birds with one stone."

Another gene linked, albeit controversially, to PD is that encoding ubiquitin carboxyl-terminal esterase L1 (UCH-L1). Kabuta *et al.* showed that I93M UCH-L1, a mutant form identified in a single PD family, interacted aberrantly with the CMA components Lamp2a, Hsc70, and Hsp90, inhibited CMA and caused an increase in alpha-synuclein levels in cell culture,[43] indicating that the aberrant interaction of mutant UCH-L1 with CMA might underlie its pathogenic role. Whether other genes liked to PD may have effects on the CMA pathway is unknown, and is expected to be the focus of future investigations.

CMA IN OTHER NEURODEGENERATIVE DISEASES

Although involvement of endosomal and lysosomal pathways in Alzheimer's Disease (AD) pathogenesis has long been put forward, especially by the Randy Nixon group,[44–46] there is a paucity of information linking CMA to AD or other beta-amyloid- or Tau-related diseases. A recent study, using an inducible neuronal tauopathy cell model made the observation that fragments of mutant Tau, but not the full-length protein, translocated to the lysosomal membrane, where they underwent restricted cleavage by cathepsin L in a CMA-dependent fashion.[47] Surprisingly, these further cleavage products that were generated by CMA caused aggregation and leakage of lysosomal proteases, likely through disruption of the lysosomal membrane. At the same time they acted as inhibitors of CMA, in a similar fashion to mutant alpha-synuclein. Thus, in this setting, CMA functioning appeared to promote protein aggregation.[47]

In another AD-related manuscript worth noting in the context of CMA, the focus was on Regulator of Calcineurin 1 (RCAN1), a negative regulator of calcineurin that is increased in Down's syndrome brains and may confer neurotoxicity. RCAN1 was convincingly shown to be degraded by both the UPS and CMA, but not macroautophagy. Given the tight regulation of RCAN1 expression in a negative feedback loop with calcineurin, the possibility exists that CMA dysfunction could lead to neurotoxicity via RCAN1 upregulation.[48]

Numerous pieces of evidence link impairments of macroautophagy to Huntington's Disease (HD), by demonstrating that mutant Huntingtin and its aggregates in particular are preferentially degraded by macroautophagy, and suggest that enhancement of macroautophagy may provide a valuable therapeutic option for this disease.[49–52] A recent manuscript has used an ingenious approach, by generating an artificial peptide containing two CMA recognition motifs fused to two copies of the polyglutamine-binding peptide 1 (QPB1) sequence, which binds preferentially to polyglutamine tracts. This molecule, termed RHQ, ameliorated human Huntingtin (N-terminal fragments of exon 1) aggregation and toxicity significantly more than QPB1 in various cellular models, through forced degradation of the Huntingtin-RHQ complex via CMA. Furthermore, AAV-mediated transduction of RHQ in transgenic mouse models of HD significantly ameliorated aggregation and inclusion formation, improved metabolic and neurobehavioral outcomes and extended lifespan compared to QPB1 alone.[53] Thus, forced participation of aggregate-prone proteins in the CMA degradation pathway through the use of such adaptor molecules may represent a novel therapeutic strategy in HD and other neurodegenerative diseases.

CONCLUSION

The involvement of CMA in physiological functions of the nervous system is just beginning to be explored. Understanding such functions may pave the way for therapeutic approaches targeted to CMA in various pathological conditions, such as neurodegenerative disorders.

ACKNOWLEDGEMENTS AND FUNDING

This work was supported by EC/People/FP7 "NEURASYN" grant to LS and by PDF postdoctoral fellowship to MX and LS.

REFERENCES

1. J. F. Dice, Peptide sequences that target cytosolic proteins for lysosomal proteolysis, *Trends Biochem Sci* **15**(8): 305–309 (1990).
2. H. L. Chiang, S. R. Terlecky, C. P. Plant and J. F. Dice, A role for a 70-kilodalton heat shock protein in lysosomal degradation of intracellular proteins, *Science* **246**(4928): 382–385 (1989).
3. A. M. Cuervo, J. F. Dice and E. Knecht, A population of rat liver lysosomes responsible for the selective uptake and degradation of cytosolic proteins, *J Biol Chem* **272**(9): 5606–5615 (1997).
4. H. Koga and A. M. Cuervo, Chaperone-mediated autophagy dysfunction in the pathogenesis of neurodegeneration, *Neurobiol Dis* **43**(1): 29–37 (2011).
5. F. A. Agarraberes and J. F. Dice, A molecular chaperone complex at the lysosomal membrane is required for protein translocation, *J Cell Sci* **114**(Pt 13): 2491–2499 (2001).
6. S. S. Wing, H. L. Chiang, A. L. Goldberg and J. F. Dice, Proteins containing peptide sequences related to Lys-Phe-Glu-Arg-Gln are selectively depleted in liver and heart, but not skeletal muscle, of fasted rats, *Biochem J* **275**(Pt 1): 165–169 (1991).
7. A. M. Cuervo and J. F. Dice, A receptor for the selective uptake and degradation of proteins by lysosomes, *Science* **273**(5274): 501–503 (1996).
8. U. Bandyopadhyay, S. Kaushik, L. Varticovski and A. M. Cuervo, The chaperone-mediated autophagy receptor organizes in dynamic protein complexes at the lysosomal membrane, *Mol Cell Biol* **28**(18): 5747–5763 (2008).
9. U. Bandyopadhyay, S. Sridhar, S. Kaushik, R. Kiffin and A. M. Cuervo, Identification of regulators of chaperone-mediated autophagy, *Mol Cell* **39**(4): 535–547 (2010).
10. A. M. Cuervo and J. F. Dice, Regulation of lamp2a levels in the lysosomal membrane, *Traffic* **1**(7): 570–583 (2000).
11. A. M. Cuervo, E. Knecht, S. R. Terlecky and J. F. Dice, Activation of a selective pathway of lysosomal proteolysis in rat liver by prolonged starvation, *Am J Physiol* **269**(5 Pt 1): C1200–C1208 (1995).
12. F. A. Agarraberes, S. R. Terlecky and J. F. Dice, An intralysosomal hsp70 is required for a selective pathway of lysosomal protein degradation, *J Cell Biol* **137**(4): 825–834 (1997).
13. A. M. Cuervo and J. F. Dice, Age-related decline in chaperone-mediated autophagy, *J Biol Chem* **275**(40): 31505–31513 (2000).
14. R. Kiffin, S. Kaushik, M. Zeng, U. Bandyopadhyay, C. Zhang, A. C. Massey, M. Martinez-Vicente and A. M. Cuervo, Altered dynamics of the lysosomal receptor for chaperone-mediated autophagy with age, *J Cell Sci* **120**(Pt 5): 782–791 (2007).

15. C. Zhang and A. M. Cuervo, Restoration of chaperone-mediated autophagy in aging liver improves cellular maintenance and hepatic function, *Nat Med* **14**(9): 959–965 (2008).
16. T. Vogiatzi, M. Xilouri, K. Vekrellis and L. Stefanis, Wild type alpha-synuclein is degraded by chaperone-mediated autophagy and macroautophagy in neuronal cells, *J Biol Chem* **283**(35): 23542–23556 (2008).
17. A. C. Massey, S. Kaushik, G. Sovak, R. Kiffin and A. M. Cuervo, Consequences of the selective blockage of chaperone-mediated autophagy, *Proc Natl Acad Sci USA* **103**(15): 5805–5810 (2006).
18. S. Kaushik, A. C. Massey, N. Mizushima and A. M. Cuervo, Constitutive activation of chaperone-mediated autophagy in cells with impaired macroautophagy, *Mol Biol Cell* **19**(5): 2179–2192 (2008).
19. Y. Wang, R. Singh, A. C. Massey, S. S. Kane, S. Kaushik, T. Grant, Y. Xiang, A. M. Cuervo and M. J. Czaja, Loss of macroautophagy promotes or prevents fibroblast apoptosis depending on the death stimulus, *J Biol Chem* **283**(8): 4766–4777 (2008).
20. C. Rothenberg, D. Srinivasan, L. Mah, S. Kaushik, C. M. Peterhoff, J. Ugolino, S. Fang, A. M. Cuervo, R. A. Nixon and M. J. Monteiro, Ubiquilin functions in autophagy and is degraded by chaperone-mediated autophagy, *Hum Mol Genet* **19**(16): 3219–3232 (2010).
21. C. Rothenberg and M. J. Monteiro, Ubiquilin at a crossroads in protein degradation pathways, *Autophagy* **6**(7): 979–980 (2010).
22. S. Kaushik and A. M. Cuervo, Methods to monitor chaperone-mediated autophagy, *Methods Enzymol* **452**: 297–324 (2009).
23. C. T. Chu, Diversity in the regulation of autophagy and mitophagy: lessons from Parkinson's disease, *Parkinsons Dis* **2011**: 789431 (2011).
24. J. F. Dice, S. R. Terlecky, H. L. Chiang, T. S. Olson, L. D. Isenman, S. R. Short-Russell, S. Freundlieb and L. J. Terlecky, A selective pathway for degradation of cytosolic proteins by lysosomes, *Semin Cell Biol* **1**(6): 449–455 (1990).
25. K. Furuta, X. L. Yang, J. S. Chen, S. R. Hamilton and J. T. August, Differential expression of the lysosome-associated membrane proteins in normal human tissues, *Arch Biochem Biophys* **365**(1): 75–82 (1999).
26. N. Mizushima and A. Kuma, Autophagosomes in GFP-LC3 transgenic mice, *Methods Mol Biol* **445**: 119–124 (2008).
27. S. K. Mak, A. L. McCormack, A. B. Manning-Bog, A. M. Cuervo and D. A. Di Monte, Lysosomal degradation of alpha-synuclein *in vivo*, *J Biol Chem* **285**(18): 13621–13629 (2010).
28. L. Alvarez-Erviti, M. C. Rodriguez-Oroz, J. M. Cooper, C. Caballero, I. Ferrer, J. A. Obeso and A. H. Schapira, Chaperone-mediated autophagy markers in Parkinson disease brains, *Arch Neurol* **67**(12): 1464–1472 (2010).
29. Y. Chu, H. Dodiya, P. Aebischer, C. W. Olanow and J. H. Kordower, Alterations in lysosomal and proteasomal markers in Parkinson's disease: relationship to alpha-synuclein inclusions, *Neurobiol Dis* **35**(3): 385–398 (2009).

30. M. Xilouri, T. Vogiatzi, K. Vekrellis, D. Park and L. Stefanis, Abberant alpha-synuclein confers toxicity to neurons in part through inhibition of chaperone-mediated autophagy, *PLoS One* **4**(5): e5515 (2009).
31. M. Martinez-Vicente, Z. Talloczy, S. Kaushik, A. C. Massey, J. Mazzulli, E. V. Mosharov, R. Hodara, R. Fredenburg, D. C. Wu, A. Follenzi, W. Dauer, S. Przedborski, H. Ischiropoulos, P. T. Lansbury, D. Sulzer and A. M. Cuervo, Dopamine-modified alpha-synuclein blocks chaperone-mediated autophagy, *J Clin Invest* **118**(2): 777–788 (2008).
32. A. M. Cuervo, L. Stefanis, R. Fredenburg, P. T. Lansbury and D. Sulzer, Impaired degradation of mutant alpha-synuclein by chaperone-mediated autophagy, *Science* **305**(5688): 1292–1295 (2004).
33. M. H. Polymeropoulos, C. Lavedan, E. Leroy, S. E. Ide, A. Dehejia, A. Dutra, B. Pike, H. Root, J. Rubenstein, R. Boyer, E. S. Stenroos, S. Chandrasekharappa, A. Athanassiadou, T. Papapetropoulos, W. G. Johnson, A. M. Lazzarini, R. C. Duvoisin, G. Di Iorio, L. I. Golbe and R. L. Nussbaum, Mutation in the alpha-synuclein gene identified in families with Parkinson's disease, *Science* **276**(5321): 2045–2047 (1997).
34. R. Kruger, W. Kuhn, T. Muller, D. Woitalla, M. Graeber, S. Kosel, H. Przuntek, J. T. Epplen, L. Schols and O. Riess, Ala30Pro mutation in the gene encoding alpha-synuclein in Parkinson's disease, *Nat Genet* **18**(2): 106–108 (1998).
35. A. B. Singleton, A. Singleton, S. Hague, J. Kachergus, M. Hulihan, T. Peuralinna, A. Dutra, R. Nussbaum, S. Lincoln, A. Crawley, M. Hanson, D. Maraganore, C. Adler, M. R. Cookson, M. Muenter, M. Baptista, D. Miller, J. Blancato, J. Hardy and K. Gwinn-Hardy, Alpha-synuclein locus triplication causes Parkinson's disease, *Science* **302**(5646): 841 (2003).
36. J. Simon-Sanchez, C. Schulte, J. M. Bras, M. Sharma, J. R. Gibbs, D. Berg, C. Paisan-Ruiz, P. Lichtner, S. W. Scholz, D. G. Hernandez, R. Kruger, M. Federoff, C. Klein, A. Goate, J. Perlmutter, M. Bonin, M. A. Nalls, T. Illig, C. Gieger, H. Houlden, M. Steffens, M. S. Okun, B. A. Racette, M. R. Cookson, K. D. Foote, H. H. Fernandez, B. J. Traynor, S. Schreiber, S. Arepalli, R. Zonozi, K. Gwinn, M. van der Brug, G. Lopez, S. J. Chanock, A. Schatzkin, Y. Park, A. Hollenbeck, J. Gao, X. Huang, N. W. Wood, D. Lorenz, G. Deuschl, H. Chen, O. Riess, J. A. Hardy, A. B. Singleton and T. Gasser, Genome-wide association study reveals genetic risk underlying Parkinson's disease, *Nat Genet* **41**(12): 1308–1312 (2009).
37. P. Ibanez, A. M. Bonnet, B. Debarges, E. Lohmann, F. Tison, P. Pollak, Y. Agid, A. Durr and A. Brice, Causal relation between alpha-synuclein gene duplication and familial Parkinson's disease, *Lancet* **364**(9440): 1169–1171 (2004).
38. M. C. Chartier-Harlin, J. Kachergus, C. Roumier, V. Mouroux, X. Douay, S. Lincoln, C. Levecque, L. Larvor, J. Andrieux, M. Hulihan, N. Waucquier, L. Defebvre, P. Amouyel, M. Farrer and A. Destee, Alpha-synuclein locus duplication as a cause of familial Parkinson's disease, *Lancet* **364**(9440): 1167–1169 (2004).
39. M. G. Spillantini, R. A. Crowther, R. Jakes, M. Hasegawa and M. Goedert, Alpha-synuclein in filamentous inclusions of Lewy bodies from Parkinson's disease and dementia with Lewy bodies, *Proc Natl Acad Sci USA* **95**(11): 6469–6473 (1998).

40. M. G. Spillantini, M. L. Schmidt, V. M. Lee, J. Q. Trojanowski, R. Jakes and M. Goedert, Alpha-synuclein in Lewy bodies, *Nature* **388**(6645): 839–840 (1997).
41. M. Xilouri, T. Vogiatzi, K. Vekrellis and L. Stefanis, Alpha-synuclein degradation by autophagic pathways: a potential key to Parkinson's disease pathogenesis, *Autophagy* **4**(7): 917–919 (2008).
42. Q. Yang, H. She, M. Gearing, E. Colla, M. Lee, J. J. Shacka and Z. Mao, Regulation of neuronal survival factor MEF2D by chaperone-mediated autophagy, *Science* **323**(5910): 124–127 (2009).
43. T. Kabuta, A. Furuta, S. Aoki, K. Furuta and K. Wada, Aberrant interaction between Parkinson disease-associated mutant UCH-L1 and the lysosomal receptor for chaperone-mediated autophagy, *J Biol Chem* **283**(35): 23731–23738 (2008).
44. B. Boland, A. Kumar, S. Lee, F. M. Platt, J. Wegiel, W. H. Yu and R. A. Nixon, Autophagy induction and autophagosome clearance in neurons: relationship to autophagic pathology in Alzheimer's disease, *J Neurosci* **28**(27): 6926–6937 (2008).
45. D. S. Yang, J. H. Lee and R. A. Nixon, Monitoring autophagy in Alzheimer's disease and related neurodegenerative diseases, *Methods Enzymol* **453**: 111–144 (2009).
46. R. A. Nixon, Autophagy, amyloidogenesis and Alzheimer disease, *J Cell Sci* **120** (Pt 23): 4081–4091 (2007).
47. Y. Wang, M. Martinez-Vicente, U. Kruger, S. Kaushik, E. Wong, E. M. Mandelkow, A. M. Cuervo and E. Mandelkow, Tau fragmentation, aggregation and clearance: the dual role of lysosomal processing, *Hum Mol Genet* **18**(21): 4153–4170 (2009).
48. H. Liu, P. Wang, W. Song and X. Sun, Degradation of regulator of calcineurin 1 (RCAN1) is mediated by both chaperone-mediated autophagy and ubiquitin proteasome pathways, *FASEB J* **23**(10): 3383–3392 (2009).
49. G. Bates, Huntingtin aggregation and toxicity in Huntington's disease, *Lancet* **361**(9369): 1642–1644 (2003).
50. S. Sarkar, B. Ravikumar, R. A. Floto and D. C. Rubinsztein, Rapamycin and mTOR-independent autophagy inducers ameliorate toxicity of polyglutamine-expanded huntingtin and related proteinopathies, *Cell Death Differ* **16**(1): 46–56 (2009).
51. A. Iwata, J. C. Christianson, M. Bucci, L. M. Ellerby, N. Nukina, L. S. Forno and R. R. Kopito, Increased susceptibility of cytoplasmic over nuclear polyglutamine aggregates to autophagic degradation, *Proc Natl Acad Sci USA* **102**(37): 13135–13140 (2005).
52. Z. H. Qin, Y. Wang, K. B. Kegel, A. Kazantsev, B. L. Apostol, L. M. Thompson, J. Yoder, N. Aronin and M. DiFiglia, Autophagy regulates the processing of amino terminal huntingtin fragments, *Hum Mol Genet* **12**(24): 3231–3244 (2003).
53. P. O. Bauer, A. Goswami, H. K. Wong, M. Okuno, M. Kurosawa, M. Yamada, H. Miyazaki, G. Matsumoto, Y. Kino, Y. Nagai and N. Nukina, Harnessing chaperone-mediated autophagy for the selective degradation of mutant huntingtin protein, *Nat Biotechnol* **28**(3): 256–263 (2010).

CHAPTER 5

MAINTAINING AUTOPHAGIC BALANCE: A ROLE FOR BRAKES

Salvatore J. Cherra, III and Charleen T. Chu

ABSTRACT

Dysregulation of macroautophagy has been implicated in multiple disease processes, from neurodegeneration to neoplasia. Basal autophagy is essential for maintaining cellular health, particularly in post-mitotic cells such as neurons, and autophagy is induced in response to a wide array of physiological and pathological stresses. Clearly, maintaining a finely tuned balance between protein/organelle degradation and reparative biosynthesis is crucial to cellular and organismic health. Moreover, autophagy induction that outstrips cellular capacities for autophagosome maturation/clearance places neurons under additional autophagic stress. While much has been learned within the last ten years about processes necessary for autophagy to proceed, our understanding of braking mechanisms to reduce or slow the degree of autophagy induction is just beginning to emerge. Mechanisms applicable to both canonical nucleation and Beclin 1-independent autophagy are reviewed in context with the pathological consequences of insufficient negative regulation in models of acute neuronal injury and Parkinson's disease.

INTRODUCTION

Macroautophagy (herein referred to as autophagy) was originally characterized as a stress response pathway induced by nutrient deprivation.[1] Although autophagy is primarily a cell survival response in yeast, its role in multicellular organisms is more complex, shaping both the architecture and the function of multiple cell and tissue types.[2] In the nervous system, autophagy has been shown to regulate neurodevelopment, synaptic function, protein aggregation and cell survival/death.[3–6] Furthermore, autophagy has been implicated in multiple stress and disease states within the nervous system, including adult and pediatric neurodegenerative diseases, brain tumors, viral encephalopathy, traumatic brain injury, and stroke.[3,7–12]

Autophagy modulation has become the focus of preclinical studies, with the major emphasis on inducing autophagy to clear aggregates and tangles that occur in many neurodegenerative diseases (see Chapters 2 and 13). Recent studies, however, suggest that autophagy induction under stress or disease contexts may also be detrimental.[13–21] Levels of degradation that outstrip the cellular capacity for organelle biogenesis, either due to high levels of autophagic flux or due to relative biosynthetic impairments observed with aging or oxidative stress,[22,23] would promote a catabolic state of degeneration. Alternatively, robust induction kinetics may overload the degradation machinery itself,[24] with obstruction of membrane trafficking pathways or expansion of β-amyloid generating compartments.[25] A growing number of studies indicate that neurons may be particularly susceptible to harmful effects of unbalanced autophagy given their highly polarized morphology[16] and metabolic dependence on mitochondria.[26] Here we will discuss mechanisms that serve to slow or limit the degree of autophagy in cells, focusing on recent discoveries of molecular interactions that suppress autophagosome formation.

OVERVIEW OF AUTOPHAGY INDUCTION

In simplified terms, autophagosome formation is dependent upon two major steps: **Nucleation**, or the integration of cellular signals to define sites of origin for autophagic membrane extension; and **Expansion/Closure** of the phagophore around the cargo destined for degradation (see Figure 1). The protein kinase complex of ULK1-Atg13-FIP200 and the lipid kinase complex of Beclin1-PIK3C3(Vps34)-Atg14L cooperate to increase the phosphatidylinositol 3-phosphate (PI(3)P) content in regions of the smooth endoplasmic reticulum.[27,28] This somehow promotes the recruitment of Atg16L, which functions to tether the Atg12-Atg5 components of the ubiquitin-like conjugation system to nascent autophagic membranes, completing the nucleation step.[29]

Expansion is dependent upon covalent attachment of microtubule associated protein 1 light chain 3 (LC3), an Atg8 homolog, to phosphatidylethanolamine, a process that may be facilitated in an E3-like manner by the Atg12-Atg5 complex.[30] Atg7 functions as the E1-like enzyme for both Atg12 and Atg8.[31] Atg8 homologs are essential not only for growth and closure of the autophagosome, but also serve as receptors to mediate the selective enrichment of tagged cargo into the growing autophagosome.[32,33]

Notably, in several cell death paradigms, an alternative mechanism that bypasses the Beclin 1-dependent nucleation step, but is still dependent upon the Atg7 conjugation system, has been described.[13,17,34–36] While it is unclear whether

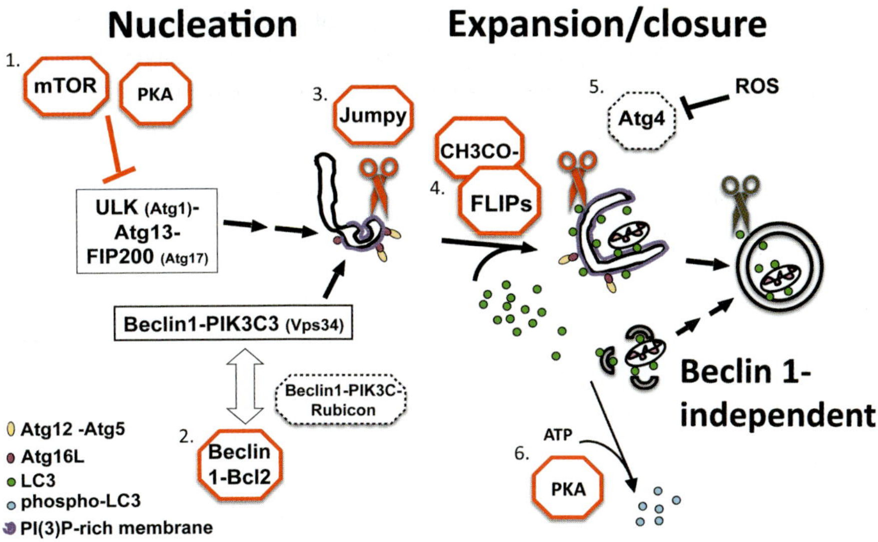

Figure 1. Mechanisms to downregulate canonical and non-canonical autophagy induction. (1) Phosphorylation of ULK1 and Atg13 by mTOR represents a major brake for autophagy induction under nutrient rich conditions. Phosphorylation of Atg1 and Atg13 by PKA also functions to inhibit autophagy induction in yeast. (2) Beclin 1 exists in several protein complexes, and addition of Rubicon to the PIK3C3 complex reduces the process of autophagy maturation. Bcl-2 is an anti-apoptotic protein that also downregulates Beclin 1-dependent autophagy. (3) The Beclin 1-PIK3C complex functions to create PI(3)P, while its phosphatase Jumpy (MTMR14) negatively regulates autophagy. (4) Acetylation of several Atg proteins has been reported. Viral FLIPs can bind Atg3, interfering with its interaction with LC3. (5) ROS-mediated inhibition of Atg4 is necessary to allow canonical autophagy to proceed, suggesting that delipidation of LC3 may also serve as a brake to autophagosome formation. (6) PKA phosphorylates LC3 Ser/Thr12, a conserved site in vertebrate species, inhibiting its participation in canonical and non-canonical injury-induced autophagy. Post-translational modification of autophagy mediators represent important potential points of up- and downregulation, allowing the autophagy machinery to rapidly fine-tune its responses to changing cellular conditions.

or not PI(3)P content is otherwise enriched, such as through decreased phosphatase activity, a consequence may be relative overactivation of autophagy, as Bcl-2 and other negative regulators that act by binding Beclin 1 [37] would not be able to regulate Beclin 1-independent autophagy induction.

Live cell imaging of hepatocyte mitophagy implicates different patterns of autophagosome formation depending upon the stimulus. With canonical nutrient deprivation-induced autophagy of mitochondria, GFP-LC3 is observed extending around mitochondria, consistent with growth from a nucleation site.[38] However, with photo-irradiation, which elicits non-canonical mitophagy, multiple GFP-LC3 puncta simultaneously appear to encircle depolarized mitochondria.[39] While these findings are currently described only in a single system, they support the hypothetical existence of alternative nucleation mechanisms, including the possibility of direct cargo recruitment of LC3 in selective autophagy. In any case, non-canonical forms of autophagy may elude several regulatory mechanisms upstream of LC3 lipidation (Figure 1). This may account for the frequent association of Beclin 1-independent autophagy with pathological situations, mediating neurite retraction or cell death.[13,17,35,40]

NEGATIVE REGULATORS OF AUTOPHAGY INDUCTION

Proteins Regulating Autophagy Induction

Mammalian target of rapamycin

The mammalian target of rapamycin (mTOR) is a highly conserved serine/threonine protein kinase that regulates cell growth, protein synthesis, cell survival, and autophagy.[41] mTOR exists in two protein complexes, mTORC1 and mTORC2. mTORC1 controls cell growth and translation by phosphorylating the 70-kDa polypeptide 1 ribosomal protein S6 kinase-1 (p70S6K) and the inhibitor of translation initiation, 4E-BP1. mTORC1, but not mTORC2, is sensitive to nutrient deprivation and inhibition by rapamycin.[42,43] Under nutrient rich conditions, mTOR is active, and cells maintain high levels of protein synthesis; however, during nutrient starvation, mTOR is inhibited, protein synthesis is reduced, and cells undergo autophagy. In addition to p70S6K and 4E-BP1, mTORC1 phosphorylates ULK1 (yeast Atg1 homolog) and Atg13 to suppress autophagy induction.[44] During starvation or rapamycin treatment, ULK1 is derepressed, allowing it to phosphorylate Atg13 and FIP200 to promote phagophore formation and the initiation of autophagy.[27]

While the precise mechanisms of how mTOR senses nutrients remain unknown, cellular energetics play a key role. Low levels of intracellular ATP caused by nutrient deprivation, hypoxia, or even mitochondrial damage may induce autophagy through AMP-activated protein kinase (AMPK).[45] mTOR is one of the major downstream targets inhibited by AMPK activation.[45] AMPK phosphorylates components of mTORC1, inhibiting mTOR activity, thereby suppressing protein translation and inducing autophagy. By integrating upstream signals (i.e. lack of nutrients or ATP reduction by consumption or cellular damage) with phagophore formation, mTOR is a major negative regulator of canonical autophagy.

Jumpy (myotubularin related protein 14)

Although phosphatidylinositol phosphorylation has been widely associated with endocytosis and endosome trafficking, it has also been suggested to play a role in autophagy initiation by demarcating sites for nascent phagophore formation.[46,47] Phosphoinositol 3-kinases (PI3K) play both stimulatory and inhibitory roles in autophagy. In particular, generation of PI(3)P by the class III PI3K (PIK3C3; Vps34 in yeast) mediates nucleation; conversely, class I PI3K products such as PI(3,4)P_2 or PI(3,4,5)P_3 suppress autophagy.[28] While the roles of PI3Ks in autophagy induction have been well-characterized, the role of phosphatases in regulating PI(3)P levels and autophagy has recently been highlighted.

Myotubularins (MTMs) are a family of lipid phosphatases that dephosphorylate PI(3)P and PI(3,5)P to regulate endocytosis and other membrane trafficking events.[48] There are 16 MTMs, but only nine of these family members possess catalytic activity.[48] While the intracellular localization and specific functions are primarily unknown, a recent study implicates Jumpy (MTMR14) as a suppressor of autophagy induction.[49] In addition to its co-localization with Golgi proteins,[50] Jumpy also localizes to phagophores.[49] Jumpy-mediated dephosphorylation of PI(3)P reduces the recruitment of WIPI-1 (yeast Atg18 homolog), a PI(3)P-binding protein essential for autophagy,[51] to phagophores.[49] Interestingly the R336Q mutation in Jumpy, which is associated with centronuclear myopathy, reduces the catalytic activity of Jumpy and abrogates its ability to suppress autophagy.[49,50] These studies indicate that a balance between lipid kinase and lipid phosphatase activity is required for proper autophagic function, and suggest that dysregulated autophagy may contribute to centronuclear myopathy.

Bcl-2 (B cell CLL/lymphoma-2)

In recent years, the interaction between autophagy mediators and apoptosis mediators has been highlighted in multiple systems in the regulation of cell death/survival. One important mechanistic link between these pathways is the interaction of antiapoptotic B cell CLL/lymphoma-2 (*BCL-2*) family members with Beclin 1 (Atg6 in yeast).[52] In addition to Bcl-2, Beclin 1 interacts with PIK3C3/Vps34 to stimulate autophagosome formation.[28] Under basal conditions Bcl-2 proteins compete for binding to Beclin 1, suppressing autophagosome formation.[37,52,53] Regulation of the bioavailability of Beclin 1 to participate with PIK3C3 in autophagy regulatory complexes is believed to serve a rheostat-like function to regulate the degree of autophagy elicited. Furthermore, mutations that disrupt the interaction between Bcl-2 and Beclin 1 are sufficient to induce autophagy,[54] suggesting that Bcl-2 proteins basally inhibit autophagy induction. Some of the cell survival effects attributable to Bcl-2 may include inhibition of "autophagic cell death" as well as of apoptosis.

FLICE-inhibitory proteins

Similar to Bcl-2, the viral FLICE-inhibitory protein (vFLIP) was initially discovered as an inhibitor of apoptosis.[55] In addition to vFLIP, there are three cellular FLIPs (cFLIP). $cFLIP_L$ contains two death-effector domains and a catalytically inactive carboxy-terminal caspase domain.[56] $cFLIP_S$, the short form, and $cFLIP_R$, isolated from the Raji B-cell line, contain two death-effector domains, but completely lack the caspase domain.[57] FLIPs primarily inhibit apoptosis by competing with caspase 8 for recruitment to death receptors.[57]

In addition to apoptosis inhibition, a recent study showed that FLIPs also suppress autophagy induction.[58] Both viral and cellular FLIPs suppress starvation- and rapamycin-induced autophagy by competing with LC3 for binding to Atg3.[58] By preventing LC3 from interacting with Atg3, FLIPs prevent LC3 lipidation, an essential step for autophagosome formation. Like other autophagy regulators, cFLIPs are sensitive to cellular growth conditions.[57] With cFLIPs transcriptionally regulated by multiple growth factor signaling pathways, they may aid cells in switching between growth and maintenance phases.

Post-Translational Modifications that Regulate Autophagy Induction

In addition to the pathways discussed above, direct post-translational modifications of autophagy proteins present the possibility of rapid, fine-tuned autophagic responses to various stimuli. Beyond regulation of expression levels, post-translational regulation offers the possibility of integrating information in response to multiple stresses in order to execute autophagy on a global network level rather than as a linear, stimulus-specific pathway. The additional benefit of multiple regulation points throughout autophagy execution is the promise of therapeutic targets for precise modulation of subsets of autophagy functions beneficial for a given disease context.

Proteolysis of autophagy proteins

The balance between cell survival and cell death signaling pathways typically determines the fate of the cell. As mentioned previously, crosstalk between autophagic and apoptotic proteins has been suggested as one mechanism to determine cell survival.[59] For example, cells may respond to an insult by initiating autophagy as a pro-survival mechanism; however, if the insult persists or if autophagic recycling is not sufficient to protect the cell, then the cell may undergo apoptosis, necrosis or autophagic cell death. In addition to reversible interactions between autophagic and apoptotic proteins that regulate cell fate, recent studies suggest that proteolysis may also play a role in cell fate determination. Calpain or caspase activation can inhibit autophagy by cleaving Atg5 or Beclin 1, respectively.[60–62] While these studies have shown that apoptosis induction causes cleavage of key autophagy proteins, the role of proteolytic cleavage in the regulation of autophagy in other contexts has not yet been demonstrated.

Regulation of autophagy by acetylation

In recent years, the roles of acetylation by p300 and the CREB-binding protein and deacetylation by Sirt1 have been shown to play critical roles not only within the nucleus, but also in many other facets of cell biology. The initial studies showed that increased acetylation of Atg5, Atg7, Atg12, and LC3 are associated with reduced autophagy levels.[63,64] While these studies provide some evidence that acetyltransferases and deacetylases could regulate autophagy, the precise effects of acetylation on the function of individual autophagy proteins remain to be determined. Acetylation can also target cargo such as mutant huntingtin for autophagic degradation,[65] and acetylated microtubules are essential for autophagosome delivery to autolysosomes.[66] Yet deacetylases such as HDAC6 are also important for maturation/fusion of autophagosomes in some contexts,[67] indicating complex mechanisms by which acetylation regulates early and late steps of autophagic degradation.

Regulation of autophagy induction by phosphorylation

With the identification of protein and lipid kinases such as mTOR, ULK1, AMPK and PIK3C3 as autophagy regulators, phosphorylation has become a highly studied post-translational modification in terms of autophagy induction. As previously discussed, the function of the ULK1-FIP200-Atg13 complex is negatively regulated by mTOR phosphorylation, and AMPK activity also regulates mTOR and ULK1 functions.[45,68] Furthermore, kinase-dead ULK1 prevents autophagy induction, suggesting that ULK1 kinase activity is required for autophagy induction in mammalian cells.[69] Although putative substrates for yeast Atg1 have been identified, the downstream targets responsible for Atg1/ULK1-mediated autophagy regulation remain unknown. In addition to kinases, protein phosphatases, such as PP1 and PP2A, have also been shown to regulate autophagy;[70] however, the substrates and mechanisms remain to be elucidated.

While the list of kinases that regulate autophagy is constantly growing, few Atg proteins have been identified as direct substrates. Thus, many questions remain to be resolved: Do these kinases and phosphatases directly or indirectly influence autophagosome formation, cargo recognition, or autophagosome trafficking? What are the autophagy-regulating substrates of these kinases? How do these pathways integrate to regulate cellular homeostasis through autophagy? One recent study has identified a death-associated protein kinase phosphorylation site on Beclin 1 that causes its dissociation from Bcl-2 proteins, allowing for autophagy induction.[71] An additional study has identified JNK phosphorylation sites on Bcl-2 that disrupt its interaction with Beclin 1, releasing Beclin 1 to induce autophagy.[72] These studies highlight specific phosphorylation events that modulate canonical autophagy induction by regulating proteins upstream of the PI(3)P-mediated demarcation of nascent autophagic membranes.

Regulation of LC3 by phosphorylation

In addition to the proteins that regulate autophagy through integration of signaling pathways upstream of autophagosome nucleation, two recent studies have identified direct phosphorylation sites on LC3 itself. Protein kinase C (PKC) is able to suppress starvation or rapamycin-induced autophagy in HEK293 cells. However, mutagenizing the two PKC phosphorylation sites identified on LC3 did not recapitulate the effects of PKC on autophagy, implicating involvement of other mechanisms.[73]

Protein kinase A (PKA) not only suppresses yeast autophagy through effects on Atg1, but also was recently shown to suppress both rapamycin-induced and Beclin 1-independent autophagy in mammalian cells.[74] The PKA phosphorylation site was identified at residue 12 between the two N-terminal helices of LC3.[74] Phosphomimetic mutation of this site is sufficient to recapitulate the effects of PKA, and ablation of this site facilitates LC3 puncta formation.[74] Although further investigation will be required to understand the mechanism(s) underlying the phosphoregulation of LC3, these studies highlight phosphorylation as a direct link between the core autophagy machinery and upstream homeostatic signaling pathways.

While it is likely that different phosphorylation sites might confer different properties to LC3 function, it is also possible that some effects are cell type specific. Many studies have focused upon delineating conserved autophagy regulation mechanisms, either among various species or cell types. Yet, very few have investigated whether autophagy regulation varies between cell types. In particular, neurons may be refractory to some common autophagy inducers,[75] although this remains controversial. Moreover, cells that are metabolically dependent upon mitochondria may be resistant to protonophore-induced mitophagy.[76] PKA phosphorylation of LC3 effectively suppresses autophagy induction in neurons and neuronal cell lines,[74] preventing the pathologic dendritic remodeling induced in parkinsonian models involving mutant leucine-rich repeat kinase 2 or sublethal doses of the complex I inhibitor 1-methyl-4-phenylpyridinium. It is possible that LC3 phosphorylation may provide neurons with additional regulatory steps for autophagy induction. LC3-I is very highly expressed in brain tissues compared to other tissues,[77,78] implicating the possibility of neuron-specific roles for different pools of LC3.

AUTOPHAGY DOWNREGULATING MECHANISMS IN RELATION TO CANONICAL AND NON-CANONICAL INDUCTION PATHWAYS

As discussed above, the canonical pathway of autophagy induction involves starvation-induced inactivation of mTOR concurrent with activation of the class III PI3 kinase PIK3C3 in complex with Beclin 1, which serves to nucleate the site of autophagosome membrane growth. In addition, pharmacologic, RNAi, and live imaging studies support the existence of additional mechanisms for injury-induced autophagy that bypass the requirement for Beclin 1-PIK3C3 activity.[17,35] Interestingly, these latter stimuli are often associated with autophagic degeneration or cell death, presumably due to dysregulation of braking mechanisms. Emerging data on braking mechanisms for autophagy that operate at different steps of nucleation and autophagosome formation are summarized in Figure 1.

(1) Phosphorylation of ULK1 and Atg13 at inhibitory sites by mTOR represents a major brake for autophagy induction under nutrient rich conditions. In particular, phosphorylation of ULK1 Ser 757 disrupts the interaction between ULK1 and AMPK.[79] Likewise, PKA has also been shown to phosphorylate Atg1 and Atg13 (S344, S437, S581) to inhibit autophagy induction in yeast.[80]

(2) A key step in identifying the site from which autophagic membranes will extend is the production of PI(3)P by the activity of the Beclin 1-class III PI3 kinase complex. Beclin 1 can exist in several complexes, and competition by Bcl2 for Beclin binding has been proposed to act in a rheostat function to limit autophagy induction to levels compatible with cell survival.[81] Notably,

addition of Rubicon to the complex reduces the process of autophagy maturation,[82,83] but whether or not it may also affect induction due to effects on PI3 kinase activity remain unknown.

(3) As PI(3)P membrane content serves to recruit autophagy initiators to the site, dephosphorylation of the phospholipid would also limit autophagy induction. The phosphatidylinositol 3-phosphate phosphatase Jumpy (MTMR14) negatively regulates autophagy, and loss of this function is associated with the disease centronuclear myopathy.[49]

(4) There are several emerging mechanisms that may act directly on the autophagy lipidation machinery. As discussed above, viral FLIPs as well as their cellular counterparts have been reported to bind Atg3, interfering with its interaction with LC3. Post-translational regulation of Atg proteins involved directly in the ubiquitin-like conjugation system provides additional points of regulation for both canonical and Beclin 1-independent forms of autophagy. Acetylation of Atg 5, 7, 12 and LC3 has been reported, which results in autophagy downregulation, although the precise sites of modification remain unknown.

(5) Atg4 functions to (a) cleave pro-LC3 to expose the C-terminal glycine necessary for autophagy and (b) to remove (delipidate) LC3 from the surface of autophagosomes, presumably allowing for reutilization. While Atg4 would thus increase availability of LC3 for autophagy induction, the observation that ROS-mediated inhibition of Atg4 is necessary to allow canonical autophagy to proceed[84] suggests that delipidation of LC3 prior to completion of the autophagosome may also serve as a brake to its formation.

(6) LC3 can be directly phosphorylated by PKA at Ser/Thr12, a conserved site in vertebrate species, inhibiting its participation in injury-induced autophagy.[74] Phosphomimetic mutants of LC3 protect against mutant LRRK2-mediated neurite degeneration and MPP^+ toxicity, two models of Beclin 1-independent autophagy.[74]

The discovery of post-translational modulation of autophagy proteins represents an important advance in our understanding of autophagy regulation as a whole. In addition to mechanisms that regulate induction, phosphorylation, acetylation, limited proteolysis, and ubiquitination would serve to rapidly regulate the bioavailability of pools of LC3 that are poised to undergo autophagy. Given the additional role of LC3 in binding autophagy cargo either directly or via cargo adaptors,[33,85,86] post-translational modifications may also serve to fine tune turnover of mitochondria, protein aggregates and other cargo during selective autophagy. The ability to modulate the tone of induced autophagy, rather than completely inhibiting it, is undoubtedly critical for maintaining the proper balance for this important process.

ROLES OF AUTOPHAGY INHIBITION DURING DISEASE IN THE NERVOUS SYSTEM

As the role of autophagy in the nervous system, especially during injury or disease, appears to be variable and context dependent, so too are the effects of autophagy inhibition. Under certain disease states, autophagic degradation may be directly inhibited. Decreased levels of Beclin 1 have been observed in aged human patients and in early stages of Alzheimer's disease.[87,88] In Huntington's disease mutant huntingtin is able to sequester Beclin 1.[88] Similarly, herpes simplex virus-1, which can cause encephalopathy, also suppresses autophagy by inhibiting Beclin 1 function.[89] Furthermore, genetic ablation of *Atg5* or *Atg7* in the nervous system of mice leads to neurodegeneration.[90,91] In addition to inhibiting the early steps of autophagy, lysosomal storage diseases, Alzheimer's disease, and Neimann-Pick disease have each been shown to decrease degradative flux through the lysosome.[92–95] In mucolipidosis type IV, autophagy induction by starvation is not affected, but there is delayed clearance of autolysosomes following re-feeding of the cells.[96] Collectively, these studies illustrate how failure of autophagy induction or clearance may play a role in neurodegeneration associated with multiple diseases (see also Section 2 of this volume). One possible therapy could be the alleviation of other braking mechanisms to rescue deficiencies in autophagy induction and to re-establish basal autophagy levels. For diseases in which lysosomal degradation is attenuated, mechanisms to increase the capacity for lysosomal proteolysis may be a potential therapy to restore lysosomal degradation.[97]

While insufficient autophagic degradation may be responsible for some forms of neurodegeneration, dysregulated induction may also lead to neuronal injury. Hypoxia-ischemia-, seizure-, or trauma-induced brain injuries have been associated with elevations in autophagy.[8,9,98] Furthermore, excitotoxicity, which develops during such injuries, elicits autophagosome accumulation in degenerating neurites.[99] While these studies often do not distinguish increased induction from decreased maturation/degradation, the **autophagic stress** produced from an imbalance between induction and degradation may ultimately lead to the demise of injured neurons.[24] Treatment for such neuronal injury may involve enhancing molecular brakes to slow autophagy induction, thereby allowing lysosomal degradation to keep pace in clearing autophagosomal contents. As lysosomal function often declines with age, restoring degradative capacity through lysosome-directed therapies can also alleviate autophagic stress, and such a strategy has been used in an animal model of AD.[97] These observations indicate that by ameliorating the deficiencies in autophagic recycling, one may be able to restore homeostasis and function to neurons, potentially preventing disease progression.

Even among different models for the same disease, there have been apparently conflicting studies that suggest that either autophagy inhibition or activation may be beneficial.[8,25,100–102] While the desired manipulation for a disease may at first appear obvious, other disease factors require consideration. For example, a common finding among different forms of PD is protein aggregation, most notably in the form of Lewy bodies. Although Lewy body formation and alpha-synuclein aggregates suggest the possibility of insufficient autophagy, recent studies suggest that macroautophagy induction even under these circumstances could be detrimental.[15] Moreover, autophagy that is induced to clear alpha-synuclein aggregates may fail in the disease context due to impaired bioenergetics, lysosomal function or regenerative biosynthesis[22,23,97,103] (see also Chapter 12).

In several parkinsonian models, both genetic and toxin-based, robust levels of macroautophagy appear to be detrimental for function or survival[13,15–17,74,104] (see also Chapter 17). Some of this may be attributed to autophagic stress from concurrent lysosomal dysfunction,[105] although genetic ablation of autophagy induction indicates that downregulating induction is neuroprotective.[13,15,16] Interestingly, several of these injuries induce autophagy in a PIK3C3-independent manner, suggesting escape from canonical, Beclin 1-associated regulatory mechanisms.[13,17] Beclin 1-independent autophagy also appears to mediate cell death in non-neuronal cells.[35] Moreover, inhibiting autophagy is protective in models that elicit ERK2-dependent mitophagy,[17,106] suggesting that excessive degradation of mitochondria may be harmful to neuronal cell types. Perhaps this accounts for differences observed between transformed cell lines and primary neurons in the ability of uncoupling drugs to induce parkin-mediated mitophagy.[76,107] Ultimately, the outcome of autophagy induction may rest upon other metabolic factors within the cell. These may include declining lysosomal capacities, microtubule function, and bioenergetic factors discussed above, as well as the biosynthetic capacity of the cell to replace degraded structures.[7] Notably, impaired transcriptional responses and decreased mitochondrial biogenesis have been observed in oxidative models of aging and disease.[22,23]

Due to the critical role of basal autophagy in maintaining neuron health,[90,91] identification of targets that can suppress induction of Beclin 1-independent autophagy while sparing basal autophagy may comprise an interesting therapeutic strategy. Understanding the differences in autophagy regulation between canonically regulated autophagy, with multiple braking systems that modulate PI(3)P content, and the potentially deleterious mechanisms of Beclin

1-independent autophagy, which escapes such regulation, could lead to the identification of therapeutic targets for acute brain injuries and neurodegenerative diseases. While the mechanisms underlying Beclin 1-independent autophagy remain unknown, several stimuli involve mitochondrial perturbation or generation of reactive oxygen species.

One final note is that neurons are post-mitotic and must survive for more than half a century, unlike other cell types that can execute cell death pathways to obviate malfunctions in protein or organelle homeostasis. From a genetic viewpoint, cell-type specific (dys)regulation could underlie mechanisms that cause familial forms of neurodegenerative diseases, despite expression of mutant proteins throughout the entire body. To this end, it may be speculated that neurons modulate key homeostatic pathways, including apoptosis and autophagy, differently than other cell types.

CONCLUSION

Autophagy is a highly conserved, tightly regulated cellular homeostasis pathway required for the maintenance and function of the central nervous system. Dysregulation or deficiencies in the autophagy pathway may underlie neuronal dysfunction and death in multiple brain diseases, including AD, HD, PD, prion disease, stroke, trauma or viral encephalitis. Further understanding of the mechanisms that regulate the autophagy pathway, both shared and unique to neurons, may provide insight to preserve or restore neuronal health in the face of disease. Importantly, the study of autophagy regulation in the nervous system has highlighted the beneficial role of basal autophagy in disease prevention and led to the discovery of non-canonical autophagy. It is clear that modulatory factors that regulate or limit the degree of autophagy induced in response to injury are critically important in cellular homeostasis, just as brakes and low gear are essential for preventing excitotoxicity.[108]

REFERENCES

1. D. J. Klionsky and S. D. Emr, Autophagy as a regulated pathway of cellular degradation, *Science* **290**: 1717–1721 (2000).
2. F. Cecconi and B. Levine, The role of autophagy in mammalian development: cell makeover rather than cell death, *Dev Cell* **15**: 344–357 (2008).
3. B. Boland and R. A. Nixon, Neuronal macroautophagy: from development to degeneration, *Mol Aspects Med* **27**: 503–519 (2006).
4. K. E. Larsen and D. Sulzer, Autophagy in neurons: a review, *Histol Histopathol* **17**: 897–908 (2002).

5. E. D. Plowey and C. T. Chu, Synaptic dysfunction in genetic models of Parkinson's disease: a role for autophagy? *Neurobiol Dis* **43**: 60–67 (2011).
6. A. M. Rowland, J. E. Richmond, J. G. Olsen, D. H. Hall and B. A. Bamber, Presynaptic terminals independently regulate synaptic clustering and autophagy of GABAA receptors in *Caenorhabditis elegans, J Neurosci* **26**: 1711–1720 (2006).
7. S. J. Cherra and C. T. Chu, Autophagy in neuroprotection and neurodegeneration: a question of balance, *Future Neurol* **3**: 309–323 (2008).
8. M. Koike, M. Shibata, M. Tadakoshi, K. Gotoh, M. Komatsu, S. Waguri, N. Kawahara, K. Kuida, S. Nagata, E. Kominami, K. Tanaka and Y. Uchiyama, Inhibition of autophagy prevents hippocampal pyramidal neuron death after hypoxic-ischemic injury, *Am J Pathol* **172**: 454–469 (2008).
9. Y. Lai, R. W. Hickey, Y. Chen, H. Bayir, M. L. Sullivan, C. T. Chu, P. M. Kochanek, C. E. Dixon, L. W. Jenkins, S. H. Graham, S. C. Watkins and R. S. Clark, Autophagy is increased after traumatic brain injury in mice and is partially inhibited by the antioxidant gamma-glutamylcysteinyl ethyl ester, *J Cereb Blood Flow Metab* **28**: 540–550 (2008).
10. K. C. Walls, B. J. Klocke, P. Saftig, M. Shibata, Y. Uchiyama, K. A. Roth and J. J. Shacka, Altered regulation of phosphatidylinositol 3-kinase signaling in cathepsin D-deficient brain, *Autophagy* **3**: 222–229 (2007).
11. D. Zhou, E. Masliah and S. A. Spector, Autophagy is increased in postmortem brains of persons with HIV-1-associated encephalitis, *J Infect Dis* **203**: 1647–1657 (2011).
12. J. M. Levy and A. Thorburn, Modulation of pediatric brain tumor autophagy and chemosensitivity, *J Neurooncol* **106**: 281–290 (2012).
13. E. D. Plowey, S. J. Cherra, 3rd, Y. J. Liu and C. T. Chu, Role of autophagy in G2019S-LRRK2-associated neurite shortening in differentiated SH-SY5Y cells, *J Neurochem* **105**: 1048–1056 (2008).
14. L. Du, R. W. Hickey, H. Bayir, S. C. Watkins, V. A. Tyurin, F. Guo, P. M. Kochanek, L. W. Jenkins, J. Ren, G. Gibson, C. T. Chu, V. E. Kagan and R. S. Clark, Starving neurons show sex difference in autophagy, *J Biol Chem* **284**: 2383–2396 (2009).
15. M. Xilouri, T. Vogiatzi, K. Vekrellis, D. Park and L. Stefanis, Abberant alpha-synuclein confers toxicity to neurons in part through inhibition of chaperone-mediated autophagy, *PLoS ONE* **4**: e5515 (2009).
16. H. C. Cheng, S. R. Kim, T. F. Oo, T. Kareva, O. Yarygina, M. Rzhetskaya, C. Wang, M. During, Z. Talloczy, K. Tanaka, M. Komatsu, K. Kobayashi, H. Okano, N. Kholodilov and R. E. Burke, Akt suppresses retrograde degeneration of dopaminergic axons by inhibition of macroautophagy, *J Neurosci* **31**: 2125–2135 (2011).
17. J. H. Zhu, C. Horbinski, F. Guo, S. Watkins, Y. Uchiyama and C. T. Chu, Regulation of autophagy by extracellular signal-regulated protein kinases during 1-methyl-4-phenylpyridinium-induced cell death, *Am J Pathol* **170**: 75–86 (2007).
18. L. Chakrabarti, J. Eng, N. Ivanov, G. A. Garden and A. R. La Spada, Autophagy activation and enhanced mitophagy characterize the Purkinje cells of pcd mice prior to neuronal death, *Mol Brain* **2**: 24 (2009).

19. K. C. Choi, S. H. Kim, J. Y. Ha, S. T. Kim and J. H. Son, A novel mTOR activating protein protects dopamine neurons against oxidative stress by repressing autophagy related cell death, *J Neurochem* **112**: 366–376 (2010).
20. C. Samara and N. Tavernarakis, Autophagy and cell death in *Caenorhabditis elegans*, *Curr Pharm Des* **14**: 97–115 (2008).
21. H. Zhu, P. Tannous, J. L. Johnstone, Y. Kong, J. M. Shelton, J. A. Richardson, V. Le, B. Levine, B. A. Rothermel and J. A. Hill, Cardiac autophagy is a maladaptive response to hemodynamic stress, *J Clin Invest* **117**: 1782–1793 (2007).
22. E. M. Chalovich, J. H. Zhu, J. Caltagarone, R. Bowser and C. T. Chu, Functional repression of cAMP response element in 6-hydroxydopamine-treated neuronal cells, *J Biol Chem* **281**: 17870–17881 (2006).
23. J. H. Zhu, A. Gusdon, H. Cimen, B. Van Houten, E. Koc and C. T. Chu, Impaired mitochondrial biogenesis contributes to depletion of functional mitochondria in chronic MPP+ toxicity, *Cell Death Dis* **3**: e312 (2012).
24. C. T. Chu, Autophagic stress in neuronal injury and disease, *J Neuropathol Exp Neurol* **65**: 423–432 (2006).
25. W. H. Yu, A. M. Cuervo, A. Kumar, C. M. Peterhoff, S. D. Schmidt, J. H. Lee, P. S. Mohan, M. Mercken, M. R. Farmery, L. O. Tjernberg, Y. Jiang, K. Duff, Y. Uchiyama, J. Naslund, P. M. Mathews, A. M. Cataldo and R. A. Nixon, Macroautophagy — a novel beta-amyloid peptide-generating pathway activated in Alzheimer's disease, *J Cell Biol* **171**: 87–98 (2005).
26. A. M. Gusdon and C. T. Chu, To eat or not to eat: neuronal metabolism, mitophagy and Parkinson's disease, *Antioxid Redox Signal* **14**: 1979–1987 (2011).
27. N. Hosokawa, T. Hara, T. Kaizuka, C. Kishi, A. Takamura, Y. Miura, S. Iemura, T. Natsume, K. Takehana, N. Yamada, J. L. Guan, N. Oshiro and N. Mizushima, Nutrient-dependent mTORC1 association with the ULK1-Atg13-FIP200 complex required for autophagy, *Mol Biol Cell* **20**: 1981–1991 (2009).
28. A. Petiot, E. Ogier-Denis, E. F. Blommaart, A. J. Meijer and P. Codogno, Distinct classes of phosphatidylinositol 3′-kinases are involved in signaling pathways that control macroautophagy in HT-29 cells, *J Biol Chem* **275**: 992–998 (2000).
29. Z. Xie and D. J. Klionsky, Autophagosome formation: core machinery and adaptations, *Nat Cell Biol* **9**: 1102–1109 (2007).
30. C. He and D. J. Klionsky, Regulation mechanisms and signaling pathways of autophagy, *Annu Rev Genet* **43**: 67–93 (2009).
31. Y. Ohsumi, Molecular dissection of autophagy: two ubiquitin-like systems, *Nat Rev Mol Cell Biol* **2**: 211–216 (2001).
32. V. Kirkin, T. Lamark, Y. S. Sou, G. Bjorkoy, J. L. Nunn, J. A. Bruun, E. Shvets, D. G. McEwan, T. H. Clausen, P. Wild, I. Bilusic, J. P. Theurillat, A. Øvervatn, T. Ishii, Z. Elazar, M. Komatsu, I. Dikic and T. Johansen, A role for NBR1 in autophagosomal degradation of ubiquitinated substrates, *Mol Cell* **33**: 505–516 (2009).
33. S. Pankiv, T. H. Clausen, T. Lamark, A. Brech, J. A. Bruun, H. Outzen, A. Øvervatn, G. Bjørkøy and T. Johansen, p62/SQSTM1 binds directly to Atg8/LC3 to facilitate

degradation of ubiquitinated protein aggregates by autophagy, *J Biol Chem* **282**: 24131–24145 (2007).
34. C. T. Chu, J. Zhu and R. Dagda, Beclin 1-independent pathway of damage-induced mitophagy and autophagic stress: implications for neurodegeneration and cell death, *Autophagy* **3**: 663–666 (2007).
35. F. Scarlatti, R. Maffei, I. Beau, P. Codogno and R. Ghidoni, Role of non-canonical Beclin 1-independent autophagy in cell death induced by resveratrol in human breast cancer cells, *Cell Death Differ* **15**: 1318–1329 (2008).
36. K. S. Yee, S. Wilkinson, J. James, K. M. Ryan and K. H. Vousden, PUMA- and Bax-induced autophagy contributes to apoptosis, *Cell Death Differ* **16**: 1135–1145 (2009).
37. S. Pattingre, A. Tassa, X. Qu, R. Garuti, X. H. Liang, N. Mizushima, M. Packer, M. D. Schneider and B. Levine, Bcl-2 antiapoptotic proteins inhibit Beclin 1-dependent autophagy, *Cell* **122**: 927–939 (2005).
38. I. Kim and J. J. Lemasters, Mitochondrial degradation by autophagy (mitophagy) in GFP-LC3 transgenic hepatocytes during nutrient deprivation, *Am Physiol* **300**: C308–317 (2011).
39. I. Kim and J. J. Lemasters, Mitophagy selectively degrades individual damaged mitochondria after photoirradiation, *Antioxid Redox Signal* **14**: 1919–1928 (2011).
40. S. Tian, J. Lin, J. Zhou, X. Wang, Y. Li, X. Ren, W. Yu, W. Zhong, J. Xiao, F. Sheng, Y. Chen, C. Jin, S. Li, Z. Zheng and B. Xia, Beclin 1-independent autophagy induced by a Bcl-X(L)/Bcl-2 targeting compound, Z18, *Autophagy* **6:** 1032–1041 (2010).
41. S. Wullschleger, R. Loewith and M. N. Hall, TOR signaling in growth and metabolism, *Cell* **124**: 471–484 (2006).
42. R. Loewith, E. Jacinto, S. Wullschleger, A. Lorberg, J. L. Crespo, D. Bonenfant, W. Oppliger, P. Jenoe and M. N. Hall, Two TOR complexes, only one of which is rapamycin sensitive, have distinct roles in cell growth control, *Mol Cell* **10**: 457–468 (2002).
43. X. F. Zheng, D. Florentino, J. Chen, G. R. Crabtree and S. L. Schreiber, TOR kinase domains are required for two distinct functions, only one of which is inhibited by rapamycin, *Cell* **82**: 121–130 (1995).
44. M. Mehrpour, A. Esclatine, I. Beau and P. Codogno, Overview of macroautophagy regulation in mammalian cells, *Cell Res* **20**: 748–762 (2010).
45. R. J. Shaw, LKB1 and AMP-activated protein kinase control of mTOR signalling and growth, *Acta Physiol (Oxford, England)* **196**: 65–80 (2009).
46. E. F. Blommaart, U. Krause, J. P. Schellens, H. Vreeling-Sindelarova and A. J. Meijer, The phosphatidylinositol 3-kinase inhibitors wortmannin and LY294002 inhibit autophagy in isolated rat hepatocytes, *Eur J Biochem* **243**: 240–246 (1997).
47. A. Simonsen and S. A. Tooze, Coordination of membrane events during autophagy by multiple class III PI3-kinase complexes, *J Cell Biol* **186**: 773–782 (2009).
48. I. Vergne and V. Deretic, The role of PI3P phosphatases in the regulation of autophagy, *FEBS Lett* **584**: 1313–1318 (2010).

49. I. Vergne, E. Roberts, R. A. Elmaoued, V. Tosch, M. A. Delgado, T. Proikas-Cezanne, J. Laporte and V. Deretic, Control of autophagy initiation by phosphoinositide 3-phosphatase jumpy, *EMBO J* **28**: 2244–2258 (2009).
50. V. Tosch, H. M. Rohde, H. Tronchere, E. Zanoteli, N. Monroy, C. Kretz, N. Dondaine, B. Payrastre, J. L. Mandel and J. Laporte, A novel PtdIns3P and PtdIns(3,5)P2 phosphatase with an inactivating variant in centronuclear myopathy, *Hum Mol Genet* **15**: 3098–3106 (2006).
51. K. Obara, T. Sekito, K. Niimi and Y. Ohsumi, The Atg18-Atg2 complex is recruited to autophagic membranes via phosphatidylinositol 3-phosphate and exerts an essential function, *J Biol Chem* **283**: 23972–23980 (2008).
52. X. H. Liang, L. K. Kleeman, H. H. Jiang, G. Gordon, J. E. Goldman, G. Berry, B. Herman and B. Levine, Protection against fatal Sindbis virus encephalitis by beclin, a novel Bcl-2-interacting protein, *J Virol* **72**: 8586–8596 (1998).
53. M. C. Maiuri, G. Le Toumelin, A. Criollo, J. C. Rain, F. Gautier, P. Juin, E. Tasdemir, G. Pierron, K. Troulinaki, N. Tavernarakis, J. A. Hickman, O. Geneste and G. Kroemer, Functional and physical interaction between Bcl-X(L) and a BH3-like domain in Beclin-1, *EMBO J* **26**: 2527–2539 (2007).
54. A. Oberstein, P. D. Jeffrey and Y. Shi, Crystal structure of the Bcl-XL-Beclin 1 peptide complex: Beclin 1 is a novel BH3-only protein, *J Biol Chem* **282**: 13123–13132 (2007).
55. M. Thome, P. Schneider, K. Hofmann, H. Fickenscher, E. Meinl, F. Neipel, C. Mattmann, K. Burns, J. L. Bodmer, M. Schroter, C. Scaffidi, P. H. Krammer, M. E. Peter and J. Tschopp, Viral FLICE-inhibitory proteins (FLIPs) prevent apoptosis induced by death receptors, *Nature* **386**: 517–521 (1997).
56. J. Tschopp, M. Irmler and M. Thome, Inhibition of fas death signals by FLIPs, *Curr Opin Immunol* **10**: 552–558 (1998).
57. R. C. Budd, W. C. Yeh and J. Tschopp, cFLIP regulation of lymphocyte activation and development, *Nat Rev* **6**: 196–204 (2006).
58. J. S. Lee, Q. Li, J. Y. Lee, S. H. Lee, J. H. Jeong, H. R. Lee, H. Chang, F. C. Zhou, S. J. Gao, C. Liang and J. U. Jung, FLIP-mediated autophagy regulation in cell death control, *Nat Cell Biol* **11**: 1355–1362 (2009).
59. A. Eisenberg-Lerner, S. Bialik, H. U. Simon and A. Kimchi, Life and death partners: apoptosis, autophagy and the cross-talk between them, *Cell Death Differ* **16**: 966–975 (2009).
60. S. Yousefi, R. Perozzo, I. Schmid, A. Ziemiecki, T. Schaffner, L. Scapozza, T. Brunner and H. U. Simon, Calpain-mediated cleavage of Atg5 switches autophagy to apoptosis, *Nat Cell Biol* **8**: 1124–1132 (2006).
61. D. H. Cho, Y. K. Jo, J. J. Hwang, Y. M. Lee, S. A. Roh and J. C. Kim, Caspase-mediated cleavage of ATG6/Beclin-1 links apoptosis to autophagy in HeLa cells, *Cancer Lett* **274**: 95–100 (2009).
62. S. Luo and D. C. Rubinsztein, Apoptosis blocks Beclin 1-dependent autophagosome synthesis: an effect rescued by Bcl-xL, *Cell Death Differ* **17**: 268–277 (2010).

63. I. H. Lee and T. Finkel, Regulation of autophagy by the p300 acetyltransferase, *J Biol Chem* **284**: 6322–6328 (2009).
64. I. H. Lee, L. Cao, R. Mostoslavsky, D. B. Lombard, J. Liu, N. E. Bruns, M. Tsokos, F. W. Alt and T. Finkel, A role for the NAD-dependent deacetylase Sirt1 in the regulation of autophagy, *Proc Natl Acad Sci USA* **105**: 3374–3379 (2008).
65. H. Jeong, F. Then, T. J. Melia, Jr., J. R. Mazzulli, L. Cui, J. N. Savas, C. Voisine, P. Paganetti, N. Tanese, A. C. Hart, A. Yamamoto and D. Krainc, Acetylation targets mutant huntingtin to autophagosomes for degradation, *Cell* **137**: 60–72 (2009).
66. R. Xie, S. Nguyen, W. L. McKeehan and L. Liu, Acetylated microtubules are required for fusion of autophagosomes with lysosomes, *BMC Cell Biol* **11**: 89 (2010).
67. J. Y. Lee, H. Koga, Y. Kawaguchi, W. Tang, E. Wong, Y. S. Gao, U. B. Pandey, S. Kaushik, E. Tresse, J. Lu, J. P. Taylor, A. M. Cuervo and T. P. Yao, HDAC6 controls autophagosome maturation essential for ubiquitin-selective quality-control autophagy, *EMBO J* **29**: 969–980 (2010).
68. D. F. Egan, D. B. Shackelford, M. M. Mihaylova, S. R. Gelino, R. A. Kohnz, W. Mair, D. S. Vasquez, A. Joshi, D. M. Gwinn, R. Taylor, J. M. Asara, J. Fitzpatrick, A. Dillin, B. Viollet, M. Kundu, M. Hansen and R. J. Shaw, Phosphorylation of ULK1 (hATG1) by AMP-activated protein kinase connects energy sensing to mitophagy, *Science* **331:** 456–461 (2011).
69. E. Y. Chan, A. Longatti, N. C. McKnight and S. A. Tooze, Kinase-inactivated ULK proteins inhibit autophagy via their conserved C-terminal domains using an Atg13-independent mechanism, *Mol Cell Biol* **29**: 157–171 (2009).
70. I. Holen, P. B. Gordon and P. O. Seglen, Inhibition of hepatocytic autophagy by okadaic acid and other protein phosphatase inhibitors, *Eur J Biochem* **215**: 113–122 (1993).
71. E. Zalckvar, H. Berissi, L. Mizrachy, Y. Idelchuk, I. Koren, M. Eisenstein, H. Sabanay, R. Pinkas-Kramarski and A. Kimchi, DAP-kinase-mediated phosphorylation on the BH3 domain of beclin 1 promotes dissociation of beclin 1 from Bcl-XL and induction of autophagy, *EMBO Rep* **10**: 285–292 (2009).
72. Y. Wei, S. Pattingre, S. Sinha, M. Bassik and B. Levine, JNK1-mediated phosphorylation of Bcl-2 regulates starvation-induced autophagy, *Mol Cell* **30**: 678–688 (2008).
73. H. Jiang, D. Cheng, W. Liu, J. Peng and J. Feng, Protein kinase C inhibits autophagy and phosphorylates LC3, *Biochem Biophys Res Commun* **395**: 471–476 (2010).
74. S. J. Cherra, 3rd, S. M. Kulich, G. Uechi, M. Balasubramani, J. Mountzouris, B. W. Day and C. T. Chu, Regulation of the autophagy protein LC3 by phosphorylation, *J Cell Biol* **190**: 533–539 (2010).
75. A. S. Tsvetkov, J. Miller, M. Arrasate, J. S. Wong, M. A. Pleiss and S. Finkbeiner, A small-molecule scaffold induces autophagy in primary neurons and protects against toxicity in a Huntington disease model, *Proc Natl Acad Sci USA* **107**: 16982–16987 (2010).

76. V. S. Van Laar, B. Arnold, S. J. Cassady, C. T. Chu, E. A. Burton and S. B. Berman, Bioenergetics of neurons inhibit the translocation response of Parkin following rapid mitochondrial depolarization, *Hum Mol Genet* **20**: 927–940 (2011).
77. S. S. Mann and J. A. Hammarback, Molecular characterization of light chain 3. A microtubule binding subunit of MAP1A and MAP1B, *J Biol Chem* **269**: 11492–11497 (1994).
78. I. Tanida, T. Ueno and E. Kominami, LC3 conjugation system in mammalian autophagy, *Int J Biochem Cell Biol* **36**: 2503–2518 (2004).
79. J. Kim, M. Kundu, B. Viollet and K. L. Guan, AMPK and mTOR regulate autophagy through direct phosphorylation of Ulk1, *Nat Cell Biol* **13**: 132–141 (2011).
80. J. S. Stephan, Y. Y. Yeh, V. Ramachandran, S. J. Deminoff and P. K. Herman, The Tor and PKA signaling pathways independently target the Atg1/Atg13 protein kinase complex to control autophagy, *Proc Natl Acad Sci USA* **106**: 17049–17054 (2009).
81. B. Levine, S. Sinha and G. Kroemer, Bcl-2 family members: dual regulators of apoptosis and autophagy, *Autophagy* **4**: 600–606 (2008).
82. K. Matsunaga, T. Saitoh, K. Tabata, H. Omori, T. Satoh, N. Kurotori, I. Maejima, K. Shirahama-Noda, T. Ichimura, T. Isobe, S. Akira, T. Noda and T. Yoshimori, Two Beclin 1-binding proteins, Atg14L and Rubicon, reciprocally regulate autophagy at different stages, *Nat Cell Biol* **11**: 385–396 (2009).
83. Y. Zhong, Q. J. Wang, X. Li, Y. Yan, J. M. Backer, B. T. Chait, N. Heintz and Z. Yue, Distinct regulation of autophagic activity by Atg14L and Rubicon associated with Beclin 1-phosphatidylinositol-3-kinase complex, *Nat Cell Biol* **11**: 468–476 (2009).
84. R. Scherz-Shouval, E. Shvets, E. Fass, H. Shorer, L. Gil and Z. Elazar, Reactive oxygen species are essential for autophagy and specifically regulate the activity of Atg4, *EMBO J* **26**: 1749–1760 (2007).
85. T. Lamark, V. Kirkin, I. Dikic and T. Johansen, NBR1 and p62 as cargo receptors for selective autophagy of ubiquitinated targets, *Cell Cycle* **8**: 1986–1990 (2009).
86. L. Liu, D. Feng, G. Chen, M. Chen, Q. Zheng, P. Song, Q. Ma, C. Zhu, R. Wang, W. Qi, L. Huang, P. Xue, B. Li, X. Wang, H. Jin, J. Wang, F. Yang, P. Liu, Y. Zhu, S. Sui and Q. Chen, Mitochondrial outer-membrane protein FUNDC1 mediates hypoxia-induced mitophagy in mammalian cells, *Nat Cell Biol* **14**: 177–185 (2012).
87. F. Pickford, E. Masliah, M. Britschgi, K. Lucin, R. Narasimhan, P. A. Jaeger, S. Small, B. Spencer, E. Rockenstein, B. Levine and T. Wyss-Coray, The autophagy-related protein beclin 1 shows reduced expression in early Alzheimer disease and regulates amyloid beta accumulation in mice, *J Clin Invest* **118**: 2190–2199 (2008).
88. M. Shibata, T. Lu, T. Furuya, A. Degterev, N. Mizushima, T. Yoshimori, M. MacDonald, B. Yankner and J. Yuan, Regulation of intracellular accumulation of mutant Huntingtin by Beclin 1, *J Biol Chem* **281**: 14474–14485 (2006).
89. A. Orvedahl, D. Alexander, Z. Talloczy, Q. Sun, Y. Wei, W. Zhang, D. Burns, D. A. Leib and B. Levine, HSV-1 ICP34.5 confers neurovirulence by targeting the Beclin 1 autophagy protein, *Cell Host Microbe* **1**: 23–35 (2007).

90. M. Komatsu, S. Waguri, T. Chiba, S. Murata, J. Iwata, I. Tanida, T. Ueno, M. Koike, Y. Uchiyama, E. Kominami and K. Tanaka, Loss of autophagy in the central nervous system causes neurodegeneration in mice, *Nature* **441**: 880–884 (2006).
91. T. Hara, K. Nakamura, M. Matsui, A. Yamamoto, Y. Nakahara, R. Suzuki-Migishima, M. Yokoyama, K. Mishima, I. Saito, H. Okano and N. Mizushima, Suppression of basal autophagy in neural cells causes neurodegenerative disease in mice, *Nature* **441**: 885–889 (2006).
92. M. Koike, M. Shibata, S. Waguri, K. Yoshimura, I. Tanida, E. Kominami, T. Gotow, C. Peters, K. von Figura, N. Mizushima, P. Saftig and Y. Uchiyama, Participation of autophagy in storage of lysosomes in neurons from mouse models of neuronal ceroid-lipofuscinoses (Batten disease), *Am J Pathol* **167**: 1713–1728 (2005).
93. C. Settembre, A. Fraldi, L. Jahreiss, C. Spampanato, C. Venturi, D. Medina, R. de Pablo, C. Tacchetti, D. C. Rubinsztein and A. Ballabio, A block of autophagy in lysosomal storage disorders, *Hum Mol Genet* **17**: 119–129 (2008).
94. C. D. Pacheco, R. Kunkel and A. P. Lieberman, Autophagy in Niemann-Pick C disease is dependent upon Beclin-1 and responsive to lipid trafficking defects, *Hum Mol Genet* **16**: 1495–1503 (2007).
95. R. A. Nixon, J. Wegiel, A. Kumar, W. H. Yu, C. Peterhoff, A. Cataldo and A. M. Cuervo, Extensive involvement of autophagy in Alzheimer disease: an immuno-electron microscopy study, *J Neuropathol Exp Neurol* **64**: 113–122 (2005).
96. J. J. Jennings, Jr., J. H. Zhu, Y. Rbaibi, X. Luo, C. T. Chu and K. Kiselyov, Mitochondrial aberrations in mucolipidosis type IV, *J Biol Chem* **281**: 39041–39050 (2006).
97. D. S. Yang, P. Stavrides, P. S. Mohan, S. Kaushik, A. Kumar, M. Ohno, S. D. Schmidt, D. Wesson, U. Bandyopadhyay, Y. Jiang, M. Pawlik, C. M. Peterhoff, A. J. Yang, D. A. Wilson, P. St George-Hyslop, D. Westaway, P. M. Mathews, E. Levy, A. M. Cuervo and R. A. Nixon, Reversal of autophagy dysfunction in the TgCRND8 mouse model of Alzheimer's disease ameliorates amyloid pathologies and memory deficits, *Brain* **134**: 258–277 (2011).
98. J. J. Shacka, J. Lu, Z. L. Xie, Y. Uchiyama, K. A. Roth and J. Zhang, Kainic acid induces early and transient autophagic stress in mouse hippocampus, *Neurosci Lett* **414**: 57–60 (2007).
99. Q. J. Wang, Y. Ding, D. S. Kohtz, N. Mizushima, I. M. Cristea, M. P. Rout, B. T. Chait, Y. Zhong, N. Heintz and Z. Yue, Induction of autophagy in axonal dystrophy and degeneration, *J Neurosci* **26**: 8057–8068 (2006).
100. S. Carloni, G. Buonocore and W. Balduini, Protective role of autophagy in neonatal hypoxia-ischemia induced brain injury, *Neurobiol Dis* **32**: 329–339 (2008).
101. V. Vingtdeux, L. Giliberto, H. Zhao, P. Chandakkar, Q. Wu, J. E. Simon, E. M. Janle, J. Lobo, M. G. Ferruzzi, P. Davies and P. Marambaud, AMP-activated protein kinase signaling activation by resveratrol modulates amyloid-beta peptide metabolism, *J Biol Chem* **285**: 9100–9113 (2010).

102. S. J. Cherra, 3rd, R. K. Dagda and C. T. Chu, Review: autophagy and neurodegeneration: survival at a cost?, *Neuropathol Appl Neurobiol* **36**: 125–132 (2010).
103. W. H. Yu, B. Dorado, H. Y. Figueroa, L. Wang, E. Planel, M. R. Cookson, L. N. Clark and K. E. Duff, Metabolic activity determines efficacy of macroautophagic clearance of pathological oligomeric alpha-synuclein, *Am J Pathol* **175**: 736–747 (2009).
104. C. Gomez-Santos, I. Ferrer, A. F. Santidrian, M. Barrachina, J. Gil and S. Ambrosio, Dopamine induces autophagic cell death and alpha-synuclein increase in human neuroblastoma SH-SY5Y cells, *J Neurosci Res* **73**: 341–350 (2003).
105. B. Dehay, J. Bove, N. Rodriguez-Muela, C. Perier, A. Recasens, P. Boya and M. Vila, Pathogenic lysosomal depletion in Parkinson's disease, *J Neurosci* **30**: 12535–12544 (2010).
106. R. K. Dagda, J. Zhu, S. M. Kulich and C. T. Chu, Mitochondrially localized ERK2 regulates mitophagy and autophagic cell stress: implications for Parkinson's disease, *Autophagy* **4**: 770–782 (2008).
107. D. Narendra, A. Tanaka, D. F. Suen and R. J. Youle, Parkin is recruited selectively to impaired mitochondria and promotes their autophagy, *J Cell Biol* **183**: 795–803 (2008).
108. Q. Chang and L. J. Martin, Glycine receptor channels in spinal motoneurons are abnormal in a transgenic mouse model of amyotrophic lateral sclerosis, *J Neurosci* **31**: 2815–2827 (2011).

SECTION 2

AUTOPHAGY AND NEUROLOGICAL DISEASES

CHAPTER 6

AUTOPHAGY AND ITS CROSS-TALK WITH CELL DEATH IN NEURAL DEVELOPMENT

Sabrina Di Bartolomeo and Francesco Cecconi

ABSTRACT

The study of autophagy-deficient model systems has made it possible to appreciate the importance of a correctly functioning autophagic machinery in the development of both lower and higher eukaryotes. Autophagy is crucial to embryogenesis, potentially acting in parallel with apoptosis to remodel tissues. An increase in autophagic cells is observed during embryonic stages characterized by massive cell elimination, such as the complex events of fly and worm metamorphosis. Moreover, autophagy probably protects cells during metabolic stress and nutrient paucity, which both occur during the tissue remodeling of embryogenesis. Several lines of evidence also highlight the importance of autophagy in later stages of embryonic and postnatal development. Thus, it can be concluded that there is a close interplay between autophagy and the processes of cell death, proliferation and differentiation, and that correct central nervous system (CNS) development depends upon an intact autophagic machinery.

INTRODUCTION

Autophagy is a lysosome-mediated degradative process by which eukaryotes recycle organelles and long-lived proteins. Macroautophagy,[1] in particular, represents a complex process involving the formation of membrane-surrounded structures, called autophagosomes, which engulf the substrates and deliver them to lysosomes for degradation by moving along cytoskeletal structures.[2,3] Many autophagy-related genes (Atg) have been discovered and characterized in yeast,[4,5] and some of them are conserved in higher eukaryotes. In the multicellular organisms, specific Atg genes, together with their regulators, control different stages of autophagosome formation and maturation. *Atg1, Atg6, Ambra1* and *Vps34*, among others, are involved in the formation of the early autophagosome membrane;[6] *Atg5, Atg7, 12, 16*, and others allow the maturation of the autophagosome membrane.[2] A large body of evidence has recently emerged to support the crucial role played by autophagy during the development of a plethora of organisms, including *Arabidopsis thaliana, Drosophila discoideum, D. melanogaster, Caenorhabditis elegans* up to mammals.[7] Autophagy, in fact, has recently been proposed as a crucial mechanism for the cellular remodeling that occurs during the development of multicellular organisms.[8,9] During embryogenesis, cells undergo various processes, including proliferation, cell death and differentiation, thus allowing the embryo to transform into an adult organism. The exact mechanism by which autophagy contributes to embryogenesis has not been clarified, but it is generally thought that autophagy could have a dual role, both in adaptation to the stress and in cell elimination occurring during the process. In particular, cells require the appropriate tools to rapidly modify their protein content and morphology in order to respond to intrinsic and external stimuli. Moreover, damaged or unnecessary organelles and cells have to be cleared from the transforming tissues.

In this chapter, we describe some specific examples of autophagy occurrence and their relevance during the embryogenesis of eukaryotes, including discussion of the role of autophagy in vertebrate CNS development.

THE ROLE OF AUTOPHAGY IN THE DEVELOPMENT OF EUKARYOTES

Autophagy in *Drosophila melanogaster* Development

The fruit fly *D. melanogaster* develops through the three distinct stages: larva, pupa and adult insect, each characterized by its own body plan. During this complex metamorphosis, a constant supply of nutrients is required to provide the energy necessary for growth, metabolism and survival; autophagy is thought to play an important role in such a process. Analogous to what is observed in yeast, worms and mice, *atg* inactivation may result in severe phenotypes in fruit flies. Mutation of *atg18* and *atg6* is lethal at the larval stage, indicating that these genes are essential for development through this stage.[10] *Atg1* expression is also crucial for development as *atg1* mutants show reduced larval viability, and those that do survive cannot develop beyond pupa.[10] *Atg7*-defective mutant flies, on the other hand, are fully viable, fertile and do not present major morphological defects during metamorphosis; they only show an attenuation of autophagy. Emerging *atg7* mutant adults, however, are hypersensitive to stress and have a reduced life span, potentially related to accumulation of ubiquitin-rich aggregates in degenerating neurons.[11] Moreover, when the *atg1* gene is disrupted in *Tor*-deficient flies, which are characterized by a reduced cell size and a decreased viability, there is an increasing severity of their phenotypes that results in embryonic lethality. These findings clearly indicate that autophagy is required to maintain cell size, viability and normal metabolism.[10,11]

There is considerable evidence that autophagy is induced in several *D. melanogaster* tissues as a normal physiological response to the rise of hormones which occurs during metamorphosis and is responsible for the cellular reorganization necessary to prepare for the maturation of adult tissues. In particular, the steroid hormone 20-hydroxyecdysone (ecdysone) that increases at the end of the larval stage defines the length of the development period and controls larval molting and metamorphosis during the fruit fly's life cycle.[9,12] The larval growth is arrested when normal levels of ecdysone are disrupted.[13] It has been observed that the mRNA levels of some *atg* genes increase following the ecdysone burst in degenerating larval structures, in the fat body,[14] in the intestine and in salivary glands.[15] The larval fat body is a nutrient storage organ analogous to the vertebrate liver, from which glycogen and lipids are mobilized to imaginal discs in response to amino acid starvation.[10] In particular, Rusten *et al.* clearly demonstrated that ecdysone signaling leads to a reduction of phosphatidylinositol-3,4,5-tris-phosphate (PIP3) levels by downregulating phosphoinositide (PI) 3-kinase (PI3K) activity in the fat body.[14] This phenomenon is thought to be the mechanism for induction of autophagy. It has also been demonstrated that autophagy, but not apoptosis, is essential in ecdysone-induced degradation of dying midgut cells; therefore this degradation does not occur in flies devoid of *atg1, atg2* or *atg18*,[16,17] and it is delayed in *atg7* mutants.[11]

The salivary gland is a larval tissue that is degraded within a few hours during metamorphosis, showing the morphological characteristics of "type II cell death," as autophagy is sometimes referred to.[18] Cell death in the salivary gland is induced by an increase in ecdysone levels ten to 12 hours after puparium formation (apf), and is maintained at high level for up to 16 hours apf. The presence of autophagosomes and absence of phagocytosis in the cytosol of these cells has been observed.[14] Remarkably, the lack of *atg* genes, including *atg1, atg2, atg3, atg6, atg7, atg8, atg12* and *atg18*, leads to a failure to degrade the fly's salivary glands. In addition, the overexpression of *atg1* results in autophagy induction and is sufficient to induce caspase-independent cell death.[16] Recently, it has been observed that the dynein light chain 1 (*ddlc1*), a component of the dynein motor complex, is required for cell death of salivary glands. In the salivary glands of *ddlc1* mutants, caspases are activated but autophagy is attenuated, thus indicating an involvement of *ddlc1* in regulating autophagy but not apoptosis in dying cells.[19] The mechanism by which *ddlc1* regulates autophagy remains unclear; nonetheless, overexpression of *atg1* is sufficient to rescue salivary gland degradation in mutant flies. Notably, studies conducted in mammalian cells have demonstrated the importance of the dynein motor complex for the transport of autophagosomes along microtubules.[3] However, as dynein heavy chain mutations do not result in an impairment of autophagy in salivary gland,[17] it cannot be excluded that *ddlc1* could also regulate autophagy in a motor-independent manner. During the *D. melanogaster* development a parallel induction of autophagy and apoptosis has also been observed in the *Drosophila* ovary during the two stages of oogenesis named germarium and mid-oogenesis, these representing two nutrient status checkpoints. It has been demonstrated that autophagy occurs in degenerating midstage egg chambers and also in germaria of nutrient deprived *D. melanogaster*.[20] Genetic inactivation of *atg7* results in a significant decrease of autophagy in dying midstage egg chambers and in germaria of starved flies, further supporting the induction of autophagy during these stages of oogenesis.[20] In some aspects, cell death taking place during *D. melanogaster* oogenesis is similar to that observed in larval salivary glands. Both ovary and salivary gland contain peculiar large and polyploid cells, and the entire tissues undergo cell death simultaneously.

Finally, it has recently been observed that both autophagy and caspases are required for the disruption of the extraembryonic membrane amnioserosa (AS) during embryogenesis.[21,22] Although it is known that the elimination of the AS involves caspase-dependent nuclear fragmentation, tissue dissassembly and engulfment by phagocytic macrophages, the degenerating structure shows autophagic hallmarks during the final stages of embryogenesis.

In conclusion, during *Drosophila* development, autophagy is induced in response to different stimuli in several tissues such as fat body, intestine, salivary glands and ovary and cooperates with canonical apoptosis to remodel developing tissues. In light of all the evidence reported, it appears clear that apoptosis and autophagy work in parallel in the disruption of larval tissues during fly development.

Autophagy in *Caenorhabditis elegans* Development

The role of autophagy in development has also been studied in the nematode *C. elegans,* in which many of the components of the autophagic machinery are conserved. Similar to other eukaryotic organisms, different autophagy genes play distinct roles during worm embryogenesis, with some of these genes showing partial redundancy to each other. For example, worms deficient in the yeast *Atg6* homolog *bec-1* exhibit a lethal phenotype and die before or during the first larval stage, showing increased vacuolization and molting defects.[23] In addition to the involvement in viability, *bec-1* activity also seems to be involved in fertility since mutant worms that reach adulthood are sterile. A lethal phenotype is also observed in worms carrying a mutation in the *C. elegans* ortholog of Class III PI3K *Vps34* (*let-512*), and in *atg8 (lgg-1)* or *atg18 (F41E6,13)* knockdown worms.[24,25] In contrast to *D. melanogaster atg1* mutants, *C. elegans* devoid of the atg1 ortholog *unc-51* develop into mature organisms although they exhibit axonal defects that impair motor neuron function,[26] similarly to what observed in vertebrates, as will be illustrated later.

It is known that *C. elegans* larvae respond to stressful conditions, such as starvation, high population density or increased temperature by arresting development at the third larval stage (L3). This is referred to as dauer diapause.[27] Dauer larvae undergo a number of metabolic and morphological changes, as distinct from normal L3 larvae, which allow long-term survival. These changes include increased intestinal fat storage, pharyngeal, intestinal and hypodermal constriction, total body elongation, and the formation of a thick cuticle that seals them from the environment. If external conditions improve, dauer larvae can resume reproductive development, reach the adult stage, and have a normal lifespan.[27] When the food supply is limited, autophagy activation allows cells to generate a pool of amino acids sufficient for the neosynthesis of proteins essential for survival. Although the regulation of dauer development has been characterized, the cellular pathways involved in dauer morphogenesis are less well understood. In dauer larvae with a *daf-2* (the insulin-like receptor) mutation there is an increase of autophagy in hypodermal precursor cells.[25] The depletion of autophagic activity in *daf-2* mutants in which *bec-1, unc-51, atg7, atg8 and atg18* are also individually knocked down results in a defect in dauer formation, with worms dying within a few days. It is these studies that help demonstrate that autophagy acts downstream of the insulin/IGF pathway and is required for normal dauer morphogenesis.

Very recently a novel role has been proposed for autophagy in degrading the P granules in somatic cells during the early stages of embryo segmentation in *C. elegans*.[28] P granules are cytosolic aggregates containing both proteins and RNA. Initially they are expressed in all the cells of the worm blastula, but they quickly redistribute exclusively into the germ cell, thus determining its fate as a precursor of the germinal lineage. According to this observation, autophagy should play a relevant function in orchestrating cell lineage differentiation, as also postulated in mammalian cells.[29]

Interestingly, autophagy-related processes are involved in the clustering of neurotransmitter receptors during the neuronal development of *C. elegans*. First, the axonal outgrowth along the antero-posterior axis is impaired in *unc-51* worm mutants.[30] Second, the degradation of GABA receptors that are internalized in the absence of presynaptic inputs in muscle cells seems to be specifically mediated by autophagy. Notably, the mammalian $GABA_A$-receptor associated protein Gabarap is an ortholog of the yeast autophagy protein Atg8. Receptor internalization and degradation is associated with autophagy since receptors co-localized both with the Atg8 ortholog LGG-1 and with BEC-1 in autophagosomes of non-innervated muscles. Trafficking of receptors from cell surface to autophagosomes likely occurs *via* the endocytic pathway since blocking endocytosis prevents receptors from aggregating in autophagosomes.[31,32]

Autophagy During the Vertebrate Development

Autophagy activation has been observed in a few specific steps during the vertebrate embryogenesis.[9] It has recently been demonstrated, for example, that autophagy is upregulated after egg fertilization in mouse, and that this process is essential for preimplantation development.[33] A significant increase of autophagosomes can be observed in fertilized oocytes within a few hours following fertilization. After fertilization, the increase of autophagy is accompanied by the mTOR substrate S6 kinase dephosphorylation, possibly as a consequence of the Ca^{2+} wave following fertilization.[33] Interestingly, autophagy is suppressed from the late one-cell to the middle two-cell stage, and is reactivated after the late two-cell stage. To understand whether activation of autophagy is critical for early embryogenesis, some models of autophagy-deficient embryos have also been developed. For example, it is known that *Atg5*$^{-/-}$ mice survive embryogenesis, but die soon after birth due to an inability to feed and overcome the critical starvation perinatal period.[34] In the early Atg5–/– embryos, the small quantity of cytoplasmic Atg5 protein accumulated into maternal oocytes before meiosis, is probably sufficient to support autophagy induction; in order to study the role of Atg5 in early development, oocytes completely deficient for this protein have been generated.[33–35] When an *Atg5*$^{-}$ sperm fertilizes an *Atg5*-deficient oocyte, the embryonic development stops at the eight-cell stage, thus demonstrating that the Atg5-dependent autophagy pathway is essential for the very early developmental stages.

Induction of autophagy at the very early stages of development of a mouse embryo may be responsible for both degradation of maternal proteins and macromolecules and for generation of a pool of free amino acids available for zygotic protein synthesis. In vertebrates, oocytes accumulate maternal mRNAs and proteins during oogenesis.[36] These RNAs are responsible for the synthesis of proteins necessary for oocyte maturation and egg activation after fertilization, thus allowing the transition from egg to zygote. After fertilization, however, maternal proteins have to be degraded and the mRNAs encoded by the zygotic genome start to be translated. Concomitantly, the cellular organelles are remodeled by processes that have not yet been completely characterized. Since autophagy is considered a bulk degradation process, it represents a good candidate for eliminating the zygotic maternal material.

Another example of autophagy occurrence during embryogenesis can be observed during the cavitation of the early embryo. During the gastrulation phase, generation of a proamniotic cavity depends on apoptosis activation and removal of the inner ectodermal cells.[37] It has been demonstrated that two pro-autophagic genes, *beclin 1* and *Atg5*, are required for the clearance of cells that have died during the cavitation process. Embryos devoid of Beclin 1, exhibited a developmental delay culminating in severely reduced body size and lethality at embryonic day 7 (E7.5).[38] More recently, it has been demonstrated that aggregates of inner-cell-mass-derived embryonic stem cells (embryoid bodies, EB) obtained from Atg5- or Beclin 1-defective embryos fail to cavitate due to accumulation of dead cells.[39] Such accumulation depends not on an increase in cell death in these autophagy-defective cells, but on a defect in the engulfment of dead corpses by phagocytic cells. Autophagy-deficient EBs fail, in fact, to express the "eat-me" (phosphatidylserine exposure) and the "come-get-me" (lysophosphatidylcholine release) signals necessary for the phagocytic cells to recognize and engulf apoptotic bodies.[39] Autophagy-dependent ATP production seems to be necessary for completion of the apoptotic program during development; the engulfment defects observed in these models are associated with low cellular ATP levels. As phosphatidylserine (PS) exposure characterizes both apoptotic cells and healthy phagocytes, it is possible that phagocytic functions are also impaired. A massive accumulation of apoptotic bodies, due to the lack of PS exposure, has also been observed in the embryonic retina and lungs of *Atg5* null mice[39] and during the early stages of retinal development (embryonic day 5, E5) in a chick model following pharmacological inhibition of autophagy.[40] Intriguingly, autophagy inhibition does not result in failure to expose PS and accumulation of cell corpses in later stages of retinal development (E9), thus demonstrating that autophagy is essential for PS exposure only for some forms of morphogenic cell death. As PS exposure is an evolutionarily conserved mechanism by which phagocytes recognize and engulf apoptotic cells, it could also be the phenomenon responsible for accumulation of embryonic cell corpses observed in *bec-1* null *C. Elegans*.[23]

Therefore, autophagy can be defined as a crucial event in the removal of cell corpses accumulated during the morphogenetic apoptosis which leads to embryo cavitation and neural retina formation.

The importance of autophagy has also been demonstrated during organogenesis and postnatal development. Besides the crucial role played by autophagy for a correct CNS development (discussed in the last paragraph of this chapter), it has recently been demonstrated that cardiogenesis, hematopoiesis and osteogenesis all require an intact autophagic machinery. In particular, an autophagy defect leads to cardiac hypertrophy and dysfunction in adult mouse, although the exact pathologic function of autophagy in embryogenesis has not been yet identified.[41] A constitutive autophagy probably represents a homeostatic mechanism necessary for cardiomyocites remodeling that maintain cardiac size and function, whereas an increase of its activity protects the heart from hemodynamic stress.

In studying the development of blood cells, a reduction in lymphocytes has been observed in Atg5- and Atg7-deficient immune cells.[1,42] Atg5- and Atg7-devoid immune cells are more prone to apoptosis, and accumulation of mitochondria and reactive oxygen species. In this sense, autophagy should guarantee a stringent quality and quantity control of organelles. Removal of organelles, for example, is crucial for erythroblasts that first differentiate into anucleate reticulocytes and then mature into organelle-devoid erytrocytes. Atg7-deprived erythroblasts fail to differentiate, and show the typical hallmarks of apoptosis and mitochondria accumulation.

Skeletogenesis offers another striking example of autophagy occurrence during organogenesis. Autophagy is, in fact, upregulated in chondrocytes to enable them to survive environmental stress represented by the hypoxic cartilagineous structure in which the cells are embedded. Under such conditions, chondrocytes terminally differentiate and synthesize extracellular matrix, allowing vascularization and initiation of osteogenesis. Autophagic chondrocytes that encounter a newly vascularized nutrient-rich matrix can stop self-digestion and contribute to osteogenesis; otherwise the excess of autophagy might culminate as a severe cannibalistic response that eliminates these cells from the tissue.[43–45]

It has been recently observed that autophagy-related proteins could play a role in the specification of the left-right axis. In vertebrates the left-right axis specification depends in part on a flow of the morphogen Sonic hedgehog (Shh) in the left cells of the node during gastrulation that induces a gene expression in the left part of the primitive streak. The Shh asymmetric concentration is perhaps due to ciliary movements of the node cells.[46] It has observed that the *UVRAG* gene, which encodes another component of the Beclin 1/Vps34 complex,[47] is mutated in a human case of abnormal left-right axis formation, resulting in heterotaxy and other malformations.[48] The UVRAG-mutated phenotype could be related to an alteration of Shh distribution in the node. Interestingly, a similar left-right axis dysregulation has been identified in a Drosophila mutated in the UVRAG ortholog gene.[49] Otherwise, since UVRAG is also involved in endocytic pathway,[50] it cannot be excluded that its role on axis specification actually depends on endocytosis rather than on autophagy regulation.

Common Features of the Phenotypes in Different Organisms

A large body of evidence supports the important role played by autophagy during the development of eukaryotes, but the exact mechanisms by which autophagy contributes to embryonic phenotypes are not clear. Both in invertebrate and vertebrate organisms, it is generally thought that autophagy plays an essential dual role both in cell elimination, and in adaptation to stress phenomena occurring during morphogenesis. First, autophagy seems to be crucial during embryogenesis by acting in tissue remodeling in parallel with apoptosis. An increase of autophagic cells is, in fact, observed in the embryonic stages characterized by massive cell elimination as, for example, metamorphosis, glandular and fat body degeneration in insects and lumen formation in vertebrates.

Moreover, autophagic processes probably protect cells during metabolic stress and nutrient paucity that both occur during tissue remodeling such as in the storage tissues during fly metamorphosis or in the mature chondrocytes, which attempt to survive hypoxic conditions during mammalian osteogenesis. Developing cells have to adapt constantly and quickly to both intrinsic and environmental changes in order to survive and to differentiate.

In vertebrates, autophagy defects can be lethal for the animal if the mutated gene is involved in the early stages of development, or can lead to severe phenotypes if the mutation affects later stages. The varying phenotypes of mice harboring mutations in the different *atg* genes may reflect intrinsic differences in the developmental effects of genes that function at different stages of autophagy. On the other hand, autophagosome maturation is regulated by families of factors which are redundant for a long period of embryonic life, but which play a crucial role throughout later stages of development and immediately after birth. Moreover, the possibility that autophagy genes might play autophagy-independent roles cannot be excluded in order to explain the evident variability of phenotypes observed.

THE ROLE OF AUTOPHAGY IN THE DEVELOPMENT OF THE NERVOUS SYSTEM

The Biogenesis of the Nervous System

The nervous system originates from the ectodermal germ layer, one of the three primitive embryonic tissues.[51] During the process of neurulation, a specific portion of the dorsal ectoderm called the neural plate gives rise to the neural tube that will form the brain and spinal cord.[52] Failure to close the neural tube is the cause of severe to lethal defects in humans; the lack of closure of the tube's anterior end results in a anencephaly, while the failure to close the posterior part of the tube causes "spina bifida," the severity of which is related to the extent of the neural part exposed. It has been observed that human neural tube closure depends on a complex interplay between genetic and environmental factors. Among others, the genes *Pax3*, *sonic hedgehog* (*Shh*) and *open brain* are crucial for a proper formation of the neural tube, although dietary factors also play an important role in this regard.

Once closed, the neural tube undergoes differentiation phenomena leading to the generation of a mature nervous system. The ballooning of the early brain and its rapid expansion are caused by the positive pressure exerted on the walls of the neural tube by the fluid forming within. Subsequently, along the antero-posterior axis, the primary vesicles further differentiate into telencephalon, which goes on to originate the cerebral hemispheres (cerebrum), olfactory lobes and hippocampus and diencephalon which then form the hypothalamus, thalamus and retina. The caudal hindbrain originates pons and cerebellum and the *medulla oblungata* posteriorly. The entire hindbrain develops a segmental pattern essential for the specification of the nerve position along the body. The different segments, called rhombomeres, subsequently specify different ganglia, whose axons form nerves.

Besides following an antero-posterior axis, the nervous system also develops along a dorsal-ventral axis.[52] The dorso-ventral polarity of the neural tube is determined by the surrounding tissues; signals coming from epidermis are responsible for the specification of the tube's dorsal region whereas signals from the notochord specify the ventral part. Analogously, Sonic hedgehog protein (Shh), initially expressed and secreted by the notochord, induces the nearby cells of the neural tube to differentiate into the floor plate. The cells of the floor plate start, in turn, to express Shh, thus generating a protein gradient along the ventral-dorsal axis.

Both adult cerebrum and cerebellum are organized into interacting layers which present a typical composition in neuronal and glial cells. As the spatial organization is critical for the proper brain function, each new cell has to migrate and differentiate in the proper way and place.[53] In the adult brain of vertebrates and in some invertebrates, neurogenesis still occurs in specific areas of the brain dependently on the presence of mitotic stem cells. Both apoptosis and autophagy are involved in progenitor elimination, as the inhibition of these processes promotes long-term survival of these cells. During adult neurogenesis, stem cell differentiation seems to be strictly related to autophagy; it has been observed that glioma progenitor cells are characterized by a lower autophagic activity in comparison with neural stem cells. The reduced autophagy level in glioma progenitor cells, which may be due to the loss of *PTEN* gene, is correlated with an impairment of their differentiation; stimulating autophagy with rapamycin treatment promotes glial differentiation.[54]

Autophagy Genes Involved in Neurogenesis

As mentioned above, CNS development is strongly influenced by the autophagic process. Data in recent literature show that the expression of some autophagy-regulating genes is crucial in different steps of neurogenesis from the early neurulation steps to the maturation of specific neuronal cells, as illustrated below and in Table 1.

Atg5- or *Atg7*-deficient mice die a few hours after birth because they are unable to survive the physiological perinatal starvation. The suckling incapability observed in these mice may also suggest the presence of a defect in CNS functions. By using conditional knockout mouse aimed at studying the role of these genes in specific tissues, it has been neatly demonstrated that *Atg5* and *Atg7* genes are necessary for motor function, and that their deficiency results in a progressive accumulation of ubiquitinated proteins, starting from E15.5, that may be associated with eventual neurodegeneration.[55–57] Moreover, *Atg7* inactivation in cerebellar Purkinje cells results in degeneration of axon terminals, similar to what has been observed in Ulk1-deficient neural cells.[57] As above mentioned, it has been reported in both mouse and chick models that *Atg5* also plays a crucial role during the formation of the neural portion of the retina.[40]

Another example of the role played by autophagy in the CNS development comes from observations done on Ulk1/2-defective mice. Ulk1, the mammalian ortholog of the yeast Atg1, is a protein kinase involved in the early steps of autophagosome formation by complexing with and activating other Atg proteins.[58] In addition to the above-mentioned role of Ulk1 in *C. elegans* development, it has been seen that the inactivation of *Ulk1* by retroviral infection of a dominant negative mutant in immature cerebellar granule cells and by RNA silencing in spinal

Table 1. Autophagy deficient mouse models associated with embryonic phenotype in the CNS.

Protein	Role of autophagy	Embryonic pattern expression	Type of mutation	Embryonic phenotype in the CNS	Reference(s)
Atg5	Autophagosome elongation	Expressed by e11.0 in neuronal and glial precursors	Conditional allele: $Atg5^{tm1Myok}$ combined w/Tg(Nestin-Cre)	Ubiquitin-positive inclusions in the brain of e15.0 embryos and postnatal mice; perinatal death	55
Atg7	Autophagosome elongation	Expressed by e10.5	Conditional allele: $Atg7^{tm1T\ 1.\ chi}$ combined w/Tg (Nestin-Cre)	Progressive deficits in motor function and degeneration of the axon terminal during postnatal brain development; perinatal death	56
Ulk1	Autophagosome nucleation	Transcript expressed in embryonic cerebellum, cerebral cortex, olfactory bulb and hippocampus starting by e16	Retroviral GFP vector containing K46R and K39T dominant negative Ulk1 cDNAs expressed in cultured granule cells Ulk1 and Ulk2 shRNA inpSUPER vector system expressed in embryonic DRG neurons	Impairment of axon outgrowth and branching in cultured granule cells and sensory neurons	59, 60
Ambra1	Autophagosome nucleation	Detectable by e7.0 in the nervous sytem to become ubiquitous later on	Gene trap allele: $Ambra1^{Gt(pGT1.8geo)Fcec}$ The gene trap insertion causes a 459 aa C-terminal truncation leading to a fusion transcript	Excess of cell proliferation and cell death in embryonic brain; neural tube defects; embryonic lethality	62

neurons results in the impairment of axon outgrowth.[59,60] The formation of the neuronal circuits depends, in fact, on the outgrowth of neurites between neuronal populations, such as granule and Purkinje neurons in cerebellum. To make connections, each developing neural cell extends its own axon toward the target cell by means of a structure called the "growth cone." The growth cone integrates signals coming from the surrounding environment and from cells and guides the axon in the proper direction. It has been observed that Ulk1 starts to be expressed at E16 in cerebellar granule cells of mammals and that both Ulk1 and Ulk2 proteins localize to vesicular structures in the growth cones of mouse spinal sensory neurons.[60] The genetic silencing of both Ulk1 and Ulk2 leads to excessive axon arborization and impairment of axon elongation. In this sense, Ulk1/2 proteins regulate filopodia extension and branching of sensory axons. The forced expression of the dominant negative form of Ulk1 also impairs the differentiation of cerebellar neurons.[60]

Moreover, Ulk1/2 are implicated in endocytosis and vesicle trafficking mechanisms, as the genetic downregulation of these genes causes a block of endocytosis of nerve growth factor.[40,60] Recent studies carried out in Drosophila indicate for UNC51/Atg1 fly homolog a role in vesicles transport along axons by regulating the interaction between synaptic vesicles and kinesin motor complex.[61] UNC51/Atg1 fly homolog of Ulk1 interacts and phosphorylates UNC-76 — an adaptor protein of kinesin heavy chain — thus enabling it to bind the synaptic vesicle protein synaptotagmin-1. The UNC51-mediated phosphorylation of UNC-76 is essential for axonal transport.[61]

Based on this observation, it is emerging that Ulk1 activity at the growth cones is crucial for the formation of fibers and the proper progression of CNS development. Since the mechanisms by which Ulk1 controls the process as described above are not completely clarified, it remains to be shown whether the role of Ulk1 in neurodevelopment depends on its function in regulating autophagy.

Studies conducted on the autophagy regulator Ambra1 clearly demonstrate the influence of autophagy on CNS development in the early phases of neurogenesis. Although highly conserved in vertebrates, no orthologs of Ambra1 have been identified in lower eukaryotes. Ambra1 forms a multimolecular complex with Beclin1 which regulates the Beclin1-Vps34 kinase activity. Ambra1 promotes the early steps of autophagosome formation both in basal and induced autophagy.[6,62] Under basal conditions, the Ambra1/Beclin 1 complex is recruited to the cytoskeletal motor protein dynein machinery through direct binding to dynein light chains 1/2. When autophagy is induced by environmental stimuli, such as amino acid deprivation or rapamycin treatment, Ulk1 kinase promotes the dissociation of Ambra1/Beclin 1 from the dynein complex, thus enabling execution of the autophagic program.[65] During embryogenesis, Ambra1 is highly expressed in the CNS, especially in the neural plate, its expression later becoming ubiquitous. Mouse embryos devoid in *Ambra1* die at E16.5, showing pronounced defects in the CNS, such as hyperproliferation of the neuroepithelium, midbrain and hindbrain exencephaly and defective closure of the neural tube.[62] *Ambra1*-devoid neuroepithelium at E11.5 is highly enriched for ubiquitin, similarly to what is described for neurons devoid of *Atg5* at E15.5.[55,62,63] During the neural tube differentiation, the brain patterning of *Ambra1* mutant embryos appears dysregulated, with the morphogen Shh diffused in the tube floor plate rather than concentrated in the notochord, which is considered a bona fide signaling center in neurogenesis (Figure 1).[62,64] In normal development of CNS, the ventral part of the neural tube, highly enriched in Ambra1 protein, may be responsible for producing and exposing specific receptors on the membrane, responsible for the restriction of Shh espression or degradation in that specific area. Thus, Ambra1 might regulate the proper receptor/ligand interaction and determine a cellular makeover in neural tube areas more subject to embryogenetic rearrangements. The strong neuronal phenotype observed in Ambra1 mutants suggests the existence of a tissue-specific regulation of autophagy. A distinct regulation of the basal autophagosome formation could take place in specific tissues, due to the variety of signaling networks responsible for their differentiation during embryogenesis.

In the neural tissue of *Ambra1* mutants an excess of apoptotic cells is also present, thus demonstrating the existence of a strict relationship between autophagy, apoptosis and cell proliferation. Different from Ambra1–/– embryo, Beclin 1–/– embryo die at E7.5, precluding the study of its role in CNS development, even though these proteins are members of the same functional complex. Such a difference can be explained by the fact that Ambra1 is highly expressed only in the nervous system during early embryogenesis and only in vertebrates, whereas Beclin 1 is a ubiquitously expressed and evolutionarily conserved gene. In addition to autophagy, these two proteins may play specific roles in other cellular processes crucial to the development, such as cell proliferation and trafficking. In light of this evidence, Ambra1 seems to be essential for the control of cell proliferation and ensuring cell survival during CNS development.

CONCLUSION

Since its first definition in the early 1960s, by Christian De Duve and colleagues, autophagy has been often associated with cell death processes. The availability of a number of animal and cellular models of autophagy deficiency, and the identification of the major genes involved in its regulation, have opened up a new

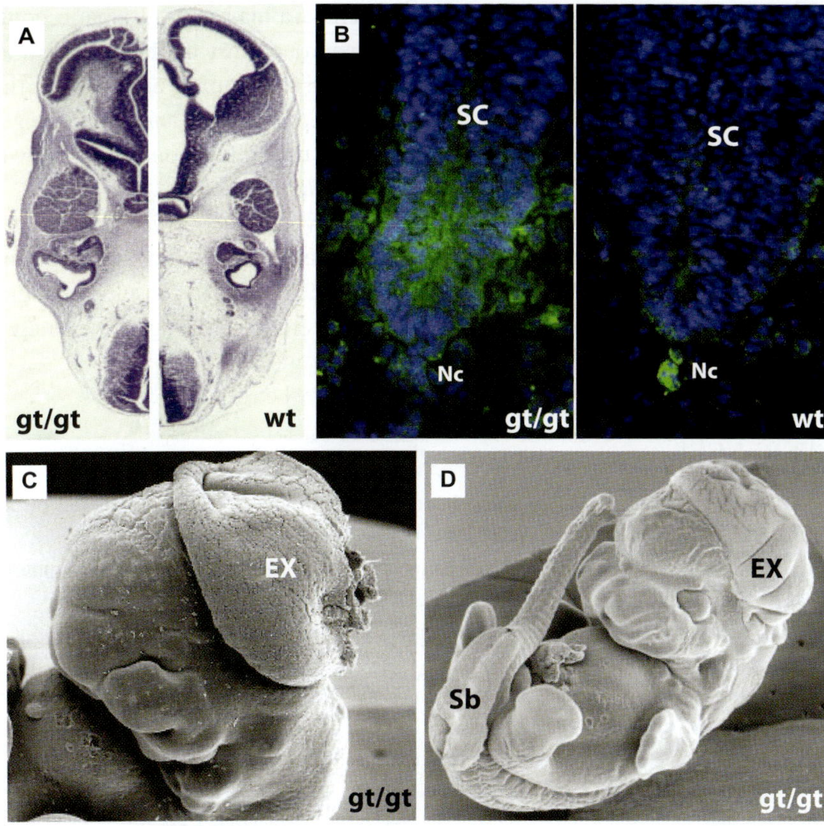

Figure 1. Neural phenotype of Ambra1 mutants. **(A)** Representative images showing a cross-section of E12.5 *Ambra1*$^{gt/gt}$ (gt/gt, left panel) and wt (wt, right panel) mouse embryos. Note the absence of a normal ventricular system, and the relative overgrowth of the neuroepithelium. **(B)** Immunohistochemical analysis of cross-sections from E9.5 *Ambra1*$^{gt/gt}$ (gt/gt) and wild type (wt) spinal cord (SC), showing a displaced expression of the signaling morphogen sonic hedgehog (Shh) in green. (Nc, notochord.) Electron scanning microscopy analysis of E11.5 (**C**) and E12.5 (**D**) *Ambra1*$^{gt/gt}$ embryos. Note the extensive midbrain/hindbrain exencephaly (EX). Some embryos also exhibit lumbosacral spinal bifida (Sb) (D). (Excerpts from Ref. 62, originally published in *Nature* by the Nature Publishing Group.)

avenue in the field of cell biology. Autophagy is often a pro-survival process and, when the cell dies *with* autophagy, it is likely that the autophagy-dependent cell defense is not robust enough to spare its life. In a developing embryonic cell, such a pro-survival activity can be turned on in a very efficient way to rapidly change its state. For example, this activity can help the cell to switch from proliferation to differentiation, to activate neuronal growth and regulate morphogen distribution, or to control its own death. The nervous system is a paradigm in this case, since any slight modifications of the cell fate may have dramatic consequences for the embryo. Indeed, developing neurons must modulate their morphology, position, and connectivity with the other cells during a very short time window. Autophagy has also been shown to cause dendritic[66] and axonal degeneration in primary neurons *in vitro*[67] and *in vivo*;[68] thus effects on central nervous system function may be more complex than on other tissues.

By analyzing a handful of gene phenotypes, we have observed the tip of an iceberg. Autophagy might well become an important new aspect of neurodevelopment. As Conrad Hal Waddington stated in 1956, "whatever the immediate operations of the genes turn out to be, they most certainly belong to the category of developmental processes."

REFERENCES

1. W. A. Dunn, Jr., Autophagy and related mechanisms of lysosome-mediated protein degradation, *Trends Cell Biol* **4**: 139–143 (1994).
2. C. He and D. J. Klionsky, Regulation mechanisms and signaling pathways of autophagy, *Annu Rev Genet* **43**: 67–93 (2009).
3. A. Longatti and S. A. Tooze, Vesicular trafficking and autophagosome formation, *Cell Death Differ* **16**: 956–965 (2009).
4. Y. Inoue and D. J. Klionsky, Regulation of macroautophagy in *Saccharomyces cerevisiae*, *Semin Cell Dev Biol* **21**: 664–670 (2010).
5. H. Nakatogawa, K. Suzuki, Y. Kamada and Y. Ohsumi, Dynamics and diversity in autophagy mechanisms: lessons from yeast, *Nat Rev Mol Cell Biol* **10**: 458–467 (2009).
6. C. He and B. Levine, The Beclin 1 interactome, *Curr Opin Cell Biol* **22**: 140–149 (2010).
7. J. H. Doelling, J. M. Walker, E. M. Friedman, A. R. Thompson and R. D. Vierstra, The APG8/12-activating enzyme APG7 is required for proper nutrient recycling and senescence in *Arabidopsis thaliana*, *J Biol Chem* **277**: 33105–33114 (2002).
8. F. Cecconi and B. Levine, The role of autophagy in mammalian development: cell makeover rather than cell death, *Dev Cell* **15**: 344–357 (2008).
9. S. Di Bartolomeo, F. Nazio and F. Cecconi, The role of autophagy during development in higher eukaryotes, *Traffic* **11**: 1280–1289 (2010).
10. R. C. Scott, O. Schuldiner and T. P. Neufeld, Role and regulation of starvation-induced autophagy in the Drosophila fat body, *Dev Cell* **7**: 167–178 (2004).
11. G. Juhasz, B. Erdi, M. Sass and T. P. Neufeld, Atg7-dependent autophagy promotes neuronal health, stress tolerance, and longevity but is dispensable for metamorphosis in Drosophila, *Genes Dev* **21**: 3061–3066 (2007).
12. K. King-Jones, J. P. Charles, G. Lam and C. S. Thummel, The ecdysone-induced DHR4 orphan nuclear receptor coordinates growth and maturation in Drosophila, *Cell* **121**: 773–784 (2005).

13. A. Garen, L. Kauvar and J. A. Lepesant, Roles of ecdysone in Drosophila development, *Proc Natl Acad Sci USA* **74**: 5099–5103 (1977).
14. T. E. Rusten, K. Lindmo, G. Juhasz, M. Sass, P. O. Seglen, A. Brech and H. Stenmark, Programmed autophagy in the Drosophila fat body is induced by ecdysone through regulation of the PI3K pathway, *Dev Cell* **7**: 179–192 (2004).
15. C. Y. Lee and E. H. Baehrecke, Steroid regulation of autophagic programmed cell death during development, *Development* **128**: 1443–1455 (2001).
16. D. Denton, B. Shravage, R. Simin, E. H. Baehrecke and S. Kumar, Larval midgut destruction in Drosophila: not dependent on caspases but suppressed by the loss of autophagy, *Autophagy* **6**: 163–165 (2010).
17. D. Denton, B. Shravage, R. Simin, K. Mills, D. L. Berry, E. H. Baehrecke and S. Kumar, Autophagy, not apoptosis, is essential for midgut cell death in Drosophila, *Curr Biol* **19**: 1741–1746 (2009).
18. D. L. Berry and E. H. Baehrecke, Growth arrest and autophagy are required for salivary gland cell degradation in Drosophila, *Cell* **131**: 1137–1148 (2007).
19. Y. Batlevi, D. N. Martin, U. B. Pandey, C. R. Simon, C. M. Powers, J. P. Taylor and E. H. Baehrecke, Dynein light chain 1 is required for autophagy, protein clearance, and cell death in Drosophila, *Proc Natl Acad Sci USA* **107**: 742–747 (2010).
20. Y. C. Hou, S. Chittaranjan, S. G. Barbosa, K. McCall and S. M. Gorski, Effector caspase Dcp-1 and IAP protein Bruce regulate starvation-induced autophagy during *Drosophila melanogaster* oogenesis, *J Cell Biol* **182**: 1127–1139 (2008).
21. K. McCall, Eggs over easy: cell death in the Drosophila ovary, *Dev Biol* **274**: 3–14 (2004).
22. N. Mohseni, S. C. McMillan, R. Chaudhary, J. Mok and B. H. Reed, Autophagy promotes caspase-dependent cell death during Drosophila development, *Autophagy* **5**: 329–338 (2009).
23. K. Takacs-Vellai, T. Vellai, A. Puoti, M. Passannante, C. Wicky, A. Streit, A. L. Kovacs and F. Muller, Inactivation of the autophagy gene bec-1 triggers apoptotic cell death in *C. elegans*, *Curr Biol* **15**: 1513–1517 (2005).
24. L. Roggo, V. Bernard, A. L. Kovacs, A. M. Rose, F. Savoy, M. Zetka, M. P. Wymann and F. Muller, Membrane transport in *Caenorhabditis elegans*: an essential role for VPS34 at the nuclear membrane, *EMBO J* **21**: 1673–1683 (2002).
25. A. Melendez, Z. Talloczy, M. Seaman, E. L. Eskelinen, D. H. Hall and B. Levine, Autophagy genes are essential for dauer development and life-span extension in *C. elegans*, *Science* **301**: 1387–1391 (2003).
26. S. L. McIntire, G. Garriga, J. White, D. Jacobson and H. R. Horvitz, Genes necessary for directed axonal elongation or fasciculation in *C. elegans*, *Neuron* **8**: 307–322 (1992).
27. D. L. Riddle, *Genetic and Environmental Regulation of Dauer Larva Development* (Cold Spring Harbor Laboratory Press, Cold Spring Harbor, NY, 1997).
28. Y. Zhang, L. Yan, Z. Zhou, P. Yang, E. Tian, K. Zhang, Y. Zhao, Z. Li, B. Song, J. Han, L. Miao and H. Zhang, SEPA-1 mediates the specific recognition and degradation of P granule components by autophagy in *C. elegans*, *Cell* **136**: 308–321 (2009).

29. Y. Tian, Z. Li, W. Hu, H. Ren, E. Tian, Y. Zhao, Q. Lu, X. Huang, P. Yang, X. Li, A. Kovacs, L. Yu and H. Zhang, *C. elegans* screen identifies autophagy genes specific to multicellular organisms, *Cell* **141**: 1042–1055 (2010).
30. A. M. Rowland, J. E. Richmond, J. G. Olsen, D. H. Hall and B. A. Bamber, Presynaptic terminals independently regulate synaptic clustering and autophagy of GABAA receptors in *Caenorhabditis elegans*, *J Neurosci* **26**: 1711–1720 (2006).
31. D. C. Rubinsztein, The roles of intracellular protein-degradation pathways in neurodegeneration, *Nature* **443**: 780–786 (2006).
32. B. A. Bamber and A. M. Rowland, Shaping cellular form and function by autophagy, *Autophagy* **2**: 247–249 (2006).
33. S. Tsukamoto, A. Kuma, M. Murakami, C. Kishi, A. Yamamoto and N. Mizushima, Autophagy is essential for preimplantation development of mouse embryos, *Science* **321**: 117–120 (2008).
34. A. Kuma, M. Hatano, M. Matsui, A. Yamamoto, H. Nakaya, T. Yoshimori, Y. Ohsumi, T. Tokuhisa and N. Mizushima, The role of autophagy during the early neonatal starvation period, *Nature* **432**: 1032–1036 (2004).
35. N. Hosokawa, Y. Hara and N. Mizushima, Generation of cell lines with tetracycline-regulated autophagy and a role for autophagy in controlling cell size, *FEBS Lett* **580**: 2623–2629 (2006).
36. M. L. Stitzel and G. Seydoux, Regulation of the oocyte-to-zygote transition, *Science* **316**: 407–408 (2007).
37. E. Coucouvanis and G. R. Martin, Signals for death and survival: a two-step mechanism for cavitation in the vertebrate embryo, *Cell* **83**: 279–287 (1995).
38. Z. Yue, S. Jin, C. Yang, A. J. Levine and N. Heintz, Beclin 1, an autophagy gene essential for early embryonic development, is a haploinsufficient tumor suppressor, *Proc Natl Acad Sci USA* **100**: 15077–15082 (2003).
39. X. Qu, Z. Zou, Q. Sun, K. Luby-Phelps, P. Cheng, R. N. Hogan, C. Gilpin and B. Levine, Autophagy gene-dependent clearance of apoptotic cells during embryonic development, *Cell* **128**: 931–946 (2007).
40. M. A. Mellen, E. J. de la Rosa and P. Boya, The autophagic machinery is necessary for removal of cell corpses from the developing retinal neuroepithelium, *Cell Death Differ* **15**: 1279–1290 (2008).
41. A. Nakai, O. Yamaguchi, T. Takeda, Y. Higuchi, S. Hikoso, M. Taniike, *et al*, The role of autophagy in cardiomyocytes in the basal state and in response to hemodynamic stress, *Nat Med* **13**: 619–624 (2007).
42. M. Mortensen and A. K. Simon, Nonredundant role of Atg7 in mitochondrial clearance during erythroid development, *Autophagy* **6**: 423–425 (2010).
43. H. I. Roach, T. Aigner and J. B. Kouri, Chondroptosis: a variant of apoptotic cell death in chondrocytes?, *Apoptosis* **9**: 265–277 (2004).
44. I. M. Shapiro, C. S. Adams, T. Freeman and V. Srinivas, Fate of the hypertrophic chondrocyte: microenvironmental perspectives on apoptosis and survival in the epiphyseal growth plate, *Birth Defects Res C Embryo Today* **75**: 330–339 (2005).

45. V. Srinivas and I. M. Shapiro, Chondrocytes embedded in the epiphyseal growth plates of long bones undergo autophagy prior to the induction of osteogenesis, *Autophagy* **2**: 215–216 (2006).
46. N. Hirokawa, Y. Tanaka, Y. Okada and S. Takeda, Nodal flow and the generation of left-right asymmetry, *Cell* **125**: 33–45 (2006).
47. C. Liang, P. Feng, B. Ku, I. Dotan, D. Canaani, B. H. Oh and J. U. Jung, Autophagic and tumour suppressor activity of a novel Beclin1-binding protein UVRAG, *Nat Cell Biol* **8**: 688–699 (2006).
48. A. Iida, M. Emi, R. Matsuoka, E. Hiratsuka, K. Okui, H. Ohashi, J. Inazawa, Y. Fukushima, T. Imai and Y. Nakamura, Identification of a gene disrupted by inv(11) (q13.5;q25) in a patient with left-right axis malformation, *Hum Genet* **106**: 277–287 (2000).
49. C. He and A. Orvedahl, 2007 keystone symposium on autophagy in health and disease, *Autophagy* **3**: 527–536 (2007).
50. C. Liang, D. Sir, S. Lee, J. H. Ou and J. U. Jung, Beyond autophagy: the role of UVRAG in membrane trafficking, *Autophagy* **4**: 817–820 (2008).
51. S. F. Gilbert, *Developmental Biology* (Sinauer Associates Inc., MA Sunderland, MA, US, 2000), pp. 379–395.
52. J. L. Smith and G. C. Schoenwolf, Neurulation: coming to closure, *Trends Neurosci* **20**: 510–517 (1997).
53. A. Chenn and S. K. McConnell, Cleavage orientation and the asymmetric inheritance of Notch1 immunoreactivity in mammalian neurogenesis, *Cell* **82**: 631–641 (1995).
54. Y. Zhao, Q. Huang, J. Yang, M. Lou, A. Wang, J. Dong, Z. Qin and T. Zhang, Autophagy impairment inhibits differentiation of glioma stem/progenitor cells, *Brain Res* **1313**: 250–258 (2010).
55. T. Hara, K. Nakamura, M. Matsui, A. Yamamoto, Y. Nakahara, R. Suzuki-Migishima, M. Yokoyama, K. Mishima, I. Saito, H. Okano and N. Mizushima, Suppression of basal autophagy in neural cells causes neurodegenerative disease in mice, *Nature* **441**: 885–889 (2006).
56. M. Komatsu and E. Kominami, [Autophagic-lysosomal system: physiology and pathology], *Nihon Shinkei Seishin Yakurigaku Zasshi* **26**: 75–81 (2006).
57. M. Komatsu, Q. J. Wang, G. R. Holstein, V. L. Friedrich, Jr., J. Iwata, E. Kominami, B. T. Chait, K. Tanaka and Z. Yue, Essential role for autophagy protein Atg7 in the maintenance of axonal homeostasis and the prevention of axonal degeneration, *Proc Natl Acad Sci USA* **104**: 14489–14494 (2007).
58. Y. Kamada, T. Funakoshi, T. Shintani, K. Nagano, M. Ohsumi and Y. Ohsumi, Tor-mediated induction of autophagy via an Apg1 protein kinase complex, *J Cell Biol* **150**: 1507–1513 (2000).
59. T. Tomoda, J. H. Kim, C. Zhan and M. E. Hatten, Role of Unc51.1 and its binding partners in CNS axon outgrowth, *Genes Dev* **18**: 541–558 (2004).
60. X. Zhou, J. R. Babu, S. da Silva, Q. Shu, I. A. Graef, T. Oliver, T. Tomoda, T. Tani, M. W. Wooten and F. Wang, Unc-51-like kinase 1/2-mediated endocytic processes

regulate filopodia extension and branching of sensory axons, *Proc Natl Acad Sci USA* **104**: 5842–5847 (2007).

61. H. Toda, H. Mochizuki, R. Flores, 3rd, R. Josowitz, T. B. Krasieva, V. J. Lamorte, E. Suzuki, J. G. Gindhart, K. Furukubo-Tokunaga and T. Tomoda, UNC-51/ATG1 kinase regulates axonal transport by mediating motor-cargo assembly, *Genes Dev* **22**: 3292–3307 (2008).
62. G. M. Fimia, A. Stoykova, A. Romagnoli, L. Giunta, S. Di Bartolomeo, R. Nardacci, M. Corazzari, C. Fuoco, A. Ucar, P. Schwartz, P. Gruss, M. Piacentini, K. Chowdhury and F. Cecconi, Ambra1 regulates autophagy and development of the nervous system, *Nature* **447**: 1121–1125 (2007).
63. M. Komatsu, S. Waguri, T. Chiba, S. Murata, J. Iwata, I. Tanida, T. Ueno, M. Koike, Y. Uchiyama, E. Kominami and K. Tanaka, Loss of autophagy in the central nervous system causes neurodegeneration in mice, *Nature* **441**: 880–884 (2006).
64. L. Wilson and M. Maden, The mechanisms of dorsoventral patterning in the vertebrate neural tube, *Dev Biol* **282**: 1–13 (2005).
65. S. Di Bartolomeo, M. Corazzari, F. Nazio, S. Oliverio, G. Lisi, M. Antonioli, V. Pagliarini, S. Matteoni, C. Fuoco, L. Giunta, M. D'Amelio, R. Nardacci, A. Romagnoli, M. Piacentini, F. Cecconi and G. M. Fimia, The dynamic interaction of AMBRA1 with the dynein motor complex regulates mammalian autophagy, *J Cell Biol* **191**: 155–168 (2010).
66. S. J. Cherra, 3rd, S. M. Kulich, G. Uechi, M. Balasubramani, J. Mountzouris, B. W. Day and C. T. Chu, Regulation of the autophagy protein LC3 by phosphorylation, *J Cell Biol* **190**: 533–539 (2010).
67. Y. Yang, K. Fukui, T. Koike and X. Zheng, Induction of autophagy in neurite degeneration of mouse superior cervical ganglion neurons, *Eur J Neurosci* **26**: 2979–2988 (2007).
68. H. C. Cheng, S. R. Kim, T. F. Oo, T. Kareva, O. Yarygina, M. Rzhetskaya, C. Wang, M. During, Z. Talloczy, K. Tanaka, M. Komatsu, K. Kobayashi, H. Okano, N. Kholodilov and R. E. Burke, Akt suppresses retrograde degeneration of dopaminergic axons by inhibition of macroautophagy, *J Neurosci* **31**: 2125–2135 (2011).

CHAPTER 7

AUTOPHAGY IN THE RETINA: DEVELOPMENT, PHYSIOLOGY AND PATHOLOGY

Patricia Boya

ABSTRACT

The retina is a light-sensitive tissue in the vertebrate eye that detects and processes visual images. Given its accessibility within the central nervous system, the retina has become a classical model in which to study a range of processes, from neurodevelopmental biology to pathological conditions. In this chapter I review the different studies that have addressed the role of autophagy in retinal tissue from flies to human. I recapitulate the role of autophagy during retinal development, in the physiology of retinal cells, and in animal models of retinal degeneration and other diseases that have an associated retinal phenotype.

INTRODUCTION

The retina is a light-sensitive tissue in the vertebrate eye that detects and processes visual images. Lining the inner surface of the eye, the retina senses light and creates impulses that travel through the optic nerve to the brain. Given its accessibility within the central nervous system, the retina has become a classical model in which to study a range of processes, from neurodevelopmental biology to pathological conditions.

The retina is composed of multiple cell types arranged in a layered structure, as studied and described by Santiago Ramón y Cajal over 100 years ago (Figure 1). It contains three layers of nerve cell bodies and two layers of synapses. The outer nuclear layer (Figure 2) is composed by the nuclei of photoreceptors, which are light-sensitive cells of two main classes, rods and cones. These cells synapse in the outer plexiform layer with amacrine and bipolar cells whose nuclei form the inner nuclear layer. In the inner plexiform layer amacrine and bipolar cells connect with

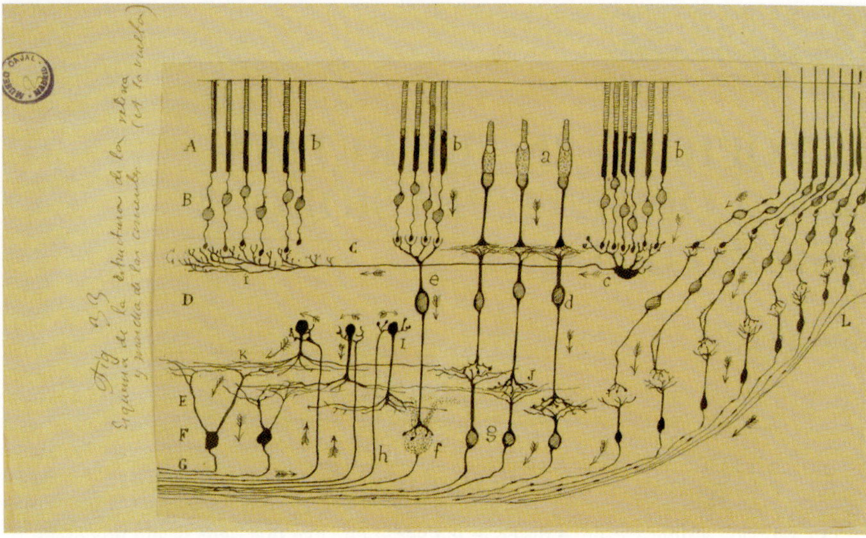

Figure 1. Schematic representation of the mammalian retina structure as depicted by Santiago Ramón y Cajal (1901). (A) Layer of rods and cones; (B) outer nuclear layer; (C) outer plexiform layer; (D) bipolar cell layer; (E) inner plexiform layer; (F) ganglion cell layer; (G) optic nerve; (L) central fossa. Reproduced with permission from the original drawing from the Cajal Institute, CSIC, Madrid, Spain.

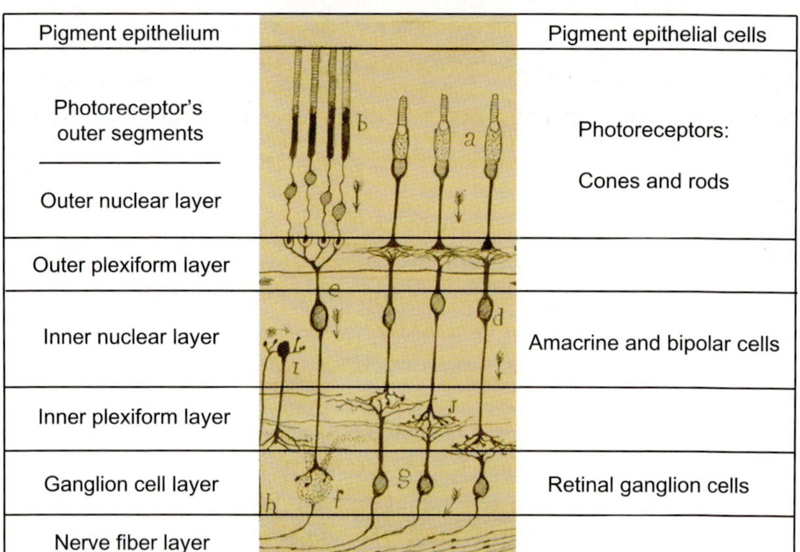

Figure 2. Mammalian retinal structure displaying the individual layers composed of cell bodies and synapses.

the retinal ganglion cells, whose axons form the optic nerve, which connects the retina to the brain. In addition to these nerve cells, the retina contains glia, known as Müller cells. Immediately outside the neuroretina, in close contact with the photoreceptor cells, are the retinal-pigmented epithelial (RPE) cells. These cells mediate the recycling of the outer segment and provide trophic support to photoreceptors.

Being part of the central nervous system, the retina is derived from the neural tube. Retinal development begins with the formation of the optic vesicle and optic cup, followed by the proliferation of the neuroepithelial cells and the generation of differentiated neurons and glia. These cells migrate to their respective retinal layers, from where they induce axon and dendrite growth. On reaching their targets, synaptogenesis occurs, followed by functional refinement.[1]

In this chapter I review the different studies that have addressed the roles of autophagy in retinal tissue from flies to human. First I describe the known role of autophagy during development, whose main function seems to be related to programmed cell death. Then I explore the few studies that have addressed the function of autophagy in the physiology of retinal cells, mainly studies on photoreceptors and RPE cells. Lastly, I summarize the studies performed in animal models of retinal degeneration and other diseases that have an associated retinal phenotype.

AUTOPHAGY AND CELL DEATH DURING RETINAL DEVELOPMENT

Autophagy plays a key role during development as evidenced by the embryonic lethality of several autophagy knockout animals like Beclin1 and Ambra1.[2,3] Although it is possible that these genes may have other roles besides autophagy these findings have increased research interest in the developmental role of autophagy. While many questions remain unanswered, it appears that autophagic processes are implicated in programmed cell death associated with the development of the nervous system.

The initial morphological classification of cell death types by Schweichel and Merker described the existence of autophagosomes in dying cells during development.[4] These authors studied vertebrate development in mice and rats by electron microscopy and classified cell death according to the role played by the lysosome in the process. Cell death was thus classified as: type I or apoptosis, type II or autophagic cell death and type III or necrosis. Though this classification system has since been replaced by more biochemically-defined categories, it demonstrates that autophagy (as defined by the morphological observation of autophagosomes) has been observed in developmental processes for more than half a century.

The retina is a well-characterized model system for the study of developmental processes. Several features make it a perfect model to study the regulation of cell differentiation and neurogenesis.[5] The retina is located outside of the brain and is thus readily accessible. It can be easily grown in organotypic cultures with defined culture medium[6] and, under such conditions, developmental processes can be accurately reproduced.[7] In addition, this highly controlled model system permits short-term pharmacological interventions.

Using retinal explants from postnatal rats, Linden and co-workers began to study the pathways implicated in programmed cell death during postnatal retinal development. In these studies, anisomycin-mediated inhibition of protein synthesis induced a form of cell death in the neuroblastic layer displaying both apoptotic and autophagic features.[8] The authors demonstrated the existence of several pathways that lead to cell demise. In one pathway simultaneous inhibition of caspases and autophagy failed to further reduce cell death in comparison to each inhibition alone, implicating autophagy and caspases in the same signaling pathway. In a parallel pathway, caspase-6-dependent mechanisms led to a TUNEL-negative form of cell death. Interestingly bongkrekic acid, an inhibitor of the mitochondrial adenine nucleotide translocator, completely inhibited cell death, whereas caspase inhibition and 3-MA (a classical inhibitor of autophagy) only partially blocked the process. Thus, in this parading cell death is dependent on mitochondrial events that trigger two different cell death cascades, one of which exhibits autophagic features.[9]

Other studies have highlighted a new role for autophagy during physiological cell death associated with retinal development. During the early phase of cell death, which affects proliferating cells,[10] autophagy is essential for cell corpse removal. Pharmacological inhibition of autophagy with 3-MA or wortmannin induces an accumulation of apoptotic cells in a well-defined area of the retinal neuroepithelium.[11] This area is the region of active differentiation where proliferating neuroblasts exit from their last cell cycle to differentiate to retinal ganglion cells (RGCs). Downregulation of autophagy blocks phosphatidylserine exposure on the surface of apoptotic cells (the principal signal for cell corpse removal) resulting in the accumulation of dead cells in the neuroepithelium. Interestingly, under these conditions autophagy is associated with the generation of energy, since 3-MA reduces the levels of ATP in organotypic retinal cultures, phenotype that is reversed by the addition of methylpyruvate, a permeable substrate for the TCA cycle that restores ATP levels, phosphatidylserine exposure and hence apoptotic cell removal (Figure 3).[11] This provision of energy for apoptotic cell clearance is however not universal, and in fact may represent an exception to the rule.[12] Indeed autophagy appears dispensable in the removal of apoptotic cells generated during other phases of retinal programmed cell death associated with chick retinal

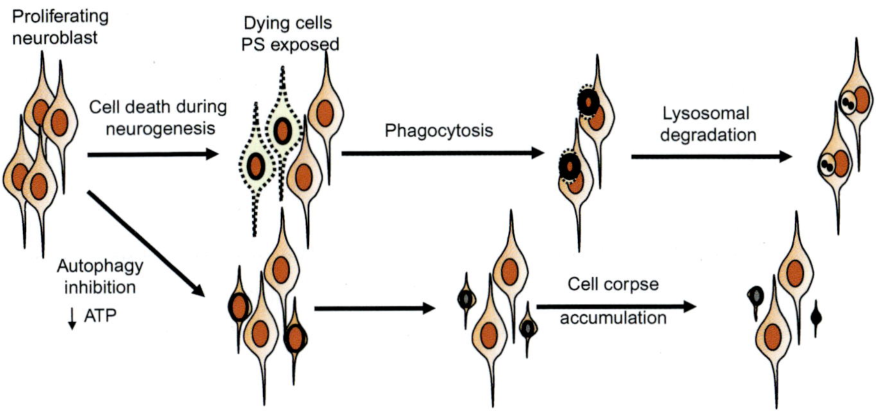

Figure 3. Autophagy is essential for cell corpse removal during programmed cell death associated with neurogenesis. Proliferating neuroblasts are affected by apoptosis and expose phosphatidylserine (PS) on their surface as a signal for neighboring cell phagocytosis. Engulfed apoptotic cells are degraded inside lysosomes. When autophagy is inhibited, ATP levels decrease, inhibiting PS expression on the surface of apoptotic cells. Dying cells accumulate in the neuroepithelium as they are no longer recognized as apoptotic.

development.[13] For example, apoptotic cell removal during morphogenetic cell death in the optic fissure is unaffected by autophagy inhibition.[13] Moreover, neurotrophic cell death of retinal ganglion cells at later developmental stages (E9) is also autophagy independent. The reasons for this differential requirement of autophagy for apoptotic cell removal are still unknown. They may be related to the energetic requirements of the cell, or may depend on the engulfing cell. In this regard at E5, microglia begin to infiltrate the retina and are only found around the optic fissure, while later at E9, their presence is detected throughout the retina.

Recently autophagy has been implicated in hyaloid vessel regression during ocular development.[14] The hyaloid vessel is a transient intraocular circulatory system that irrigates the developing eye before the formation of mature vessels in the inner retina. These hyaloid vessels undergo a complete regression as the retina matures through retinal vascularization.[15] During hyaloid vessel regression, punctate LC3 staining and caspase-3 activation is observed in endothelial cells. Ultrastructural analysis demonstrates that cells display autophagosomes and chromatin condensation. More importantly, ocular injection of rapamycin significantly enhances the regression of the hyaloid vessels, while the autophagy inhibitor 3-MA has the opposite effect. Thus autophagy contributes to endothelial cell death associated with hyaloid vessel regression during eye development although genetic evidence is lacking.

Together these findings demonstrate that autophagy participates in programmed cell death associated with retinal development (Table 1). In agreement, inhibition of autophagy produces a parallel reduction in programmed cell death. Moreover, autophagy-dependent cell death is implicated in hyaloid vessel regression during maturation of the eye. Lastly, autophagy regulates the engulfment of apoptotic cells during retinal neurogenesis by maintaining ATP levels for the developing retina. These experiments have been, however, performed in culture with pharmacological manipulations and thus further confirmation using genetic approaches are needed. In addition, whether autophagy plays additional roles in retinal development, e.g. in cell differentiation and remodeling, is unknown, and will require further studies using autophagy-deficient animals.

AUTOPHAGY IN PHYSIOLOGICAL FUNCTIONS OF THE RETINA

No alterations in the retina have been described in autophagy-deficient mutant mice that survive to adulthood. We found no structural differences in the retinal layers in Atg4B knockout mice as compared with wild-type animals. This was confirmed by the absence of any functional defects in these animals, as determined by electroretinogram (our unpublished observations). Some studies have addressed the role of autophagy in different retinal cell types — photoreceptors, cells of the retinal epithelium and ganglion cells. Here we describe the evidence outlining the physiologic role of autophagy in these different retinal cell types.

Autophagy in Photoreceptors

Photoreceptors are specialized nerve cells in the retina that capture light and trigger a chain of biological processes known as phototransduction. There are two types of photoreceptors, rods and cones, which are responsible for night vision and diurnal color vision, respectively. Retinal photoreceptors are very specialized cells. Although they possess several cytoprotective mechanisms, many pathological situations induce photoreceptor damage and cell death, causing subsequent vision defects. Perturbations in structural components, energy metabolism or in the phototransduction cascade increase the risk of photoreceptor cell death.[16]

The pioneering studies of Remé *et al.* demonstrated the presence of active autophagy in cone photoreceptors.[17] They studied the consequences of hibernation on the morphology of retinal cells. Using electron microscopy, they revealed

Table 1. Evidence of autophagy during development.

System	Species	Age	Cell type	Stimulus	Autophagy determination	Autophagy modulation	Phenotype after autophagy inhibition	Role of autophagy	References
Retinal explants	Rat	P1	Neuroblast	Protein synthesis inhibitors	EM	3-MA	Partial rescue from cell death	PCD	8
Retinal explants	Chicken	E5	Neuroblast	PCD associated with development	LC3 lipidation	3-MA	Cell corpse accumulation	Maintaining ATP levels	11,12
		E9	RGCs	Neurotrophic cell death	LC3 lipidation	3-MA	No phenotype		13
		E5	RGCs	Morphogenetic cell death	LC3 lipidation	3-MA	No phenotype		13
Retinal explants	Mouse	E13.5	RGCs	PCD associated with development		Ambra disruption	RGC mislocalization, increased cell death	Survival	Mellen, MA (unpublished)
Eye sections	Mouse	P4-P28	Hyaloid vessels	Hyaloid vessel regression	LC3 lipidation, LC3 IF, EM	3-MA/ Rapamycin	Hyaloid regression block/ acceleration	PCD in hyaloid cells	14

E, embryonic day; EM, electron microscopy; IF, immunofluorescence; P, postnatal day; PCD, programmed cell death; RGCs, retinal ganglion cells.

autophagosomes in the inner segment of rod photoreceptors in the retina under normal conditions.[17] The vacuoles were spherical and of 0.5 μm in diameter, the typical morphology of mammalian autophagosomes. In the ellipsoid area of the inner segment, a region of high mitochondrial density, evidence of mitochondrial autophagy (mitophagy) was observed. Modified cone morphology was seen in the retinas of hibernating animals, along with a reduced number of mitochondria and other organelles. In addition, an increased number of autophagosomes in comparison to non-hibernating animals was detected.[17] The increase in autophagosome number was slowly attenuated with the return of the animal to homeothermic conditions, and after five days cells displayed a similar appearance to those of non-hibernating animals. These data point to a putative role for retinal autophagy in the degradation of cell components, especially mitochondria, during hibernation.[17] Interestingly, the engulfed mitochondria in hibernating animals were round in shape, in contrast to the elongated free mitochondria. Based on these observations, the authors proposed the selective targeting of round versus elongated mitochondria. This conclusion is now supported by recent evidence indicating that mitochondrial autophagy could be influenced by mitochondrial morphology.[18]

Photoreceptor autophagy displays circadian rhythm related variations,[19] independent of animal activity and feeding/fasting status. Autophagy peaks at the middle of the light phase (in cycles of 12-hour dark/12-hour light) and is thought to play a role in the degradation of visual pigments, which enables adaptation to new light conditions.[20] An increase in the number of autophagosomes is observed when animals are switched to a brighter environment. As such, autophagy-mediated degradation has been proposed as a means of reducing the capacity for light absorption, thus protecting photoreceptor cells from light-induced damage.[20]

The only evidence that downregulation of autophagy induces degeneration in photoreceptor cells *in vivo* comes from fly studies. Like other insects *Drosophila melanogaster* possesses compound eyes, which are composed of around 800 omatidia. Each omatidium contains eight photoreceptor cells, which mediate phototransduction. The fly visual system is a very powerful tool to dissect the molecular mechanisms of neurodegeneration by genetic manipulation of selected genes in photoreceptor cells.[21] In light/dark cycles of 12-hour/12-hour, Atg7 mutant flies displayed a reduction in the number of rhabdomeres in comparison to wild-type animals. Interestingly this effect was only observed in old (20 days) but not in young flies.[22] In addition, TOR induction-induced photoreceptor loss in flies exposed to 12-hour/12-hour light/dark cycles; this effect was reduced in those maintained in darkness, indicating that degeneration is dependent on phototransduction. Interestingly, forced expression of Atg1 rescues this phenotype, suggesting that TOR induction followed by autophagy inhibition underlies this mechanism of cell degeneration.[22]

Autophagy in Retinal Pigment Epithelial Cells

Some *in vitro* studies have addressed the role of autophagy in cells of the RPE. These cells are found outside the photoreceptor layer of the neural retina, where they form a single cell layer over a basal membrane comprising one of the layers of Bruch's membrane. RPE cells are essential for the survival and correct functioning of photoreceptors. In the important process of photoreceptor disc shedding, RPE cells phagocytose aged discs of the outer segment of photoreceptor cells, which are subsequently degraded inside lysosomes. In addition, RPE cells sustain the retinal environment by supplying small molecules such as amino acids, ascorbic acid and D-glucose, while maintaining a tight barrier to choroidal blood-borne substances. RPE cell dysfunction is associated with age-related macular degeneration (AMD), one of the leading worldwide causes of blindness in the elderly.

RPE cells are postmitotic, suggesting high levels of autophagy and other detoxifying mechanisms. These cells are resistant to oxidative stress, demonstrating high autophagic flux, which enhances their ability to clear damaged cellular structures.[23] RPE cells are exposed to high levels of oxygen and metabolize a considerable amount of cell membranes from photoreceptors in the presence of constant light. These conditions favor the formation of the age pigment lipofuscin, resulting in RPE cell damage. Components of lipofuscin such as the A2-E fluorophore inhibit the lysosomal proton pump, increasing lysosomal pH and inhibiting lysosomal hydrolases.[24] *In vitro* incubation of immortalized human RPE cells with photoreceptor outer segments induces lipofuscin accumulation.[25] Peroxidated photoreceptor outer segments further exacerbate this process by inducing lysosomal dysfunction, which in turn blocks protein degradation by autophagy.[25] Interestingly, autophagy inhibition with 3-MA induces lipofuscin accumulation in ARPE-19 cells, a human transformed retinal epithelial cell line. Thus lysosomal alterations, lipofuscin accumulation and reduction of autophagy are closely related events that may contribute to RPE cell dysfunction.

The cytoprotective role of autophagy in RPE cells has been demonstrated in immortalized RPE cells (hTERT-RPE-1). Treatment with the death receptor TRAIL induces AMPK (AMP-activated protein kinase) activation, mTOR blockade and induction of autophagy, as demonstrated in h-TERT-RPE-1 GFP-LC3 cells.[26] Interestingly blockade of the pathway by RNA silencing of AMPK or TAK-1 (transforming growth factor-β-activating kinase-1) reduces autophagy activation and sensitizes cells to cell death. Thus, it is conceivable that upregulation of autophagy in RPE cells during pathologic situations may also play a similar cytoprotective role but *in vivo* evidences are still lacking.

Autophagy in Retinal Ganglion Cells

RGCs are the only projecting neurons of the retina. Their axons form the optic nerve and transmit visual information to the brain.[27] RGCs are the first neurons to differentiate in the retina, and undergo apoptotic death in a predictable time-dependent manner during development.[28] Ambra-1 deficient retinas show increased levels of programmed cell death during development, implicating autophagy as a pro-survival mechanism during RGC development. Whether autophagy is an essential process for the differentiation of these cells is currently unknown, but our preliminary studies indicate that Ambra-1 deficient retinas display alterations in the localization of RGCs in the retinas of E13.5 mice (Mellén MA, unpublished results).

In adult RGCs one of the conditions essential for survival is the presence of neurotrophic factors. Whether autophagy represents an essential mechanism for adult RGC survival is currently unknown. An *in vitro* study using RGC-5 cells (retinal ganglion cell clone-5) indicates that autophagy is upregulated after serum starvation. Autophagy induction appears to be cytoprotective as pharmacological inhibition of autophagy using 3-MA, Wortmannin or Bafilomycin A increase cell death after serum starvation in these cells.[29]

AUTOPHAGY IN RETINAL AND OPTIC NERVE PATHOLOGY

The role of autophagy in retinal diseases remains poorly understood, though recent studies have provided considerable evidence establishing this link. Here I review the latest evidence implicating autophagy in retinal degeneration.

Retinal Dystrophies

Retinal dystrophies are a subset of degenerative diseases that involve photoreceptor cell death, cones, and rods. Rod cell death is initially associated with loss of night vision. Progression of the disease leads to cell death of cone photoreceptors, resulting in complete blindness.[30] Retinitis pigmentosa (RP) is caused by mutations in at least 35 different genes that alter photoreceptor homeostasis leading to cell death. The most common mutation in RP is in the rhodopsin gene, which codes for a principal protein of the outer segment of the photoreceptor.

Several mouse models mimic the degenerative process of RP. These have been widely studied to determine the molecular mechanisms of cell death involved, and to search for new therapeutic strategies.[31] Cell death of cones and rods is characterized by apoptotic morphology and caspase activation, though cathepsins and calpains also have been implicated.[16] Some studies have described the activation of autophagy, as determined by LC3 lipidation, in models of RP.[32,33] Interestingly, the stimulus for induction and the kinetics of photoreceptor cell death varies between different RP animal models (Table 2). The *rd/rd or rd1* (from retinal degeneration) model displays a spontaneous mutation of the β-subunit of the rod-phosphodiesterase gene, resulting in high cGMP levels and calcium overload in photoreceptors. In contrast the *rds* (retinal degeneration slow) model displays a mutation in the peripherin gene that mediates structural defects in the outer segment of the rods, resulting in rod and cone degeneration. Lastly, in the light-induced degeneration model, oxidative stress is thought to induce cell death. Thus, a range of triggers for photoreceptor cell death have the ability to induce autophagosome formation. To date, however, it is unclear if autophagy plays an active role in cell demise[33] or represents a cytoprotective response. The later hypothesis may be supported by the fact that autophagy has been shown to maintain proper ATP levels.[32] In these models the mRNA expression for the glycolytic enzyme phosphofructokinase is upregulated before the photoreceptors start to degenerate.[32]

In an attempt to unravel the mechanism of photoreceptor cell death, a recent study performed microarray analyses at the peak of cone cell death in four different RP models. Several genes implicated in the insulin/mTOR pathway were upregulated in all four models, suggesting that retinas were suffering from starvation-like conditions.[34] Indeed, expression of HIF-1 and the glucose transporter GLUT1 were also upregulated at the protein level in all four models.[34] Interestingly macroautophagy, as determined by LC3 puncta using a viral vector coding for GFP-LC3, was unchanged in all models, though an increase in LAMP-2A levels suggested upregulation of chaperone-mediated autophagy. Mice treated systemically with insulin displayed a reduction in photoreceptor cell death, suggesting that cone starvation contributes to cell death in RP.[34] Interestingly, treatment with rapamycin reduced protein expression of selected opsin pigments that are also lost in mutant mice. Whether this mechanism is common to other retinal neurodegenerative situations remains unclear.

Table 2. Evidence of autophagy in *in vivo* models of retinal degeneration.

Human disease	Model	Species	Affected cell type	Autophagy determination	Autophagy modulation	Phenotype	Role of autophagy	Reference
Retinitis pigmentosa	*rd1*	Mouse	PR	LC3 lipidation			?	33
	rds	Mouse	PR	LC3 lipidation			?	
	Light-induced damage	Mouse	PR	LC3 lipidation			?	
Retinitis pigmentosa	*rd1*	Mouse	PR					34
	PDE beta	Mouse	PR		Rapamycin	Reduced expression of cone opsin		
	$Rho^{-/-}$	Mouse	PR					
	P23H	Mouse	PR					
Proteinopathies	Human huntingtin fragment over-expression, light	Fly	PR		Atg1 overex-pression	Rescue	Cytoprotective	22
	Rhodopsin mutations	Fly	PR		Atg1 overex-pression	Rescue	Cytoprotective	22
	Rhodopsin mutations	Fly	PR				Rhodopsin degradation, cytoprotective	71

(*Continued*)

Table 2. *(Continued)*

Human disease	Model	Species	Affected cell type	Autophagy determination	Autophagy modulation	Phenotype	Role of autophagy	Reference
Lysosomal storage diseases	Niemann–Pick type C1	Mouse	PR	LC3 IF, EM			?	72
	Infantile neuronal ceroid lipofuscinosis	Fly	PR				?	71
AMD	Aging	Mouse	RPE, PR	LC3 lipidation Atg12 IF, Atg12 blot			Drusen formation?	73
	Patients	Human	RPE, PR	Atg5 IF			Drusen formation?	36
Axonal damage	Optic nerve transection	Mouse	RGCs	LC3-GFP, LC3 lipidation EM, mRNA Atg genes	Rapamycin $Atg4B^{-/-}$ $Atg5^{flox}$ AAV2-Cre	Rescue Increased cell loss Increased cell loss	Cytoprotective	48
	Optic nerve crush	Rat	RGCs	EM, LC3 IF,	3-MA	Reduced axonal degeneration	Damage after acute axonal injury	49
ADOA	OPA1 deficiency	Mouse	RGCs	EM			Axonal damage?	60
Ischemia	IOP elevation	Rat	RGCs	LC3 lipidation			Role in cell death?	66
Ischemic preconditioning	Retinal artery ligation	Rat	RGCs	mRNA Atg3			Cytoprotective	67

ADOA, autosomal dominant optic atrophy; AMD, age dependent macular degeneration; IF, immunofluorescence; IOP, intraocular pressure; PR, photoreceptor; RGCs, retinal ganglion cells.

Age-Related Macular Degeneration

AMD is a bilateral retinopathy affecting the central region of the retina, the macula, and is a major cause of visual impairment in older adults. The macula is responsible for the majority of photopic vision and contains the fovea, an area with the highest concentration of cone photoreceptors, which mediate diurnal vision. The dry form of AMD, which is the most common, is associated with the formation of drusen, an extracellular accumulation of biological material in Bruch's membrane, located between the retinal pigment epithelium and the choroid. In addition, aged RPE cells accumulate lipofuscin, and are less able to degrade the outer segment, contributing to photoreceptor cell death.[35]

In cultured immortalized RPE cells mitochondrial damage induced by rotenone induced an increase in LC3 staining and the formation of Atg5-Atg12 conjugates.[36] Decreased activity of the lysosomal protein cathepsin D was also observed in ARPE-19 cells after rotenone treatment. More importantly, exosome and lysosomal markers (CD63 and Lamp2) were increased in these cells after mitochondrial damage, suggesting that exosome formation may represent a compensatory mechanism to eliminate toxic aggregates in cells in response to defective lysosomal degradation. Moreover it has been proposed that material from RPE cells acts as the source of drusen. In these cells mitochondrial damage and decreased DNA-repair mechanisms are observed as a function of age.[37] Hallmarks of autophagy such as Atg12 immunolabeling as well as increased exosomal markers are observed in RPE cells from old mice. Thus, the release of the engulfed material via exosomes to the extracellular space may play a role in drusen formation.[36] Interestingly Atg5 localizes with drusen in aged donor eyes and AMD eyes.[36] These findings could indicate that autophagy may participate in the formation of drusen in both aged mice and in patients with AMD.

Light-Induced Degeneration

Exposure to light in the presence of oxygen induces the generation of free radicals, resulting in oxidative damage to membranes and other cell components of photoreceptor cells. Prolonged exposure to bright light has been used as a model to study the molecular pathways mediating photoreceptor degeneration, implicating both caspase-dependent and -independent pathways. Light exposure is also a co-factor in many retinal diseases as observed in several animal models of photoreceptor degeneration.[38,39] Increased age aggravates the susceptibility of photoreceptors to light-induced degeneration, and autophagosomes with damaged mitochondria are observed in the degenerating cells. Whether this represents a harmful or a cytoprotective response remains to be determined.[40]

Glaucoma and Optic Nerve Damage

Glaucoma is a group of ocular diseases characterized by optic nerve damage and vision loss. It affects 70 million people and is the second leading cause of blindness in the world. The disease is characterized by a slow and progressive degeneration of RGCs and their axons. The most prominent risk factor for glaucoma is elevated intraocular pressure. Hence the majority of current treatments are aimed at lowering intraocular pressure. However, in other cases (e.g. normal tension glaucoma), damage to the optic nerve occurs despite intraocular pressures within the normal range, leading to RGC death.[41]

Primary open angle glaucoma is a slow and chronic form of glaucoma, which may be a consequence of defects in trabecular meshwork (TM) function. This damage appears to be the consequence of increased oxidative stress in TM cells. Porcine TM cells exposed to hyperoxic conditions display markers of oxidative stress and increased DNA and mitochondrial damage.[42] In addition these cells show accumulation of autofluorescent substrates such as lipofuscin. Accumulation of secondary lysosomes and autophagosomes and a reduced activity of lysosomal enzymes are also observed. These findings suggest that conditions of increased oxidative stress in TM cells results in compromised autophagosomal and lysosomal activity, resulting in the accumulation of deleterious substances like the age pigment liposfucin.[43]

RCGs are the final output neurons of the retina. Their axons form the optic nerve and transmit visual information to the brain.[27] As previously stated RCGs undergo apoptotic cell death in a predictable time-dependent manner during development and as a consequence of injury in glaucoma and other optic neuropathies.[27] RGC degeneration is often modeled by optic nerve damage, including optic nerve crush and optic nerve axotomy, which produces a slow and progressive degeneration of these central nervous system neurons.[44,45] The mechanisms of RGC death are still a subject of intense investigation, and several processes including growth factor deprivation, oxidative stress and excitotoxicity have been implicated in RGC degeneration both in glaucoma models and after optic nerve axotomy.[46] Accordingly, several neurotrophic factors and antioxidants can delay RGC loss after axotomy.[27,46]

In rats, transcription of several autophagy genes was upregulated in the retina after optic nerve transection. mRNA for Atg5, Atg7 and Atg12 was induced three days after axotomy and expression levels returned back to normal levels after seven days.[29] Moreover, isolated RGCs from axotomized retinas displayed increased expression of LC3-II and beclin expression by Western blot.[29] Based on these *in vitro* findings, the authors proposed a cytoprotective role for autophagy. Signs of increased autophagy after optic nerve axotomy are also observed in other species. Hallmarks of autophagy are observed in the ganglion cell layer following optic nerve transection in rabbits, including the presence of classical double-membrane autophagosomes and intraluminal accumulation of substances with cytosol-like appearance as well as signs of apoptosis, such as chromatin condensation.[47]

We have recently demonstrated that *in vivo* autophagy is indeed a protective mechanism for RGCs after axonal damage. Optic nerve transection (ONT) induces apoptotic RGC death that is delayed in animals after autophagy induction with rapamycin.[48] More importantly, Atg4B knockout animals, although do not manifest more damage under basal conditions in comparison to wild-type animals, show increased cell death after ONT.[48] Interestingly this cytoprotective role of autophagy seems to be specific for RGCs, since deletion of the autophagy regulator Atg5 only in RGCs, by intravitreal injection of an adeno-associated vector AAV serotype 2 (which specifically transduces RGCs), increases RGCs loss in comparison to animals injected with an empty vector.[48] This is the first *in vivo* study where both increasing and decreasing autophagy strategies have been performed in the same experimental setting. In addition, these data demonstrate for the first time a cytoprotective role of autophagy in the cell soma after axonal damage opening new perspectives for regeneration and glaucoma therapies.[49]

A recent live imaging study analyzed early events following optic nerve crush in rats.[50] The authors used an epifluorescence microscope to visualize the optic nerve for up to six hours after damage. Intravitreal injections of an adeno-associated vector coding for the fluorescent marker dsRed labeled RGCs and their axons. Thirty minutes after the lesion axonal swellings were observed and axonal degeneration began. Ultrastructural analysis demonstrated microtubule and neurofilament disorganization, mitochondrial abnormalities and LC3-labeled autophagosomes. In order to investigate the role of autophagy in axonal damage, 3-MA was injected intravitreally, resulting in a significant delay in axonal degeneration up to six hours postcrush. In addition, 3-MA injection attenuated the incidence of ultrastructural abnormalities.[51] Abnormal calcium homeostasis has been implicated in numerous diseases of the nervous system and calcium influx from the extracelullar space is associated with traumatic axonal damage. After optic nerve crush, an increase in intracellular calcium levels was observed in the axons of the optic nerve.[50] This increase mediated axonal degeneration, as calcium channel inhibitors delayed axonal degradation while incubation with the calcium ionophore A23187 accelerated axonal damage. Interestingly, decreased LC3 and p62 staining was observed after calcium channel inhibition, indicating that intra-axonal calcium levels may mediate the induction of autophagy. This study suggests a contributory role for autophagy in axonal degeneration shortly after optic nerve crush[51] as has been observed in other cell systems where autophagy appears to contribute to axonal degeneration or retraction.[52–54] Whether increased autophagic activity induces axonal damage by blocking intra-axonal transport requires further investigation.

Retinal ganglion cells are the most common target tissue affected by mitochondrial dysfunction. Optic atrophy and RGC death is a common hallmark of mitochondrial disorders such as Leber's hereditary optic neuropathy, autosomal dominant optic atrophy (ADOA) and others.[55] ADOA is the most common form of primary inherited optic neuropathy that is characterized by progressive loss of visual acuity. The fundamental pathology of this disease is a loss of RGCs.[56] OPA1 is a mitochondrial GTPase related to dynamin that is mutated in ADOA.[57,58] OPA1 is involved in mitochondria fusion, cristae remodeling and sequestration of cytochrome c inside mitochondria.[59] OPA1 deficiency decreases fusion, reduces mitochondria membrane potential and targets mitochondria for autophagy-dependent removal.[60] Mice mutant for the OPA1 protein manifest alterations in the optic nerve that resemble those seen during aging but appear earlier.[61] An increase in autophagosomes was observed in RGCs from OPA1$^{+/-}$ mice and the authors speculate that an overload in the autophagic process may underlie RGC death. Conversely, autophagy-dependent removal of damaged mitochondria may be essential for avoiding RGC death in the context of optic nerve damage. Indeed a double polymorphism in the OPA1 gene has been linked to normal tension glaucoma[62] and reduced expression of OPA1 is observed in a transcriptome array of ONT in rats and in glaucomatous mice.[63,64]

Ischemic Injury

Ischemia induces tissue damage after restriction of the blood supply. In the nervous system, ischemia is associated with excitotoxicity and neuronal cell loss. Cell death after ischemia/reperfusion occurs via caspase-dependent and -independent pathways, including autophagy and necroptosis. In a mouse model of ischemic injury in pyramidal neurons, autophagy was strongly upregulated during cell death, and ablation of the autophagy regulator Atg7 protected against neuron loss.[65]

Necroptosis is an alternative cell death pathway dependent of RIP-1 kinase characterized by necrotic features with autophagy activation that occurs in cells when apoptosis is blocked.[66] In a rat model of retinal ischemia, necroptosis with autophagic features was responsible for cell death in the ganglion cell layer and the inner cell layer. The necroptosis inhibitor Nec-1 (a cell-permeant inhibitor of RIP-1 kinase) reduced autophagy and cell death, and improved the electroretinograph response seven days after ischemia[67] indicating that in this settings autophagy and necroptosis inhibition protect from cell death after retinal ischemia.

Ischemic preconditioning can protect from subsequent ischemia through gene expression-dependent mechanism. In studies performed in rats, Atg13 mRNA expression was upregulated after ischemic preconditioning, but not during ischemia, suggesting that autophagy may mediate the cytoprotective effects of preconditioning during retinal ischemia.[68]

Retinal Proteinopathies

Autophagy is one of the primary mechanisms for the clearance and elimination of toxic and aggregated proteins in cells and its pharmacological upregulation ameliorates the symptoms of many pathological conditions associated with protein misfolding and aggregation.[69] Expression of N-terminal fragments of the human huntingtin protein containing 120 residues in *Drosophila* photoreceptor neurons induced age-dependent photoreceptor loss that was attenuated by overexpression of TSC1, TSC2, or Atg1. Thus, induction of autophagy by TOR inhibition can attenuate the cell loss associated with protein aggregation in fly models.[22]

The accumulation of rhodopsin-arrestin complexes in photoreceptor neurons induces light-dependent retinal degeneration that cannot be prevented by inhibition of caspases or other effectors of the apoptotic cascade.[70] These complexes also accumulate after deletion of the phospholipase C gene in *Drosophila*. Again, overexpression of TSC1, TSC2 or Atg1 blocked photoreceptor loss, indicating that autophagy may protect those cells against neurodegeneration.[22] This effect was attributed to the elimination of the toxic rhodopsin-arrestin complexes. These data indicate that autophagy protects neurons from toxic proteins by inducing their degradation.[22] Interestingly rhodopsin-arrestin complexes are not detected in aggregates, indicating that autophagy may be effective in eliminating soluble as well as aggregated protein complexes.

Efficient rhodopsin degradation is essential for photoreceptor cell viability, as mutations of the major rhodopsin Rh1 induce its accumulation and light-dependent retinal degeneration.[71] A recent report demonstrates that Rh1 is degraded by autophagy and inactivating mutations in *Drosophila* Atg8 induce neurodegeneration of photoreceptor cells due to Rh1 accumulation.[72] A similar phenotype is observed following inactivation of the phosphatidylserine decarboxylase (Psd) gene, which results in decreased phosphatidylethanolamine production, reduced autophagy and Rh1 accumulation in Rab7-positive vesicles. Interestingly these phenotypes were attenuated after Rab7 overexpression, pointing to an efficient crosstalk between autophagy and the endosomal pathway in preventing retinal neurodegeneration.[72]

Lysosomal Storage Diseases

Niemann–Pick type C1 is a lysosomal storage disease characterized by cholesterol accumulation due to mutations of Npc genes. This affection causes severe neurodegeneration, particularly in Purkinje neurons. Retinas of Npc1-deficient mice exhibit morphological and functional defects at two months of age. Retinal alterations include impaired retinal function in electroretinograms, accumulation of the autofluorescent age pigment lipofuscin, and photoreceptor degeneration.[73] Moreover, the membranes of photoreceptor outer segments appear disorganized, suggesting defective disk shedding. Interestingly a large number of autophagosomes and multilamellar bodies are observed in RGCs of Npc1 animals by electron microscopy, and LC3 staining is increased. Thus these data show that similar to Purkinje cells RGCs mutations in Npc1 induce alterations in intracellular lipid trafficking inducing autophagy.

Neuronal ceroid lipofuscinoses (NCLs) are a large group of autosomal recessive diseases that are classified by the age of onset. Infantile, late infantile, juvenile and adult NCLs involve vision loss, dementia and epilepsy.[74] The infantile form is produced by mutations in the palmitoyl-protein thioesterase 1 (Ppt1) gene that encodes a lysosomal protein of unknown function. Recent evidence indicates that a reduction in Ppt1 expression induces photoreceptor degeneration in the fly composite eye.[72] Interestingly, Ppt1 interacts with Psd and mutations in the latter also induce neurodegeneration via downregulation of autophagy. Further research is required to determine whether defects in autophagy may contribute to the pathology of INCL[72] and lysosomal storage diseases in general.

CONCLUDING REMARKS

Evidence from isolated cells, organotypic explants and retinas from wild-type and mutant mice outline manifold roles of autophagy in the retina. During development autophagy mainly participates in programmed cell death either as a cytoprotective response, for example by supplying ATP levels, or as an active mediator of cell demise. Under pathological circumstances, autophagy acts as a cytoprotective mechanism by degrading aggregated or mutated proteins that alter photoreceptor cell homeostasis leading to cell death, as in retinitis pigmentosa. In addition, proper autophagy seems to be essential for RPE cells as its inhibition influences the formation of lipofuscin and other age pigments that may disrupt the correct degradation and recycling of photoreceptor outer segments. Autophagy could also have a role in the formation of drusen, an extracelullar accumulation of material that has a pathogenic influence in retinal function during AMD. Following traumatic injury to the optic nerve, autophagy seems to play a role in axonal degeneration a short time after the injury although it protects retinal ganglion cells in the long term. What needs to be clarified most urgently is (1) whether autophagic flux is affected in the various pathological conditions and (2) whether lysosomes contribute to neurodegenerative functions as independent entities or whether they are simply reflecting changes in the autophagic process.

REFERENCES

1. F. J. Livesey and C. L. Cepko, Vertebrate neural cell-fate determination: lessons from the retina, *Nat Rev Neurosci* **2**: 109–118 (2001).
2. Z. Yue, S. Jin, C. Yang, A. J. Levine and N. Heintz, Beclin 1, an autophagy gene essential for early embryonic development, is a haploinsufficient tumor suppressor, *Proc Natl Acad Sci USA* **100**: 15077–15082 (2003).
3. G. Fimia, A. Stoykova, A. Romagnoli, L. Giunta, S. Di Bartolomeo, R. Nardacci, M. Corazzari, C. Fuoco, A. Ucar, P. Schwartz, P. Gruss, M. Piacentini, K. Chowdhury and F. Cecconi, Ambra1 regulates autophagy and development of the nervous system, *Nature* **447**: 1121–1125 (2007).
4. J. U. Schweichel and H. J. Merker, The morphology of various types of cell death in prenatal tissues, *Teratology* **7**: 253–266 (1973).
5. T. Marquardt and P. Gruss, Generating neuronal diversity in the retina: one for nearly all, *Trends Neurosci* **25**: 32–38 (2002).
6. E. J. de la Rosa, B. Diaz and F. de Pablo, Organoculture of the chick embryonic neuroretina, *Curr Top Dev Biol* **36**: 133–144 (1998).
7. B. Diaz, J. Serna, F. De Pablo and E. J. de la Rosa, *In vivo* regulation of cell death by embryonic (pro)insulin and the insulin receptor during early retinal neurogenesis, *Development* **127**: 1641–1649 (2000).
8. C. A. Guimaraes, M. Benchimol, G. P. Amarante-Mendes and R. Linden, Alternative programs of cell death in developing retinal tissue, *J Biol Chem* **278**: 41938–41946 (2003).
9. C. A. Guimaraes and R. Linden, Programmed cell deaths. Apoptosis and alternative death styles, *Eur J Biochem* **271**: 1638–1650 (2004).
10. E. J. de la Rosa and F. de Pablo, Cell death in early neural development: beyond the neurotrophic theory, *Trends Neurosci* **23**: 454–458 (2000).

11. M. A. Mellén, E. J. de la Rosa and P. Boya, The autophagic machinery is necessary for removal of cell corpses from the developing retinal neuroepithelium, *Cell Death Differ* **15**: 1279–1290 (2008).
12. P. Boya, M. A. Mellén and E. J. de la Rosa, How autophagy is related to programmed cell death during the development of the nervous system, *Biochem Soc Trans* **36**: 813–817 (2008).
13. M. A. Mellén, E. J. de la Rosa and P. Boya, Autophagy is not universally required for phosphatidyl-serine exposure and apoptotic cell engulfment during neural development, *Autophagy* **5**: 964–972 (2009).
14. J. H. Kim, J. H. Kim, Y. S. Yu, J. Y. Mun and K. W. Kim, Autophagy-induced regression of hyaloid vessels in early ocular development, *Autophagy* **6**: 922–928 (2010).
15. M. Saint-Geniez and P. A. D'Amore, Development and pathology of the hyaloid, choroidal and retinal vasculature, *Int J Dev Biol* **48**: 1045–1058 (2004).
16. J. Sancho-Pelluz, B. Arango-Gonzalez, S. Kustermann, F. J. Romero, T. van Veen, E. Zrenner, P. Ekstrom and F. Paquet-Durand, Photoreceptor cell death mechanisms in inherited retinal degeneration, *Mol Neurobiol* **38**: 253–269 (2008).
17. C. E. Reme and R. W. Young, The effects of hibernation on cone visual cells in the ground squirrel, *Invest Ophthalmol Vis Sci* **16**: 815–840 (1977).
18. L. C. Gomes, G. Di Benedetto and L. Scorrano, During autophagy mitochondria elongate, are spared from degradation and sustain cell viability, *Nat Cell Biol* **13**: 589–598 (2011).
19. C. Remé, B. Aeberhard and M. Schoch, Circadian rhythms of autophagy and light responses of autophagy and disc shedding in the rat retina, *J Comp Physiol A Neuroethol Sensory Neural Behav Physiol* **156**: 669–677 (1985).
20. C. E. Remé, U. Wolfrum, C. Imsand, F. Hafezi and T. P. Williams, Photoreceptor autophagy: effects of light history on number and opsin content of degradative vacuoles, *Invest Ophthalmol Vis Sci* **40**: 2398–2404 (1999).
21. E. Knust, Photoreceptor morphogenesis and retinal degeneration: lessons from Drosophila, *Curr Opin Neurobiol* **17**: 541–547 (2007).
22. T. Wang, U. Lao and B. A. Edgar, TOR-mediated autophagy regulates cell death in Drosophila neurodegenerative disease, *J Cell Biol* **186**: 703–711 (2009).
23. T. Kurz, M. Karlsson, U. T. Brunk, S. E. Nilsson and C. Frennesson, ARPE-19 retinal pigment epithelial cells are highly resistant to oxidative stress and exercise strict control over their lysosomal redox-active iron, *Autophagy* **5**: 494–501 (2009).
24. M. Bergmann, F. Schutt, F. G. Holz and J. Kopitz, Inhibition of the ATP-driven proton pump in RPE lysosomes by the major lipofuscin fluorophore A2-E may contribute to the pathogenesis of age-related macular degeneration, *FASEB J* **18**: 562–564 (2004).
25. T. U. Krohne, N. K. Stratmann, J. Kopitz and F. G. Holz, Effects of lipid peroxidation products on lipofuscinogenesis and autophagy in human retinal pigment epithelial cells, *Exp Eye Res* **90**: 465–471 (2010).

26. G. Herrero-Martin, M. Hoyer-Hansen, C. Garcia-Garcia, C. Fumarola, T. Farkas, A. Lopez-Rivas and M. Jaattela, TAK1 activates AMPK-dependent cytoprotective autophagy in TRAIL-treated epithelial cells, *EMBO J* **28**: 677–685 (2009).
27. S. Isenmann, A. Kretz and A. Cellerino, Molecular determinants of retinal ganglion cell development, survival, and regeneration, *Prog Retin Eye Res* **22**: 483–543 (2003).
28. A. I. Valenciano, P. Boya and E. J. de la Rosa, Early neural cell death: numbers and cues from the developing neuroretina, *Int J Dev Biol* **53**: 1515–1528 (2009).
29. S. H. Kim, Y. Munemasa, J. M. Kwong, J. H. Ahn, S. Mareninov, L. K. Gordon, J. Caprioli and N. Piri, Activation of autophagy in retinal ganglion cells, *J Neurosci Res* **86**: 2943–2951 (2008).
30. S. Lev, Molecular aspects of retinal degenerative diseases, *Cell Mol Neurobiol* **21**: 575–589 (2001).
31. C. Dalke and J. Graw, Mouse mutants as models for congenital retinal disorders, *Exp Eye Res* **81**: 503–512 (2005).
32. H. R. Lohr, K. Kuntchithapautham, A. K. Sharma and B. Rohrer, Multiple, parallel cellular suicide mechanisms participate in photoreceptor cell death, *Exp Eye Res* **83**: 380–389 (2006).
33. K. Kunchithapautham and B. Rohrer, Autophagy is one of the multiple mechanisms active in photoreceptor degeneration, *Autophagy* **3**: 65–66 (2007).
34. C. Punzo, K. Kornacker and C. L. Cepko, Stimulation of the insulin/mTOR pathway delays cone death in a mouse model of retinitis pigmentosa, *Nat Neurosci* **12**: 44–52 (2009).
35. S. C. Finnemann, L. W. Leung and E. Rodriguez-Boulan, The lipofuscin component A2E selectively inhibits phagolysosomal degradation of photoreceptor phospholipid by the retinal pigment epithelium, *Proc Natl Acad Sci USA* **99**: 3842–3847 (2002).
36. A. L. Wang, T. J. Lukas, M. Yuan, N. Du, M. O. Tso and A. H. Neufeld, Autophagy and exosomes in the aged retinal pigment epithelium: possible relevance to drusen formation and age-related macular degeneration, *PLoS One* **4**: e4160 (2009).
37. A. L. Wang, T. J. Lukas, M. Yuan and A. H. Neufeld, Increased mitochondrial DNA damage and down-regulation of DNA repair enzymes in aged rodent retinal pigment epithelium and choroid, *Mol Vis* **14**: 644–651 (2008).
38. D. K. Vaughan, S. F. Coulibaly, R. M. Darrow and D. T. Organisciak, A morphometric study of light-induced damage in transgenic rat models of retinitis pigmentosa, *Invest Ophthalmol Vis Sci* **44**: 848–855 (2003).
39. A. Wenzel, C. Grimm, M. Samardzija and C. Reme, Molecular mechanisms of light-induced photoreceptor apoptosis and neuroprotection for retinal degeneration, *Prog Retin Eye Res* **24**: 275–306 (2005).
40. K. I. Cho, H. Yi, A. Yeh, N. Tserentsoodol, L. Cuadrado, K. Searle, Y. Hao and P. A. Ferreira, Haploinsufficiency of RanBP2 is neuroprotective against light-elicited and age-dependent degeneration of photoreceptor neurons, *Cell Death Differ* **16**: 287–297 (2009).

41. J. Qu, D. Wang and C. L. Grosskreutz, Mechanisms of retinal ganglion cell injury and defense in glaucoma, *Exp Eye Res* **91**: 48–53 (2010).
42. P. B. Liton, Y. Lin, C. Luna, G. Li, P. Gonzalez and D. L. Epstein, Cultured porcine trabecular meshwork cells display altered lysosomal function when subjected to chronic oxidative stress, *Invest Ophthalmol Vis Sci* **49**: 3961–3969 (2008).
43. P. B. Liton, Y. Lin, P. Gonzalez and D. L. Epstein, Potential role of lysosomal dysfunction in the pathogenesis of primary open angle glaucoma, *Autophagy* **5**: 122–124 (2009).
44. M. P. Villegas-Perez, M. Vidal-Sanz, M. Rasminsky, G. M. Bray and A. J. Aguayo, Rapid and protracted phases of retinal ganglion-cell loss follow axotomy in the optic-nerve of adult-rats, *J Neurobiol* **24**: 23–36 (1993).
45. T. V. Johnson and S. I. Tomarev, Rodent models of glaucoma, *Brain Res Bull* **81**: 349–358 (2010).
46. G. Tezel, Oxidative stress in glaucomatous neurodegeneration: mechanisms and consequences, *Prog Retin Eye Res* **25**: 490–513 (2006).
47. F. Germain, E. Fernandez and P. de la Villa, Morphological signs of apoptosis in axotomized ganglion cells of the rabbit retina, *Neuroscience* **144**: 898–910 (2007).
48. N. Rodriguez-Muela, F. Germain, G. Marino, P. S. Fitze and P. Boya, Autophagy promotes survival of retinal ganglion cells after optic nerve axotomy in mice, *Cell Death Differ* **19**: 162–169 (2012).
49. N. Rodriguez-Muela and P. Boya, Axonal damage, autophagy and neuronal survival, *Autophagy* **8**: 286–288 (2012).
50. J. Knoferle, J. C. Koch, T. Ostendorf, U. Michel, V. Planchamp, P. Vutova, L. Tonges, C. Stadelmann, W. Bruck, M. Bahr and P. Lingor, Mechanisms of acute axonal degeneration in the optic nerve *in vivo*, *Proc Natl Acad Sci USA* **107**: 6064–6069 (2010).
51. J. C. Koch, J. Knoferle, L. Tonges, T. Ostendorf, M. Bahr and P. Lingor, Acute axonal degeneration *in vivo* is attenuated by inhibition of autophagy in a calcium-dependent manner, *Autophagy* **6**: 658–659 (2010).
52. Y. Yang, K. Fukui, T. Koike and X. Zheng, Induction of autophagy in neurite degeneration of mouse superior cervical ganglion neurons, *Eur J Neurosci* **26**: 2979–2988 (2007).
53. M. Komatsu, Q. J. Wang, G. R. Holstein, V. L. Friedrich, Jr., J. Iwata, E. Kominami, B. T. Chait, K. Tanaka and Z. Yue, Essential role for autophagy protein Atg7 in the maintenance of axonal homeostasis and the prevention of axonal degeneration, *Proc Natl Acad Sci USA* **104**: 14489–14494 (2007).
54. E. D. Plowey, S. J. Cherra, 3rd, Y. J. Liu and C. T. Chu, Role of autophagy in G2019S-LRRK2-associated neurite shortening in differentiated SH-SY5Y cells, *J Neurochem* **105**: 1048–1056 (2008).
55. V. Carelli, C. La Morgia, M. L. Valentino, P. Barboni, F. N. Ross-Cisneros and A. A. Sadun, Retinal ganglion cell neurodegeneration in mitochondrial inherited disorders, *Biochim Biophys Acta* **1787**: 518–528 (2009).

56. P. Heiduschka, S. Schnichels, N. Fuhrmann, S. Hofmeister, U. Schraermeyer, B. Wissinger and M. V. Alavi, Electrophysiological and histologic assessment of retinal ganglion cell fate in a mouse model for OPA1-associated autosomal dominant optic atrophy, *Invest Ophthalmol Vis Sci* **51**: 1424–1431 (2010).
57. C. Delettre, G. Lenaers, J. M. Griffoin, N. Gigarel, C. Lorenzo, P. Belenguer, L. Pelloquin, J. Grosgeorge, C. Turc-Carel, E. Perret, C. Astarie-Dequeker, L. Lasquellec, B. Arnaud, B. Ducommun, J. Kaplan and C. P. Hamel, Nuclear gene OPA1, encoding a mitochondrial dynamin-related protein, is mutated in dominant optic atrophy, *Nat Genet* **26**: 207–210 (2000).
58. C. Alexander, M. Votruba, U. E. Pesch, D. L. Thiselton, S. Mayer, A. Moore, M. Rodriguez, U. Kellner, B. Leo-Kottler, G. Auburger, S. S. Bhattacharya and B. Wissinger, OPA1, encoding a dynamin-related GTPase, is mutated in autosomal dominant optic atrophy linked to chromosome 3q28, *Nat Genet* **26**: 211–215 (2000).
59. C. Frezza, S. Cipolat, O. Martins de Brito, M. Micaroni, G. V. Beznoussenko, T. Rudka, D. Bartoli, R. S. Polishuck, N. N. Danial, B. De Strooper and L. Scorrano, OPA1 controls apoptotic cristae remodeling independently from mitochondrial fusion, *Cell* **126**: 177–189 (2006).
60. G. Twig, A. Elorza, A. J. Molina, H. Mohamed, J. D. Wikstrom, G. Walzer, L. Stiles, S. E. Haigh, S. Katz, G. Las, J. Alroy, M. Wu, B. F. Py, J. Yuan, J. T. Deeney, B. E. Corkey and O. S. Shirihai, Fission and selective fusion govern mitochondrial segregation and elimination by autophagy, *EMBO J* **27**: 433–446 (2008).
61. K. E. White, V. J. Davies, V. E. Hogan, M. J. Piechota, P. P. Nichols, D. M. Turnbull and M. Votruba, OPA1 deficiency associated with increased autophagy in retinal ganglion cells in a murine model of dominant optic atrophy, *Invest Ophthalmol Vis Sci* **50**: 2567–2571 (2009).
62. T. Aung, L. Ocaka, N. D. Ebenezer, A. G. Morris, M. Krawczak, D. L. Thiselton, C. Alexander, M. Votruba, G. Brice, A. H. Child, P. J. Francis, R. A. Hitchings, O. J. Lehmann and S. S. Bhattacharya, A major marker for normal tension glaucoma: association with polymorphisms in the OPA1 gene, *Hum Genet* **110**: 52–56 (2002).
63. M. Agudo, M. C. Perez-Marin, U. Lonngren, P. Sobrado, A. Conesa, I. Canovas, M. Salinas-Navarro, J. Miralles-Imperial, F. Hallbook and M. Vidal-Sanz, Time course profiling of the retinal transcriptome after optic nerve transection and optic nerve crush, *Mol Vis* **14**: 1050–1063 (2008).
64. W. K. Ju, K. Y. Kim, J. D. Lindsey, M. Angert, K. X. Duong-Polk, R. T. Scott, J. J. Kim, I. Kukhmazov, M. H. Ellisman, G. A. Perkins and R. N. Weinreb, Intraocular pressure elevation induces mitochondrial fission and triggers OPA1 release in glaucomatous optic nerve, *Invest Ophthalmol Vis Sci* **49**: 4903–4911 (2008).
65. M. Koike, M. Shibata, M. Tadakoshi, K. Gotoh, M. Komatsu, S. Waguri, N. Kawahara, K. Kuida, S. Nagata, E. Kominami, K. Tanaka and Y. Uchiyama, Inhibition of autophagy prevents hippocampal pyramidal neuron death after hypoxic-ischemic injury, *Am J Pathol* **172**: 454–469 (2008).

66. J. Hitomi, D. E. Christofferson, A. Ng, J. Yao, A. Degterev, R. J. Xavier and J. Yuan, Identification of a molecular signaling network that regulates a cellular necrotic cell death pathway, *Cell* **135**: 1311–1323 (2008).
67. D. M. Rosenbaum, A. Degterev, J. David, P. S. Rosenbaum, S. Roth, J. C. Grotta, G. D. Cuny, J. Yuan and S. I. Savitz, Necroptosis, a novel form of caspase-independent cell death, contributes to neuronal damage in a retinal ischemia-reperfusion injury model, *J Neurosci Res* **88**: 1569–1576 (2010).
68. B. X. Wu, A. G. Darden, M. Laser, Y. Li, C. E. Crosson, E. S. Hazard, 3rd and J. X. Ma, The rat Apg3p/Aut1p homolog is upregulated by ischemic preconditioning in the retina, *Mol Vis* **12**: 1292–1302 (2006).
69. S. Sarkar, B. Ravikumar, R. A. Floto and D. C. Rubinsztein, Rapamycin and mTOR-independent autophagy inducers ameliorate toxicity of polyglutamine-expanded huntingtin and related proteinopathies, *Cell Death Differ* **16**: 46–56 (2009).
70. C. D. Hsu, M. A. Whaley, K. Frazer, D. A. Miller, K. A. Mitchell, S. M. Adams and J. E. O'Tousa, Limited role of developmental programmed cell death pathways in Drosophila norpA retinal degeneration, *J Neurosci* **24**: 500–507 (2004).
71. P. Kurada and J. E. O'Tousa, Retinal degeneration caused by dominant rhodopsin mutations in Drosophila, *Neuron* **14**: 571–579 (1995).
72. R. Midorikawa, M. Yamamoto-Hino, W. Awano, Y. Hinohara, E. Suzuki, R. Ueda and S. Goto, Autophagy-dependent rhodopsin degradation prevents retinal degeneration in Drosophila, *J Neurosci* **30**: 10703–10719 (2010).
73. T. Claudepierre, M. Paques, M. Simonutti, I. Buard, J. Sahel, R. A. Maue, S. Picaud and F. W. Pfrieger, Lack of Niemann-Pick type C1 induces age-related degeneration in the mouse retina, *Mol Cell Neurosci* **43**: 164–176 (2010).
74. K. E. Wisniewski, N. Zhong and M. Philippart, Pheno/genotypic correlations of neuronal ceroid lipofuscinoses, *Neurology* **57**: 576–581 (2001).

CHAPTER 8

GENETIC MOUSE MODELS FOR ELUCIDATION OF AUTOPHAGY-LYSOSOMAL SYSTEMS IN NEURONS UNDER PHYSIOLOGIC AND PATHOLOGIC CONDITIONS

Masaaki Komatsu, Masato Koike, Yoshinobu Ichimura and Yasuo Uchiyama

ABSTRACT

Several lines of autophagy-deficient mice have yielded important information on the role of autophagy in central nervous system pathophysiology. These model systems are reviewed with regard to the role of autophagy in lysosomal storage diseases and in relation to apoptotic and ischemic neuronal cell death. New directions in understanding selective autophagy, including the multifaceted roles of the autophagy interacting protein p62, are discussed.

INTRODUCTION

Proteolysis in eukaryotic cells can be separated into two major pathways, one mediated by the proteasome and the other by the lysosome. These pathways are considered to function independently, but recent data show that in some cases they operate in a coordinated fashion. The proteasome, in collaboration with the sophisticated ubiquitin system used for the selection of target proteins, plays crucial roles in the rapid, aggressive degradation of not only a variety of short-lived regulatory proteins, but also abnormal proteins with aberrant structures that should be eliminated from the cells in order to maintain cellular homeostasis.[1,2] On the other hand, the lysosome is an organelle that contains a diverse array of acid hydrolases. This structure whose internal environment is acidic (pH ranging from 5.0 to 5.5) is separated from the cytosol by a limiting membrane to prevent the random destruction of cellular components. In the lysosomal pathway, degradation of plasmalemmal and extracellular proteins is mediated by endocytosis, while that of cytoplasmic components is achieved through autophagy, a process divided into three classes of pathways: macroautophagy, microautophagy, and chaperone-mediated autophagy (CMA).[3]

Microautophagy is accompanied by membrane extensions of vacuoles/lysosomes as well as invagination, in which cytoplasmic components close to the vacuoles/lysosomes are sequestered and then degraded. CMA targets specific cytosolic proteins that are trapped by the heat shock cognate protein of 70 kDa (hsc70), which are translocated into the lysosomal lumen for rapid degradation through interaction with lysosome-associated membrane protein type-2A (LAMP-2A). Macroautophagy (hereafter referred to as autophagy) is the prototype of autophagy, in which isolation membranes/phagophores engulf cytoplasmic components to produce transiently double-membrane vesicles called autophagosomes. Fusion of autophagosomes with transporting vesicles from the trans-Golgi network (TGN) and/or with lysosomes triggers the breakdown of the inner autophagosomal membrane followed by degradation of the contents (Figure 1).

Since the discovery of *ATG* (autophagy-related) genes in the yeast *Saccharomyces cerevisiae*,[4] the autophagy field has expanded explosively. Subsequently, it has become apparent that autophagy is involved in multiple physiological functions in various species.[5,6] A role of induced autophagy is to supply amino acids in response to starvation, for new protein synthesis and an energy source. *Atg* mutants in *Saccharomyces cerevisiae* are sensitive to the lethal effect of nitrogen deprivation, while they do not exhibit any phenotypes in nutrient-rich conditions.[7] Moreover, autophagy is indispensable for the physiological phenomena that accompany starvation, such as fruit body formation of slime mold, dauer formation of *Caenorhabditis elegans*, and sporulation of *Saccharomyces cerevisiae*.[5] It also plays important roles during starvation in the mouse fetus when the supply of amino acids through the placenta is suddenly cut off,[8] and in the early development of fertilized eggs under low intake of nutrients from extracellular space.[9] Besides the importance of starvation-induced autophagy, growing lines of evidence have revealed that basal autophagy, which is constitutively active at low but distinct rates under physiological conditions, plays key roles in the maintenance of cellular metabolism as well as in the quality control of cellular components including organelles.

Although there have been numerous studies concerning basic autophagy and starvation-induced autophagy in peripheral tissue cells, no clear-cut *in vivo*

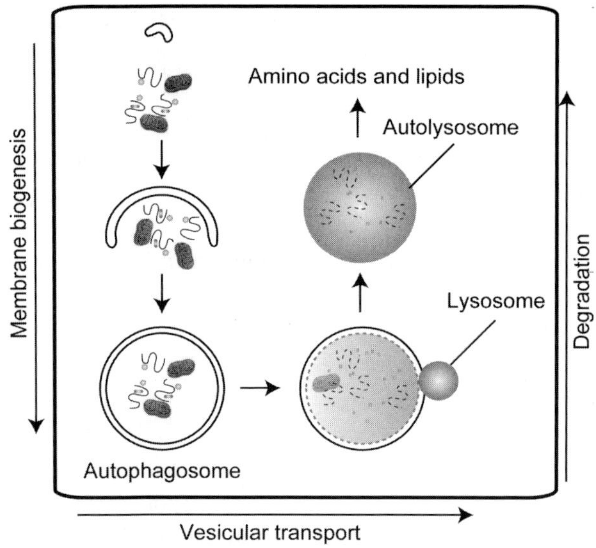

Figure 1. Schematic representation of the autophagy-lysosome system. An isolation membrane/phagophore elongates and surrounds cytoplasmic components including organelles to form the autophagosome (Membrane biogenesis). Fusion of autophagosomes with transporting vesicles from the trans-Golgi network with lysosomal enzymes and/or with lysosomes (Vesicular transport) triggers the breakdown of the inner autophagosomal membrane followed by degradation of the contents (Degradation).

study has shown autophagic responses in central nervous system (CNS) neurons after amino acids starvation. Nevertheless, the loss of Atg proteins such as Atg7 and Atg5 specifically in CNS tissue induces accumulation of ubiquitin in neurons and their death, indicating that basic autophagy is essential in the maintenance of metabolism in neurons. In addition to basic autophagy, it is very important to consider selective autophagy that is mediated by multifunctional proteins, p62 and Nbr1, together with ubiquitin and LC3, a mammalian homologue of yeast Atg8. In the first two subjects of this review, these general and specific issues of autophagy that have been obtained using various kinds of genetically manipulated animals will be involved. Moreover, neuropathological stresses such as brain ischemia, loss of lysosomal proteinases, and various neurodegenerative disorders are also known to induce autophagy in neuronal cells. In particular, it is very important to discuss on the issue of whether such autophagy induced by various disease-mediated stresses is neuroprotective or neurotoxic. Such disease-oriented autophagy will be reviewed in the last two subjects.

AUTOPHAGY DEFICIENCY IN MICE

Phenotypes of Tissue-Specific Autophagy-Deficient Mice

Mice deficient in autophagy die *in utero* (e.g. *Beclin1* knockout mice)[10,11] or die within 24 hours after birth due, at least in part, to deficient amino acid production (e.g. *Atg5*-knockout mice).[12] To investigate the physiological roles of constitutive autophagy, conditional *Atg5*- or *Atg7*- knockout mice and various tissue-specific *Atg*-knockout mice have been established and analyzed by several groups (Table 1). For example, liver-specific *Atg7*-deficient mice showed severe hepatomegaly accompanied by hepatocyte hypertrophy, resulting in severe liver injury.[12] Mice lacking *Atg5* or *Atg7* in the central nervous system exhibited behavioral deficits, such as abnormal limb-clasping reflexes and reduction of coordinated movement,[13,14] as well as massive neuronal loss in the cerebral and cerebellar cortices. Loss of *Atg5* in cardiac muscle caused cardiac hypertrophy, left ventricular dilatation and systolic dysfunction.[15] Muscle-specific deletion of *Atg7* resulted in profound muscle atrophy and an age-dependent decrease in skeletal muscle force.[16] Pancreatic B cell-specific *Atg7*-knockout mice exhibited degeneration of the islets and impaired glucose tolerance with reduced insulin secretion.[17,18]

Morphological Features Common in Various Autophagy-Deficient Cells

Electron microscopic analyses of mutant mice showed marked accumulation of swollen and deformed mitochondria in the mutant hepatocytes,[12] pancreatic B cells,[17,18] cardiac[15] and skeletal myocytes,[16] the appearance of concentric membranous structures consisting of endoplasmic reticulum or sarcoplasmic reticulum in mutant hepatocytes,[12] neuronal axons[19,20] and skeletal myocytes[16] and an increased number of peroxisomes in mutant hepatocytes.[12] In addition to the accumulation of aberrant organelles, histological analyses of tissues with defective autophagy showed the accumulation of polyubiquitylated proteins in almost all tissues, although the level varied from one region to another, forming inclusion bodies whose size and number increased with aging. Importantly, the mutant tissues, especially autophagy-deficient liver and skeletal muscle, seemed to have functionally intact proteasomes,[16,21] the impairment of which is generally known to cause abnormal ubiquitin-mediated proteolysis. Autophagy deficiency is considered to result in delays in the global turnover of cytoplasmic components, causing the accumulation of both misfolded proteins and dysregulated organelles followed by the formation of ubiquitin-positive aggregates together with accumulation of deformed organelles. In other words, basal autophagy acts as quality control machinery for cytoplasmic

Table 1. Phenotypes of *Atg*- and tissue-specific *Atg*-knockout mice.

Genotype (disrupted tissue)	Phenotypes	References
Atg3$^{-/-}$, *Atg5*$^{-/-}$, *Atg7*$^{-/-}$, *Atg9*$^{-/-}$, *Atg16L*$^{-/-}$	Neonate death and suckling defect.	8, 12, 29, 117, 118
Beclin1/Atg6$^{-/-}$	Embryonic lethal.	11
FIP200$^{-/-}$	Embryonic lethal.	119
ULK1$^{-/-}$	Embryonic lethal.	120
Atg4C$^{-/-}$	Viable and fertile.	121
LC3B$^{-/-}$	Viable and fertile.	122
GABARAP$^{-/-}$	Viable and fertile.	123
Atg5$^{flox/flox}$*:Nestin-Cre, Atg7*$^{flox/flox}$*:Nestin-Cre* (brain)	Neurodegeneration.	13, 14
FIP200$^{flox/flox}$*:Nestin-Cre* (brain)	Neurodegeneration.	124
Atg7$^{flox/flox}$*:Pcp2-Cre, Atg5*$^{flox/flox}$*:Pcp2-Cre* (Purkinje cells)	Axonal degeneration.	19, 20
Atg7$^{flox/flox}$*:Mx1-Cre, Atg7*$^{flox/flox}$*:Alb-Cre* (liver or hepatocyte)	Hepatomegaly and liver injury.	12, 21
Atg5$^{flox/flox}$*:MLCv2-Cre Atg5*$^{flox/flox}$*:MerCreMer* (cardiomyocyte)	Cardiac hypertrophy and dysfunction.	15
Atg5$^{flox/flox}$*:HSA-Cre* (skeletal muscle)	Atrophy of fast muscle.	125
Atg7$^{flox/flox}$*:Mlc1f-Cre* (skeletal muscle)	Muscle atrophy and weakness.	16
Atg7$^{flox/flox}$*:Rip-Cre* (pancreatic beta cell)	Diabetes.	17, 18
Atg5$^{flox/flox}$*:EL-Cre* (pancreatic acinar cell)	Loss of trypsinogen activation.	126
Atg5$^{flox/flox}$*:Podocin-Cre* (podocyte)	Glomerular disease.	127
Atg7$^{flox/flox}$*:Ap2-Cre* (adipocyte)	Reduced white adipose tissue due to the impaired adipogensis.	128
Atg7$^{flox/flox}$*:Vav-Cre* (hematopoietic cell)	Lethal anemia.	129
Atg5$^{flox/flox}$*:Lck-Cre, Atg7*$^{flox/flox}$*:Lck-Cre* (T cell)	Decrease in T cell survival.	130
Atg5$^{flox/flox}$*:CD19-Cre* (B cell)	Inefficient B cell development.	131
Atg5$^{flox/flox}$*:CD11c-Cre* (dendritic cell)	Defect of MHC II antigen presentation.	132
Atg5$^{flox/flox}$*:Zp3-Cre* (oocyte)	Defective early development.	9

components and is crucial for the homeostasis of various mitotic cells. Disruption of these functions of constitutive autophagy may contribute to diseases.

SELECTIVE AUTOPHAGY SUBSTRATE, p62

Selective Autophagy

The autophagosome is a globular organelle with a diameter of approximately 1 μm and a volume of 0.5×10^{-18} m^3, and it can entrap hundreds of thousands of proteins of average molecular weight. Therefore, autophagy has generally been considered a bulk protein degradation machinery that does not select the protein load. However, growing lines of evidence have shed light on other autophagy modes that selectively degrade aggregated proteins, unnecessary or damaged organelles, and invading bacterial cells. Receptor proteins (e.g. Atg19p, Atg32p, PpAtg30, SEPA-1, p62 and Nbr1) interacting with the autophagosomal membrane protein, Atg8p/LC3, and/or an adaptor protein (i.e. Atg11p) tethers potential cargo to Atg proteins essential for autophagosome formation to promote this selectivity (Table 2). Identification of these proteins imply important roles for selective autophagy in various physiological situations.

Table 2. Receptor proteins in selective autophagy.

Cargos	Cargo tags	Receptors	Atg proteins recognized by receptors	Atg8 (LC3)-recognition sequence
CVT-complex	Ape1-propeptide	Atg19p	Atg8p and Atg11p	WXXL (Atg19)
Damaged mitochondria		Atg32p	Atg8p and Atg11p	WXXI (Atg32)
Depolarized mitochondria		Nix/Bnip3L	GABARAP	WXXL (Nix/Bnip3L)
Uncoupled mitochondria	VDAC ubiquitinated by Parkin?	p62	LC3	WXXL (p62)
Peroxisomes	PpPex14p and PpPex3p	PpAtg30p	PpAtg11p	–
P granules	PGL-3	SEPA-1	LGG-1 (Atg8 homologue)	YQEL? (SEPA-1)
Ubiquitinated aggregates	Ubiquitin-chain	p62	LC3	WXXL (p62)
Ubiquitinated aggregates	Ubiquitin-chain	Nbr1	LC3, p62?	EDYIIL (Nbr1)

Atg8 Lipidation System Essential for Autophagy

Taking advantage of yeast genetics and biochemical analysis, more than 30 *ATG* genes have thus far been identified.[4,22] The molecular function of most Atg proteins is typically conserved in eukaryotes from yeasts to mammals. Among them, the ubiquitin-like protein, Atg8p, whose structure is comprised of a ubiquitin fold and two N-terminal α-helices, plays a crucial role in autophagosome formation.[23] The microtubule-associated protein 1 light chain 3 (LC3), a mammalian homolog of the yeast Atg8, is processed immediately after its synthesis by Atg4B, a cysteine protease, which exposes the C-terminal glycine residue (named LC3-I).[24] Thereafter, LC3-I is activated by Atg7 (E1) in an ATP-dependent manner and transferred to Atg3 (E2).[25,26] Subsequently, the C-terminal glycine of LC3-I conjugates to the amino group of phosphatidylethanolamine (PE) in an Atg12–Atg5•Atg16 dependent manner.[27] LC3-II is localized on the inner and outer membranes of the isolation membrane/phagophore, and is essential for membrane biogenesis and/or closure of the membrane.[28,29] LC3-II localized on the outer membrane is efficiently re-cleaved by Atg4B following the completion of autophagosome formation and recycled, whereas LC3-II present on the inner membrane is degraded together with other cellular constituents by lysosomal proteases.[30] Therefore, LC3 is frequently utilized as an autophagosome marker.[24,31]

LC3-Interacting Protein, P62

Following a comprehensive analysis of LC3-binding proteins, p62 was identified as an LC3-interacting protein.[21,32–34] Because LC3-II that is present on the inner membrane of the autophagosome is degraded, together with other cellular constituents, by lysosomal proteases, p62 trapped by LC3 is transported selectively into the autophagosome and the impaired autophagy is accompanied by the accumulation of p62.

p62 is a ubiquitously expressed cellular protein that is conserved in metazoa but not in plants and fungi, and mutations have been identified in Paget's disease in bone.[35] While it is postulated that p62 is a receptor for ubiquitinated proteins to deliver them selectively into the autophagosome (see below), it also functions as a signaling scaffold protein through interaction with various signaling proteins, such as TRAF6 and caspase-8 (Figure 2A).[36,37] Importantly, p62 has been identified as a component of inclusion bodies observed in human diseases, especially in neurodegenerative diseases (e.g. Alzheimer's disease, Parkinson's disease and amyotrophic lateral sclerosis) and in liver diseases (e.g. alcoholic hepatitis, hepatic steatosis and hepatocellular carcinoma).[38,39] It is interesting that the

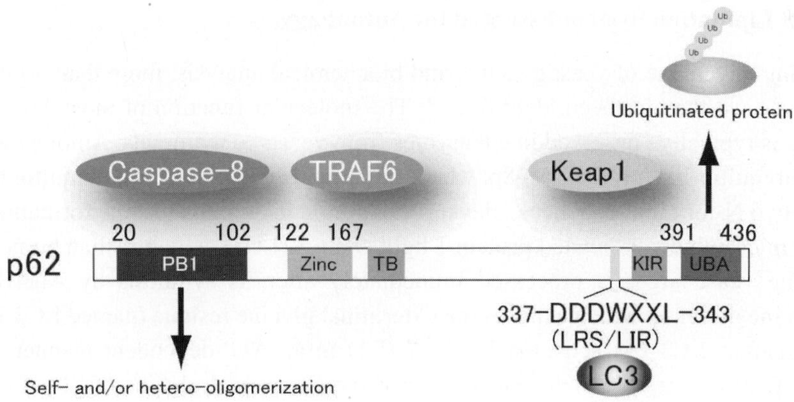

(A)

Figure 2A. Autophagy-specific substrate, p62. Domain structure of p62. PB1 mediates self-oligomerization. LC3 recognizes p62 through the LIR (LC3-interacting region)/LRS (LC3 recognition sequence). p62 interacts with Keap1 via the KIR (Keap1 interacting region). (PB1: Phox and Bem1p domain, Zinc: zinc finger domain, TB: TRAF6 binding (TB) domain, UBA: ubiquitin-associated domain.)

N-terminal PB1 domain exhibits self-aggregation, and the C-terminal ubiquitin-associated (UBA) domain can bind to ubiquitinated proteins (Figure 2A), implicating the involvement of p62 in the formation of disease-related inclusions.

Molecular Mechanism of Selective Autophagy

Autophagic degradation of p62 is severely attenuated by mutation of Phox and Bem1p (PB1) domain, which is responsible for self- and hetero-oligomerization.[33] Moreover, p62 is targeted to the autophagosome formation site on the ER in a PB1-dependent manner.[40] Furthermore, Kopito and his colleagues proposed that protein oligomerization drives autophagic substrate selection.[41] These studies indicate that oligomerization of p62 is a key event to become an autophagic substrate (Figure 2B). Furthermore, it has become evident that the LC3-interacting region, which is comprised of 11 amino acids in mouse p62 called LRS (LC3 recognition sequence), is essential for the autophagic degradation.[33,34,42,43] The sequence is almost identical to the previously reported LIR (LC3-interacting region) of human p62.[33,34,42,43] The LRS/LIR is composed of a highly conserved acidic cluster (DDD/DEE) and two hydrophobic amino acid motifs (WXXL/WXXV). The tertiary structure of the LC3-LRS complex, as determined by X-ray crystallography, indicates that the acidic cluster of LRS interacts electrostatically

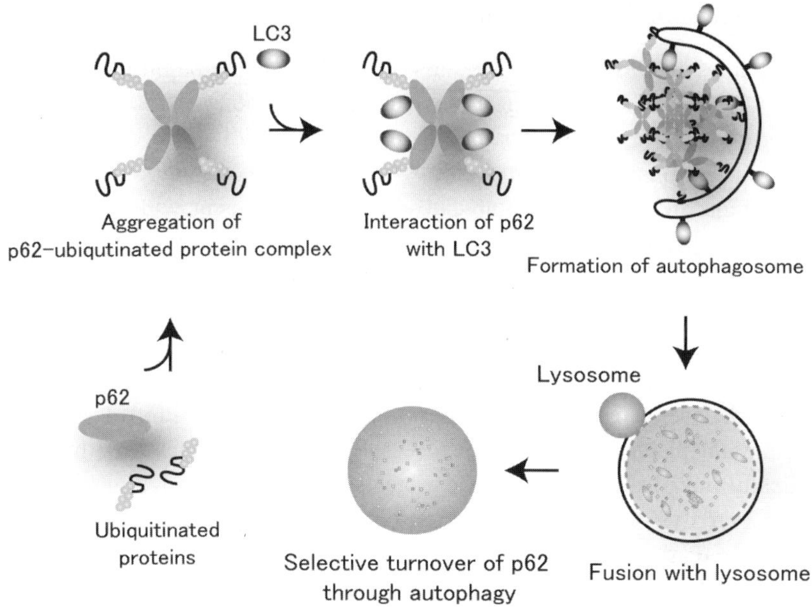

(B)

Figure 2B. A model of p62 as a receptor for ubiquitinated proteins. p62 interacts with polyubiquitinated proteins via the UBA domain. The complex is selectively sequestered in the autophagosomes through the interaction of LIR/LRS–LC3. This process is facilitated by self-oligomerization of p62 via the PB1 domains.

with the basic amino acids in the N-terminal α1 helix of LC3, and that the hydrophobic amino acids (WXXL) interact with two hydrophobic pockets formed on the molecular surface of the ubiquitin domain of LC3.[33,34,42,43] The latter hydrophobic pockets are conserved in Atg8 family proteins. Importantly, protein kinase A directly phosphorylates S12 in the α1 helix of LC3, implying that signaling pathways may be able to regulate LC3 interactions with other proteins.[44] It is of interest that the WXXL motif in Atg19p, a receptor of the cytoplasm to the vacuole targeting (Cvt) pathway, also interacts with the hydrophobic pockets on Atg8 for selective transport of Ape1.[42] Furthermore, the Atg32 and Nix/Bnip3L involved in selective degradation of the mitochondria through autophagy (mitophagy) also contain WXXL-like motifs to interact with Atg8, LC3 and GABARAP.[45,46] Therefore, the hydrophobic patch formed by β-sheets in the ubiquitin-fold core could be used as an important molecular face for substrate sorting in selective autophagy.

p62 as a Receptor for Ubiquitinated Proteins

Growing lines of evidence indicate that p62, together with ubiquitinated proteins, is transported into the autophagosomes, implying that p62 is a receptor for ubiquitinated proteins that augments their degradation in the lysosomes. Indeed, p62 and/or Nbr1, through cooperation with Alfy contribute to the autophagic degradation of ubiquitinated protein aggregates.[32,47,48] In agreement with these studies, almost all inclusions in autophagy-deficient cells are positive for ubiquitin, and p62 and/or Nbr1,[21,49] and loss of p62, greatly reduces the formation of ubiquitin inclusions in mice and flies defective in autophagy.[21,50] Moreover, p62 is also involved in the selective degradation of ubiquitinated organelles. Interestingly, recent reports have gradually revealed the relationship between Parkinson's disease and degenerated mitochondria turnover via autophagy. PINK1 and Parkin are important players in autosomal recessive Parkinsonism, as they have been genetically linked to a pathway that protects against progressive mitochondrial damage and dysfunction. Pink1, a mitochondrial kinase known to function upstream of Parkin, accumulates in the mitochondria upon treatment with an uncoupling agent, carbonyl cyanide m-chlorophenyl hydrazone (CCCP), and this accumulation is both necessary and sufficient for Parkin recruitment to the mitochondria.[51–54] Once Parkin is translocated from the cytoplasm to the mitochondria, the assembly of p62 onto the mitochondria is triggered by ubiquitination of VDAC1, which is a mitochondrial outer membrane protein.[55] Eventually, the mitochondria are enveloped selectively by autophagosomal membranes via p62-LC3 interaction. However, despite an essential role of p62 in clustering of depolarized mitochondria, additional mechanism for mitophagy might exist in cells, since elimination of mitochondria upon CCCP-treatment occurs even in *p62*-deficient cells.[56,57]

p62 as signaling hubs

Meanwhile, p62 is also involved in several signal transduction pathways through its ability to self-oligomerize.[36,37] In response to cytokine, p62 interacts with TRAF6 and then the complex oligomerizes in a p62-dependent manner, which enhances the ubiquitination of TRAF6. The resulting polyubiquitinated TRAF6 induces NF-κB activation.[58] DR4 and DR5, which are death receptors present on the cell surface, stimulate cell apoptosis through the death-inducing signaling complex (DISC). Caspase-8, which is part of the DISC, is ubiquitinated by a Cullin3-based ubiquitin ligase, and the ubiquitinated caspase-8 interacts with p62. Subsequently, the p62 promotes aggregation of the complex, leading to full

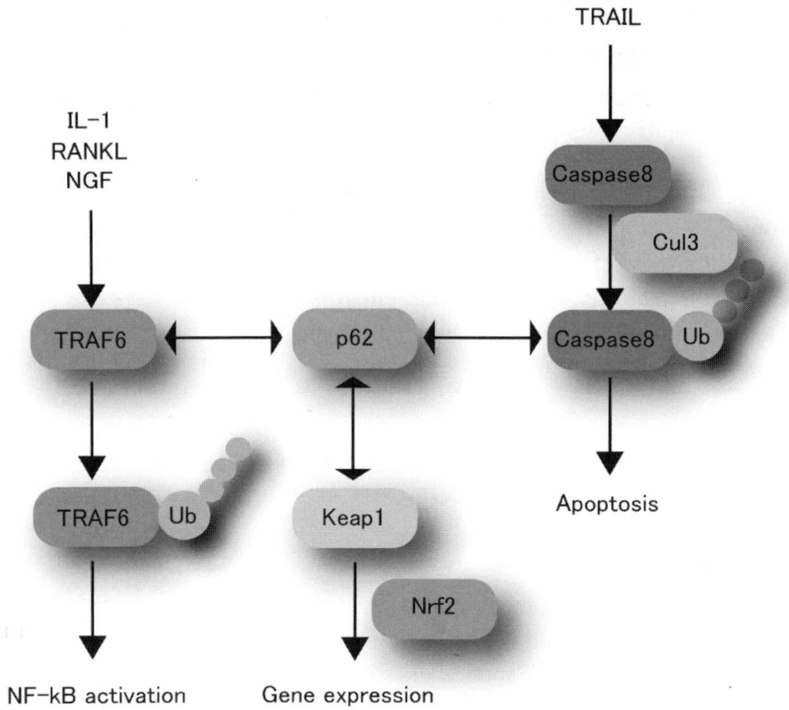

(C)

Figure 2C. The p62 signaling hub controls apoptosis, NF-κB, and stress response through the multiple interaction domains. Because the homeostatic level of p62 protein is strictly regulated by constitutive autophagy, autophagy can control the NF-κB pathway, apoptosis, and environmental stress response through the p62-degradation. (IL-1: interleukin-1, RANKL: receptor activator of NF-κB ligand, NGF: nerve growth factor, TRAF6: TNF receptor-associated factor 6, Ub: ubiquitin, NF-κB: nuclear factor-kappa B, Keap1: kelch-like ECH-associated protein 1, TRAIL: TNF related apoptosis inducing ligand, Nrf2: Nuclear factor erythroid 2-related factor 2, Cul3: cullin 3.)

activation of caspase-8 in order to execute cell death.[59] These results suggest that p62-positive aggregates function as signaling hubs that could determine whether cells survive, by activating the TRAF6-NF-κB pathway, or die by facilitating the aggregation of caspase-8 and the downstream effector caspases (Figure 2c). A novel system for the regulation of environmental stress by p62 was recently reported.[60–63] p62 interacts with the Nrf2-binding site on Keap1, a component of cullin3-based ubiquitin ligase for Nrf2.[62] The Nrf2 regulates the expression of cytoprotective genes, including antioxidant proteins and detoxification enzymes. Thus, overproduction of p62 or autophagy deficiency competes with the interaction between Nrf2 and Keap1, resulting in the stabilization of Nrf2 and transcriptional activation of Nrf2 target genes. Much like the cases of TRAF-6 and caspase-8, aggregation of p62 is needed for full activation of Nrf2.[62] Considering that the homeostatic level of p62 protein is strictly regulated by constitutive autophagy, autophagy can control the NF-κB pathway, apoptosis, and environmental stress response through the p62-degradation (Figure 2C). These defects might cause various diseases observed in tissue-specific autophagy-knockout mice. Actually, severe liver injury recognized in autophagy-deficient livers was suppressed by concomitant loss of p62.[21,32–24] Thus, there is a possibility that NF-κB pathway, apoptosis, and environmental stress response are dysregulated in human diseases accompanied by p62-accumulation, leading to the symptom onset.

AUTOPHAGY IN LYSOSOME STORAGE DISORDERS DUE TO CATHEPSIN DEFICIENCY

Lysosomal Storage Disorder

Lysosomes are the only organelle that contains various hydrolases that function in a highly acidic milieu, and degrades intracellular constituents, including organelles, into biologically active monomers such as amino acids that are recycled in the cells. Each lysosome in a cell contains more than 40 enzymes, although there are more than 70 enzymes that serve as lysosomal enzymes. The constituents of lysosomes are heterogeneous in quantity and quality, depending on the tissue or cell type. Lysosomal disorders are triggered when a particular enzyme shows insufficient activity, or is missing altogether. When this occurs, substrates of the enzyme deficit accumulate in the cells. Since neurons differ from most peripheral tissue cells, representing post-mitotic cells that function throughout life, severe mental and motor disorders are usually manifested.[64]

Neuronal Ceroid Lipofuscinoses (NCLs)

The most common inherited neurodegenerative disease in childhood is neuronal ceroid lipofuscinosis (NCL or CLN), which is the general name for a family of at least ten genetically separate neurodegenerative disorders characterized by a decline of mental and other capacities, epilepsy, and visual loss through retinal degeneration.[65] NCLs are categorized as a lysosomal storage disorder and are pathologically characterized by the accumulation of proteolipids, which may include subunit c of mitochondrial ATP synthase or sphingolipid activator proteins in the neurons.[66–69] NCLs are known to involve various types of the disease that differ from each other, depending upon the age of symptomatic onset, duration, and early onset of manifestations like seizures and blindness. Due to methodological progress, gene defects responsible for each type of NCL have been clarified and referred to as CLNs, in which ten causative genes have already been revealed.[65] However, the molecular cell biological aspect of each CLN gene product has not yet been clarified.

Cathepsin D Deficiency as CLN10

Through generation of cathepsin D (CD)-deficient mice, CD reportedly is involved in limited proteolysis rather than bulk proteolysis.[70] CD-deficient mice are born normally, but die at postnatal day (P) 26 ± 1 because of massive intestinal necrosis, thromboembolia, and lymphopenia. These mice manifest seizures and become blind after P20. It was found that the CNS neurons in CD-deficient mice demonstrate a new form of lysosomal accumulation disease with a phenotype resembling NCLs, in which subunit c of mitochondrial ATP synthase accumulates in the lysosomes of the affected neurons.[71–74] Typical morphological features of NCL neurons are granular osmiophilic deposits (GRODs) and fingerprint profiles that can also be seen in these mutant mouse neurons (Figures 3A to 3D). The fact that cathepsin B (CB) and subunit c are detected by immunoelectron microscopy in GRODs of CD-deficient neurons[71] suggest that the GRODs and fingerprint profiles are lysosomes. It has therefore been thought that the CD-deficient mouse is a novel model of NCLs that corresponds to the severe form (CLN10, OMIM:116840).[75,76]

Unlike CD in lysosomes, mice deficient in cathepsin B or cathepsin L (CB or CL) grow normally and become fertile. However, cathepsin L is known to be involved in the regulation of epithelial cell proliferation and differentiation in the skin and in bone turnover during normal development and in pathological states.[77,78] It is interesting that, when CB and CL are doubly deficit in mice, subunit c is detected as granules in the neuronal perikarya of the brain, while the localization pattern of LC3 is similar to that of subunit c in the neuronal perikarya and fibrous in the dendrites of cerebral and cerebellar cortical neurons.[73] Autophagosomes (double membrane-bound vacuoles (AV)) containing part of the cytoplasm are frequently detected by electron microscopy in the CNS neurons of CD-deficient (Figures 3B to 3D) and CB/CL-double deficient mice that are near the terminal stage of the disease (near P14).[73] It has been shown that AVs undergo stepwise maturation by fusing with endosomes and/or lysosomes.[79] Early and late AVs in this maturation process are distinguishable by electron microscopy:[80,81] nascent or immature AVs encircled by endoplasmic reticulum (ER)-like

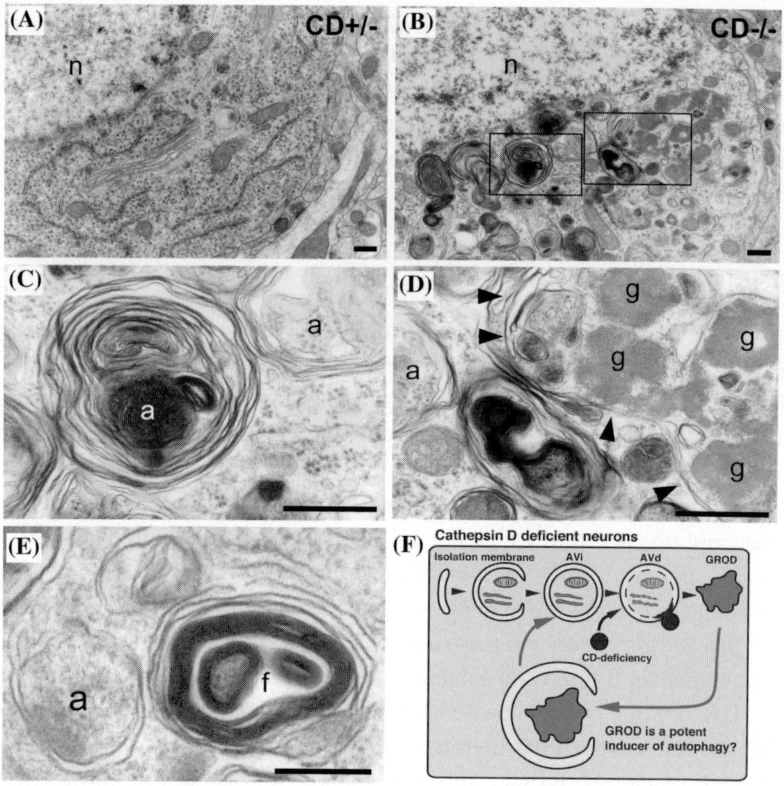

Figure 3. (A to E) Electron micrographs of cerebral cortical neurons in brains from cathepsin D deficient (CD–/–) (B to E) and control-littermate mice (A) at P24. The boxed areas in (B) are shown in C and D. In the perikarya of a cerebral cortical neuron from CD–/– mice (B to E), granular osmiophilic deposits (GRODs) (g) abundantly accumulate, and autophagosome-like structures (a) encircled by multi-layered smooth membrane saccules together with part of the cytoplasm are clearly visible. In addition, a typical fingerprint profile (f) is also observed in a CD–/– neuron (E). Note that GRODs are enwrapped together with part of the cytoplasm by double-membrane structures (arrowheads). (n: nucleus, bars indicate 1 μm. (F) A schematic illustration for a hypothetical model of the "pathologically activated" autophagic pathway in CD–/– neurons in which GRODs act as a potent inducer of autophagy. In neurons from CD–/– mice, even after autophagosomes receive lysosomal enzymes via transporting vesicles from the trans-Golgi network or lysosomes, certain substrates of cathepsin D would remain undigested and accumulate in lysosomal compartments. The result leads to the generation of GRODs, which can be detected in nascent autophagosomes, indicating that such GRODs may in turn induce autophagy, accelerating the sequestration of the GRODs themselves by autophagy.

membrane saccules contain part of the morphologically-intact cytoplasm (AVi, nascent autophagosomes) and mature AVs encircled by a single membrane possess partially degraded, but identifiable, cytoplasmic materials and structures (AVd, autolysosomes). It is interesting that lysosome-like structures, such as dense bodies, GRODs, and autophagosomes in CD-deficient and CB/CL-deficient neurons are often surrounded, along with part of the cytoplasm, by membrane saccules (Figure 3D). When electron microscopic data are evaluated by morphometry, the volume densities of GRODs and AVs (AVi and AVd) in CD-deficient neurons increase days after birth. It is worthy to note that half of the AVi counted possess GRODs and 20% of GRODs are detected in AVi. In addition to these morphological findings, it is also evident that cytoplasmic LC3 is largely converted to its membrane-bound form,[73] indicating that autophagosome formation or defect in completion of autophagic degradation occurs frequently in these mutant neurons. These lines of evidence suggest that the presence of GRODs and AVs with undigested materials such as GROD-like inclusions may be a potent inducer of autophagy in neuronal cells (Figure 3F). At present, however, it remains largely unknown whether interaction of LC3 with p62-ubiquitinated proteins is required for intake of GRODs into autophagosomes.

LAMP2 and Autophagy

The lysosome-associated membrane proteins-1 and -2 (LAMP1 and LAMP2) are highly glycosylated membrane proteins that represent well-known marker proteins of lysosomes. Although LAMP2-deficiency is different from the loss of lysosomal cathepsins, increased autophagosomes have also been found in various tissue cells of LAMP2-deficient mice.[82] In such mice, autophagic degradation of long-lived proteins, which is usually conducted in lysosomes, is severely impaired in hepatocytes, while cardiac myocytes are ultrastructurally abnormal and heart contractility is severely reduced. These results indicate that LAMP2 plays important roles in the maintenance of lysosomal functions and is critical for autophagy in peripheral tissue cells. A deficiency of LAMP2 in humans is known to cause Danon's disease, which is associated with the accumulation of autophagic material in striated myocytes.[82,83] Moreover, since subunit c of mitochondrial ATP synthase accumulates in the lysosomes of NCL neurons, it is suspected that autophagy, a pathway that regulates mitochondrial turnover, may be impaired in NCL neurons;[84] however, it would be reasonable to think that accumulation of subunit c could be attributed to the impairment of degradation in lysosomes rather than to increased sequestration by autophagy. In homozygous mutant Cln3 knockin mice, autophagy is highly induced, but isolated autophagic vacuoles and lysosomes from the mice are immature.[84] Thus, impairment of lysosomal functions due to the loss of lysosomal cathepsins and LAMP2, and mutant Cln3 knockin facilitates autophagosome accumulation, resulting in lysosomal storage disorder.

Until recently, however, what signaling is essential for autophagosome formation remained largely unknown. As stated above, in conditional *Atg7*-knockout mice, numerous ubiquitinated aggregates are detected in the cytosol of hepatocytes or CNS neurons with the presence of functional proteasomes.[12–14] These results indicate that protein ubiquitination may serve as a signal to the autophagic process, in addition to the proteasomal pathway. The presence of ubiquitin aggregates is one of the common pathological characteristics of neurodegenerative diseases including lysosomal storage disorders.[85–87] As explained in the former section, genetic ablation of p62 suppresses the appearance of ubiquitin-positive protein aggregates in hepatocytes and neurons, indicating that p62 plays an important role in inclusion body formation, although the pathologic process associated with autophagic deficiency is cell type-specific.[21,49] At present, it remains largely unknown whether such protein-protein interaction plays a key role in the accumulation of lysosomes in NCLs.

CELL DEATH AND AUTOPHAGY

Cell Death and Lysosomes

Based on observations of ultrastructural changes in the metamorphosis-related degeneration of insect intersegmental muscles, Beaulaton and Lockshin have shown that autophagy is responsible for the degradation of mitochondria, glycogen particles, ribosomes, and other organized sarcoplasmic structures, but not for the dissolution of myofilaments, which appears to be independent of lysosomal activity.[88] Therefore, they suggested that lysosomal hydrolases are liberated into the sarcoplasm, leading to the destruction of myofibrils and cell death. Recent research has found that the degradation of myofilaments is conducted by the ubiquitin proteasome system in cooperation with calpain.[89] As Beaulation and Lockshin[88] observed by electron microscopy, in the degradation of intersegmental muscular components shortly after ecdysis to the moth, proteolysis by the autophagy/lysosomal system, in cooperation with the ubiquitin-proteasome system and calpain, serves the adaptive function of providing amino acids for the non-feeding adult insect. Therefore, it may be difficult to speculate that excess autophagy induces metamorphosis-related cell death of intersegmental muscles.

Before Peter Clarke[90] reviewed the death modes of developing CNS neurons, it was shown that there are three different cell death types during development:[91,92]

(1) cell death in which lysosomes do not play any detectable role in the degeneration of the cells, except for heterophagocytosis (apoptotic); (2) cell death that is largely conducted by its own lysosomes (autophagocytosis); and (3) cells that are destroyed without lysosomes playing any detectable role (non-lysosomal vesiculate). Numerous studies have shown the presence of autophagy-related cell death in various CNS tissues, peripheral tissues, and cultured cells.[93–100] Since 3-methyladenine (3-MA) was identified as an inhibitor of autophagy,[101,102] cell death accompanied by autophagy and inhibited by 3-MA has been suggested to be autophagic cell death.[99,103] L298 fibroblastic cells are known to die by apoptosis when treated with tumor necrosis factor, ceramide, oxidants, or irradiation. Such apoptotic cell death is mediated by activation of the caspase cascade. However, Z-VAD-fmk (benzyloxycarbonylvalyl-alanyl–aspartic acid (O-methyl)–fluoromethylketone), a pan-caspase inhibitor, does not prevent this apoptosis of L298 cells.[104] Treatment with Z-VAD-fmk also induces death in various cell lines, such as U937 monocytoid cells, RAW 264.7 macrophage cells, and primary mouse macrophages. Yu *et al.*[100] have shown that Z-VAD-fmk-induced cell death is prevented by 3-MA or wortmannin. Subsequently, they have tried to find ways to directly inhibit autophagy, because 3-MA and wortmannin are general inhibitors of phosphatidylinositol-3 (PI3) kinases, which may affect both autophagy and non-apoptotic cell death. They found that the mRNA knockdown of *Atg7* or *Beclin-1*, which are essential for autophagy, prevents Z-VAD-fmk-induced cell death. Therefore, potent inhibitors of apoptosis, like Z-VAD-fmk, may have unanticipated effects on autophagic cell death.[100]

Autophagic neuron death

As discussed above, autophagy is induced during starvation, differentiation, and normal growth control to maintain homeostasis and survival.[8,12,105] However, it is also involved in neurodegenerative disorders.[73,106–108] In fact, autophagy is highly induced in CA1 pyramidal neurons of the gerbil hippocampus after brief forebrain ischemia and such damaged pyramidal neurons undergo delayed neuronal death.[109] The induction of autophagy, assessed using LC3 as an autophagic marker, has also been shown in neonatal and adult mouse cortex, hippocampus, and striatum after hypoxic/ischemic (H/I) brain injury.[110–116] To answer the question of whether autophagy is neuroprotective or neurotoxic after H/I injury, this H/I injury-induced neuron death was examined using autophagy-disabled neonatal mice, and caspase-3-deficient and caspase-activated DNase (CAD)-deficient mice.[112] The data can be summarized into three primary conclusions. First, H/I injury-induced pyramidal neuron death in the neonatal hippocampus occurs mostly in a caspase 3-independent

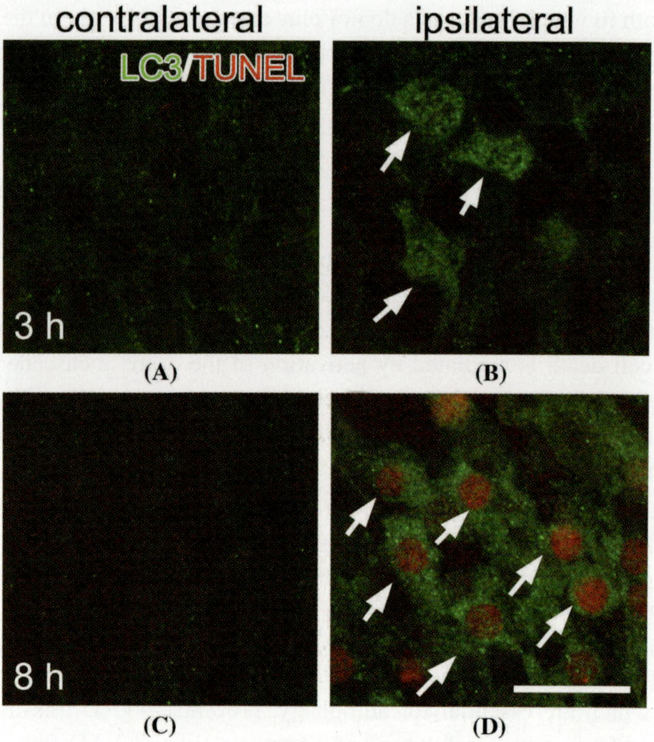

Figure 4. H/I injury induces autophagy in hippocampal pyramidal neurons of wild-type mouse brains. (A to D). Immunostaining of LC3 (green) in the contralateral (A and C) and ipsilateral (B and D) hippocampi at three (A and B) and eight (C and D) hours after H/I injury. Arrows show intense granular immunoreactivity for LC3 in the pyramidal neurons of the ipsilateral hippocampus, the nuclei of which are positively stained for TUNEL (red) at eight hours after H/I injury. (Bar indicates 10 μm.)

manner. In this model, caspase 3-independent neuron death is accompanied by DNA ladder formation that is mediated by an unknown DNase other than CAD. Second, H/I injury induces autophagy in neonatal hippocampal pyramidal neurons (Figure 4), while this pyramidal neuron death is prevented by *Atg7* deficiency. Finally, pyramidal neuron death in the adult hippocampus after H/I injury is caspase-independent and accompanied by autophagosome formation. In particular, the morphological features of these degenerating adult neurons resemble the features of type 2 neuron death, as defined by Clarke.[90] These data provide direct evidence for autophagy-induced neuron death following neonatal mouse H/I brain injury.[112] Even though these results were obtained using mice deficient in *Atg7* specifically in CNS tissue, it remains largely unknown how the loss of *Atg7* prevents H/I injury-induced pyramidal neuron death in the neonatal hippocampus.[112–114] It is, therefore, very important to understand the mechanism by which pyramidal neurons regulate the two opposite downstream effects of autophagy, survival and death, after H/I insult.

CONCLUDING REMARKS

It is difficult to detect autophagy in CNS neurons in response to starvation. Importance of autophagy in CNS tissue is understood when autophagy-deficient animals are analyzed in the tissue. Accumulation of ubiquitinated protein aggregates appear in CNS neurons of such animals, indicating that ubiquitin signal is involved in not only ubiquitin-proteasome but also autophagy-lysosomal systems. Indeed, p62, a multifunctional protein that has binding regions for both ubiquitin and LC3 is degraded through autophagy in lysosomes. The binding of p62 or Nbr1 with ubiquitin may further bind LC3 that is localized on the isolation membrane of autophagosomes; this may explain selective intake of p62 into autophagosomes that become autolysosomes after fusion with lysosomes. At present, it remains largely unknown whether ubiquitin signal is contributed to the degradation of proteins in lysosomes of neurons. However, since ubiquitin signal is present in accumulated lysosomes of neurons suffering from lysosomal storage disorders including NCLs, protein conjugation of ubiquitin, p62, and LC3 may play a crucial role in the production of autophagosomes/autolysosomes. Auotphagy is also induced in various types of neuronal disorders including H/I brain injury. As stated above, autophagy is essential for the maintenance of metabolism, leading to neuronal survival. However, our data using Atg7-deficient neonatal mice specifically in CNS tissue provide direct evidence for autophagy-induced neuron death following neonatal mouse H/I brain injury.

References

1. A. L. Goldberg, Protein degradation and protection against misfolded or damaged proteins, *Nature* **426**: 895–899 (2003).
2. A. Hershko and A. Ciechanover, The ubiquitin system, *Annu Rev Biochem* **67**: 425–479 (1998).
3. N. Mizushima, B. Levine, A. M. Cuervo and D. J. Klionsky, Autophagy fights disease through cellular self-digestion, *Nature* **451**: 1069–1075 (2008).
4. H. Nakatogawa, K. Suzuki, Y. Kamada and Y. Ohsumi, Dynamics and diversity in autophagy mechanisms: lessons from yeast, *Nat Rev Mol Cell Biol* **10**: 458–467 (2009).
5. B. Levine and D. J. Klionsky, Development by self-digestion: molecular mechanisms and biological functions of autophagy, *Dev Cell* **6**: 463–477 (2004).

6. B. Levine and G. Kroemer, Autophagy in the pathogenesis of disease, *Cell* **132**: 27–42 (2008).
7. M. Tsukada and Y. Ohsumi, Isolation and characterization of autophagy-defective mutants of *Saccharomyces cerevisiae*, *FEBS Lett* **333**: 169–174 (1993).
8. A. Kuma, M. Hatano, M. Matsui, A. Yamamoto, H. Nakaya, T. Yoshimori, Y. Ohsumi, T. Tokuhisa and N. Mizushima, The role of autophagy during the early neonatal starvation period, *Nature* **432**: 1032–1036 (2004).
9. S. Tsukamoto, A. Kuma, M. Murakami, C. Kishi, A. Yamamoto and N. Mizushima, Autophagy is essential for preimplantation development of mouse embryos, *Science* **321**: 117–120 (2008).
10. X. Qu, J. Yu, B. Bhagat, N. Furuya, H. Hibshoosh, A. Troxel, J. Rosen, E. L. Eskelinen, N. Mizushima, Y. Ohsumi, G. Cattoretti and B. Levine, Promotion of tumorigenesis by heterozygous disruption of the beclin 1 autophagy gene, *J Clin Invest* **112**: 1809–1820 (2003).
11. Z. Yue, S. Jin, C. Yang, A. J. Levine and N. Heintz, Beclin 1, an autophagy gene essential for early embryonic development, is a haploinsufficient tumor suppressor, *Proc Natl Acad Sci USA* **100**: 15077–15082 (2003).
12. M. Komatsu, S. Waguri, T. Ueno, J. Iwata, S. Murata, I. Tanida, J. Ezaki, N. Mizushima, Y. Ohsumi, Y. Uchiyama, E. Kominami, K. Tanaka and T. Chiba, Impairment of starvation-induced and constitutive autophagy in Atg7-deficient mice, *J Cell Biol* **169**: 425–434 (2005).
13. T. Hara, K. Nakamura, M. Matsui, A. Yamamoto, Y. Nakahara, R. Suzuki-Migishima, M. Yokoyama, K. Mishima, I. Saito, H. Okano and N. Mizushima, Suppression of basal autophagy in neural cells causes neurodegenerative disease in mice, *Nature* **441**: 885–889 (2006).
14. M. Komatsu, S. Waguri, T. Chiba, S. Murata, J. Iwata, I. Tanida, T. Ueno, M. Koike, Y. Uchiyama, E. Kominami and K. Tanaka, Loss of autophagy in the central nervous system causes neurodegeneration in mice, *Nature* **441**: 880–884 (2006).
15. A. Nakai, O. Yamaguchi, T. Takeda, Y. Higuchi, S. Hikoso, M. Taniike, S. Omiya, I. Mizote, Y. Matsumura, M. Asahi, K. Nishida, M. Hori, N. Mizushima and K. Otsu, The role of autophagy in cardiomyocytes in the basal state and in response to hemodynamic stress, *Nat Med* **13**: 619–624 (2007).
16. E. Masiero, L. Agatea, C. Mammucari, B. Blaauw, E. Loro, M. Komatsu, D. Metzger, C. Reggiani, S. Schiaffino and M. Sandri, Autophagy is required to maintain muscle mass, *Cell Metab* **10**: 507–515 (2009).
17. C. Ebato, T. Uchida, M. Arakawa, M. Komatsu, T. Ueno, K. Komiya, K. Azuma, T. Hirose, K. Tanaka, E. Kominami, R. Kawamori, Y. Fujitani and H. Watada, Autophagy is important in islet homeostasis and compensatory increase of beta cell mass in response to high-fat diet, *Cell Metab* **8**: 325–332 (2008).
18. H. S. Jung, K. W. Chung, J. Won Kim, J. Kim, M. Komatsu, K. Tanaka, Y. H. Nguyen, T. M. Kang, Y. H. Yoon, J. W. Kim, Y. T. Jeong, M. S. Han, M. K. Lee, K. W. Kim,

J. Shin and M. S. Lee, Loss of autophagy diminishes pancreatic beta cell mass and function with resultant hyperglycemia, *Cell Metab* **8**: 318–324 (2008).

19. M. Komatsu, Q. J. Wang, G. R. Holstein, V. L. Friedrich Jr., J. Iwata, E. Kominami, B. T. Chait, K. Tanaka and Z. Yue, Essential role for autophagy protein Atg7 in the maintenance of axonal homeostasis and the prevention of axonal degeneration, *Proc Natl Acad Sci USA* **104**: 14489–14494 (2007).
20. J. Nishiyama, E. Miura, N. Mizushima, M. Watanabe and M. Yuzaki, Aberrant membranes and double-membrane structures accumulate in the axons of Atg5-null Purkinje cells before neuronal death, *Autophagy* **3**: 591–596 (2007).
21. M. Komatsu, S. Waguri, M. Koike, Y. S. Sou, T. Ueno, T. Hara, N. Mizushima, J. Iwata, J. Ezaki, S. Murata, J. Hamazaki, Y. Nishito, S. Iemura, T. Natsume, T. Yanagawa, J. Uwayama, E. Warabi, H. Yoshida, T. Ishii, A. Kobayashi, M. Yamamoto, Z. Yue, Y. Uchiyama, E. Kominami and K. Tanaka, Homeostatic levels of p62 control cytoplasmic inclusion body formation in autophagy-deficient mice, *Cell* **131**: 1149–1163 (2007).
22. Z Xie and D. J. Klionsky, Autophagosome formation: core machinery and adaptations, *Nat Cell Biol* **9**: 1102–1109 (2007).
23. H. Nakatogawa, Y. Ichimura and Ohsumi Y, Atg8, a ubiquitin-like protein required for autophagosome formation, mediates membrane tethering and hemifusion, *Cell* **130**: 165–178 (2007).
24. Y. Kabeya, N. Mizushima, T. Ueno, A. Yamamoto, T. Kirisako, T. Noda, E. Kominami, Y. Ohsumi and T. Yoshimori, LC3, a mammalian homologue of yeast Apg8p, is localized in autophagosome membranes after processing, *EMBO J* **19**: 5720–5728 (2000).
25. I. Tanida, E. Tanida-Miyake, T. Ueno and E. Kominami, The human homolog of *Saccharomyces cerevisiae* Apg7p is a protein-activating enzyme for multiple substrates including human Apg12p, GATE-16, GABARAP, and MAP-LC3, *J Biol Chem* **276**: 1701–1706 (2001).
26. I. Tanida, E. Tanida-Miyake, M. Komatsu, T. Ueno and E. Kominami, Human Apg3p/Aut1p homologue is an authentic E2 enzyme for multiple substrates, GATE-16, GABARAP, and MAP-LC3, and facilitates the conjugation of hApg12p to hApg5p, *J Biol Chem* **277**: 13739–13744 (2002).
27. N. Fujita, T. Itoh, H. Omori, M. Fukuda, T. Noda and T. Yoshimori, The Atg16L complex specifies the site of LC3 lipidation for membrane biogenesis in autophagy, *Mol Biol Cell* **19**: 2092–2100 (2008).
28. N. Fujita, M. Hayashi-Nishino, H. Fukumoto, H. Omori, A. Yamamoto, T. Noda and T. Yoshimori, An Atg4B mutant hampers the lipidation of LC3 paralogues and causes defects in autophagosome closure, *Mol Biol Cell* **19**: 4651–4659 (2008).
29. Y. S. Sou, S. Waguri, J. Iwata, T. Ueno, T. Fujimura, T. Hara, N. Sawada, A. Yamada, N. Mizushima, Y. Uchiyama, E. Kominami, K. Tanaka and M. Komatsu, The Atg8 conjugation system is indispensable for proper development of autophagic isolation membranes in mice, *Mol Biol Cell* **19**: 4762–4775 (2008).

30. I. Tanida, N. Minematsu-Ikeguchi, T. Ueno and E. Kominami, Lysosomal turnover, but not a cellular level, of endogenous LC3 is a marker for autophagy, *Autophagy* **1**: 84–91 (2005).
31. N. Mizushima, A. Yamamoto, M. Matsui, T. Yoshimori and Y. Ohsumi, *In vivo* analysis of autophagy in response to nutrient starvation using transgenic mice expressing a fluorescent autophagosome marker, *Mol Biol Cell* **15**: 1101–1111 (2004).
32. G. Bjørkøy, T. Lamark, A. Brech, H. Outzen, M. Perander, A. Øvervatn, H. Stenmark and T. Johansen, p62/SQSTM1 forms protein aggregates degraded by autophagy and has a protective effect on huntingtin-induced cell death, *J Cell Biol* **171**: 603–614 (2005).
33. Y. Ichimura, T. Kumanomidou, Y. S. Sou, T. Mizushima, J. Ezaki, T. Ueno, E. Kominami, T. Yamane, K. Tanaka and M. Komatsu, Structural basis for sorting mechanism of p62 in selective autophagy, *J Biol Chem* **283**: 22847–22857 (2008).
34. S. Pankiv, T. H. Clausen, T. Lamark, A. Brech, J. A. Bruun, H. Outzen, A. Øvervatn, G. Bjørkøy and T. Johansen, p62/SQSTM1 binds directly to Atg8/LC3 to facilitate degradation of ubiquitinated protein aggregates by autophagy, *J Biol Chem* **282**: 24131–24145 (2007).
35. N. Laurin, J. P. Brown, J. Morissette and V. Raymond, Recurrent mutation of the gene encoding sequestosome 1 (SQSTM1/p62) in Paget disease of bone, *Am J Hum Genet* **70**: 1582–1588 (2002).
36. J. Moscat and M. T. Diaz-Meco, p62 at the crossroads of autophagy, apoptosis, and cancer, *Cell* **137**: 1001–1004 (2009).
37. J. Moscat, M. T. Diaz-Meco, A. Albert and S. Campuzano, Cell signaling and function organized by PB1 domain interactions, *Mol Cell* **23**: 631–640 (2006).
38. C. Stumptner, A. Fuchsbichler, H. Heid, K. Zatloukal and H. Denk, Mallory body — a disease-associated type of sequestosome, *Hepatology* **35**: 1053–1062 (2002).
39. K. Zatloukal, C. Stumptner, A. Fuchsbichler, H. Heid, M. Schnoelzer, L. Kenner, R. Kleinert, M. Prinz, A. Aguzzi and H. Denk, p62 is a common component of cytoplasmic inclusions in protein aggregation diseases, *Am J Pathol* **160**: 255–263 (2002).
40. E. Itakura and N. Mizushima, p62 Targeting to the autophagosome formation site requires self-oligomerization but not LC3 binding, *J Cell Biol* **192**: 17–27 (2011).
41. B. E. Riley, S. E. Kaiser, T. A. Shaler, A. C. Ng, T. Hara, M. S. Hipp, K. Lage, R. J. Xavier, K. Y. Ryu, K. Taguchi, M. Yamamoto, K. Tanaka, N. Mizushima, M. Komatsu and R. R. Kopito, Ubiquitin accumulation in autophagy-deficient mice is dependent on the Nrf2-mediated stress response pathway: a potential role for protein aggregation in autophagic substrate selection, *J Cell Biol* **191**: 537–552 (2010).
42. N. N. Noda, H. Kumeta, H. Nakatogawa, K. Satoo, W. Adachi, J. Ishii, Y. Fujioka , Y. Ohsumi and F. Inagaki, Structural basis of target recognition by Atg8/LC3 during selective autophagy, *Genes Cells* **13**: 1211–1218 (2008).
43. E. Shvets, E. Fass, R. Scherz-Shouval and Z. Elazar, The N-terminus and Phe52 residue of LC3 recruit p62/SQSTM1 into autophagosomes, *J Cell Sci* **121**: 2685–2695 (2008).

44. S. J. Cherra 3rd, S. M. Kulich, G. Uechi, M. Balasubramani, J. Mountzouris, B. W. Day and C. T. Chu, Regulation of the autophagy protein LC3 by phosphorylation, *J Cell Biol* **190**: 533–539 (2010).
45. I. Novak, V. Kirkin, D. G. McEwan, J. Zhang, P. Wild, A. Rozenknop, V. Rogov, F. Löhr, D. Popovic, A. Occhipinti, A. S. Reichert, J. Terzic, V. Dötsch, P. A. Ney and I. Dikic, Nix is a selective autophagy receptor for mitochondrial clearance, *EMBO Rep.* **11**: 45–51 (2010).
46. K. Okamoto, N. Kondo-Okamoto and Y. Ohsumi, Mitochondria-anchored receptor Atg32 mediates degradation of mitochondria via selective autophagy, *Dev Cell* **17**: 87–97 (2009).
47. M. Filimonenko, P. Isakson, K. D. Finley, M. Anderson, H. Jeong, T. J. Melia, B. J. Bartlett, K. M. Myers, H. C. Birkeland, T. Lamark, D. Krainc, A. Brech, H. Stenmark, A. Simonsen and A. Yamamoto, The selective macroautophagic degradation of aggregated proteins requires the PI3P-binding protein Alfy, *Mol Cell* **38**: 265–279 (2010).
48. V. Kirkin, T. Lamark, Y. S. Sou, G. Bjørkøy, J. L. Nunn, J. A. Bruun, E. Shvets, D. G. McEwan, T. H. Clausen, P. Wild, I. Bilusic, J. P. Theurillat, A. Øvervatn, T. Ishii, Z. Elazar, M. Komatsu, I. Dikic and T. Johansen, A role for NBR1 in autophagosomal degradation of ubiquitinated substrates, *Mol Cell* **33**: 505–516 (2009).
49. V. Kirkin, T. Lamark, T. Johansen and I. Dikic, NBR1 cooperates with p62 in selective autophagy of ubiquitinated targets, *Autophagy* **5**: 732–733. (2009)
50. I. P. Nezis, A. Simonsen, A. P. Sagona, K. Finley, S. Gaumer, D. Contamine, T. E. Rusten, H. Stenmark and A. Brech, Ref(2)P, the *Drosophila melanogaster* homologue of mammalian p62, is required for the formation of protein aggregates in adult brain, *J Cell Biol* **180**: 1065–1071 (2008).
51. N. Matsuda, S. Sato, K. Shiba, K. Okatsu, K. Saisho, C. A. Gautier, Y. S. Sou, S. Saiki, S. Kawajiri, F. Sato, M. Kimura, M. Komatsu, N. Hattori and K. Tanaka, PINK1 stabilized by mitochondrial depolarization recruits Parkin to damaged mitochondria and activates latent Parkin for mitophagy, *J Cell Biol* **189**: 211–221 (2010).
52. D. Narendra, A. Tanaka, D. F. Suen and R. J. Youle, Parkin is recruited selectively to impaired mitochondria and promotes their autophagy, *J Cell Biol* **183**: 795–803 (2008).
53. D. P. Narendra, S. M. Jin, A. Tanaka, D. F. Suen, C. A. Gautier, J. Shen, M. R. Cookson and R. J. Youle, PINK1 is selectively stabilized on impaired mitochondria to activate Parkin, *PLoS Biol* **8**: e1000298 (2010).
54. C. Vives-Bauza, C. Zhou, Y. Huang, M. Cui, R. L. de Vries, J. Kim, J. May, M. A. Tocilescu, W. Liu, H. S. Ko, J. Magrané, D. J. Moore, V. L. Dawson, R. Grailhe, T. M. Dawson, C. Li, K. Tieu and S. Przedborski, PINK1-dependent recruitment of Parkin to mitochondria in mitophagy, *Proc Natl Acad Sci USA* **107**: 378–383 (2010).
55. S. Geisler, K. M. Holmström, D. Skujat, F. C. Fiesel, O. C. Rothfuss, P. J. Kahle and W. Springer, PINK1/Parkin-mediated mitophagy is dependent on VDAC1 and p62/SQSTM1, *Nat Cell Biol* **12**: 119–131 (2010).

56. D. Narendra, L. A. Kane, D. N. Hauser, I. M. Fearnley and R. J. Youle, p62/SQSTM1 is required for Parkin-induced mitochondrial clustering but not mitophagy; VDAC1 is dispensable for both, *Autophagy* **6**: 1090–1106 (2010).
57. K. Okatsu, K. Saisho, M. Shimanuki, K. Nakada, H. Shitara, Y. S. Sou, M. Kimura, S. Sato, N. Hattori, M. Komatsu, K. Tanaka and N. Matsuda, p62/SQSTM1 cooperates with Parkin for perinuclear clustering of depolarized mitochondria, *Genes Cells* **15**: 887–900 (2010).
58. A. Duran, J. F. Linares, A. S. Galvez, K. Wikenheiser, J. M. Flores, M. T. Diaz-Meco and J. Moscat, The signaling adaptor p62 is an important NF-kappaB mediator in tumorigenesis, *Cancer Cell* **13**: 343–354 (2008).
59. Z. Jin, Y. Li, R. Pitti, D. Lawrence, V. C. Pham, J. R. Lill and A. Ashkenazi, Cullin3-based polyubiquitination and p62-dependent aggregation of caspase-8 mediate extrinsic apoptosis signaling, *Cell* **137**: 721–735 (2009).
60. W. Fan, Z. Tang, D. Chen, D. Moughon, X. Ding, S. Chen, M. Zhu and Q. Zhong, Keap1 facilitates p62-mediated ubiquitin aggregate clearance via autophagy, *Autophagy* **6**: 614–621 (2010).
61. A. Jain, T. Lamark, E. Sjøttem, K. B. Larsen, J. A. Awuh, A. Øvervatn, M. McMahon, J. D. Hayes and T. Johansen, p62/SQSTM1 is a target gene for transcription factor NRF2 and creates a positive feedback loop by inducing antioxidant response element-driven gene transcription, *J Biol Chem* **285**: 2576–22591 (2010).
62. M. Komatsu, H. Kurokawa, S. Waguri, K. Taguchi, A. Kobayashi, Y. Ichimura, Y. S. Sou, I. Ueno, A. Sakamoto, K. I. Tong, M. Kim, Y. Nishito, S. Iemura, T. Natsume, T. Ueno, E. Kominami, H. Motohashi, K. Tanaka and M. Yamamoto, The selective autophagy substrate p62 activates the stress responsive transcription factor Nrf2 through inactivation of Keap1, *Nat Cell Biol* **12**: 213–223 (2010).
63. A. Lau, X. J. Wang, F. Zhao, V. F. Villeneuve, T. Wu, T. Jiang, Z. Sun, E. White and D. D. Zhang, A noncanonical mechanism of Nrf2 activation by autophagy deficiency: direct interaction between Keap1 and p62, *Mol Cell Biol* **30**: 3275–3285 (2010).
64. M. Jeyakumar, R. A. Dwek, T. D. Butters and F. M. Platt, Storage solutions: treating lysosomal disorders of the brain, *Nat Rev Neurosci* **6**: 713–725 (2005).
65. A. Kohlschütter and A. Schulz, Towards understanding the neuronal ceroid lipofuscinoses, *Brain Dev* **31**: 499–502 (2009).
66. I. M. Fearnley, J. E. Walker, R. D. Martinus, R. D. Jolly, K. B. Kirkland, G. J. Shaw and D. N. Palmer, The sequence of the major protein stored in ovine ceroid lipofuscinosis is identical with that of the dicyclohexylcarbodiimide-reactive proteolipid of mitochondrial ATP synthase, *Biochem J* **268**: 751–758 (1990).
67. N. A. Hall, B. D. Lake, N. N. Dewji and A. D. Patrick, Lysosomal storage of subunit c of mitochondrial ATP synthase in Batten's disease (ceroid-lipofuscinosis), *Biochem J* **275**(Pt1): 269–272 (1991).
68. E. Kominami, J. Ezaki, D. Muno, K. Ishido, T. Ueno and L. S. Wolfe, Specific storage of subunit c of mitochondrial ATP synthase in lysosomes of neuronal ceroid lipofuscinosis (Batten's disease), *J Biochem (Tokyo)* **111**: 278–282 (1992).

69. D. N. Palmer, R. D. Martinus, S. M. Cooper, G. G. Midwinter, J. C. Reid and R. D. Jolly, Ovine ceroid lipofuscinosis. The major lipopigment protein and the lipid-binding subunit of mitochondrial ATP synthase have the same NH2-terminal sequence, *J Biol Chem* **264**: 5736–5740 (1989).
70. P. Saftig, M. Hetman, W. Schmahl, K. Weber, L. Heine, H. Mossmann, A. Köster, B. Hess, M. Evers and K. von Figura, Mice deficient for the lysosomal proteinase cathepsin D exhibit progressive atrophy of the intestinal mucosa and profound destruction of lymphoid cells, *EMBO J* **14**: 3599–3608 (1995).
71. M. Koike, H. Nakanishi, P. Saftig, J. Ezaki, K. Isahara, Y. Ohsawa, W. Schulz-Schaeffer, T. Watanabe, S. Waguri, S. Kametaka, M. Shibata, K. Yamamoto, E. Kominami, C. Peters, K. von Figura and Y. Uchiyama. Cathepsin D deficiency induces lysosomal storage with ceroid lipofuscin in mouse CNS neurons, *J Neurosci* **20**: 6898–6906 (2000).
72. M. Koike, M. Shibata, Y. Ohsawa, H. Nakanishi, T. Koga, S. Kametaka, S. Waguri, T. Momoi, E. Kominami, C. Peters, K. von Figura, P. Saftig and Y. Uchiyama, Involvement of two different cell death pathways in retinal atrophy of cathepsin D-deficient mice, *Mol Cell Neurosci* **22**: 146–161 (2003).
73. M. Koike, M. Shibata, S. Waguri, K. Yoshimura, I. Tanida, E. Kominami, T. Gotow, C. Peters, K. von Figura, N. Mizushima, P. Saftig and Y. Uchiyama, Participation of autophagy in storage of lysosomes in neurons from mouse models of neuronal ceroid-lipofuscinoses (Batten disease), *Am J Pathol* **167**: 1713–1728 (2005).
74. H. Nakanishi, J. Zhang, M. Koike, T. Nishioku, Y. Okamoto, E. Kominami, K. von Figura, C. Peters, K. Yamamoto, P. Saftig and Y. Uchiyama, Involvement of nitric oxide released from microglia-macrophages in pathological changes of cathepsin D-deficient mice, *J Neurosci* **21**: 7526–7533 (2001).
75. E. Siintola, S. Partanen, P. Strömme, A. Haapanen, M. Haltia, J. Maehlen, A. E. Lehesjoki and J. Tyynelä, Cathepsin D deficiency underlies congenital human neuronal ceroid-lipofuscinosis, *Brain* **129**: 1438–1445 (2006).
76. R. Steinfeld, K. Reinhardt, K. Schreiber, M. Hillebrand, R. Kraetzner, W. Bruck, P. Saftig and J. Gartner, Cathepsin D deficiency is associated with a human neurodegenerative disorder, *Am J Hum Genet* **78**: 988–998 (2006).
77. W. Potts, J. Bowyer, H. Jones, D. Tucker, A. J. Freemont, A. Millest, C. Martin, W. Vernon, D. Neerunjun, G. Slynn, F. Harper and R. Maciewicz, Cathepsin L-deficient mice exhibit abnormal skin and bone development and show increased resistance to osteoporosis following ovariectomy, *Int J Exp Pathol* **85**: 85–96 (2004).
78. W. Roth, J. Deussing, V. A. Botchkarev, M. Pauly-Evers, P. Saftig, A. Hafner, P. Schmidt, W. Schmahl, J. Scherer, I. Anton-Lamprecht, K. von Figura, R. Paus and C. Peters, Cathepsin L deficiency as molecular defect of furless: hyperproliferation of keratinocytes and pertubation of hair follicle cycling, *FASEB J* **14**: 2075–2086 (2000).
79. W. A. Dunn Jr., Autophagy and related mechanisms of lysosome-mediated protein degradation, *Trends Cell Biol* **4**: 139–143 (1994).

80. W. A. Dunn Jr., Studies on the mechanisms of autophagy: maturation of the autophagic vacuole, *J Cell Biol* **110**: 1935–1945 (1990).
81. W. Liou, H. J. Geuze, M. J. Geelen and J. W. Slot, The autophagic and endocytic pathways converge at the nascent autophagic vacuoles, *J Cell Biol* **136**: 61–70 (1997).
82. Y. Tanaka, G. Guhde, A. Suter, E. L. Eskelinen, D. Hartmann, R. Lüllmann-Rauch, P. M. Janssen, J. Blanz, K. von Figura and P. Saftig, Accumulation of autophagic vacuoles and cardiomyopathy in LAMP-2-deficient mice, *Nature* **406**: 902–906 (2000).
83. I. Nishino, J. Fu, K. Tanji, T. Yamada, S. Shimojo, T. Koori, M. Mora, J. E. Riggs, S. J. Oh, Y. Koga, C. M. Sue, A. Yamamoto, N. Murakami, S. Shanske, E. Byrne, E. Bonilla, I. Nonaka, S. DiMauro and M. Hirano, Primary LAMP-2 deficiency causes X-linked vacuolar cardiomyopathy and myopathy (Danon disease), *Nature* **406**: 906–910 (2000).
84. Y. Cao, J. A. Espinola, E. Fossale, A. C. Massey, A. M. Cuervo, M. E. MacDonald and S. L. Cotman, Autophagy is disrupted in a knock-in mouse model of juvenile neuronal ceroid lipofuscinosis, *J Biol Chem* **281**: 20483–20493 (2006).
85. H. C. Ardley, C. C. Hung and P. A. Robinson, The aggravating role of the ubiquitin-proteasome system in neurodegeneration, *FEBS Lett* **579**: 571–576 (2005).
86. C. Settembre, A. Fraldi, L. Jahreiss, C. Spampanato, C. Venturi, D. Medina, R. de Pablo, C. Tacchetti, D. C. Rubinsztein and A. Ballabio, A block of autophagy in lysosomal storage disorders, *Hum Mol Genet* **17**: 119–129 (2008).
87. S. S. Zhan, K. Beyreuther and H. P. Schmitt, Neuronal ubiquitin and neurofilament expression in different lysosomal storage disorders, *Clin Neuropathol* **11**: 251–255 (1992).
88. J. Beaulaton and R. A. Lockshin, Ultrastructural study of the normal degeneration of the intersegmental muscles of *Anthereae polyphemus* and *Manduca sexta* (Insecta, Lepidoptera) with particular reference of cellular autophagy, *J Morphol* **154**: 39–57 (1977).
89. D. E. Goll, G. Neti, S. W. Mares and V. F. Thompson, Myofibrillar protein turnover: the proteasome and the calpains, *J Anim Sci* **86**: E19–E35 (2007)
90. P. G. Clarke, Developmental cell death: morphological diversity and multiple mechanisms, *Anat Embryol (Berl)* **181**: 195–213 (1990).
91. J. U. Schweichel, Electron microscopic studies on the degradation of the apical ridge during the development of limbs in rat embryos, *Z Anat Entwicklungsgesch* **136**: 192–203 (1972).
92. J. U. Schweichel and H. J. Merker, The morphology of various types of cell death in prenatal tissues, *Teratology* **7**: 253–266 (1973).
93. W. Bursch, The autophagosomal-lysosomal compartment in programmed cell death, *Cell Death Differ* **8**: 569–581 (2001).
94. N. Canu, R. Tufi, A. L. Serafino, G. Amadoro, M. T. Ciotti and P. Calissano, Role of the autophagic-lysosomal system on low potassium-induced apoptosis in cultured cerebellar granule cells, *J Neurochem* **92**: 1228–1242 (2005).
95. K. Isahara, Y. Ohsawa, S. Kanamori, M. Shibata, S. Waguri, N. Sato, T. Gotow, T. Watanabe, T. Momoi, K. Urase, E. Kominami and Y. Uchiyama, Regulation of a

novel pathway for cell death by lysosomal aspartic and cysteine proteinases, *Neuroscience* **91**: 233–249 (1999).

96. Y. Ohsawa, K. Isahara, S. Kanamori, M. Shibata, S. Kametaka, T. Gotow, T. Watanabe, E. Kominami and Y. Uchiyama, An ultrastructural and immunohistochemical study of PC12 cells during apoptosis induced by serum deprivation with special reference to autophagy and lysosomal cathepsins, *Arch Histol Cytol* **61**: 395–403 (1998).
97. M. Shibata, S. Kanamori, K. Isahara, Y. Ohsawa, A. Konishi, S. Kametaka, T. Watanabe, S. Ebisu, K. Ishido, E. Kominami and Y. Uchiyama, Participation of cathepsins B and D in apoptosis of PC12 cells following serum deprivation, *Biochem Biophys Res Commun* **251**: 199–203 (1998).
98. A. Telbisz, A. L. Kovács and Z. Somosy, Influence of X-ray on the autophagic-lysosomal system in rat pancreatic acini, *Micron* **33**: 143–151 (2002).
99. Y. Uchiyama, Autophagic cell death and its execution by lysosomal cathepsins, *Arch Histol Cytol* **64**: 233–246 (2001).
100. L.Yu, A. Alva, H. Su, P. Dutt, E. Freundt, S. Welsh, E. H. Baehrecke and M. J. Lenardo, Regulation of an ATG7-beclin 1 program of autophagic cell death by caspase-8, *Science* **304**: 1500–1502 (2004).
101. P. B. Gordon and P. O. Seglen, 6-substituted purines: a novel class of inhibitors of endogenous protein degradation in isolated rat hepatocytes, *Arch Biochem Biophys* **217**: 282–294 (1982).
102. P. O. Seglen and P. B. Gordon, 3-Methyladenine: specific inhibitor of autophagic/lysosomal protein degradation in isolated rat hepatocytes, *Proc Natl Acad Sci USA* **79**: 1889–1892 (1982).
103. W. Bursch, A. Ellinger, H. Kienzl, L. Török, S. Pandey, M. Sikorska, R. Walker and R. S. Hermann, Active cell death induced by the anti-estrogens tamoxifen and ICI 164 384 in human mammary carcinoma cells (MCF-7) in culture: the role of autophagy, *Carcinogenesis* **17**: 1595–1607 (1996).
104. W. Fiers, R. Beyaert, W. Declercq and P. Vandenabeele, More than one way to die: apoptosis, necrosis and reactive oxygen damage, *Oncogene* **18**: 7719–7730 (1999).
105. T. Shintani and D. J. Klionsky, Autophagy in health and disease: a double-edged sword, *Science* **306**: 990–995 (2004).
106. C. T. Chu, Autophagic stress in neuronal injury and disease, *J Neuropathol Exp Neurol* **65**: 423–432 (2006).
107. R. A. Nixon, Autophagy in neurodegenerative disease: friend, foe or turncoat?, *Trends Neurosci* **29**: 528–535 (2006).
108. J. H. Zhu, C. Horbinski, F. Guo, S. Watkins, Y. Uchiyama Y and C. T. Chu, Regulation of autophagy by extracellular signal-regulated protein kinases during 1-methyl-4-phenylpyridinium-induced cell death, *Am J Pathol* **170**: 75–86 (2007).
109. T. Nitatori, N. Sato, S. Waguri, Y. Karasawa, H. Araki, K. Shibanai, E. Kominami and Y. Uchiyama, Delayed neuronal death in the CA1 pyramidal cell layer of the gerbil hippocampus following transient ischemia is apoptosis, *J Neurosci* **15**: 1001–1011 (1995).

110. F. Adhami, G. Liao, Y. M. Morozov, A. Schloemer, V. J. Schmithorst, J. N. Lorenz, R. S. Dunn, C. V. Vorhees, M. Wills-Karp, J. L. Degen, R. J. Davis, N. Mizushima, P. Rakic, B. J. Dardzinski, S. K. Holland, F. R. Sharp and C. Y. Kuan, Cerebral ischemia-hypoxia induces intravascular coagulation and autophagy, *Am J Pathol* **169**: 566–583 (2006).
111. F. Adhami, A. Schloemer and C. Y. Kuan, The roles of autophagy in cerebral ischemia, *Autophagy* **3**: 42–44 (2007).
112. M. Koike, M. Shibata, M. Tadakoshi, K. Gotoh, M. Komatsu, S. Waguri, N. Kawahara, K. Kuida, S. Nagata, E. Kominami, K. Tanaka and Y. Uchiyama, Inhibition of autophagy prevents hippocampal pyramidal neuron death after hypoxic-ischemic injury, *Am J Pathol* **172**: 454–569 (2008).
113. Y. Uchiyama, M. Koike and M. Shibata, Autophagic neuron death in neonatal brain ischemia/hypoxia, *Autophagy* **4**: 404–408 (2008).
114. Y. Uchiyama, M. Koike, M. Shibata and M. Sasaki, Autophagic neuron death, *Methods Enzymol* **453**: 33–51 (2009).
115. C. Zhu, X. Wang, F. Xu, B. A. Bahr, M. Shibata, Y. Uchiyama, H. Hagberg and K. Blomgren, The influence of age on apoptotic and other mechanisms of cell death after cerebral hypoxia-ischemia, *Cell Death Differ* **12**: 162–176 (2005).
116. C. Zhu, F. Xu, X. Wang, M. Shibata, Y. Uchiyama, K. Blomgren and H. Hagberg, Different apoptotic mechanisms are activated in male and female brains after neonatal hypoxia-ischaemia, *J Neurochem* **96**: 1016–1027 (2006).
117. T. Saitoh, N. Fujita, M. H. Jang, S. Uematsu, B. G. Yang, T. Satoh, H. Omori, T. Noda, N. Yamamoto, M. Komatsu, K. Tanaka, T. Kawai, T. Tsujimura, O. Takeuchi, T. Yoshimori and S. Akira, Loss of the autophagy protein Atg16L1 enhances endotoxin-induced IL-1beta production, *Nature* **456**: 264–268 (2008).
118. T. Saitoh, N. Fujita, T. Hayashi, K. Takahara, T. Satoh, H. Lee, K. Matsunaga, S. Kageyama, H. Omori, T. Noda, N. Yamamoto, T. Kawai, K. Ishii, O. Takeuchi, T. Yoshimori and S. Akira, Atg9a controls dsDNA-driven dynamic translocation of STING and the innate immune response, *Proc Natl Acad Sci USA* **106**: 20842–20846 (2009).
119. B. Gan, X. Peng, T. Nagy, A. Alcaraz, H. Gu and J. L. Guan, Role of FIP200 in cardiac and liver development and its regulation of TNFalpha and TSC-mTOR signaling pathways, *J Cell Biol* **175**: 121–133 (2006).
120. M. Kundu, T. Lindsten, C. Y. Yang, J. Wu, F. Zhao, J. Zhang, M. A. Selak, P. A. Ney and C. B. Thompson, Ulk1 plays a critical role in the autophagic clearance of mitochondria and ribosomes during reticulocyte maturation, *Blood* **112**: 1493–1502 (2008).
121. G. Mariño, N. Salvador-Montoliu, A. Fueyo, E. Knecht, N. Mizushima and C. López-Otín, Tissue-specific autophagy alterations and increased tumorigenesis in mice deficient in Atg4C/autophagin-3, *J Biol Chem* **282**: 18573–18583 (2007).
122. G. M. Cann, C. Guignabert, L. Ying, N. Deshpande, J. M. Bekker, L. Wang, B. Zhou and M. Rabinovitch, Developmental expression of LC3alpha and beta: absence

of fibronectin or autophagy phenotype in LC3beta knockout mice, *Dev Dyn* **237**: 187–195 (2008).

123. G. A. O'Sullivan, M. Kneussel, Z. Elazar and H. Betz, GABARAP is not essential for GABA receptor targeting to the synapse, *Eur J Neurosci* **22**: 2644–2648 (2005).
124. C. C. Liang, C. Wang, X. Peng, B. Gan and J. L. Guan, Neural-specific deletion of FIP200 leads to cerebellar degeneration caused by increased neuronal death and axon degeneration, *J Biol Chem* **285**: 3499–3509 (2010).
125. N. Raben, V. Hill, L. Shea, S. Takikita, R. Baum, N. Mizushima, E. Ralston, P. Plotz, Suppression of autophagy in skeletal muscle uncovers the accumulation of ubiquitinated proteins and their potential role in muscle damage in Pompe disease, *Hum Mol Genet* **17**: 3897–3908 (2008).
126. D. Hashimoto, M. Ohmuraya, M. Hirota, A. Yamamoto, K. Suyama, S. Ida, Y. Okumura, E. Takahashi, H. Kido, K. Araki, H. Baba, N. Mizushima and K. Yamamura, Involvement of autophagy in trypsinogen activation within the pancreatic acinar cells, *J Cell Biol* **181**: 1065–1072 (2008).
127. B. Hartleben, M. Gödel, C. Meyer-Schwesinger, S. Liu, T. Ulrich, S. Köbler, T. Wiech, F. Grahammer, S. J. Arnold, M. T. Lindenmeyer, C. D. Cohen, H. Pavenstädt, D. Kerjaschki, N. Mizushima, A. S. Shaw, G. Walz and T. B. Huber. Autophagy influences glomerular disease susceptibility and maintains podocyte homeostasis in aging mice, *J Clin Invest* **120**: 1084–1096 (2010).
128. R. Singh, Y. Xiang, Y. Wang, K. Baikati, A. M. Cuervo, Y. K. Luu, Y. Tang, J. E. Pessin, G. J. Schwartz and M. J. Czaja, Autophagy regulates adipose mass and differentiation in mice, *J Clin Invest* **119**: 3329–3339 (2009).
129. M. Mortensen, D. J. Ferguson, M. Edelmann, B. Kessler, K. J. Morten, M. Komatsu and A. K. Simon, Loss of autophagy in erythroid cells leads to defective removal of mitochondria and severe anemia *in vivo*, *Proc Natl Acad Sci USA* **107**: 832–837 (2010).
130. H. H. Pua, J. Guo, M. Komatsu and Y. W. He, Autophagy is essential for mitochondrial clearance in mature T lymphocytes, *J Immunol* **182**: 4046–4055 (2009).
131. B. C. Miller, Z. Zhao, L. M. Stephenson, K. Cadwell, H. H. Pua, H. K. Lee, N. Mizushima, A. Iwasaki, H. W. He, W. Swat and H. W. Virgin 4th, The autophagy gene ATG5 plays an essential role in B lymphocyte development, *Autophagy* **4**: 309–314 (2008).
132. H. K. Lee, L. M. Mattei, B. E. Steinberg, P. Alberts, Y. H. Lee, A. Chervonsky, N. Mizushima, S. Grinstein and A. Iwasaki, *In vivo* requirement for Atg5 in antigen presentation by dendritic cells, *Immunity* **32**: 227–239 (2010).

CHAPTER 9

AUTOPHAGY IN AMYOTROPHIC LATERAL SCLEROSIS

Jozsef Gal and Haining Zhu

ABSTRACT

Amyotrophic lateral sclerosis (ALS, Lou Gehrig's disease) is a neurodegenerative disease that primarily affects motor neurons. Mutations in multiple genes have been found to cause the familial forms of ALS. Research has been primarily focused on the familial ALS cases using animal and cellular models. The etiology of the disease is still not fully understood despite the intensive research. Increasing amount of evidence suggests that the perturbation or insufficiency of cellular protein turnover, especially autophagy, plays an important role in ALS pathogenesis. In this chapter, we review the involvement of impaired protein turnover by autophagy in ALS and discuss the role of familial ALS mutations in causing such autophagic impairment.

INTRODUCTION

Amyotrophic lateral sclerosis, also known as Lou Gehrig's disease or ALS is a progressive and fatal neurodegenerative disease that primarily affects motor neurons.[1] Severe muscle wasting resulting from the loss of innervation at the neuromuscular junction usually leads to paralysis and respiratory failure. No effective treatment is currently available for ALS. The only FDA-approved drug Riluzole provides limited benefit slowing the progression of the disease. It is likely that a large number of motor neurons had already been lost by the time of the clinical presentation of the disease.

Approximately 90% of all ALS cases are sporadic (sALS) with no clear involvement of a genetic component based on the family history of a given patient. Mutations in multiple genes have been found to cause the familial forms of ALS (fALS). Such gene mutations sometimes are found in sALS cases although the causal relationship between mutations and sALS has not been proven in many cases. The clinical symptoms of sALS and fALS are usually indistinguishable. Among the fALS patients, the actual presentation of symptoms, i.e. site and age of onset, progression and severity could be influenced by the implicated gene and even by the nature of specific mutations.[2,3]

At the cellular level, a hallmark of the disease is the appearance of pathologic protein inclusions in the remaining motor neurons and accompanying glial cells.[4,5] The misfolding and aggregation of proteins can be induced by many factors such as oxidative damage, modifications or mutations. Protein aggregation starts with misfolding of individual protein molecules and formation of soluble protein oligomers, and proceeds towards less and less soluble high molecular weight structures. Accumulation of misfolded proteins in cells further reflects a failure of cellular protein turnover. Whereas such large structures are clearly pathologic and might interfere with multiple cellular processes, they may also represent a mechanism to gather smaller aggregates or oligomers in a less harmful form.[6–8] Thus, it remains to be determined whether the large microscopically visible protein inclusions (e.g. aggresomes) might be detrimental or protective.

In recent years, a significant amount of research in ALS has been focused on understanding the protein aggregation pathology. Protein degradation pathways play a critical role in removing misfolded proteins and preventing protein aggregation. Autophagy is one of the two major protein degradation pathways. We review the accumulating evidence for the involvement of autophagy in ALS in this chapter.

PROTEIN TURNOVER PATHWAYS AND THEIR INVOLVEMENT IN ALS

The two major known protein degradation pathways in mammalian cells are the ubiquitin-proteasome system (UPS) and the macroautophagy-lysosome pathway (called autophagy in the rest of the chapter), which function in a highly coordinated manner.[9] The proteasome can degrade soluble substrates but is limited in its ability to degrade protein oligomers and aggregates.[10] Autophagy can degrade soluble proteins, protein aggregates and organelles. The substrate spectra of the proteasome and autophagy somewhat overlap, thus they can compensate for each other's failure to a certain extent. Aggregation-prone proteins can cause an impairment of UPS,[10–13] so their removal primarily proceeds through the autophagy pathway.[9] It was also reported that the inhibition of autophagy compromises the ubiquitin-proteasome pathway.[14]

The cytoplasmic inclusions in ALS are usually ubiquitin positive.[4,15–25] However, accumulation of ubiquitinated proteins in patient samples can reflect either the failure of UPS, or autophagy or both. In fact, in the case of neurodegenerative diseases with extensive protein aggregation, a general failure of both systems is conceivable.

Alterations of UPS in ALS

The status of UPS in ALS is somewhat ambiguous. Most studies reported mild-to-severe proteasomal inhibition or deficiency in ALS[26–31] although upregulation was also reported.[32] We commonly observed a mild increase of the proteolytical activity of proteasomes in cells transfected with fALS mutants of SOD1 (F. Zhang and H. Zhu, unpublished). This paradox situation is probably caused by different measurement methods, different stages in the pathology and different model systems. Earlier studies measuring proteolytic activities without sufficient consideration of regulatory mechanisms need to be revisited. Moreover, an initial adaptive upregulation and an eventual failure of UPS is conceivable. A recent study using an *in vivo* UPS activity reporter mouse model demonstrated a mild decrease of UPS activity in motor neurons of the symptomatic ALS mice.[26]

Alterations of Autophagy in ALS

Recent studies have shown that it is critical to maintain autophagic activity in the central nervous system. Deletion of the key autophagy gene *atg5* or *atg7* in neurons in transgenic mice resulted in neurodegeneration and the accumulation of polyubiquitinated proteins and ubiquitin-immunoreactive inclusions.[33,34] The ubiquitin positivity of the inclusions suggested that either UPS was insufficient to compensate for the failure of autophagy, or that these ubiquitinated proteins did not constitute proteasomal substrates.

The involvement of autophagy in ALS has been reported but its functional role in the disease etiology is less clear. The analysis of post-mortem spinal cord samples from sALS and fALS patients showed increased levels of polyubiquitin and autophagy markers Beclin 1 and the lipidated form of microtubule associated protein light chain 3 (LC3-II) in the majority of the samples, suggesting autophagy involvement in the disease.[35] The protein inclusions in ALS are often immunopositive for ubiquitin and p62 (an adaptor protein for autophagy substrates), both of which are also found in protein inclusions in the *atg5* and *atg7* knockout mice. The following section will discuss the role of p62 and LC3 in ALS in the context of autophagy. Moreover, alterations in autophagy were also reported in multiple fALS models caused by different gene mutations and will be discussed in detail in the corresponding sections.

ROLE OF P62 AND LC3 IN AUTOPHAGY AND IN ALS

P62-Positive Protein Inclusions in ALS

The p62 protein, also called sequestosome 1 or SQSTM1, can be found in pathological inclusions in various cases of ALS. It is common that ubiquitin is also found in p62-positive inclusions.[36,37] P62 immunopositive inclusions were reported in familial ALS caused by CHMP2B mutation,[15,38,39] ANG/angiogenin mutation,[17] FIG4 mutation,[25] and more recently in the TDP-43[19,20] and FUS inclusions.[22,23,40] We reported that p62 co-localized with SOD1- and ubiquitin-positive inclusions in the G93A SOD1 transgenic ALS mouse model.[16]

Since p62-positive inclusions have been observed in many subsets of ALS patients, it becomes tempting to speculate whether mutations in p62 could possibly be responsible for a subset of familial ALS. Mutations in p62 have been implicated in Paget's disease of bone, in which protein inclusions are also a pathological feature.[41] There are no published reports yet on p62 mutations linked to fALS.[a]

In addition to ALS, p62 is found in protein inclusions in a number of aggregation diseases.[42–45] However, p62 is not always co-localized with ubiquitin in pathologic protein inclusions. In frontotemporal lobar degeneration with motor neuron disease (FTLD-MND) that shows overlapping symptoms with ALS, p62 was reported to co-localize with TDP-43 inclusions, but not with ubiquitin.[46] The result suggests that p62 may participate in disease-related processes independent of polyubiquitin.

The Mechanistic Role of p62 in Autophagy

The p62 protein plays a critical role in autophagy as a cargo receptor. The domain structure of p62 is illustrated in Figure 1. The N-terminal PB1 domain can facilitate homo-oligomerization. The C-terminal ubiquitin association (UBA) domain can interact with polyubiquitin chains.[16,47] A separate sequence named LC3 interaction region (LIR) is responsible for interacting with LC3 (see below) and is critical to its ability to shuttle substrates to autophagosomes for degradation.[48–50] The depletion of p62 inhibited autophagic degradation of aggregation-prone polyglutamine-expanded huntingtin inclusions while p62 protected cells from cell death induced by polyglutamine-expanded huntingtin.[46] Vice versa, inhibition of

[a]Since publication of this chapter, there has been a new study reporting p62 mutations in ALS. [F. Fecto, J. Yan , S. P. Vemula, E. Liu, Y. Yang *et al.*, SQSTM1 mutations in familial and sporadic amytrophic lateral sclerosis, *Arch Neurol* **68**: 1440–1446 (2011)].

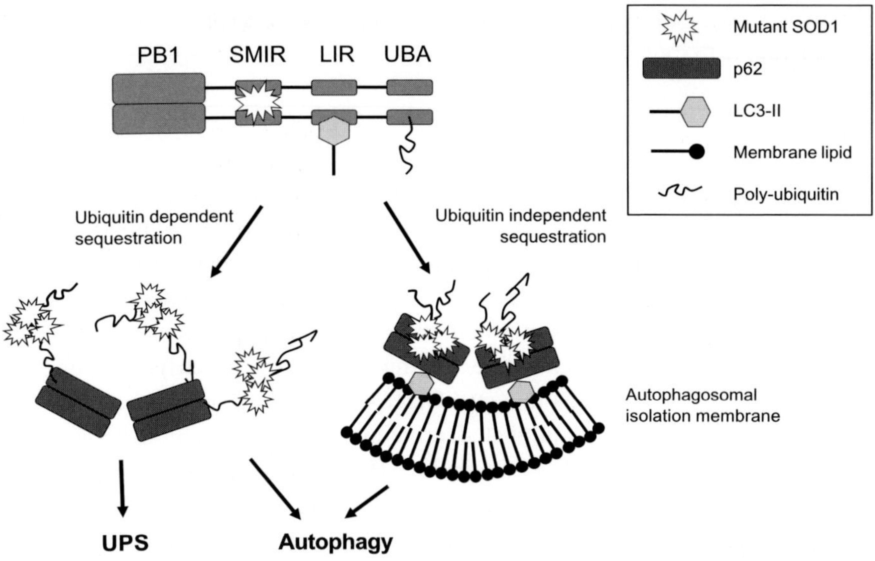

Figure 1. The domain structure of p62 and its functional role as an adaptor targeting misfolded proteins for degradation. P62 recognizes mutant SOD1 via ubiquitin-dependent and -independent mechanisms. The ubiquitin-dependent mechanism requires the UBA domain, and the ubiquitin-independent mechanism requires the PB1 domain and the SMIR motif. P62 can target proteins to both UPS and autophagy, the latter depending on the p62-LC3 interaction mediated by the LIR motif.

autophagy caused elevated levels of p62 and induced more and larger p62 inclusions.[47,51,52]

The steady-state p62 level has recently been used as a marker of autophagy degradation activity. For instance, an elevated level of p62 would be interpreted as autophagy inhibition or failure.[52,53] This involves a critical assumption that p62 biosynthesis is not itself regulated. It has been reported that p62 can be induced at the transcriptional level by various stresses including oxidative stress[54] and proteasome inhibition.[55] Thus, caution should be exercised when using p62 levels as a marker of autophagy activity. In the case of SOD1-mediated fALS, we found significant, age-dependent accumulation of the p62 protein in the G93A SOD1 transgenic mice.[16] However, only a moderate trend of elevated p62 mRNA levels with marginal statistical significance was found in the spinal cords of ALS mice using real-time PCR (J. Gal and H. Zhu, unpublished). The p62 mRNA level was only approximately 20% higher in the G93A mice than in the WT SOD1 transgenic mice at the end stage of the disease. The moderate elevation of the p62 mRNA level did not match the accumulation of the p62 protein. Thus, the evidence suggested that the accumulation of p62 in ALS mice was probably due mostly to slowing of its turnover, which was indicative of a suppressed autophagy degradation activity.

The Mechanisms of p62 Recognizing Misfolded Proteins

We have reported that SOD1-positive inclusions in the spinal cord motor neurons in the G93A SOD1 transgenic mice were also positive for p62 and ubiquitin.[16] In addition, p62 was co-immunoprecipitated with fALS mutants but not with WT SOD1 from both transgenic mouse spinal cord extracts and transfected cells.[16,56]

Since p62 is a critical adaptor protein targeting proteins to autophagic degradation, it is important to better define the mechanism by which p62 recognizes autophagic substrates such as mutant SOD1. It has been generally believed that p62 recognizes the polyubiquitinated proteins via the interaction between its UBA domain and the polyubiquitin chain.[57] However, to the best of our knowledge, ubiquitination of SOD1 has not been demonstrated to be required for its autophagic degradation. Interestingly, the UBA domain of p62 was dispensable for the p62-mutant SOD1 interaction, suggesting that p62 can recognize fALS mutants of SOD1 via an ubiquitin-independent mechanism.[16,56] We further identified two distinct regions within p62 that were essential to its binding to mutant SOD1: the N-terminal PB1 domain and a separate internal region we termed as "SOD1 mutant interaction region" or SMIR (Figure 1). The PB1 domain is required to ensure oligomeric status of p62. The SMIR is the region that interacted with mutant SOD1 and several conserved amino acid residues within SMIR were also demonstrated to be critical to the interaction. The SMIR and LIR (LC3 interaction region) were also independent of each other. Moreover, the existence of mutant SOD1 in acidic autolysosomes decreased in cells lacking p62, suggesting that p62 indeed can function as an adaptor between mutant SOD1 and the autophagy machinery.[55]

Although we have only demonstrated the ubiquitin-independent interaction mechanism between p62 and the fALS mutants of SOD1, we speculate that this may represent a more general mechanism recognizing other misfolded or aggregated autophagic substrates. This hypothesis remains to be tested in future studies.

Role of LC3 in Autophagy and in ALS

LC3 is a mammalian homolog of the yeast Atg8, a protein critical to autophagosome biosynthesis. LC3 is proteolytically processed with its C-terminus removed, resulting in the LC3-I form with a C-terminal glycine residue. This glycine residue can subsequently be conjugated to phosphatidylethanolamine resulting in the LC3-II form. LC3-I is cytosolic, whereas LC3-II is associated with autophagosome membranes.[58] Multiple autophagy (Atg) genes are involved in mediating or regulating the conjugation process. The accumulation of LC3-II can be considered as an indicator of alterations in autophagy, but itself is not an unequivocal marker of autophagy activation. Either upregulated autophagosome formation or suppression of autophagic degradation of LC3-II could cause a higher LC3-II level.[53]

The appearance of LC3-positive puncta under microscope has been used as surrogate measurement of autophagosomes. However, it is noted that LC3-positive puncta do not necessarily imply autophagosome as LC3 may be incorporated into protein aggregates unbounded by membrane.[59,60] Additional evidence of autophagosomes using electron microscopy, or demonstration that the C-terminal glycine is necessary for puncta formation, would better support the conclusion.

The presence of LC3 in inclusions has been reported in multiple cases of ALS, including FUS-positive basophilic inclusions,[40] the G93A SOD1 mouse model[61,62] and neurons and astrocytes of mice carrying a mutation in FIG4.[25,63,64] Moreover, genetic ablation of ALS2/alsin in SOD1 H46R mice resulted in enhanced accumulation of LC3-II.[21] Details regarding how LC3 is implicated in various subsets of ALS will be discussed in appropriate sections in the chapter.

THE INVOLVEMENT OF AUTOPHAGY IN FAMILIAL ALS

Mutations in multiple genes have been linked to familial ALS. Strikingly, most familial ALS gene mutations were demonstrated to either result in an abundant aggregation-prone protein that requires degradation by autophagy (e.g. SOD1 and TDP-43) or affect the autophagy-lysosome pathway (e.g. ALS2/alsin, CHMP2B, VAPB, FIG4 and dynactin). We review the evidence for the involvement of autophagy in the familial forms of ALS below. Figure 2 illustrates how fALS gene mutations may relate to the autophagy pathway. In the case of ALS patients, only post-mortem autopsy data representing end-stage are available in most studies. Most of the mechanistic studies were performed using cellular and transgenic animal models.

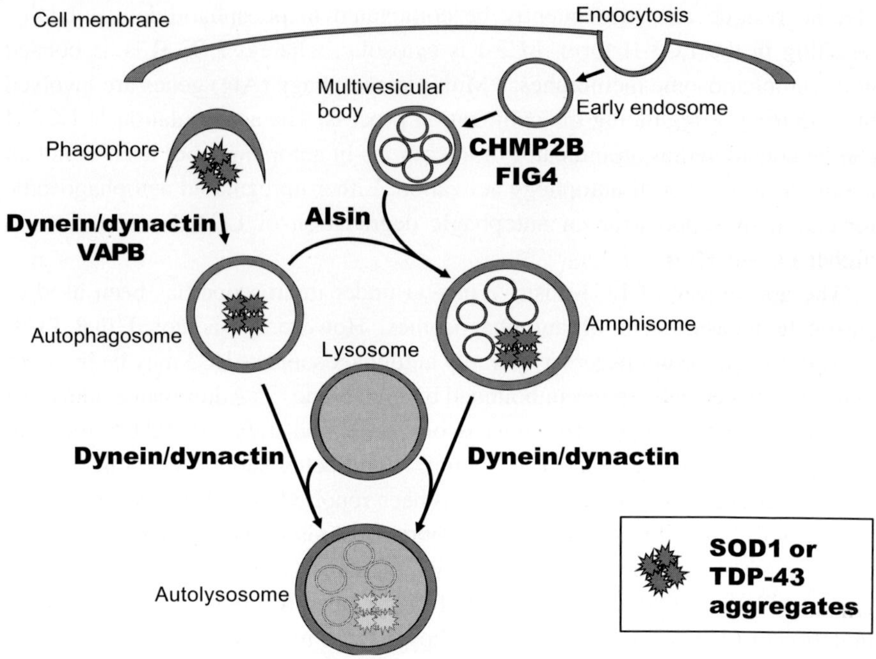

Figure 2. A schematic diagram of the autophagosome-endosome-lysosome system and its lesions in ALS. Autophagic substrates (e.g. SOD1 or TDP43 aggregates) can be recognized and engulfed by autophagosomes. Autophagosomes can either directly fuse with lysosomes to form autolysosomes, or fuse with late endosomes/multivesicular bodies to form amphisomes that can subsequently fuse with lysosomes. Protein aggregates are degraded in autolysosomes. The proteins, whose mutations cause various subsets of familial ALS, are shown in bold along the pathway.

ALS1/SOD1

Mutant SOD1 aggregation in fALS

The copper-zinc superoxide dismutase (SOD1) was the first gene whose mutations were associated with familial ALS.[65,66] Since then, well over 100 SOD1 mutations were found to cause familial ALS, most of which cause disease with dominant inheritance.[1,3,67] SOD1 is a homodimeric metalloenzyme catalyzing the dismutation of superoxide anion to molecular oxygen and hydrogen peroxide. SOD1 knockout mice did not develop ALS-like symptoms[68] while mutant SOD1 transgenic mice did.[69] Thus, a gain-of-function hypothesis has been prevailing in the field, although the nature of the gained toxicity remains debatable.

Aggregation is a common feature of many SOD1 mutants.[4,70–72] The mechanism by which the fALS mutations cause aggregation has been studied extensively.[3,67,73] Detailed studies have also reported that the mutations can cause perturbed metal ion binding,[74–76] increased susceptibility to intramolecular disulfide bond reduction,[76] architectural destabilization,[77–79] oxidative damage,[80] delayed post-translational folding kinetics,[81] increased hydrophobicity,[82] and decreased half-life of the mutants.[83] Moreover, the mutations can destabilize SOD1 monomers, weaken the dimer interface, or both.[84,85] Our laboratory and others also demonstrated formation of soluble oligomers of fALS mutant SOD1 using crosslinking technique.[86,87]

SOD1 immunopositive inclusions were detected in spinal cord anterior horn cells of familial ALS patients with SOD1 mutations.[88] The most abundant component of the protein inclusions in mutant SOD1 transgenic mouse models was SOD1.[71,72,89–93] SOD1 aggregation started in the presymptomatic stage and the levels of accumulating insoluble SOD1 often correlated with disease symptoms in transgenic mice. More interestingly, a "SOD1 exposed dimer interface" (SEDI) antibody was developed to specifically recognize those SOD1 conformations in which the native dimer is disrupted or misfolded. The SEDI antibody detected misfolded SOD1 primarily within degenerating motor neurons prior to the onset of symptoms.[85] Another set of misfolded SOD1-specific antibodies were developed and were used to demonstrate that the hydrophobic soluble form of misfolded mutant SOD1 was enriched in the spinal cord. The levels of such misfolded species in the different mutant SOD1 transgenic mouse models inversely correlated with lifespan.[94,95]

The role of autophagy in mutant SOD1-mediated fALS

Since the degradation of protein aggregates is primarily the task of autophagy, insufficient autophagic removal of SOD1 aggregates is a likely contributor to the pathogenesis. It has been reported that WT and mutant SOD1 are degraded by both UPS and autophagy.[96] The contributions of autophagy and proteasome to the degradation of mutant SOD1 are comparable. Autophagy inhibition increased SOD1 levels in both the detergent-soluble and -insoluble fractions.[96] More interestingly, the study also showed that autophagy inhibition induced mutant SOD1-mediated cell death.

In G93A SOD1 transgenic mice, an increased level of LC3-II was reported at the symptomatic stage compared with non-transgenic or WT SOD1 transgenic mice.[61,62] However, as discussed earlier, an increased level of LC3-II is not sufficient evidence for autophagy activation. The ratio of phosphorylated mTOR/

Ser[2448] immunopositive motor neurons to total motor neurons was decreased in the G93A mice, suggesting involvement of the mTOR pathway in the regulation of mutant SOD1-induced autophagy.[62]

It was reported that the aggregate formation and detergent insolubility of G93A SOD1 were significantly decreased in the presence of millimolar concentrations of trehalose in NSC-34 cells,[97] possibly due to the ability of trehalose to induce autophagy.[98] A recent study reported that the small heat shock protein HspB8 may facilitate the autophagic turnover of mutant SOD1 in neurons.[99]

Autophagic degradation of mutant SOD1 in cell types other than motor neurons has also been examined. A transgenic mouse model expressing G93A SOD1 selectively in skeletal muscle was generated and was found to develop progressive muscle atrophy, alterations in the contractile apparatus and mitochondrial dysfunction.[100] The dephosphorylated/activated form of FoxO3 (a transcription factor regulating autophagy[101]) accumulated in the G93A muscle. Autophagy-related targets downstream of FoxO3, such as cathepsin L (a lysosomal enzyme), Bnip3 (a Bcl2-related protein involved in mitophagy) and LC3-II, were upregulated in G93A atrophic muscles. Inhibition of autophagy by LC3 knockdown rescued muscle atrophy. Moreover, the autophagy of endocytosed mutant SOD1 was also demonstrated in microglia,[102] the cell type that may contribute to the neuro-inflammation component of ALS.

ALS2/Alsin

Mutations in the gene encoding ALS2/alsin are responsible for an autosomal recessive familial form of ALS.[103,104] Alsin was implicated in endosomal dynamics.[105] Expression of alsin protected motor neuronal cells from mutant SOD1 toxicity and alsin interacted with mutant SOD1, but not with WT SOD1, via the RhoGEF domain.[106] A recent study reported that genetic ablation of ALS2 in H46R SOD1 transgenic mice aggravated the disease and enhanced the accumulation of insoluble high molecular weight SOD1, polyubiquitinated proteins, p62 and LC3-II.[21] ALS2 was co-localized with LC3 and p62 and partly with SOD1 on autophagosome/endosome hybrid compartments. Loss of ALS2 significantly lowered the lysosome-dependent clearance of LC3 and p62 in cultured cells. It was proposed that ALS2 plays a role in endolysosomal trafficking, particularly in the fusion between endosomes and autophagosomes.

ALS10/TDP-43

The TAR DNA binding protein of 43 kDa (TDP-43) was identified as a major protein in the ubiquitin-positive inclusions in frontotemporal lobar degeneration and ALS.[107–109] Several TDP-43 mutations, primarily clustering in the C-terminal glycine-rich domain, were found in both familial and sporadic ALS cases (see review[110]). The pathogenic TDP-43 mutations, which exhibit autosomal dominant inheritance, were found to make the protein more prone to aggregation.[111,112]

TDP-43 is commonly found in the cytoplasmic inclusions in the non-SOD1-mediated familial and sporadic ALS cases, but TDP-43 is typically absent or scarce in SOD1 fALS.[17–20,46,113,114] TDP-43 is mostly localized to the nuclei of healthy cells. Pathologic TDP-43 in cytoplasmic inclusions is hyper-phosphorylated and ubiquitinated and can undergo proteolytic cleavage generating C-terminal fragments.[115,116] A study using three-dimensional deconvolution imaging showed that TDP-43 might not be the major ubiquitination target in the inclusions.[117] The expression of pathologic TDP-43 C-terminal fragments in cultured cells resulted in the formation of cytoplasmic inclusions that were ubiquitinated and phosphorylated, recapitulating the TDP-43 proteinopathy. However, it remains to be determined whether the C-terminal fragments are the toxic species causing the disease.

TDP-43 is subject to autophagic degradation. Functional multivesicular bodies (MVBs), which are required for the autophagic clearance of protein aggregates, were also required for the clearance of TDP-43.[38] Autophagy inhibition using 3-methyladenine (3-MA) in cultured cells increased the accumulation of the pathological C-terminal fragments of TDP-43 whereas autophagy activation using rapamycin reduced their level.[118]

Ubiquitin-like protein ubiquilin 1 contributes to the aggregation of TDP-43 and likely plays a role in its autophagosomal targeting. Polyubiquitinated TDP-43 interacted with ubiquilin 1 and overexpression of ubiquilin 1 recruited TDP-43 to detergent-resistant cytoplasmic aggregates that were also LC3-positive.[119] Overexpression of even WT human TDP-43 in *Drosophila* motor neurons led to motor dysfunction. Co-expression of ubiquilin 1 reduced steady-state TDP-43 expression, but intriguingly enhanced the severity of TDP-43 phenotypes.[120] In addition, small heat shock protein B8 (HspB8) was also reported to promote autophagic removal of insoluble and aggregated species of a C-terminally truncated, aggregation-prone version of TDP-43.[99]

Recently, two reports implicated both the ubiquitin-proteasome system and autophagy in the degradation of TDP-43.[121,122] Both the proteasome inhibitor lactacystin and the autophagy inhibitor 3-MA suppressed TDP-43 degradation and enhanced inclusion formation. The effect of 3-MA appeared to be more potent. Combined lactacystin and 3-MA treatment resulted in robust inclusion formation.[121] The degradation of the C-terminal 25 kDa fragment implicated in ALS pathogenesis was also found to be under the control of the proteasome and autophagy.[122]

ALS8/VAPB

A mutation in the vesicle-trafficking protein VAPB was reported to cause autosomal dominant late-onset spinal muscular atrophy and ALS.[123] VAPB plays a critical role in the regulation of the unfolded protein response (UPR). There are three major UPR pathways (PERK, ATF6 and IRE1/XBP1) in mammalian cells,[124] and VAPB is implicated in the modulation of two of them (ATF6[125] and IRE1/XBP1[126]). UPR and ER stress have been reported in sporadic and familial ALS.[127] A subset of motor neurons, which are especially vulnerable in ALS, are also prone to UPR and ER stress and induce microglial activation and selective axonal degeneration.[128] Moreover, ER stress is known to activate autophagy.[9]

The X-box-binding protein-1 (XBP-1) is a key unfolded protein response (UPR) transcription factor that regulates genes involved in protein folding and quality control. VAPB knockdown attenuated the IRE1/XBP1-mediated UPR to induced ER stress, and the P56S fALS mutation in VAPB suppressed UPR in a dominant-negative fashion.[126] Surprisingly, mutant SOD1 transgenic mice with a specific deletion of XBP-1 in the nervous system were more resistant to developing the disease. XBP-1 deficiency increased autophagy in motor neurons and reduced accumulation of mutant SOD1 aggregates in the spinal cord.[35] The study also showed a marked activation of both UPR and autophagy in the post-mortem spinal cord samples from sALS and fALS patients.[35]

CHMP2B and the ESCRT Pathway in ALS

Mutations in the endosomal sorting complexes required for transport (ESCRT) III-complex subunit CHMP2B were first implicated in frontotemporal dementia,[129] then also in familial ALS.[15] ESCRTs are required to sort integral membrane proteins into intralumenal vesicles of the multivesicular body (MVB). Autophagosomes can fuse directly with lysosomes, but they can also fuse with MVBs to form amphisomes that can subsequently fuse with lysosomes.[130] This provides a mechanism for degrading endocytic material. It was shown that autophagic degradation was inhibited in cells depleted of ESCRT subunits and in cells expressing CHMP2B mutants. In ESCRT-depleted cells, the formation of autolysosomes was inhibited, autophagosomes and amphisomes accumulated, membrane-free aggregates formed and TDP-43 accumulated in p62- and ubiquitin-positive cytoplasmic inclusions.[38] Moreover, a recent microarray analysis of motor neurons from CHMP2B cases[64] showed downregulation of Ulk1 (mammalian homologue of Atg1), a key protein in autophagy activation. The study also showed that overexpression of fALS mutant CHMP2B in cultured cells caused large cytoplasmic vacuoles, aberrant lysosomal localization and impairment of autophagy.[64] The above studies suggest that autophagy plays a role in mutant CHMP2B-mediated fALS.

The Dynein/Dynactin Complex

Mutations in dynactin (DCTN1) are a cause of familial ALS (lower motor neuron disease or LMND) and are also possibly involved in the pathogenesis of sporadic cases.[131–133] Mutagenesis studies showed that mutations in the cytoplasmic dynein heavy chain (e.g. *Legs at odd angles (Loa)* and *Cramping 1 (Cra1)* mutations) resulted in motor neuron degeneration and ALS-like symptoms.[134] A targeted disruption of the dynein-dynactin complex by dynamitin overexpression in transgenic mice resulted in inhibition of retrograde axonal transport and late-onset progressive motor neuron degeneration.[135] The impairment of axonal transport, in which the dynein-dynactin complex plays a critical role, is an early presymptomatic event in ALS patients and animal models.[136–145]

Interestingly, the pathogenic processes caused by familial ALS mutants of SOD1 and the dynein/dynactin complex intersect. We and others found that ALS mutants of SOD1 interacted with dynein[86,146] and that the interaction disrupted dynein-mediated retrograde transport[145,147] and contributed to the formation of protein inclusions.[86,148] However, for unclear reasons, the ALS-like symptoms of the G93A mice were surprisingly ameliorated on the *Loa* or the *Cra1* background[136,149,150] (see review[151]).

The dynein/dynactin-mediated retrograde transport is necessary for autophagy. Autophagosomes or related structures were shown to be transported to and fused with lysosomes to form autolysosomes in which protein degradation takes place; the process relies on dynein/dynactin.[152–154] In addition, dynein/dynactin may also play important roles in the early steps of autophagosome formation.[155] Overexpression of dynamitin or microinjection of anti-dynein intermediate chain antibody perturbed the dynein-dependent distribution of endosomes and lysosomes.[156] Decreased dynein function impairs autophagic clearance of protein aggregates through impairment of the autophagosome-lysosome fusion.[157,158] In *Drosophila*, dynein light chain 1 mutants attenuated autophagy, caused accumulation of p62 and ubiquitin-positive inclusions in neurons and decreased larval motility and synaptic boutons, recapitulating many traits of neurodegenerative diseases.[159] Importantly, cell death associated with autophagy and vesicular trafficking defects were reported in a mutant dynactin transgenic mouse model of ALS.[160]

FIG4

FIG4/SAC3 is a phosphoinositide 5-phosphatase that regulates the level of $PI(3,5)P_2$, a signaling lipid mediating retrograde trafficking of endosomal vesicles to the *trans*-Golgi network.[161,162] FIG4 mutations were first found to cause a severe form of Charcot-Marie-Tooth disease (CMT4J) that impacts both sensory and motor neurons.[163] Non-synonymous mutants of FIG4 were later found in ALS11 and primary lateral sclerosis (PLS) patients, and heterozygosity for a deleterious allele of FIG4 appears to be a risk factor for ALS and PLS.[63]

A recent publication established a role for FIG4 and $PI(3,5)P_2$ in the regulation of autophagy in the mammalian central nervous system.[25] The levels of p62 and LC3-II were elevated in the $FIG4^{-/-}$ brain with no evidence for autophagy activation, indicating impairment of autophagic degradation. Cytoplasmic inclusion bodies found in the degenerating regions of the $FIG4^{-/-}$ brain were positive for p62 and ubiquitin but negative for TDP-43. The results establish a clear connection between FIG4 mutation, autophagy dysfunction and neurodegeneration.

Fused in Sarcoma (FUS)

Fused in sarcoma (FUS) mutations have been recently reported to be responsible for ALS6.[164,165] We and others reported that the fALS mutations in FUS caused abnormal accumulation and inclusions of the protein in the cytoplasm.[166–168] Our unpublished results using inhibitors suggest that the ubiquitin-proteasome system may play a more important role in FUS degradation than autophagy does (J. Zhang and H. Zhu, unpublished results).

THE POSSIBLE INVOLVEMENT OF HDAC6 IN AUTOPHAGY IN ALS

Histone deacetylase 6 (HDAC6) protein is an important regulator of aggresome formation[169] and autophagy.[170,171] HDAC6 is also a regulatory link between the UPS and autophagy pathways.[9,172] Although it is conceivable that HDAC6 may play a role in protein homeostasis in ALS, no studies have been reported yet. A recently published paper provided an unexpected link between the HDAC6-mediated protein aggregation-degradation pathway and the fALS genes involved in RNA metabolism. TDP-43 and FUS were found to associate with HDAC6 mRNA and silencing of either TDP-43 or FUS reduced the level of the HDAC6 mRNA.[173] It remains to be further studied how HDAC6 may participate in the pathology of the disease.

FUTURE DIRECTIONS

Emerging evidence support that autophagy likely plays a role in the pathology of ALS, but many questions remain to be addressed in future studies. We specifically discuss autophagic degradation of mitochondria and potential therapeutic benefit of modulating autophagy.

Mitophagy in ALS

Mitophagy, the autophagic removal of damaged mitochondria, has been implicated in multiple neurodegenerative diseases.[174] Although mitophagy is essential, excessive mitophagy can cause collateral damage.[175] Mitophagy has been particularly implicated in Parkinson's disease (see review[176]). Mitochondrial abnormalities have been well established in ALS.[177,178] It is conceivable that damaged mitochondria need to be degraded and/or recycled, and mitophagy as the major mechanism for mitochondrial degradation could play a role in ALS. Recent studies showed that treatment with lithium, which can induce autophagic activity, normalized mitochondrial size and increased the number of normal mitochondria in G93A SOD1 transgenic mouse spinal cord and in cellular models.[179] Lithium and rapamycin also appeared to stimulate the biogenesis of mitochondria.[180] These results suggest that upregulation of mitophagy can remove damaged mitochondria and improve the pool of functional mitochondria.

Therapeutic Promises of Autophagy Modulation

It was reported that Li(+)-carbonate significantly enhanced motor neuron survival, motor function and survival time in ALS model mice.[181] Additional studies found that lithium attenuated disease progression in ALS patients and animal models likely by activating autophagy.[179,180,182] Lithium is believed to activate autophagy by inhibiting inositol monophosphatase and decreasing the level of the autophagy suppressing molecule IP3.[183,184] In the G93A SOD1 transgenic mice, lithium delayed disease onset and augmented life span along with activation of autophagy. Lithium facilitated the clearance of ubiquitin and SOD1 aggregates and increased the number of autophagic vacuoles. Lithium was also reported to activate autophagy in motor neurons in organotypic slice cultures of spinal cord based on the accumulation of autophagic vacuoles and the enhanced LC3 immunostaining.[185] However, follow-up studies found no therapeutic or neuroprotective effect of lithium in mutant SOD1 transgenic mice.[186,187] Clinical trials testing lithium carbonate[188] or a combination of lithium and Riluzole[189] found no evidence of neuroprotection by lithium. Other compounds have been shown to modulate autophagy activity, it remains to be further studied whether and how modulation of autophagy activity could be utilized for therapeutic intervention of ALS for the benefit of the patients.

CONCLUSION REMARKS

There is convincing evidence for the involvement of autophagy in the etiology of ALS. The p62-and ubiquitin-positive protein inclusions found in neuronal and glial cells closely resemble the inclusions seen in autophagy-impaired cells. The familial ALS mutations of SOD1 and TDP-43 appear to induce abundant aggregated substrates for autophagy. In other inherited forms of ALS caused by mutations in CHMP2B, Alsin, dynactin, VAPB and FIG4, alterations or impairment of autophagy or related pathways were also demonstrated. This chapter provides a comprehensive review on the role of autophagy in familial ALS.

It is likely that the modulation of autophagy may offer an alternative therapeutic avenue for the treatment of ALS and other neurodegenerative diseases. The trials with lithium as an autophagy inducer have resulted in mixed results. The specificity of the modulation of autophagy remains a concern because of the possibility of unintended degradation of off-target substrates, thus deserving additional careful examination. It is necessary to further refine our understanding of the mechanisms of autophagic targeting and degradation in ALS. Additional studies are needed to develop therapeutic approaches specifically enhancing the autophagic clearance of aggregated proteins.

ACKNOWLEDGMENTS

This study was in part supported by the NIH grants R01NS049126 and R21AG032567 to Dr. Haining Zhu. The support from NIH/NCRR Center of Biomedical Research Excellence in the Molecular Basis of Human Disease

(COBRE, P20RR020171) is acknowledged. We thank Dr. Qingjun Wang for critically reading and commenting the manuscript.

REFERENCES

1. L. C. Wijesekera and P. N. Leigh, Amyotrophic lateral sclerosis, *Orphanet J Rare Dis* **4**: 3 (2009).
2. S. Millecamps, F. Salachas, C. Cazeneuve, P. Gordon, B. Bricka, A. Camuzat, L. Guillot-Noel, O. Russaouen, G. Bruneteau, P. F. Pradat *et al.*, SOD1, ANG, VAPB, TARDBP, and FUS mutations in familial amyotrophic lateral sclerosis: genotype-phenotype correlations, *J Med Genet* **47**: 554–560 (2010).
3. J. S. Valentine, P. A. Doucette and S. Zittin Potter, Copper-zinc superoxide dismutase and amyotrophic lateral sclerosis, *Annu Rev Biochem* **74**: 563–593 (2005).
4. M. Watanabe, M. Dykes-Hoberg, V. C. Culotta, D. L. Price, P. C. Wong and J. D. Rothstein, Histological evidence of protein aggregation in mutant SOD1 transgenic mice and in amyotrophic lateral sclerosis neural tissues, *Neurobiol Dis* **8**: 933–941 (2001).
5. Y. S. Piao, K. Wakabayashi, A. Kakita, M. Yamada, S. Hayashi, T. Morita, F. Ikuta, K. Oyanagi and H. Takahashi, Neuropathology with clinical correlations of sporadic amyotrophic lateral sclerosis: 102 autopsy cases examined between 1962 and 2000, *Brain Pathol* **13**: 10–22 (2003).
6. C. A. Ross and M. A. Poirier, Opinion: what is the role of protein aggregation in neurodegeneration?, *Nat Rev Mol Cell Biol* **6**: 891–898 (2005).
7. M. Arrasate, S. Mitra, E. S. Schweitzer, M. R. Segal and S. Finkbeiner, Inclusion body formation reduces levels of mutant huntingtin and the risk of neuronal death, *Nature* **431**: 805–810 (2004).
8. R. R. Kopito, Aggresomes, inclusion bodies and protein aggregation, *Trends Cell Biol* **10**: 524–530 (2000).
9. W. X. Ding and X. M. Yin, Sorting, recognition and activation of the misfolded protein degradation pathways through macroautophagy and the proteasome, *Autophagy* **4**: 141–150 (2008).
10. E. J. Bennett, N. F. Bence, R. Jayakumar and R. R. Kopito, Global impairment of the ubiquitin-proteasome system by nuclear or cytoplasmic protein aggregates precedes inclusion body formation, *Mol Cell* **17**: 351–365 (2005).
11. J. N. Keller, E. Dimayuga, Q. Chen, J. Thorpe, J. Gee and Q. Ding, Autophagy, proteasomes, lipofuscin, and oxidative stress in the aging brain, *Int J Biochem Cell Biol* **36**: 2376–2391 (2004).
12. N. F. Bence, R. M. Sampat and R. R. Kopito, Impairment of the ubiquitin-proteasome system by protein aggregation, *Science* **292**: 1552–1555 (2001).
13. D. H. Perlmutter, The cellular response to aggregated proteins associated with human disease, *J Clin Invest* **110**: 1219–1220 (2002).

14. V. I. Korolchuk, A. Mansilla, F. M. Menzies and D. C. Rubinsztein, Autophagy inhibition compromises degradation of ubiquitin-proteasome pathway substrates, *Mol Cell* **33**: 517–527 (2009).
15. N. Parkinson, P. G. Ince, M. O. Smith, R. Highley, G. Skibinski, P. M. Andersen, K. E. Morrison, H. S. Pall, O. Hardiman, J. Collinge *et al.*, ALS phenotypes with mutations in CHMP2B (charged multivesicular body protein 2B), *Neurology* **67**: 1074–1077 (2006).
16. J. Gal, A. L. Strom, R. Kilty, F. Zhang and H. Zhu, p62 accumulates and enhances aggregate formation in model systems of familial amyotrophic lateral sclerosis, *J Biol Chem* **282**: 11068–11077 (2007).
17. D. Seilhean, C. Cazeneuve, V. Thuries, O. Russaouen, S. Millecamps, F. Salachas, V. Meininger, E. Leguern and C. Duyckaerts, Accumulation of TDP-43 and alpha-actin in an amyotrophic lateral sclerosis patient with the K17I ANG mutation, *Acta Neuropathol* **118**: 561–573 (2009).
18. I. R. Mackenzie, E. H. Bigio, P. G. Ince, F. Geser, M. Neumann, N. J. Cairns, L. K. Kwong, M. S. Forman, J. Ravits, H. Stewart *et al.*, Pathological TDP-43 distinguishes sporadic amyotrophic lateral sclerosis from amyotrophic lateral sclerosis with SOD1 mutations, *Ann Neurol* **61**: 427–434 (2007).
19. S. Maekawa, P. N. Leigh, A. King, E. Jones, J. C. Steele, I. Bodi, C. E. Shaw, T. Hortobagyi and S. Al-Sarraj, TDP-43 is consistently co-localized with ubiquitinated inclusions in sporadic and Guam amyotrophic lateral sclerosis but not in familial amyotrophic lateral sclerosis with and without SOD1 mutations, *Neuropathology* **29**: 672–683 (2009).
20. H. Braak, A. Ludolph, D. R. Thal and K. Del Tredici, Amyotrophic lateral sclerosis: dash-like accumulation of phosphorylated TDP-43 in somatodendritic and axonal compartments of somatomotor neurons of the lower brainstem and spinal cord, *Acta Neuropathol* **120**: 67–74 (2010).
21. S. Hadano, A. Otomo, R. Kunita, K. Suzuki-Utsunomiya, A. Akatsuka, M. Koike, M. Aoki, Y. Uchiyama, Y. Itoyama and J. E. Ikeda, Loss of ALS2/Alsin exacerbates motor dysfunction in a SOD1 H46R-expressing mouse ALS model by disturbing endolysosomal trafficking, *PLoS One* **5**: e9805 (2010).
22. H. X. Deng, H. Zhai, E. H. Bigio, J. Yan, F. Fecto, K. Ajroud, M. Mishra, S. Ajroud-Driss, S. Heller, R. Sufit *et al.*, FUS-immunoreactive inclusions are a common feature in sporadic and non-SOD1 familial amyotrophic lateral sclerosis, *Ann Neurol* **67**: 739–748 (2010).
23. T. Tateishi, T. Hokonohara, R. Yamasaki, S. Miura, H. Kikuchi, A. Iwaki, H. Tashiro, H. Furuya, Y. Nagara, Y. Ohyagi *et al.*, Multiple system degeneration with basophilic inclusions in Japanese ALS patients with FUS mutation, *Acta Neuropathol* **119**: 355–364 (2010).
24. D. G. Munoz, M. Neumann, H. Kusaka, O. Yokota, K. Ishihara, S. Terada, S. Kuroda and I. R. Mackenzie, FUS pathology in basophilic inclusion body disease, *Acta Neuropathol* **118**: 617–627 (2009).

25. C. J. Ferguson, G. M. Lenk and M. H. Meisler, Defective autophagy in neurons and astrocytes from mice deficient in PI(3,5)P2, *Hum Mol Genet* **18**: 4868–4878 (2009).
26. C. Cheroni, M. Marino, M. Tortarolo, P. Veglianese, S. De Biasi, E. Fontana, L. V. Zuccarello, C. J. Maynard, N. P. Dantuma and C. Bendotti, Functional alterations of the ubiquitin-proteasome system in motor neurons of a mouse model of familial amyotrophic lateral sclerosis, *Hum Mol Genet* **18**: 82–96 (2009).
27. K. Puttaparthi, C. Wojcik, B. Rajendran, G. N. DeMartino and J. L. Elliott, Aggregate formation in the spinal cord of mutant SOD1 transgenic mice is reversible and mediated by proteasomes, *J Neurochem* **87**: 851–860 (2003).
28. M. Urushitani, J. Kurisu, K. Tsukita and R. Takahashi, Proteasomal inhibition by misfolded mutant superoxide dismutase 1 induces selective motor neuron death in familial amyotrophic lateral sclerosis, *J Neurochem* **83**: 1030–1042 (2002).
29. S. Allen, P. R. Heath, J. Kirby, S. B. Wharton, M. R. Cookson, F. M. Menzies, R. E. Banks and P. J. Shaw, Analysis of the cytosolic proteome in a cell culture model of familial amyotrophic lateral sclerosis reveals alterations to the proteasome, antioxidant defenses, and nitric oxide synthetic pathways, *J Biol Chem* **278**: 6371–6383 (2003).
30. D. H. Hyun, M. Lee, B. Halliwell and P. Jenner, Proteasomal inhibition causes the formation of protein aggregates containing a wide range of proteins, including nitrated proteins, *J Neurochem* **86**: 363–373 (2003).
31. E. Kabashi, J. N. Agar, D. M. Taylor, S. Minotti and H. D. Durham, Focal dysfunction of the proteasome: a pathogenic factor in a mouse model of amyotrophic lateral sclerosis, *J Neurochem* **89**: 1325–1335 (2004).
32. K. Aquilano, G. Rotilio and M. R. Ciriolo, Proteasome activation and nNOS downregulation in neuroblastoma cells expressing a Cu, Zn superoxide dismutase mutant involved in familial ALS, *J Neurochem* **85**: 1324–1335 (2003).
33. T. Hara, K. Nakamura, M. Matsui, A. Yamamoto, Y. Nakahara, R. Suzuki-Migishima, M. Yokoyama, K. Mishima, I. Saito, H. Okano *et al.*, Suppression of basal autophagy in neural cells causes neurodegenerative disease in mice, *Nature* **441**: 885–889 (2006).
34. M. Komatsu, S. Waguri, T. Chiba, S. Murata, J. Iwata, I. Tanida, T. Ueno, M. Koike, Y. Uchiyama, E. Kominami *et al.*, Loss of autophagy in the central nervous system causes neurodegeneration in mice, *Nature* **441**: 880–884 (2006).
35. C. Hetz, P. Thielen, S. Matus, M. Nassif, F. Court, R. Kiffin, G. Martinez, A. M. Cuervo, R. H. Brown and L. H. Glimcher, XBP-1 deficiency in the nervous system protects against amyotrophic lateral sclerosis by increasing autophagy, *Genes Dev* **23**: 2294–2306 (2009).
36. T. Nakano, K. Nakaso, K. Nakashima and E. Ohama, Expression of ubiquitin-binding protein p62 in ubiquitin-immunoreactive intraneuronal inclusions in amyotrophic lateral sclerosis with dementia: analysis of five autopsy cases with broad clinicopathological spectrum, *Acta Neuropathol* **107**: 359–364 (2004).

37. Y. Mizuno, M. Amari, M. Takatama, H. Aizawa, B. Mihara and K. Okamoto, Immunoreactivities of p62, an ubiqutin-binding protein, in the spinal anterior horn cells of patients with amyotrophic lateral sclerosis, *J Neurol Sci* **249**: 13–18 (2006).
38. M. Filimonenko, S. Stuffers, C. Raiborg, A. Yamamoto, L. Malerod, E. M. Fisher, A. Isaacs, A. Brech, H. Stenmark and A. Simonsen, Functional multivesicular bodies are required for autophagic clearance of protein aggregates associated with neurodegenerative disease, *J Cell Biol* **179**: 485–500 (2007).
39. T. E. Rusten and A. Simonsen, ESCRT functions in autophagy and associated disease, *Cell Cycle* **7**: 1166–1172 (2008).
40. T. Matsuoka, N. Fujii, A. Kondo, A. Iwaki, T. Hokonohara, H. Honda, K. Sasaki, S. O. Suzuki and T. Iwaki, An autopsied case of sporadic adult-onset amyotrophic lateral sclerosis with FUS-positive basophilic inclusions, *Neuropathology* **31**: 71–76 (2011).
41. M. H. Helfrich and L. J. Hocking, Genetics and aetiology of Pagetic disorders of bone, *Arch Biochem Biophys* **473**: 172–182 (2008).
42. E. Kuusisto, A. Salminen and I. Alafuzoff, Ubiquitin-binding protein p62 is present in neuronal and glial inclusions in human tauopathies and synucleinopathies, *Neuroreport* **12**: 2085–2090 (2001).
43. K. Zatloukal, C. Stumptner, A. Fuchsbichler, H. Heid, M. Schnoelzer, L. Kenner, R. Kleinert, M. Prinz, A. Aguzzi and H. Denk, p62 Is a common component of cytoplasmic inclusions in protein aggregation diseases, *Am J Pathol* **160**: 255–263 (2002).
44. U. Nagaoka, K. Kim, N. R. Jana, H. Doi, M. Maruyama, K. Mitsui, F. Oyama and N. Nukina, Increased expression of p62 in expanded polyglutamine-expressing cells and its association with polyglutamine inclusions, *J Neurochem* **91**: 57–68 (2004).
45. E. Kuusisto, A. Salminen and I. Alafuzoff, Early accumulation of p62 in neurofibrillary tangles in Alzheimer's disease: possible role in tangle formation, *Neuropathol Appl Neurobiol* **28**: 228–237 (2002).
46. M. Hiji, T. Takahashi, H. Fukuba, H. Yamashita, T. Kohriyama and M. Matsumoto, White matter lesions in the brain with frontotemporal lobar degeneration with motor neuron disease: TDP-43-immunopositive inclusions co-localize with p62, but not ubiquitin, *Acta Neuropathol* **116**: 183–191 (2008).
47. G. Bjørkøy, T. Lamark, A. Brech, H. Outzen, M. Perander, A. Øvervatn, H. Stenmark and T. Johansen, p62/SQSTM1 forms protein aggregates degraded by autophagy and has a protective effect on huntingtin-induced cell death, *J Cell Biol* **171**: 603–614 (2005).
48. S. Pankiv, T. H. Clausen, T. Lamark, A. Brech, J. A. Bruun, H. Outzen, A. Øvervatn, G. Bjørkøy and T. Johansen, p62/SQSTM1 binds directly to Atg8/LC3 to facilitate degradation of ubiquitinated protein aggregates by autophagy, *J Biol Chem* **282**: 24131–24145 (2007).
49. E. Shvets, E. Fass, R. Scherz-Shouval and Z. Elazar, The N-terminus and Phe52 residue of LC3 recruit p62/SQSTM1 into autophagosomes, *J Cell Sci* **121**: 2685–2695 (2008).

50. Y. Ichimura, T. Kumanomidou, Y. S. Sou, T. Mizushima, J. Ezaki, T. Ueno, E. Kominami, T. Yamane, K. Tanaka and M. Komatsu, Structural basis for sorting mechanism of p62 in selective autophagy, *J Biol Chem* **283**: 22847–22857 (2008).
51. M. Komatsu, S. Waguri, M. Koike, Y. S. Sou, T. Ueno, T. Hara, N. Mizushima, J. Iwata, J. Ezaki, S. Murata *et al.*, Homeostatic levels of p62 control cytoplasmic inclusion body formation in autophagy-deficient mice, *Cell* **131**: 1149–1163 (2007).
52. Z. Yue, Regulation of neuronal autophagy in axon: implication of autophagy in axonal function and dysfunction/degeneration, *Autophagy* **3**: 139–141 (2007).
53. D. J. Klionsky, H. Abeliovich, P. Agostinis, D. K. Agrawal, G. Aliev, D. S. Askew, M. Baba, E. H. Baehrecke, B. A. Bahr, A. Ballabio *et al.*, Guidelines for the use and interpretation of assays for monitoring autophagy in higher eukaryotes, *Autophagy* **4**: 151–175 (2008).
54. T. Ishii, T. Yanagawa, T. Kawane, K. Yuki, J. Seita, H. Yoshida and S. Bannai, Murine peritoneal macrophages induce a novel 60-kDa protein with structural similarity to a tyrosine kinase p56lck-associated protein in response to oxidative stress, *Biochem Biophys Res Commun* **226**: 456–460 (1996).
55. E. Kuusisto, T. Suuronen and A. Salminen, Ubiquitin-binding protein p62 expression is induced during apoptosis and proteasomal inhibition in neuronal cells, *Biochem Biophys Res Commun* **280**: 223–228 (2001).
56. J. Gal, A. L. Strom, D. M. Kwinter, R. Kilty, J. Zhang, P. Shi, W. Fu, M. W. Wooten and H. Zhu, Sequestosome 1/p62 links familial ALS mutant SOD1 to LC3 via an ubiquitin-independent mechanism, *J Neurochem* **111**: 1062–1073 (2009).
57. P. K. Kim, D. W. Hailey, R. T. Mullen and J. Lippincott-Schwartz, Ubiquitin signals autophagic degradation of cytosolic proteins and peroxisomes, *Proc Natl Acad Sci USA* **105**: 20567–20574 (2008).
58. Y. Kabeya, N. Mizushima, T. Ueno, A. Yamamoto, T. Kirisako, T. Noda, E. Kominami, Y. Ohsumi and T. Yoshimori, LC3, a mammalian homologue of yeast Apg8p, is localized in autophagosome membranes after processing, *EMBO J* **19**: 5720–5728 (2000).
59. A. Kuma, M. Matsui and N. Mizushima, LC3, an autophagosome marker, can be incorporated into protein aggregates independent of autophagy: caution in the interpretation of LC3 localization, *Autophagy* **3**: 323–328 (2007).
60. E. Shvets and Z. Elazar, Autophagy-independent incorporation of GFP-LC3 into protein aggregates is dependent on its interaction with p62/SQSTM1, *Autophagy* **4**: 1054–1056 (2008).
61. L. Li, X. Zhang and W. Le, Altered macroautophagy in the spinal cord of SOD1 mutant mice, *Autophagy* **4**: 290–293 (2008).
62. N. Morimoto, M. Nagai, Y. Ohta, K. Miyazaki, T. Kurata, M. Morimoto, T. Murakami, Y. Takehisa, Y. Ikeda, T. Kamiya *et al.*, Increased autophagy in transgenic mice with a G93A mutant SOD1 gene, *Brain Res* **1167**: 112–117 (2007).

63. C. Y. Chow, J. E. Landers, S. K. Bergren, P. C. Sapp, A. E. Grant, J. M. Jones, L. Everett, G. M. Lenk, D. M. McKenna-Yasek, L. S. Weisman *et al.*, Deleterious variants of FIG4, a phosphoinositide phosphatase, in patients with ALS, *Am J Hum Genet* **84**: 85–88 (2009).
64. L. E. Cox, L. Ferraiuolo, E. F. Goodall, P. R. Heath, A. Higginbottom, H. Mortiboys, H. C. Hollinger, J. A. Hartley, A. Brockington, C. E. Burness *et al.*, Mutations in CHMP2B in lower motor neuron predominant amyotrophic lateral sclerosis (ALS), *PLoS One* **5**: e9872 (2010).
65. D. R. Rosen, T. Siddique, D. Patterson, D. A. Figlewicz, P. Sapp, A. Hentati, D. Donaldson, J. Goto, J. P. O'Regan, H. X. Deng *et al.*, Mutations in Cu/Zn superoxide dismutase gene are associated with familial amyotrophic lateral sclerosis, *Nature* **362**: 59–62 (1993).
66. H. X. Deng, A. Hentati, J. A. Tainer, Z. Iqbal, A. Cayabyab, W. Y. Hung, E. D. Getzoff, P. Hu, B. Herzfeldt, R. P. Roos *et al.*, Amyotrophic lateral sclerosis and structural defects in Cu, Zn superoxide dismutase, *Science* **261**: 1047–1051 (1993).
67. M. Chattopadhyay and J. S. Valentine, Aggregation of copper-zinc superoxide dismutase in familial and sporadic ALS, *Antioxid Redox Signal* **11**: 1603–1614 (2009).
68. A. G. Reaume, J. L. Elliott, E. K. Hoffman, N. W. Kowall, R. J. Ferrante, D. F. Siwek, H. M. Wilcox, D. G. Flood, M. F. Beal, R. H. Brown, Jr. *et al.*, Motor neurons in Cu/Zn superoxide dismutase-deficient mice develop normally but exhibit enhanced cell death after axonal injury, *Nat Genet* **13**: 43–47 (1996).
69. M. E. Gurney, H. Pu, A. Y. Chiu, M. C. Dal Canto, C. Y. Polchow, D. D. Alexander, J. Caliendo, A. Hentati, Y. W. Kwon, H. X. Deng *et al.*, Motor neuron degeneration in mice that express a human Cu, Zn superoxide dismutase mutation, *Science* **264**: 1772–1775 (1994).
70. J. Wang, H. Slunt, V. Gonzales, D. Fromholt, M. Coonfield, N. G. Copeland, N. A. Jenkins and D. R. Borchelt, Copper-binding-site-null SOD1 causes ALS in transgenic mice: aggregates of non-native SOD1 delineate a common feature, *Hum Mol Genet* **12**: 2753–2764 (2003).
71. J. A. Johnston, M. J. Dalton, M. E. Gurney and R. R. Kopito, Formation of high molecular weight complexes of mutant Cu, Zn-superoxide dismutase in a mouse model for familial amyotrophic lateral sclerosis, *Proc Natl Acad Sci USA* **97**: 12571–12576 (2000).
72. L. I. Bruijn, M. K. Houseweart, S. Kato, K. L. Anderson, S. D. Anderson, E. Ohama, A. G. Reaume, R. W. Scott and D. W. Cleveland, Aggregation and motor neuron toxicity of an ALS-linked SOD1 mutant independent from wild-type SOD1, *Science* **281**: 1851–1854 (1998).
73. P. J. Hart, Pathogenic superoxide dismutase structure, folding, aggregation and turnover, *Curr Opin Chem Biol* **10**: 131–138 (2006).
74. J. P. Crow, J. B. Sampson, Y. Zhuang, J. A. Thompson and J. S. Beckman, Decreased zinc affinity of amyotrophic lateral sclerosis-associated superoxide dis-

mutase mutants leads to enhanced catalysis of tyrosine nitration by peroxynitrite, *J Neurochem* **69**: 1936–1944 (1997).

75. J. J. Goto, H. Zhu, R. J. Sanchez, A. Nersissian, E. B. Gralla, J. S. Valentine and D. E. Cabelli, Loss of *in vitro* metal ion binding specificity in mutant copper-zinc superoxide dismutases associated with familial amyotrophic lateral sclerosis, *J Biol Chem* **275**: 1007–1014 (2000).
76. A. Tiwari and L. J. Hayward, Familial amyotrophic lateral sclerosis mutants of copper/zinc superoxide dismutase are susceptible to disulfide reduction, *J Biol Chem* **278**: 5984–5992 (2003).
77. R. M. Cardoso, M. M. Thayer, M. DiDonato, T. P. Lo, C. K. Bruns, E. D. Getzoff and J. A. Tainer, Insights into Lou Gehrig's disease from the structure and instability of the A4V mutant of human Cu, Zn superoxide dismutase, *J Mol Biol* **324**: 247–256 (2002).
78. M. A. Hough, J. G. Grossmann, S. V. Antonyuk, R. W. Strange, P. A. Doucette, J. A. Rodriguez, L. J. Whitson, P. J. Hart, L. J. Hayward, J. S. Valentine *et al.*, Dimer destabilization in superoxide dismutase may result in disease-causing properties: structures of motor neuron disease mutants, *Proc Natl Acad Sci USA* **101**: 5976–5981 (2004).
79. E. L. Shipp, F. Cantini, I. Bertini, J. S. Valentine and L. Banci, Dynamic properties of the G93A mutant of copper-zinc superoxide dismutase as detected by NMR spectroscopy: implications for the pathology of familial amyotrophic lateral sclerosis, *Biochemistry* **42**: 1890–1899 (2003).
80. R. Rakhit, P. Cunningham, A. Furtos-Matei, S. Dahan, X. F. Qi, J. P. Crow, N. R. Cashman, L. H. Kondejewski and A. Chakrabartty, Oxidation-induced misfolding and aggregation of superoxide dismutase and its implications for amyotrophic lateral sclerosis, *J Biol Chem* **277**: 47551–47556 (2002).
81. C. K. Bruns and R. R. Kopito, Impaired post-translational folding of familial ALS-linked Cu, Zn superoxide dismutase mutants, *EMBO J* **26**: 855–866 (2007).
82. A. Tiwari, Z. Xu and L. J. Hayward, Aberrantly increased hydrophobicity shared by mutants of Cu, Zn-superoxide dismutase in familial amyotrophic lateral sclerosis, *J Biol Chem* **280**: 29771–29779 (2005).
83. D. R. Borchelt, M. K. Lee, H. S. Slunt, M. Guarnieri, Z. S. Xu, P. C. Wong, R. H. Brown, Jr., D. L. Price, S. S. Sisodia and D. W. Cleveland, Superoxide dismutase 1 with mutations linked to familial amyotrophic lateral sclerosis possesses significant activity, *Proc Natl Acad Sci USA* **91**: 8292–8296 (1994).
84. M. J. Lindberg, R. Bystrom, N. Boknas, P. M. Andersen and M. Oliveberg, Systematically perturbed folding patterns of amyotrophic lateral sclerosis (ALS)-associated SOD1 mutants, *Proc Natl Acad Sci USA* **102**: 9754–9759 (2005).
85. R. Rakhit, J. Robertson, C. Vande Velde, P. Horne, D. M. Ruth, J. Griffin, D. W. Cleveland, N. R. Cashman and A. Chakrabartty, An immunological epitope selective for pathological monomer-misfolded SOD1 in ALS, *Nat Med* **13**: 754–759 (2007).

86. A. L. Strom, P. Shi, F. Zhang, J. Gal, R. Kilty, L. J. Hayward and H. Zhu, Interaction of amyotrophic lateral sclerosis (ALS)-related mutant copper-zinc superoxide dismutase with the dynein-dynactin complex contributes to inclusion formation, *J Biol Chem* **283**: 22795–22805 (2008).
87. F. Zhang and H. Zhu, Intracellular conformational alterations of mutant SOD1 and the implications for fALS-associated SOD1 mutant induced motor neuron cell death, *Biochim Biophys Acta* **1760**: 404–414 (2006).
88. N. Shibata, K. Asayama, A. Hirano and M. Kobayashi, Immunohistochemical study on superoxide dismutases in spinal cords from autopsied patients with amyotrophic lateral sclerosis, *Dev Neurosci* **18**: 492–498 (1996).
89. S. Sasaki, H. Warita, T. Murakami, N. Shibata, T. Komori, K. Abe, M. Kobayashi and M. Iwata, Ultrastructural study of aggregates in the spinal cord of transgenic mice with a G93A mutant SOD1 gene, *Acta Neuropathol* **109**: 247–255 (2005).
90. M. Basso, T. Massignan, G. Samengo, C. Cheroni, S. De Biasi, M. Salmona, C. Bendotti and V. Bonetto, Insoluble mutant SOD1 is partly oligoubiquitinated in amyotrophic lateral sclerosis mice, *J Biol Chem* **281**: 33325–33335 (2006).
91. B. F. Shaw, H. L. Lelie, A. Durazo, A. M. Nersissian, G. Xu, P. K. Chan, E. B. Gralla, A. Tiwari, L. J. Hayward, D. R. Borchelt *et al.*, Detergent-insoluble aggregates associated with amyotrophic lateral sclerosis in transgenic mice contain primarily full-length, unmodified superoxide dismutase-1, *J Biol Chem* **283**: 8340–8350 (2008).
92. J. Wang, G. Xu, H. Li, V. Gonzales, D. Fromholt, C. Karch, N. G. Copeland, N. A. Jenkins and D. R. Borchelt, Somatodendritic accumulation of misfolded SOD1-L126Z in motor neurons mediates degeneration: alphaB-crystallin modulates aggregation, *Hum Mol Genet* **14**: 2335–2347 (2005).
93. D. Bergemalm, K. Forsberg, V. Srivastava, K. S. Graffmo, P. M. Andersen, T. Brannstrom, G. Wingsle and S. L. Marklund, Superoxide dismutase-1 and other proteins in inclusions from transgenic amyotrophic lateral sclerosis model mice, *J Neurochem* **114**: 408–418 (2010).
94. P. Zetterstrom, H. G. Stewart, D. Bergemalm, P. A. Jonsson, K. S. Graffmo, P. M. Andersen, T. Brannstrom, M. Oliveberg and S. L. Marklund, Soluble misfolded subfractions of mutant superoxide dismutase-1s are enriched in spinal cords throughout life in murine ALS models, *Proc Natl Acad Sci USA* **104**: 14157–14162 (2007).
95. K. Forsberg, P. A. Jonsson, P. M. Andersen, D. Bergemalm, K. S. Graffmo, M. Hultdin, J. Jacobsson, R. Rosquist, S. L. Marklund and T. Brannstrom, Novel antibodies reveal inclusions containing non-native SOD1 in sporadic ALS patients, *PLoS One* **5**: e11552 (2010).
96. T. Kabuta, Y. Suzuki and K. Wada, Degradation of amyotrophic lateral sclerosis-linked mutant Cu, Zn-superoxide dismutase proteins by macroautophagy and the proteasome, *J Biol Chem* **281**: 30524–30533 (2006).

97. C. Gomes, C. Escrevente and J. Costa, Mutant superoxide dismutase 1 overexpression in NSC-34 cells: effect of trehalose on aggregation, TDP-43 localization and levels of co-expressed glycoproteins, *Neurosci Lett* **475**: 145–149 (2010).
98. S. Sarkar and D. C. Rubinsztein, Small molecule enhancers of autophagy for neurodegenerative diseases, *Mol Biosyst* **4**: 895–901 (2008).
99. V. Crippa, D. Sau, P. Rusmini, A. Boncoraglio, E. Onesto, E. Bolzoni, M. Galbiati, E. Fontana, M. Marino, S. Carra *et al.*, The small heat shock protein B8 (HspB8) promotes autophagic removal of misfolded proteins involved in amyotrophic lateral sclerosis (ALS), *Hum Mol Genet* **19**: 3440–3456 (2010).
100. G. Dobrowolny, M. Aucello, E. Rizzuto, S. Beccafico, C. Mammucari, S. Boncompagni, S. Belia, F. Wannenes, C. Nicoletti, Z. Del Prete *et al.*, Skeletal muscle is a primary target of SOD1G93A-mediated toxicity, *Cell Metab* **8**: 425–436 (2008).
101. C. Mammucari, G. Milan, V. Romanello, E. Masiero, R. Rudolf, P. Del Piccolo, S. J. Burden, R. Di Lisi, C. Sandri, J. Zhao *et al.*, FoxO3 controls autophagy in skeletal muscle *in vivo*, *Cell Metab* **6**: 458–471 (2007).
102. F. Meissner, K. Molawi and A. Zychlinsky, Mutant superoxide dismutase 1-induced IL-1beta accelerates ALS pathogenesis, *Proc Natl Acad Sci USA* **107**: 13046–13050 (2010).
103. S. Hadano, C. K. Hand, H. Osuga, Y. Yanagisawa, A. Otomo, R. S. Devon, N. Miyamoto, J. Showguchi-Miyata, Y. Okada, R. Singaraja *et al.*, A gene encoding a putative GTPase regulator is mutated in familial amyotrophic lateral sclerosis 2, *Nat Genet* **29**: 166–173 (2001).
104. Y. Yang, A. Hentati, H. X. Deng, O. Dabbagh, T. Sasaki, M. Hirano, W. Y. Hung, K. Ouahchi, J. Yan, A. C. Azim *et al.*, The gene encoding alsin, a protein with three guanine-nucleotide exchange factor domains, is mutated in a form of recessive amyotrophic lateral sclerosis, *Nat Genet* **29**: 160–165 (2001).
105. A. Otomo, S. Hadano, T. Okada, H. Mizumura, R. Kunita, H. Nishijima, J. Showguchi-Miyata, Y. Yanagisawa, E. Kohiki, E. Suga *et al.*, ALS2, a novel guanine nucleotide exchange factor for the small GTPase Rab5, is implicated in endosomal dynamics, *Hum Mol Genet* **12**: 1671–1687 (2003).
106. K. Kanekura, Y. Hashimoto, T. Niikura, S. Aiso, M. Matsuoka and I. Nishimoto, Alsin, the product of ALS2 gene, suppresses SOD1 mutant neurotoxicity through RhoGEF domain by interacting with SOD1 mutants, *J Biol Chem* **279**: 19247–19256 (2004).
107. T. Arai, M. Hasegawa, H. Akiyama, K. Ikeda, T. Nonaka, H. Mori, D. Mann, K. Tsuchiya, M. Yoshida, Y. Hashizume *et al.*, TDP-43 is a component of ubiquitin-positive tau-negative inclusions in frontotemporal lobar degeneration and amyotrophic lateral sclerosis, *Biochem Biophys Res Commun* **351**: 602–611 (2006).
108. M. Neumann, D. M. Sampathu, L. K. Kwong, A. C. Truax, M. C. Micsenyi, T. T. Chou, J. Bruce, T. Schuck, M. Grossman, C. M. Clark *et al.*, Ubiquitinated TDP-43 in frontotemporal lobar degeneration and amyotrophic lateral sclerosis, *Science* **314**: 130–133 (2006).

109. L. K. Kwong, M. Neumann, D. M. Sampathu, V. M. Lee and J. Q. Trojanowski, TDP-43 proteinopathy: the neuropathology underlying major forms of sporadic and familial frontotemporal lobar degeneration and motor neuron disease, *Acta Neuropathol* **114**: 63–70 (2007).
110. G. S. Pesiridis, V. M. Lee and J. Q. Trojanowski, Mutations in TDP-43 link glycine-rich domain functions to amyotrophic lateral sclerosis, *Hum Mol Genet* **18**: R156–R162 (2009).
111. B. S. Johnson, D. Snead, J. J. Lee, J. M. McCaffery, J. Shorter and A. D. Gitler, TDP-43 is intrinsically aggregation-prone, and amyotrophic lateral sclerosis-linked mutations accelerate aggregation and increase toxicity, *J Biol Chem* **284**: 20329–20339 (2009).
112. T. Arai, M. Hasegawa, T. Nonoka, F. Kametani, M. Yamashita, M. Hosokawa, K. Niizato, K. Tsuchiya, Z. Kobayashi, K. Ikeda *et al.*, Phosphorylated and cleaved TDP-43 in ALS, FTLD and other neurodegenerative disorders and in cellular models of TDP-43 proteinopathy, *Neuropathology* **30**: 170–181 (2010).
113. C. F. Tan, H. Eguchi, A. Tagawa, O. Onodera, T. Iwasaki, A. Tsujino, M. Nishizawa, A. Kakita and H. Takahashi, TDP-43 immunoreactivity in neuronal inclusions in familial amyotrophic lateral sclerosis with or without SOD1 gene mutation, *Acta Neuropathol* **113**: 535–542 (2007).
114. E. L. Tudor, C. M. Galtrey, M. S. Perkinton, K. F. Lau, K. J. De Vos, J. C. Mitchell, S. Ackerley, T. Hortobagyi, E. Vamos, P. N. Leigh *et al.*, Amyotrophic lateral sclerosis mutant vesicle-associated membrane protein-associated protein-B transgenic mice develop TAR-DNA-binding protein-43 pathology, *Neuroscience* **167**: 774–785 (2010).
115. L. M. Igaz, L. K. Kwong, Y. Xu, A. C. Truax, K. Uryu, M. Neumann, C. M. Clark, L. B. Elman, B. L. Miller, M. Grossman *et al.*, Enrichment of C-terminal fragments in TAR DNA-binding protein-43 cytoplasmic inclusions in brain but not in spinal cord of frontotemporal lobar degeneration and amyotrophic lateral sclerosis, *Am J Pathol* **173**: 182–194 (2008).
116. L. M. Igaz, L. K. Kwong, A. Chen-Plotkin, M. J. Winton, T. L. Unger, Y. Xu, M. Neumann, J. Q. Trojanowski and V. M. Lee, Expression of TDP-43 C-terminal fragments *in vitro* recapitulates pathological features of TDP-43 proteinopathies, *J Biol Chem* **284**: 8516–8524 (2009).
117. T. Sanelli, S. Xiao, P. Horne, J. Bilbao, L. Zinman and J. Robertson, Evidence that TDP-43 is not the major ubiquitinated target within the pathological inclusions of amyotrophic lateral sclerosis, *J Neuropathol Exp Neurol* **66**: 1147–1153 (2007).
118. A. Caccamo, S. Majumder, J. J. Deng, Y. Bai, F. B. Thornton and S. Oddo, Rapamycin rescues TDP-43 mislocalization and the associated low molecular mass neurofilament instability, *J Biol Chem* **284**: 27416–27424 (2009).
119. S. H. Kim, Y. Shi, K. A. Hanson, L. M. Williams, R. Sakasai, M. J. Bowler and R. S. Tibbetts, Potentiation of amyotrophic lateral sclerosis (ALS)-associated TDP-43 aggregation by the proteasome-targeting factor, ubiquilin 1, *J Biol Chem* **284**: 8083–8092 (2009).

120. K. A. Hanson, S. H. Kim, D. A. Wassarman and R. S. Tibbetts, Ubiquilin modifies TDP-43 toxicity in a Drosophila model of amyotrophic lateral sclerosis (ALS), *J Biol Chem* **285**: 11068–11072 (2010).
121. M. Urushitani, T. Sato, H. Bamba, Y. Hisa and I. Tooyama, Synergistic effect between proteasome and autophagosome in the clearance of polyubiquitinated TDP-43, *J Neurosci Res* **88**: 784–797 (2010).
122. X. Wang, H. Fan, Z. Ying, B. Li, H. Wang and G. Wang, Degradation of TDP-43 and its pathogenic form by autophagy and the ubiquitin-proteasome system, *Neurosci Lett* **469**: 112–116 (2010).
123. A. L. Nishimura, M. Mitne-Neto, H. C. Silva, A. Richieri-Costa, S. Middleton, D. Cascio, F. Kok, J. R. Oliveira, T. Gillingwater, J. Webb *et al.*, A mutation in the vesicle-trafficking protein VAPB causes late-onset spinal muscular atrophy and amyotrophic lateral sclerosis, *Am J Hum Genet* **75**: 822–831 (2004).
124. J. Wu and R. J. Kaufman, From acute ER stress to physiological roles of the unfolded protein response, *Cell Death Differ* **13**: 374–384 (2006).
125. C. Gkogkas, S. Middleton, A. M. Kremer, C. Wardrope, M. Hannah, T. H. Gillingwater and P. Skehel, VAPB interacts with and modulates the activity of ATF6, *Hum Mol Genet* **17**: 1517–1526 (2008).
126. K. Kanekura, I. Nishimoto, S. Aiso and M. Matsuoka, Characterization of amyotrophic lateral sclerosis-linked P56S mutation of vesicle-associated membrane protein-associated protein B (VAPB/ALS8), *J Biol Chem* **281**: 30223–30233 (2006).
127. K. Kanekura, H. Suzuki, S. Aiso and M. Matsuoka, ER stress and unfolded protein response in amyotrophic lateral sclerosis, *Mol Neurobiol* **39**: 81–89 (2009).
128. S. Saxena, E. Cabuy and P. Caroni, A role for motoneuron subtype-selective ER stress in disease manifestations of FALS mice, *Nat Neurosci* **12**: 627–636 (2009).
129. G. Skibinski, N. J. Parkinson, J. M. Brown, L. Chakrabarti, S. L. Lloyd, H. Hummerich, J. E. Nielsen, J. R. Hodges, M. G. Spillantini, T. Thusgaard *et al.*, Mutations in the endosomal ESCRTIII-complex subunit CHMP2B in frontotemporal dementia, *Nat Genet* **37**: 806–808 (2005).
130. P. B. Gordon and P. O. Seglen, Prelysosomal convergence of autophagic and endocytic pathways, *Biochem Biophys Res Commun* **151**: 40–47 (1988).
131. C. Munch, A. Rosenbohm, A. D. Sperfeld, I. Uttner, S. Reske, B. J. Krause, R. Sedlmeier, T. Meyer, C. O. Hanemann, G. Stumm *et al.*, Heterozygous R1101K mutation of the DCTN1 gene in a family with ALS and FTD, *Ann Neurol* **58**: 777–780 (2005).
132. C. Munch, R. Sedlmeier, T. Meyer, V. Homberg, A. D. Sperfeld, A. Kurt, J. Prudlo, G. Peraus, C. O. Hanemann, G. Stumm *et al.*, Point mutations of the p150 subunit of dynactin (DCTN1) gene in ALS, *Neurology* **63**: 724–726 (2004).

133. I. Puls, C. Jonnakuty, B. H. LaMonte, E. L. Holzbaur, M. Tokito, E. Mann, M. K. Floeter, K. Bidus, D. Drayna, S. J. Oh *et al.*, Mutant dynactin in motor neuron disease, *Nat Genet* **33**: 455–456 (2003).
134. M. Hafezparast, R. Klocke, C. Ruhrberg, A. Marquardt, A. Ahmad-Annuar, S. Bowen, G. Lalli, A. S. Witherden, H. Hummerich, S. Nicholson *et al.*, Mutations in dynein link motor neuron degeneration to defects in retrograde transport, *Science* **300**: 808–812 (2003).
135. B. H. LaMonte, K. E. Wallace, B. A. Holloway, S. S. Shelly, J. Ascano, M. Tokito, T. Van Winkle, D. S. Howland and E. L. Holzbaur, Disruption of dynein/dynactin inhibits axonal transport in motor neurons causing late-onset progressive degeneration, *Neuron* **34**: 715–727 (2002).
136. D. Kieran, M. Hafezparast, S. Bohnert, J. R. Dick, J. Martin, G. Schiavo, E. M. Fisher and L. Greensmith, A mutation in dynein rescues axonal transport defects and extends the life span of ALS mice, *J Cell Biol* **169**: 561–567 (2005).
137. A. C. Breuer and M. B. Atkinson, Fast axonal transport alterations in amyotrophic lateral sclerosis (ALS) and in parathyroid hormone (PTH)-treated axons, *Cell Motil Cytoskeleton* **10**: 321–330 (1988).
138. A. C. Breuer, M. P. Lynn, M. B. Atkinson, S. M. Chou, A. J. Wilbourn, K. E. Marks, J. E. Culver and E. J. Fleegler, Fast axonal transport in amyotrophic lateral sclerosis: an intra-axonal organelle traffic analysis, *Neurology* **37**: 738–748 (1987).
139. J. F. Collard, F. Cote and J. P. Julien, Defective axonal transport in a transgenic mouse model of amyotrophic lateral sclerosis, *Nature* **375**: 61–64 (1995).
140. E. Perlson, G. B. Jeong, J. L. Ross, R. Dixit, K. E. Wallace, R. G. Kalb and E. L. Holzbaur, A switch in retrograde signaling from survival to stress in rapid-onset neurodegeneration, *J Neurosci* **29**: 9903–9917 (2009).
141. S. Sasaki and M. Iwata, Impairment of fast axonal transport in the proximal axons of anterior horn neurons in amyotrophic lateral sclerosis, *Neurology* **47**: 535–540 (1996).
142. H. Warita, Y. Itoyama and K. Abe, Selective impairment of fast anterograde axonal transport in the peripheral nerves of asymptomatic transgenic mice with a G93A mutant SOD1 gene, *Brain Res* **819**: 120–131 (1999).
143. T. L. Williamson and D. W. Cleveland, Slowing of axonal transport is a very early event in the toxicity of ALS-linked SOD1 mutants to motor neurons, *Nat Neurosci* **2**: 50–56 (1999).
144. B. Zhang, P. Tu, F. Abtahian, J. Q. Trojanowski and V. M. Lee, Neurofilaments and orthograde transport are reduced in ventral root axons of transgenic mice that express human SOD1 with a G93A mutation, *J Cell Biol* **139**: 1307–1315 (1997).
145. P. Shi, A. L. Strom, J. Gal and H. Zhu, Effects of ALS-related SOD1 mutants on dynein- and KIF5-mediated retrograde and anterograde axonal transport, *Biochim Biophys Acta* **1802**: 707–716 (2010).
146. F. Zhang, A. L. Strom, K. Fukada, S. Lee, L. J. Hayward and H. Zhu, Interaction between familial amyotrophic lateral sclerosis (ALS)-linked SOD1 mutants and the dynein complex, *J Biol Chem* **282**: 16691–16699 (2007).

147. L. A. Ligon, B. H. LaMonte, K. E. Wallace, N. Weber, R. G. Kalb and E. L. Holzbaur, Mutant superoxide dismutase disrupts cytoplasmic dynein in motor neurons, *Neuroreport* **16**: 533–536 (2005).
148. J. A. Johnston, M. E. Illing and R. R. Kopito, Cytoplasmic dynein/dynactin mediates the assembly of aggresomes, *Cell Motil Cytoskeleton* **53**: 26–38 (2002).
149. A. M. El-Kadi, V. Bros-Facer, W. Deng, A. Philpott, E. Stoddart, G. Banks, G. S. Jackson, E. M. Fisher, M. R. Duchen, L. Greensmith *et al.*, The legs at odd angles (Loa) mutation in cytoplasmic dynein ameliorates mitochondrial function in SOD1G93A mouse model for motor neuron disease, *J Biol Chem* **285**: 18627–18639 (2010).
150. M. Teuchert, D. Fischer, B. Schwalenstoecker, H. J. Habisch, T. M. Bockers and A. C. Ludolph, A dynein mutation attenuates motor neuron degeneration in SOD1(G93A) mice, *Exp Neurol* **198**: 271–274 (2006).
151. A. L. Strom, J. Gal, P. Shi, E. J. Kasarskis, L. J. Hayward and H. Zhu, Retrograde axonal transport and motor neuron disease, *J Neurochem* **106**: 495–505 (2008).
152. S. Kimura, T. Noda and T. Yoshimori, Dynein-dependent movement of autophagosomes mediates efficient encounters with lysosomes, *Cell Struct Funct* **33**: 109–122 (2008).
153. L. Jahreiss, F. M. Menzies and D. C. Rubinsztein, The itinerary of autophagosomes: from peripheral formation to kiss-and-run fusion with lysosomes, *Traffic* **9**: 574–587 (2008).
154. K. Katsumata, J. Nishiyama, T. Inoue, N. Mizushima, J. Takeda and M. Yuzaki, Dynein- and activity-dependent retrograde transport of autophagosomes in neuronal axons, *Autophagy* **6**: 378–385 (2010).
155. M. Yamamoto, S. O. Suzuki and M. Himeno, The effects of dynein inhibition on the autophagic pathway in glioma cells, *Neuropathology* **30**: 1–6 (2010).
156. J. K. Burkhardt, C. J. Echeverri, T. Nilsson and R. B. Vallee, Overexpression of the dynamitin (p50) subunit of the dynactin complex disrupts dynein-dependent maintenance of membrane organelle distribution, *J Cell Biol* **139**: 469–484 (1997).
157. B. Ravikumar, A. Acevedo-Arozena, S. Imarisio, Z. Berger, C. Vacher, C. J. O'Kane, S. D. Brown and D. C. Rubinsztein, Dynein mutations impair autophagic clearance of aggregate-prone proteins, *Nat Genet* **37**: 771–776 (2005).
158. D. C. Rubinsztein, B. Ravikumar, A. Acevedo-Arozena, S. Imarisio, C. J. O'Kane and S. D. Brown, Dyneins, autophagy, aggregation and neurodegeneration, *Autophagy* **1**: 177–178 (2005).
159. Y. Batlevi, D. N. Martin, U. B. Pandey, C. R. Simon, C. M. Powers, J. P. Taylor and E. H. Baehrecke, Dynein light chain 1 is required for autophagy, protein clearance, and cell death in Drosophila, *Proc Natl Acad Sci USA* **107**: 742–747 (2010).
160. F. M. Laird, M. H. Farah, S. Ackerley, A. Hoke, N. Maragakis, J. D. Rothstein, J. Griffin, D. L. Price, L. J. Martin and P. C. Wong, Motor neuron disease occurring in a mutant dynactin mouse model is characterized by defects in vesicular trafficking, *J Neurosci* **28**: 1997–2005 (2008).
161. A. C. Rutherford, C. Traer, T. Wassmer, K. Pattni, M. V. Bujny, J. G. Carlton, H. Stenmark and P. J. Cullen, The mammalian phosphatidylinositol 3-phosphate

5-kinase (PIKfyve) regulates endosome-to-TGN retrograde transport, *J Cell Sci* **119**: 3944–3957 (2006).

162. Y. Zhang, S. N. Zolov, C. Y. Chow, S. G. Slutsky, S. C. Richardson, R. C. Piper, B. Yang, J. J. Nau, R. J. Westrick, S. J. Morrison *et al.*, Loss of Vac14, a regulator of the signaling lipid phosphatidylinositol 3,5-bisphosphate, results in neurodegeneration in mice, *Proc Natl Acad Sci USA* **104**: 17518–17523 (2007).
163. C. Y. Chow, Y. Zhang, J. J. Dowling, N. Jin, M. Adamska, K. Shiga, K. Szigeti, M. E. Shy, J. Li, X. Zhang *et al.*, Mutation of FIG4 causes neurodegeneration in the pale tremor mouse and patients with CMT4J, *Nature* **448**: 68–72 (2007).
164. T. J. Kwiatkowski, Jr., D. A. Bosco, A. L. Leclerc, E. Tamrazian, C. R. Vanderburg, C. Russ, A. Davis, J. Gilchrist, E. J. Kasarskis, T. Munsat *et al.*, Mutations in the FUS/TLS gene on chromosome 16 cause familial amyotrophic lateral sclerosis, *Science* **323**: 1205–1208 (2009).
165. C. Vance, B. Rogelj, T. Hortobagyi, K. J. De Vos, A. L. Nishimura, J. Sreedharan, X. Hu, B. Smith, D. Ruddy, P. Wright *et al.*, Mutations in FUS, an RNA processing protein, cause familial amyotrophic lateral sclerosis type 6, *Science* **323**: 1208–1211 (2009).
166. J. Gal, J. Zhang, D. M. Kwinter, J. Zhai, H. Jia, J. Jia and H. Zhu, Nuclear localization sequence of FUS and induction of stress granules by ALS mutants, *Neurobiol Aging* **32**: 2323.e27–2323.e40.
167. D. A. Bosco, N. Lemay, H. K. Ko, H. Zhou, C. Burke, T. J. Kwiatkowski, Jr., P. Sapp, D. McKenna-Yasek, R. H. Brown, Jr. and L. J. Hayward, Mutant FUS proteins that cause amyotrophic lateral sclerosis incorporate into stress granules, *Hum Mol Genet* **19**: 4160–4175 (2010).
168. D. Dormann, R. Rodde, D. Edbauer, E. Bentmann, I. Fischer, A. Hruscha, M. E. Than, I. R. Mackenzie, A. Capell, B. Schmid *et al.*, ALS-associated fused in sarcoma (FUS) mutations disrupt Transportin-mediated nuclear import, *EMBO J* **29**: 2841–2857.
169. Y. Kawaguchi, J. J. Kovacs, A. McLaurin, J. M. Vance, A. Ito and T. P. Yao, The deacetylase HDAC6 regulates aggresome formation and cell viability in response to misfolded protein stress, *Cell* **115**: 727–738 (2003).
170. A. Iwata, B. E. Riley, J. A. Johnston and R. R. Kopito, HDAC6 and microtubules are required for autophagic degradation of aggregated huntingtin, *J Biol Chem* **280**: 40282–40292 (2005).
171. J. Y. Lee, H. Koga, Y. Kawaguchi, W. Tang, E. Wong, Y. S. Gao, U. B. Pandey, S. Kaushik, E. Tresse, J. Lu *et al.*, HDAC6 controls autophagosome maturation essential for ubiquitin-selective quality-control autophagy, *EMBO J* **29**: 969–980 (2010).
172. U. B. Pandey, Z. Nie, Y. Batlevi, B. A. McCray, G. P. Ritson, N. B. Nedelsky, S. L. Schwartz, N. A. DiProspero, M. A. Knight, O. Schuldiner *et al.*, HDAC6 rescues neurodegeneration and provides an essential link between autophagy and the UPS, *Nature* **447**: 859–863 (2007).

173. S. H. Kim, N. Shanware, M. J. Bowler and R. S. Tibbetts, Amylotrophic lateral sclerosis-associated proteins TDP-43 and FUS/TLS function in a common biochemical complex to coregulate HDAC6 mRNA, *J Biol Chem* **285**: 34097–34105 (2010).
174. H. Chen and D. C. Chan, Mitochondrial dynamics — fusion, fission, movement, and mitophagy — in neurodegenerative diseases, *Hum Mol Genet* **18**: R169–176 (2009).
175. D. C. Rubinsztein, The roles of intracellular protein-degradation pathways in neurodegeneration, *Nature* **443**: 780–786 (2006).
176. C. T. Chu, A pivotal role for PINK1 and autophagy in mitochondrial quality control: implications for Parkinson disease, *Hum Mol Genet* **19**: R28–R37 (2010).
177. P. Shi, J. Gal, D. M. Kwinter, X. Liu and H. Zhu, Mitochondrial dysfunction in amyotrophic lateral sclerosis, *Biochim Biophys Acta* **1802**: 45–51 (2010).
178. P. Shi, Y. Wei, J. Zhang, J. Gal and H. Zhu, Mitochondrial dysfunction is a converging point of multiple pathological pathways in amyotrophic lateral sclerosis, *J Alzheimers Dis* **20**(Suppl 2): S311–S324 (2010).
179. F. Fornai, P. Longone, L. Cafaro, O. Kastsiuchenka, M. Ferrucci, M. L. Manca, G. Lazzeri, A. Spalloni, N. Bellio, P. Lenzi *et al.*, Lithium delays progression of amyotrophic lateral sclerosis, *Proc Natl Acad Sci USA* **105**: 2052–2057 (2008).
180. F. Fornai, P. Longone, M. Ferrucci, P. Lenzi, C. Isidoro, S. Ruggieri and A. Paparelli, Autophagy and amyotrophic lateral sclerosis: the multiple roles of lithium, *Autophagy* **4**: 527–530 (2008).
181. J. H. Shin, S. I. Cho, H. R. Lim, J. K. Lee, Y. A. Lee, J. S. Noh, I. S. Joo, K. W. Kim and B. J. Gwag, Concurrent administration of Neu2000 and lithium produces marked improvement of motor neuron survival, motor function, and mortality in a mouse model of amyotrophic lateral sclerosis, *Mol Pharmacol* **71**: 965–975 (2007).
182. L. Pasquali, P. Longone, C. Isidoro, S. Ruggieri, A. Paparelli and F. Fornai, Autophagy, lithium, and amyotrophic lateral sclerosis, *Muscle Nerve* **40**: 173–194 (2009).
183. A. Criollo, M. C. Maiuri, E. Tasdemir, I. Vitale, A. A. Fiebig, D. Andrews, J. Molgo, J. Diaz, S. Lavandero, F. Harper *et al.*, Regulation of autophagy by the inositol trisphosphate receptor, *Cell Death Differ* **14**: 1029–1039 (2007).
184. S. Sarkar, R. A. Floto, Z. Berger, S. Imarisio, A. Cordenier, M. Pasco, L. J. Cook and D. C. Rubinsztein, Lithium induces autophagy by inhibiting inositol monophosphatase, *J Cell Biol* **170**: 1101–1111 (2005).
185. J. Caldero, N. Brunet, O. Tarabal, L. Piedrafita, M. Hereu, V. Ayala and J. E. Esquerda, Lithium prevents excitotoxic cell death of motoneurons in organotypic slice cultures of spinal cord, *Neuroscience* **165**: 1353–1369 (2010).
186. C. Pizzasegola, I. Caron, C. Daleno, A. Ronchi, C. Minoia, M. T. Carri and C. Bendotti, Treatment with lithium carbonate does not improve disease progression in two different strains of SOD1 mutant mice, *Amyotroph Lateral Scler* **10**: 221–228 (2009).

CHAPTER 10

AUTOPHAGY FAILURE IN ALZHEIMER'S DISEASE AND LYSOSOMAL STORAGE DISORDERS: A COMMON PATHWAY TO NEURODEGENERATION?

Devin M. Wolfe and Ralph A. Nixon

ABSTRACT

In many lysosomal storage disorders (LSDs), neurodegeneration and progressive cognitive decline are prominent features, reflecting the particularly critical role lysosomes play in neuronal function and survival. While neurodegenerative LSDs develop at the opposite end of the age spectrum from Alzheimer's disease, the neuropathology of several includes lesions, such as neurofibrillary tangles, otherwise typical of aging related dementias, including AD. Many LSDs also share with AD a characteristic form of neuritic dystrophy in which autophagic vacuoles are the predominant organelle accumulating within axon and dendritic swellings. In both AD and a growing number of LSDs, the basis for this dystrophy, and possibly other common neurodegenerative pathology, seems to be a defect in neuronal autophagy and most often impaired lysosomal proteolysis. In this chapter, we discuss the various ways that the pathobiology of these different disorders may inform each other and may suggest common strategies for therapeutic intervention.

INTRODUCTION

Alzheimer's disease (AD), the most common form of neurodegenerative dementia, is characterized by progressive cognitive decline beginning in later life (65+) although familial AD (FAD) may have an early onset in the 50's, 40's or even 30's. Lysosomal storage disorders (LSDs), by contrast, are a less common group of metabolic disorders varying in symptomatology, progression, and age of onset, with most appearing during embryonic or early postnatal development. Unlike Alzheimer's disease, the major pathological consequences in lysosomal storage disorders may present in a number of organs, although the brain is very often prominently involved.[1,2] Moreover, while rare forms of AD may be caused by autosomal dominant mutations in amyloid-β precursor protein and presenilins 1 and 2, AD is predominantly "sporadic," albeit influenced by many genes, including most significantly allelic variation of the Apolipoprotein E gene.[3] Lysosomal storage disorders are all monogenically inherited and most of the causative genetic mutations are autosomal recessive.[4,5] Despite these disparate etiologies and symptom profiles, AD and LSDs exhibit significant parallels in their pathobiology, a commonality that current research suggests stems in large part from defects in autophagy, the major cellular pathway for the degradation of long-lived proteins and organelles.

Autophagy is important for the survival and function of all cell types, and is of particular importance in neurons.[6] Its failure leading to neurodegenerative consequences in AD and in a number of lysosomal storage disorders including: Pompe disease, Niemann–Pick type C (NPC), the neuronal ceroid lipofuscinoses (NCLs), mucolipidosis type IV (MLIV), GM1 and GM2-gangliosidosis and multiple sulphatase deficiency[7–11] has reinforced the significance of lysosomal system dysfunction in the pathogenesis of these and possibly other neurodegenerative diseases. In this chapter, we will review our current understanding of autophagy in the progression of AD and of LSDs that have a predominant neurological component and we will consider how its dysfunction can promote severe neurodegeneration in these disorders. The sharing of diverse neuropathological features between AD and some LSDs, as discussed here, yields new clues about pathophysiological mechanisms in AD, which remain incompletely understood.

AUTOPHAGY IN THE PATHOGENESIS OF ALZHEIMER'S DISEASE

Macroautophagy (hereafter autophagy) is an essential cellular pathway by which long-lived proteins and whole organelles are sequestered, trafficked and delivered to the lysosome for degradation. Its initiation, regulation and progression have been thoroughly described in a number of excellent reviews[12–14] (Figure 1). Autophagy has been shown to influence pathogenesis in a broad spectrum of diseases, including cancer, heart disease, infectious diseases, and importantly neurodegenerative disorders.[15–17] As such, autophagy dysfunction has been studied as a cause and consequence of many of these diseases, and its regulation has become a target of therapeutic interventions for several human disorders.[18–20] Varying degrees of altered autophagy function are evident in many

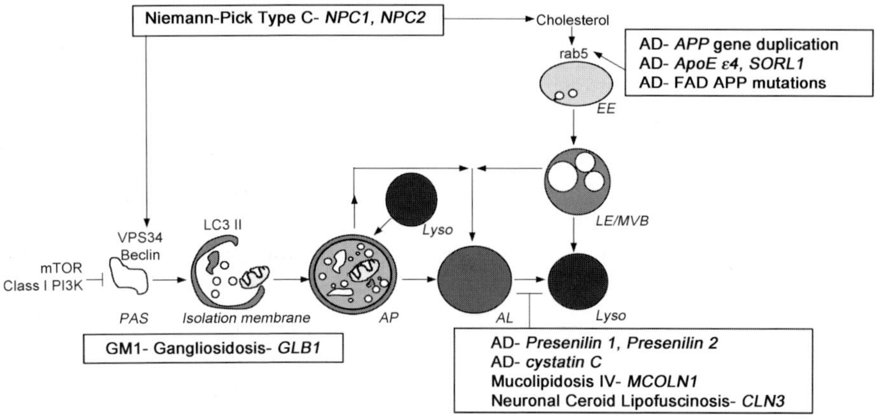

Figure 1. The autophagic pathway and its involvement in Alzheimer's disease and lysosomal storage disorders. Neuronal autophagy operates rapidly to clear unwanted organelles and long-lived proteins from the cell. Induction regulation occurs upstream of the autophagy machinery via the action of Class I PI3-kinase and mTOR kinase. Autophagy initiation is regulated via the coordinated action of at least five protein complexes which together subserve functions similar to the yeast "pre-autophagosomal structure (PAS)." Among these complexes is the VPS34 (class III PI3K)/Beclin 1 complex which has been implicated in several disorders, including Niemann-Pick type C. Initiation begins with omegasome formation from the ER and other membrane sources (not depicted). The omegasome coordinates formation of the isolation membrane, where engulfment of cytosolic material occurs. Via the action of LC3 and its associated conjugation system, the isolation membrane is elongated and eventually closed to form the autophagosome (AP). This mature autophagosome then fuses with a lysosome (LY) to form an autolysosome (AL) where its contents are degraded by lysosomal hydrolytic enzymes, in an acidic environment, and the resulting degradation products, i.e. amino acids, can then be recycled by the cell (not shown). Substrates internalized by endocytosis within early endosomes (EE) and delivered to the late endosome/multivesicular body (LE/MVB) can then reach lysosomes for degradation directly or by entering the autophagy pathway by fusing with the autophagosome to form an amphisome — an intermediate in the maturation to the autolysosome. In the case of AD and many lysosomal storage disorders, the normal flux of this pathway from initiation to degradation is altered, resulting in the accumulation of various vacuolar intermediates in the autophagy pathway. Alzheimer's disease can impact the homeostasis of the lysosome via dysregulation of its lumenal pH, which delays the proteolysis of cargo in autolysosomes and amphisomes leading to accumulation of these autophagic intermediates in the cell. Similar proteolytic defects have been observed in the Neuronal Ceroid Lipofuscinoses and Mucolipidosis IV. Conversely, Niemann–Pick type C and GM1-gangliosidosis can enhance the induction of autophagy. NPC and AD also can alter the endocytic pathway via Rab5, resulting in an increased rate of endocytosis and enlarged endosomes, potentially increasing the degradative burden on the lysosome.

neurodegenerative diseases, including Alzheimer's disease, Parkinson's disease, Huntington's disease and a number of the primary lysosomal storage disorders although the nature of these alterations and their pathogenic significance in most of these disorders is not well understood.[21] The autophagy-related pathology in Alzheimer's disease has been considered uniquely extensive among late age onset neurodegenerative diseases and its prominence and nature have drawn comparisons to certain LSDs.

Since their description over a century ago, senile plaques and neurofibrillary tangles have served as the pathognmonic neuropathological lesions of Alzheimer's disease.[22] Aggregated forms of the peptide amyloid β and the protein tau are prominent proteinaceous components of the senile plaque and tangle, respectively, and the pathogenic roles of these proteins collecting inside and outside of neurons (Figure 2) are considered significant albeit still controversial. A second defining feature of the senile plaque is a plexus of "dystrophic neurites" (DNs)[23] — grossly swollen segments of axons and dendrites that form a network within and around the senile plaque, which for this reason is often called the "neuritic plaque." DNs are also scattered widely throughout the neuropil of affected brain regions (Figure 2). Although commonly detected by their increased content of hyperphosphorylated tau, DNs in AD almost exclusively contain autophagy related organelles. Ultrastructural images of DNs in AD brain show the bulk of the cytoplasm in these swellings to be replaced by single or double-membrane limited autophagic vacuoles filled with incompletely digested waste proteins and membrane constituents.[24] Although the presence of many double membrane-limited vacuoles in DNs suggests an autophagosome origin of these structures, the amorphous electron-dense cellular debris within the lumens of most vacuoles implies at least some degree of fusion of lysosomes or late endosomes with autophagosomes in DNs and limited substrate hydrolysis. Subsequent proteolytic clearance, however, is markedly impaired within AVs, including in many cases the digestion of the inner membrane of the former autophagosome.[24] This morphological pattern associated with proteolytic clearance deficits is quite distinct from the collections of electron dense lysosomes that appear when autophagy is experimentally over-induced.[25] These observations have suggested that the defect lies in the penultimate step of autophagy, autolysosomal maturation and substrate proteolysis. The pattern in AV accumulation in AD is also distinct from that in dystrophic neurites in many other neurodegenerative conditions where different admixtures of vesicular and fibrous organelles may collect and reflect major differences in mechanisms leading to transport failure and organelle accumulation in these disorders.[26]

The accumulation of autolysosomes during autophagy resulting from failure of lysosomal proteolysis in AD is strongly supported by other observations. Similar types of AVs develop in DNs when lysosomal cathepsin activity is inhibited by deleting specific cathepsins or blocking their activity.[27,28] Indeed, inhibiting lysosomal proteolysis in primary cortical neurons in culture either with cathepsin inhibitors or by reducing the extent of lysosomal acidification was recently shown

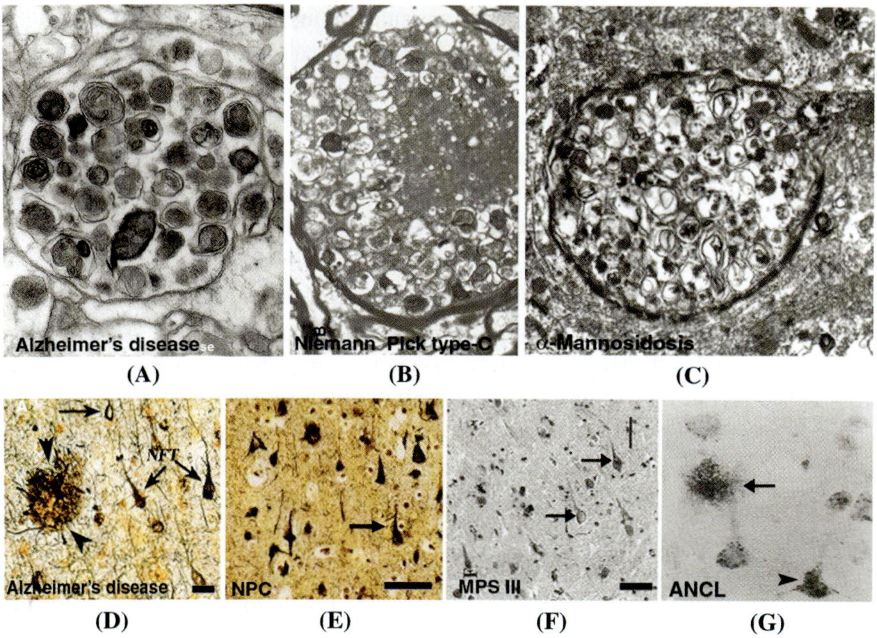

Figure 2. Alzheimer's Disease pathology is present in multiple lysosomal storage disorders. **(A)** A transmission EM image of a dystrophic neurite in cross-section in the brain of an AD patient. These neurites and their characteristic accumulation of autophagic intermediates and other vesicles are also seen in neurons from patients with the lysosomal storage disorders Niemann–Pick type C (reprinted and modified from Ref. 114, with permission from Elsevier) **(B)** and α-Mannosidosis (modified and reprinted from Ref. 115, with permission from Elsevier) **(C)**. AD also presents with the formation of senile plaques and neurofibrillary tangles **(D)**. Neurofibrillary tangles also occur in Niemann–Pick type C (modified and reprinted from Ref. 114, with permission from Elsevier) **(E)** and Mucopolysaccharidosis III (modified and reprinted from Ref. 116) **(F)** which are indicated by arrows. In the case of Adult-onset Neuronal Ceroid Lipofuscinosis, senile plaques formed by accumulation of APP and/or its metabolites are shown as immunoreactivity with an anti-amyloid antibody (4G-8) in the neocortex **(G)**. The altered processing of APP is a feature of some LSDs, such as NPC and NCL (depicted here). Monoclonal antibodies to Aβ epitopes label intracellular punctate structures in some neurons (arrowhead) as well as extracellular deposits of diffuse Aβ immunoreactivity (arrow) (G modified and reprinted with kind permission from Springer Science + Business Media, Ref. 63, Figure 5B). (Bars = 50 μM in D and F and 30 μM in F.)

to impede selectively the transport of autophagy- and lysosome-related vesicular compartments and cause them to accumulate as the principal organelles within DNs. Moreover, the dystrophic swellings also acquired additional Alzheimer-like features, including cytoskeletal hyperphosphorylation.[29] In an APP transgenic model of AD where impairment in the lysosomal clearance of autophagic substrates was demonstrated, partially restoring the proteolytic efficiency of the lysosomes by genetically deleting the endogenous cysteine cathepsin inhibitor, cystatin B, substantially ameliorated the neuropathology and cognitive deficits that usually develop in this model.[30] Similar results have been seen using additional approaches to enhance cysteine cathepsin activities.[31,32] Finally, genetic evidence in AD has directly linked mutations of presenilin 1, the most common cause of early onset FAD, to lysosomal dysfunction. In cells from individuals with familial AD caused by mutations of PS1 and in cell and animal models involving deletion of this gene, acidification of lysosomes is impeded due to impaired PS1-mediated maturation of the lysosomal proton pump, vATPase, and its assembly on lysosomes. Consequently, lysosomal hydrolase activity is significantly decreased, ultimately resulting in a severe retardation in the lysosome's ability to degrade incoming substrates, including those transported via the endocytic pathway.[33] PS1 mutations in FAD promote autophagy pathology and neuritic dystrophy in patients, and also greatly accelerate the onset of neuritic dystrophy and amyloid-related pathologies when introduced into mutant APP transgenic models of FAD. Interestingly, overexpression of FAD-mutant APP alone induces late appearance of autophagic pathology similar to that seen at much earlier ages in models overexpressing both mutated PS1 and APP constructs,[12,34] suggesting that mutant APP and/or its multiple aberrant cleavage products, such as Aβ, can also adversely affect cellular autophagy.[35]

While autophagy is predominantly thought of as a method for the removal of proteins and organelles that are no longer essential, recent work has demonstrated that the autophagic pathway is also critical for clearing abnormally aggregated proteins, including Aβ and tau.[36] Stimulating autophagy pharmacologically, for example, greatly reduces the accumulation of either Aβ or tau in AD mouse models.[37–39] Although the role of autophagic defects in the development of dystrophic neurites seems clear, a question remains as to whether defective autophagy directly triggers the aggregation of Aβ and tau, or whether elevated Aβ and tau aggregation cause defective autophagy, or both. It is most likely that the development of autophagy failure in AD is multifactorial as both genetic factors and environmental factors seem to play a role. In addition to factors that interrupt the autophagic and lysosomal machinery, other influences may include an increased delivery of normal or damaged substrates to the autophagy system, which may then overwhelm an already partially impaired system. For example, one of the first pathological occurrences in AD are endosome anomalies reflecting an increase in the rate of endocytosis.[40–42] This rab5-dependent pathological response is promoted by cholesterol,[43,44] ApoE E4, a cholesterol transporting protein,[45] and β-cleaved APP, a cholesterol binding protein.[46] In cell model systems, rab5 upregulation leads to increased delivery of substrates to lysosomes and also endosomal accumulation of substrates and diminished lysosomal degradation rates.[47] Because a considerable proportion of endocytic cargo reaches lysosomes via amphisomes of the autophagic pathway,[48,49] this may also impede autophagy efficiency. Importantly, endocytosis is the mechanism by which secreted Aβ re-enters cells, giving rise to a mechanism whereby accelerated endocytosis promotes increased delivery of Aβ to amphisomes and lysosomes, contributing to autophagic dysfunction[50] which in turn could lead to impaired Aβ degradation, more Aβ aggregation, and propagation of this toxic cascade.[51]

The connection between autophagic-lysosomal defects and tau hyperphosphorylation and fibril formation also seems apparent although the mechanism is unclear. A recent study suggests that similar to the accumulation of Aβ, tau aggregation may be promoted by defective autophagy.[52] Interestingly, this study demonstrated an exacerbation in mutant tau aggregation when autophagy was inhibited with 3-methyladenine,[52] implicating autophagic dysfunction in the aggregation of tau fibrils. Conversely, induction of autophagy with rapamycin in a triple-transgenic mouse model of AD, overexpressing mutant forms of APP, PS1, and tau, reduces the accumulation of tau in neurons.[53] Other studies have also implicated chaperone mediated autophagy (CMA), in the pathogenesis of Parkinson's, Huntington's and in the tauopathy of AD.[54–56] CMA differs significantly from macroautophagy, in that there is no double membrane that engulfs the material to be transported, and instead soluble cytosolic proteins bind cytosolic chaperones and are then transported to the lysosome, where they are translocated across the lysosomal membrane into the lumen for digestion.[57] In the case of AD, it has been suggested that mutant tau may disrupt this CMA pathway and contribute to neuronal degeneration. While mutant tau is targeted to the lysosome, its translocation across the membrane is affected, leaving the N-terminus accessible to proteolytic cleavage. These liberated fragments are ultimately toxic, oligomerizing in the lysosome, blocking further CMA and resulting in more widespread lysosomal dysfunction.[58]

Given the importance of neuronal autophagy in recycling long lived proteins, as well as in degrading defective organelles and proteins, lysosome dysfunction has long been linked to mechanisms of neurodegeneration. Aged and unstable mitochondria signal clearance by autophagy thereby diminishing the chances of their initiating apoptosis.[59] Indeed, the failure to target and clear damaged mitochondria through mitophagy is considered a major contributing factor to neuronal cell death in Parkinson's disease. It is possible that in AD, the failure to efficiently remove specific toxic proteins, including not only Aβ and tau but also activated caspases[60] and other pro-apoptotic molecules, contributes to lowered neuronal survival in AD. Accumulation of incompletely digested proteins and lipids and their stasis and oxidation within uncleared AVs opens the door to their interacting with lysosomal membranes directly to increase membrane permeabilization and leakage of hydrolytic enzymes into the cytoplasm.[61] Rapid release of these hydrolases after a catastrophic loss of membrane integrity of a lysosome could initiate cellular necrosis.[61,62] Conversely, the slower release of lysosomal cathepsins into the cellular mileau induces a number of signaling pathways which can initiate an apoptotic response.[61]

LYSOSOMAL STORAGE DISORDERS AND AUTOPHAGY

At the other end of the age spectrum from AD are lysosomal storage disorders, a family of ~50 different metabolic disorders among which are some of the most common *pediatric* neurodegenerative disorders. As implied in their categorization, all of these diseases are characterized by the bulk storage within lysosomes of incompletely digested materials that may include aggregated proteins, lipids, heavy metals or other macromolecules.[1] Although multiple physiological systems are usually affected, many present with a severe neurological component. Even despite the marked difference in age of onset between AD and most LSDs, which may itself significantly influence the regional neuropathological pictures emerging in these disorders, intriguing overlaps have been noted in the neuropathology of certain LSDs. These overlaps include, besides severe neurodegeneration, lesions that have heretofore been considered to be defining hallmarks of AD, such as paired helical filaments of tau and neurofibrillary tangles, AD-like neuritic dystrophy featuring selective AV accumulations, and increased amyloidogenic APP processing leading to diffuse extracellular deposition of Aβ deposits, albeit not true neuritic plaque development.[4,9,63] Understanding how autophagic defects common to AD and certain LSDs drive this neuropathology should provide a clearer picture as to roles the lysosomal system plays in AD pathogenesis and could play in therapeutic interventions.

Niemann–Pick type C (NPC) is a prototypical lysosomal storage disorder as 95% of NPC cases stem from loss of function mutations in a single gene, NPC1.[64] About 5% of cases are caused by mutations in a second gene, NPC2. Although the exact function of these proteins remains unclear, they are both resident endosomal-lysosomal proteins involved in cholesterol transport. NPC1 contains a putative sterol sensing domain common to many proteins involved in maintaining cholesterol transport.[65] The major lysosomal storage material in NPC1 cells consists of unesterified cholesterol and sphingolipids.[66,67] NPC symptomatology can vary significantly with neurological involvement increasingly prominent in later onset cases.[68] Age of onset ranges from *in utero* to mid teens, and in very rare cases, adulthood; NPC is characterized by progressive neurodegeneration, enlargement of the liver and spleen, and results ultimately in death.[69] While the clinical presentation diverges substantially from AD, there are a number of striking commonalities between the two disorders on a cellular level. These include the formation of neurofibrillary tangles associated with tau hyper-phosphorylation, neuroaxonal dystrophy, AD-like endosome swelling, and progressive neuronal loss (Figure 2). Importantly, dysregulation of the autophagic pathway has been clearly identified in NPC.[70–75]

Initial investigation regarding the role of autophagic dysfunction in NPC uncovered that, similarly to AD, there is a significant accumulation of neuronal AVs in an NPC mouse model.[76] These findings were further substantiated by a number of additional studies utilizing other NPC models, which has made NPC one of the most thoroughly studied LSDs in regards to impaired autophagy.[9,77] While the accumulation of undigested and unprocessed autophagic vesicles in neurons of the diseased brain is the same between NPC and AD, there are some differences in the mechanism of AV accumulation. As previously mentioned, AV accumulation in AD appears to be the result of defective lysosomal proteolysis rather than a pathological induction of autophagy.[50] In NPC, however, evidence suggests that the AV accumulation in neurons results at least in part from hyper-active autophagy *induction*, likely mediated via the class III-PI3K/Beclin 1 complex.[9] This excessive induction alters the flux of the autophagic pathway leading to a build-up of autophagic intermediates similar to that seen in AD.[24] These observations do not exclude the additional possibility that proteolytic steps of autophagy are also impeded. In this regard, autolysosomal clearance of autophagic substrates is normally exceptionally efficient even during strong autophagy induction.[25] Thus, the prominent AV accumulation in NPC implies that AV clearance may be less efficient than normal. Moreover, NPC1 displays a number of other similarities to AD that may result from defective autophagy, including tau hyperphosphorylation and aggregation, as well as elevated levels of APP cleavage products βCTF, Aβ42 and Aβ40 and accumulation of these species in characteristically enlarged endosomal-lysosomal compartments, as seen in early AD.[78,79]

The connection between NPC and AD through the important role of cholesterol illustrates further mechanistic overlaps between these two conditions. Inheritance of the e4 allele of Apolipoprotein E (APOE), a protein that is a key regulator of neuronal cholesterol transport, promotes disease progression in both AD[45] and NPC.[80] Importantly, other risk genes for AD are also related to cholesterol handling and the role of cholesterol in APP metabolism.[81,82] It is therefore not surprising, given the critical importance of NPC1 in cellular cholesterol trafficking, that the endosome dysfunction highly characteristic of AD is also seen in NPC[83] and seems to be tightly linked to the pathobiology by which cholesterol acts pathogenically in both diseases, linking endosome dysfunction, APP processing, and possibly endosomal-lysosomal impairment. Specifically, elevated intracellular cholesterol caused by overexpressed ApoE4 has been shown to result in elevated βCTF levels,[44,84] levels that are also significantly elevated in NPC, and in increased delivery of Aβ42 to the lysosome, a phenomenon observed in both AD and NPC.[44,79,84]

Autophagic impairment has been reported in a number of other LSDs with a severe neurodegenerative phenotype, including the most common of the pediatric neurodegenerative conditions, the neuronal ceroid lipofuscinoses. Like NPC, disease onset and rate of progression in this family of disorders varies widely. Each individual NCL stems from mutations in a different gene.[5,85] Interestingly, while the clinical symptoms are significantly different from those in AD, they are similar within each subtype of NCL, with blindness, seizures, cognitive and motor decline being the predominant symptoms.[5,86,87] Again if we delve into the cellular pathology, NCL and AD display a significant level of overlap, including defective autophagy, altered lysosomal function, altered APP processing and neurodegeneration.[7,63,88,89] Nearly all of the NCLs, including the juvenile form (JNCL) which is the most common NCL, accumulate a range of autophagic substrates including subunit C of the mitochondrial ATPase in their lysosomes,[90] suggesting impaired turnover of the protein. Interestingly in the neurons of JNCL brains, there is not only lysosomal storage of mitochondrial ATPase subunit C, but also a number of membranous inclusions that suggest improper vesicular trafficking and/or fusion in the cell.[91,92] In one study, autophagic induction and AV accumulation were monitored in a knockin CLN3 mouse model where it was shown that the marker LC3-II was elevated and mTOR was downregulated. Furthermore ultrastructural analysis identified accumulated autophagic intermediates, suggesting a block in autophagosome maturation similar to that in AD.[7] Given the breadth of research that has suggested that acidification of lysosomes in CLN3–/– cells is compromised,[40,88,93] one could imagine a situation that closely mimics AD, where altered lysosomal pH inhibits proteolysis and results in an aggregation of AVs. While a role for amyloid deposition in the NCLs has never been firmly established, evidence of altered processing in JNCL is suggested by the increased cellular levels of APP and/or its metabolites in the lysosome-related membranous inclusions characteristic of these diseases (Figure 2).[63] This suggests that, as in AD, autophagic/lysosomal dysfunction may increase the delivery of APP-rich organelles for autophagy or block the subsequent autophagic processing of APP.[50]

Additional LSDs with demonstrable autophagic dysfunction include the GM1 and GM2-gangliosidosis and mucolipidosis type IV.[94–96] Mucolipidosis type IV (MLIV) is another devastating pediatric neurodegenerative disorder, caused by mutations in TRPML1 (or Mucolipin 1/Mcoln1), a late endosomal/lysosomal protein.[97–99] Symptomatically, MLIV is characterized by hypotonia, progressive mental and motor decline and visual deterioration,[100] a pathogenic progression which is again markedly different from the clinical presentation of AD. Functionally, Mcoln1, like other members of the TRP family, is an integral membrane protein that likely functions as an outwardly rectifying ion channel. The selectivity of Mcoln1 remains undetermined. Although it likely functions predominantly as a Ca^{2+} transporter,[101] Mcoln1 may also be permeable to K^+ and Na^+,[102] as well as H^+.[103] While the exact function of this protein remains unclear, there is a profound effect on autophagy when it is mutated.[8]

Like all LSDs, the earliest observable pathology for MLIV are lysosomal storage bodies present in many tissues, in this case storage of mucopholysaccharides and lipids.[104,105] Recent analysis of neurons from mucolipin 1 null mice demonstrated an increased diameter of lysotracker positive vesicles, suggesting an enlarged lysosome or other acidic organelle. Further ultrastructural investigation of these positive vesicles demonstrated the presence of numerous cytoplasmic inclusion bodies, which when coupled with LC3 analysis were determined to be autophagic intermediates.[8,11] Interestingly, there is also significant overlap between AD and MLIV as to the mechanism of defective autophagy, as MLIV cells, similar to the PS1 null cells reported in Lee *et al.* (2010), exhibit a significantly altered lysosomal pH.[103] Coupled with additional experiments by Curcio-Morelli *et al.* which demonstrate a lack of increased autophagic induction and studies demonstrating that there is both a lack of defective AV/lysosome fusion[106] and measurable delay in AV degradation,[94] it is possible that proteolytic activity deficits, like those observed in PS1–/– cells, could exist in MLIV. These dysfunctional lysosomes, as has been discussed, can therefore pathogenically link the neurodegeneration in two very distinct diseases to similar defects in a common pathway.

GM1-gangliosidosis is yet another lysosomal storage disease with a significant autophagic component. Arising from mutations and decreases in activity of the lysosomal acid ß-galactosidase (GLB1), GM1-gangliosidosis is characterized the storage of the ganglioside GM1 in the lysosome.[107] Like the NCLs, GM1-gangliosidosis can be characterized by the age of onset, with an infantile, juvenile and adult variant, although unlike the NCLs, each variant is caused by mutations in the same gene. GM1-gangliosidosis is characterized by progressive central nervous system dysfunction, enlargement of internal organs and skeletal dysplasia.[107]

Like many of the LSDs, GM1-gangliosidosis presents on the cellular level with characteristic substrate storage, as well as lamellar and membranous cellular inclusions,[108] similar to those observed in the dystrophic neurites of AD neurons.[24] A recent study demonstrated that, in β-galactosidase null mouse brains, there is a significant elevation of LC-3 and Beclin 1, suggesting that autophagy is likely altered in this disease.[96] Further work by this group determined that mTOR signaling was enhanced, suggesting that similar to the case of NPC, autophagic induction is overactive and results in AV accumulation in the brains of these mice. Notably, ganglioside metabolism is suspected to have a role in AD pathogenesis, in light of observations that GM1, the same sphingolipid stored in GM1-gangliosidosis, can directly bind Aβ and facilitate its structural stabilization into beta-pleated sheets. GM1 mice also exhibit aggregation of APP in neurons.[109] Moreover, several studies have demonstrated specific alterations in ganglioside metabolism in the AD brain,[110,111] including reductions of ganglioside levels in many brain regions.[112] Moreover, a recent study demonstrated that the accumulation of sphingolipids stimulates gamma-secretase activity, which in turn increases the production of both intracellular and secreted Aβ.[113]

It is possible that many other LSDs not mentioned here share common pathologies with AD, including amyloid processing defects, tau hyperphosphorylation and neurodegeneration, and investigations of this possibility are currently active. These and other ongoing investigations will soon clarify the full consequences of autophagic/lysosomal disruption and its significance for disease progression and possible therapeutic intervention in AD and LSDs.

CONCLUSION

Alzheimer's disease is the most common form of dementia in the aging population worldwide and its treatment is still limited to symptom management. While there continues to be ample debate regarding the relative roles of Aβ and tau and their toxicity in the progression of AD, the role of autophagic dysfunction in this disease is becoming increasingly clear, both in terms of promoting accumulation of these suspected pathogenic proteins and having other primary effects on neuronal survival and disease progression.[24,51] In our efforts to further expand this understanding, incorporating knowledge about autophagic dysfunction in other neurodegenerative diseases will provide further insight into the etiology of AD. The fact that neurodegenerative diseases with such disparate clinical and pathological presentations display defects in a common pathway of vital importance to neuron survival is a crucial starting point for innovative studies aimed at uncovering the basic genesis of AD and identifying new therapeutic targets in the treatment of these and other neurodegenerative diseases.

ACKNOWLEDGEMENTS

The authors would like to thank Dr. Asok Kumar for usage of his EM of AD neurons (Figure 2A), as well as Nicole Piorkowski and Corrinne Peterhoff for their assistance in manuscript and figure preparation. Research from the Nixon laboratories is supported by the National Institute on Aging.

REFERENCES

1. S. U. Walkley, Cellular pathology of lysosomal storage disorders, *Brain Pathol* **8**: 175–193 (1998).
2. L. B. Jardim, M. M. Villanueva, C. F. de Souza and C. B. Netto, Clinical aspects of neuropathic lysosomal storage disorders, *J Inherit Metab Dis* **33**: 315–329 (2010).
3. W. J. Strittmatter and A. D. Roses, Apolipoprotein E and Alzheimer's disease, *Annu Rev Neurosci* **19**: 53–77 (1996).
4. M. L. Schultz, L. Tecedor, M. Chang and B. L. Davidson, Clarifying lysosomal storage diseases, *Trends Neurosci* **34**: 401–410 (2011).
5. J. M. Weimer, E. Kriscenski-Perry, Y. Elshatory and D. A. Pearce, The neuronal ceroid lipofuscinoses: mutations in different proteins result in similar disease, *Neuromolecular Med* **1**: 111–124 (2002).
6. T. Hara, K. Nakamura, M. Matsui, A. Yamamoto, Y. Nakahara, R. Suzuki-Migishima, M. Yokoyama, K. Mishima, I. Saito, H. Okano and N. Mizushima, Suppression of basal autophagy in neural cells causes neurodegenerative disease in mice, *Nature* **441**: 885–889 (2006).
7. Y. Cao, J. A. Espinola, E. Fossale, A. C. Massey, A. M. Cuervo, M. E. MacDonald and S. L. Cotman, Autophagy is disrupted in a knock-in mouse model of juvenile neuronal ceroid lipofuscinosis, *J Biol Chem* **281**: 20483–20493 (2006).
8. C. Curcio-Morelli, F. A. Charles, M. C. Micsenyi, Y. Cao, B. Venugopal, M. F. Browning, K. Dobrenis, S. L. Cotman, S. U. Walkley and S. A. Slaugenhaupt, Macroautophagy is defective in mucolipin-1-deficient mouse neurons, *Neurobiol Dis* **40**: 370–377 (2010).
9. C. D. Pacheco, R. Kunkel and A. P. Lieberman, Autophagy in Niemann-Pick C disease is dependent upon Beclin-1 and responsive to lipid trafficking defects, *Hum Mol Genet* **16**: 1495–1503 (2007).
10. T. Fukuda, L. Ewan, M. Bauer, R. J. Mattaliano, K. Zaal, E. Ralston, P. H. Plotz and N. Raben, Dysfunction of endocytic and autophagic pathways in a lysosomal storage disease, *Ann Neurol* **59**: 700–708 (2006).

11. J. J. Jennings, Jr., J. H. Zhu, Y. Rbaibi, X. Luo, C. T. Chu and K. Kiselyov, Mitochondrial aberrations in mucolipidosis Type IV, *J Biol Chem* **281**: 39041–39050 (2006).
12. D. S. Yang, A. Kumar, P. Stavrides, J. Peterson, C. M. Peterhoff, M. Pawlik, E. Levy, A. M. Cataldo and R. A. Nixon, Neuronal apoptosis and autophagy cross talk in aging PS/APP mice, a model of Alzheimer's disease, *Am J Pathol* **173**: 665–681 (2008).
13. N. Mizushima, Autophagy: process and function, *Genes Dev* 21: 2861–2873 (2007).
14. D. J. Klionsky and A. Nemchenko, There is more to autophagy than induction: regulating the roller coaster, *Autophagy* **7**: 801–802 (2011).
15. B. Levine and G. Kroemer, Autophagy in the pathogenesis of disease, *Cell* **132**: 27–42 (2008).
16. T. Shintani and D. J. Klionsky, Autophagy in health and disease: a double-edged sword, *Science* 306: 990–995 (2004).
17. J. Huang and D. J. Klionsky, Autophagy and human disease, *Cell Cycle* **6**: 1837–1849 (2007).
18. Y. Gorzalczany, Y. Gilad, D. Amihai, I. Hammel, R. Sagi-Eisenberg and O. Merimsky, Combining an EGFR directed tyrosine kinase inhibitor with autophagy-inducing drugs: a beneficial strategy to combat non-small cell lung cancer, *Cancer Lett* **310**: 207–215 (2011).
19. J. Bove, M. Martinez-Vicente and M. Vila, Fighting neurodegeneration with rapamycin: mechanistic insights, *Nat Rev Neurosci* **12**: 437–452 (2011).
20. A. Nemchenko, M. Chiong, A. Turer, S. Lavandero and J. A. Hill, Autophagy as a therapeutic target in cardiovascular disease, *J Mol Cell Cardiol* **51**: 584–593 (2011).
21. M. Y. Heng, D. K. Duong, R. L. Albin, S. J. Tallaksen-Greene, J. M. Hunter, M. J. Lesort, A. Osmand, H. L. Paulson and P. J. Detloff, Early autophagic response in a novel knock-in model of Huntington disease, *Hum Mol Genet* **19**: 3702–3720 (2010).
22. A. Alzheimer, H. Forstl and R. Levy, On certain peculiar diseases of old age, *Hist Psychiatry* **2**: 71–101 (1991).
23. T. Simchowicz, Histologische Studien über die senile Demenz. In: *Histologie and Histopathologische Arbeiten über die Grosshirnrinde*, Vol. IV, eds. F. Nissl and A. Alzheimer (Fischer, Jena, 1911), pp. 267–444.
24. R. A. Nixon, J. Wegiel, A. Kumar, W. H. Yu, C. Peterhoff, A. Cataldo and A. M. Cuervo, Extensive involvement of autophagy in Alzheimer disease: an immuno-electron microscopy study, *J Neuropathol Exp Neurol* **64**: 113–122 (2005).
25. B. Boland, A. Kumar, S. Lee, F. M. Platt, J. Wegiel, W. H. Yu and R. A. Nixon, Autophagy induction and autophagosome clearance in neurons: relationship to autophagic pathology in Alzheimer's disease, *J Neurosci* **28**: 6926–6937 (2008).
26. S. Y. Lee and R. A. Nixon, Primary lysosomal dysfunction causes cargo-specific deficits of axonal transport leading to Alzheimer's-like neuritic dystrophy, *Autophagy* **7**: 1562–1563 (2011).

27. U. Felbor, B. Kessler, W. Mothes, H. H. Goebel, H. L. Ploegh, R. T. Bronson and B. R. Olsen, Neuronal loss and brain atrophy in mice lacking cathepsins B and L, *Proc Natl Acad Sci USA* **99**: 7883–7888 (2002).
28. M. Koike, M. Shibata, S. Waguri, K. Yoshimura, I. Tanida, E. Kominami, T. Gotow, C. Peters, K. von Figura, N. Mizushima, P. Saftig and Y. Uchiyama, Participation of autophagy in storage of lysosomes in neurons from mouse models of neuronal ceroid-lipofuscinoses (Batten disease), *Am J Pathol* **167**: 1713–1728 (2005).
29. S. Lee, Y. Sato and R. A. Nixon, Lysosomal proteolysis inhibition selectively disrupts axonal transport of degradative organelles and causes an Alzheimer's-like axonal dystrophy, *J Neurosci* **31**: 7817–7830 (2011).
30. D. S. Yang, P. Stavrides, P. S. Mohan, S. Kaushik, A. Kumar, M. Ohno, S. D. Schmidt, D. Wesson, U. Bandyopadhyay, Y. Jiang, M. Pawlik, C. M. Peterhoff, A. J. Yang, D. A. Wilson, P. St George-Hyslop, D. Westaway, P. M. Mathews, E. Levy, A. M. Cuervo and R. A. Nixon, Reversal of autophagy dysfunction in the TgCRND8 mouse model of Alzheimer's disease ameliorates amyloid pathologies and memory deficits, *Brain* **134**: 258–277 (2011).
31. D. Butler, J. Hwang, C. Estick, A. Nishiyama, S. S. Kumar, C. Baveghems, H. B. Young-Oxendine, M. L. Wisniewski, A. Charalambides and B. A. Bahr, Protective effects of positive lysosomal modulation in Alzheimer's disease transgenic mouse models, *PLoS One* **6**: e20501 (2011).
32. B. Sun, Y. Zhou, B. Halabisky, I. Lo, S. H. Cho, S. Mueller-Steiner, N. Devidze, X. Wang, A. Grubb and L. Gan, Cystatin C-cathepsin B axis regulates amyloid beta levels and associated neuronal deficits in an animal model of Alzheimer's disease, *Neuron* **60**: 247–257 (2008).
33. J. H. Lee, W. H. Yu, A. Kumar, S. Lee, P. S. Mohan, C. M. Peterhoff, D. M. Wolfe, M. Martinez-Vicente, A. C. Massey, G. Sovak, Y. Uchiyama, D. Westaway, A. M. Cuervo and R. A. Nixon, Lysosomal proteolysis and autophagy require presenilin 1 and are disrupted by Alzheimer-related PS1 mutations, *Cell* **141**: 1146–1158 (2010).
34. L. Spokes, T. Jickells, K. Weston, B. G. Gustafsson, M. Johnsson, B. Liljebladh, D. Conley, C. Ambelas-Skjodth, J. Brandt, J. Carstensen, T. Christiansen, L. Frohn, G. Geernaert, O. Hertel, B. Jensen, C. Lundsgaard, S. Markager, W. Martinsen, B. Moller, B. Pedersen, K. Sauerberg, L. L. Sorensen, C. C. Hasager, A. M. Sempreviva, S. C. Pryor, S. W. Lund, S. Larsen, M. Tjernstrom, G. Svensson and M. Zagar, MEAD: an interdisciplinary study of the marine effects of atmospheric deposition in the Kattegat, *Environ Pollut* **140**: 453–462 (2006).
35. D. Ling, H. J. Song, D. Garza, T. P. Neufeld and P. M. Salvaterra, Abeta42-induced neurodegeneration via an age-dependent autophagic-lysosomal injury in Drosophila, *PLoS One* **4**: e4201 (2009).
36. A. Williams, L. Jahreiss, S. Sarkar, S. Saiki, F. M. Menzies, B. Ravikumar and D. C. Rubinsztein, Aggregate-prone proteins are cleared from the cytosol by autophagy: therapeutic implications, *Curr Top Dev Biol* **76**: 89–101 (2006).

37. Z. Berger, B. Ravikumar, F. M. Menzies, L. G. Oroz, B. R. Underwood, M. N. Pangalos, I. Schmitt, U. Wullner, B. O. Evert, C. J. O'Kane and D. C. Rubinsztein, Rapamycin alleviates toxicity of different aggregate-prone proteins, *Hum Mol Genet* **15**: 433–442 (2006).
38. S. Sarkar, J. E. Davies, Z. Huang, A. Tunnacliffe and D. C. Rubinsztein, Trehalose, a novel mTOR-independent autophagy enhancer, accelerates the clearance of mutant huntingtin and alpha-synuclein, *J Biol Chem* **282**: 5641–5652 (2007).
39. P. Spilman, N. Podlutskaya, M. J. Hart, J. Debnath, O. Gorostiza, D. Bredesen, A. Richardson, R. Strong and V. Galvan, Inhibition of mTOR by rapamycin abolishes cognitive deficits and reduces amyloid-beta levels in a mouse model of Alzheimer's disease, *PLoS One* **5**: e9979 (2010).
40. W. H. Yu, A. M. Cuervo, A. Kumar, C. M. Peterhoff, S. D. Schmidt, J. H. Lee, P. S. Mohan, M. Mercken, M. R. Farmery, L. O. Tjernberg, Y. Jiang, K. Duff, Y. Uchiyama, J. Naslund, P. M. Mathews, A. M. Cataldo and R. A. Nixon, Macroautophagy — a novel Beta-amyloid peptide-generating pathway activated in Alzheimer's disease, *J Cell Biol* **171**: 87–98 (2005).
41. Y. Jiang, K. A. Mullaney, C. M. Peterhoff, S. Che, S. D. Schmidt, A. Boyer-Boiteau, S. D. Ginsberg, A. M. Cataldo, P. M. Mathews and R. A. Nixon, Alzheimer's-related endosome dysfunction in Down syndrome is Abeta-independent but requires APP and is reversed by BACE-1 inhibition, *Proc Natl Acad Sci USA* **107**: 1630–1635
42. A. M. Cataldo, J. L. Barnett, S. A. Berman, J. Li, S. Quarless, S. Bursztajn, C. Lippa and R. A. Nixon, Gene expression and cellular content of cathepsin D in Alzheimer's disease brain: evidence for early up-regulation of the endosomal-lysosomal system, *Neuron* **14**: 671–680 (1995).
43. Z. S. Ji, R. D. Miranda, Y. M. Newhouse, K. H. Weisgraber, Y. Huang and R. W. Mahley, Apolipoprotein E4 potentiates amyloid beta peptide-induced lysosomal leakage and apoptosis in neuronal cells, *J Biol Chem* **277**: 21821–21828 (2002).
44. J. C. Cossec, A. Simon, C. Marquer, R. X. Moldrich, C. Leterrier, J. Rossier, C. Duyckaerts, Z. Lenkei and M. C. Potier, Clathrin-dependent APP endocytosis and Abeta secretion are highly sensitive to the level of plasma membrane cholesterol, *Biochim Biophys Acta* **1801**: 846–852 (2010).
45. A. M. Cataldo, C. M. Peterhoff, J. C. Troncoso, T. Gomez-Isla, B. T. Hyman and R. A. Nixon, Endocytic pathway abnormalities precede amyloid beta deposition in sporadic Alzheimer's disease and Down syndrome: differential effects of APOE genotype and presenilin mutations, *Am J Pathol* **157**: 277–286 (2000).
46. A. J. Beel, C. K. Mobley, H. J. Kim, F. Tian, A. Hadziselimovic, B. Jap, J. H. Prestegard and C. R. Sanders, Structural studies of the transmembrane C-terminal domain of the amyloid precursor protein (APP): does APP function as a cholesterol sensor?, *Biochemistry* **47**: 9428–9446 (2008).
47. A. M. Cataldo, P. M. Mathews, A. B. Boiteau, L. C. Hassinger, C. M. Peterhoff, Y. Jiang, K. Mullaney, R. L. Neve, J. Gruenberg and R. A. Nixon, Down syndrome

fibroblast model of Alzheimer-related endosome pathology: accelerated endocytosis promotes late endocytic defects, *Am J Pathol* **173**: 370–384 (2008).
48. C. M. Fader and M. I. Colombo, Autophagy and multivesicular bodies: two closely related partners, *Cell Death Differ* **16**: 70–78 (2009).
49. P. B. Gordon and P. O. Seglen, Prelysosomal convergence of autophagic and endocytic pathways, *Biochem Biophys Res Commun* **151**: 40–47 (1988).
50. R. A. Nixon and D. S. Yang, Autophagy failure in Alzheimer's disease-locating the primary defect, *Neurobiol Dis* **43**: 38–45 (2011).
51. R. A. Nixon and D. S. Yang, Autophagy failure in Alzheimer's disease-locating the primary defect, *Neurobiol Dis* **43**: 38–45 (2011).
52. S. I. Kim, W. K. Lee, S. S. Kang, S. Y. Lee, M. J. Jeong, H. J. Lee, S. S. Kim, G. V. Johnson and W. Chun, Suppression of autophagy and activation of glycogen synthase kinase 3beta facilitate the aggregate formation of tau, *Korean J Physiol Pharmacol* **15**: 107–114 (2011).
53. A. Caccamo, S. Majumder, A. Richardson, R. Strong and S. Oddo, Molecular interplay between mammalian target of rapamycin (mTOR), amyloid-beta, and Tau: effects on cognitive impairments, *J Biol Chem* **285**: 13107–13120 (2010).
54. A. M. Cuervo, L. Stefanis, R. Fredenburg, P. T. Lansbury and D. Sulzer, Impaired degradation of mutant α-synuclein by chaperone-mediated autophagy, *Science* **305**: 1292–1295 (2004).
55. P. O. Bauer, A. Goswami, H. K. Wong, M. Okuno, M. Kurosawa, M. Yamada, H. Miyazaki, G. Matsumoto, Y. Kino, Y. Nagai and N. Nukina, Harnessing chaperone-mediated autophagy for the selective degradation of mutant huntingtin protein, *Nat Biotechnol* **28**: 256–263 (2010).
56. Y. Wang, J. C. Groppe, J. Wu, T. Ogawa, G. Mues, R. N. D'Souza and H. Kapadia, Pathogenic mechanisms of tooth agenesis linked to paired domain mutations in human PAX9, *Hum Mol Genet* **18**: 2863–2874 (2009).
57. E. Arias and A. M. Cuervo, Chaperone-mediated autophagy in protein quality control, *Curr Opin Cell Biol* **23**: 184–189 (2011).
58. Y. Wang, M. Martinez-Vicente, U. Kruger, S. Kaushik, E. Wong, E. M. Mandelkow, A. M. Cuervo and E. Mandelkow, Tau fragmentation, aggregation and clearance: the dual role of lysosomal processing, *Hum Mol Genet* **18**: 4153–4170 (2009).
59. A. Terman, B. Gustafsson and U. T. Brunk, Autophagy, organelles and ageing, *J Pathol* **211**: 134–143 (2007).
60. D. S. Yang, J. H. Lee and R. A. Nixon, Monitoring autophagy in Alzheimer's disease and related neurodegenerative diseases, *Methods Enzymol* 453: 111–144 (2009).
61. G. Kroemer and M. Jaattela, Lysosomes and autophagy in cell death control, *Nat Rev Cancer* **5**: 886–897 (200 5).
62. P. Syntichaki and N. Tavernarakis, The biochemistry of neuronal necrosis: rogue biology? *Nat Rev Neurosci* **4**: 672–684 (2003).
63. K. E. Wisniewski, D. Maslinska, T. Kitaguchi, K. S. Kim, H. H. Goebel and M. Haltia, Topographic heterogeneity of amyloid β-protein epitopes in brains with

various forms of neuronal ceroid lipofuscinoses suggesting defective processing of amyloid precursor protein, *Acta Neuropathol* **80**: 26–34 (1990).
64. E. D. Carstea, J. A. Morris, K. G. Coleman, S. K. Loftus, *et al.*, Niemann-Pick C1 disease gene: homology to mediators of cholesterol homeostasis, *Science* **277**: 228–231 (1997).
65. Y. A. Ioannou, The structure and function of the Niemann-Pick C1 protein, *Mol Genet Metab* **71**: 175–181 (2000).
66. L. Liscum, R. M. Ruggiero and J. R. Faust, The intracellular transport of low density lipoprotein-derived cholesterol is defective in Niemann-Pick type C fibroblasts, *J Cell Biol* **108**: 1625–1636 (1989).
67. D. te Vruchte, E. Lloyd-Evans, R. J. Veldman, D. C. Neville, R. A. Dwek, F. M. Platt, W. J. van Blitterswijk and D. J. Sillence, Accumulation of glycosphingolipids in Niemann-Pick C disease disrupts endosomal transport, *J Biol Chem* **279**: 26167–26175 (2004).
68. T. Yamamoto, H. Ninomiya, M. Matsumoto, Y. Ohta, E. Nanba, Y. Tsutsumi, K. Yamakawa, G. Millat, M. T. Vanier, P. G. Pentchev and K. Ohno, Genotype-phenotype relationship of Niemann-Pick disease type C: a possible correlation between clinical onset and levels of NPC1 protein in isolated skin fibroblasts, *J Med Genet* **37**: 707–712 (2000).
69. M. T. Vanier and G. Millat, Niemann-Pick disease type C, *Clin Genet* **64**: 269–281 (2003).
70. C. Xie, D. K. Burns, S. D. Turley and J. M. Dietschy, Cholesterol is sequestered in the brains of mice with Niemann-Pick type C disease but turnover is increased, *J Neuropathol Exp Neurol* **59**: 1106–1117 (2000).
71. M. Zervas, K. Dobrenis and S. U. Walkley, Neurons in Niemann-Pick disease type C accumulate gangliosides as well as unesterified cholesterol and undergo dendritic and axonal alterations, *J Neuropathol Exp Neurol* **60**: 49–64 (2001).
72. S. Takikita, T. Fukuda, I. Mohri, T. Yagi and K. Suzuki, Perturbed myelination process of premyelinating oligodendrocyte in Niemann-Pick type C mouse, *J Neuropathol Exp Neurol* **63**: 660–673 (2004).
73. B. Bu, J. Li, P. Davies and I. Vincent, Deregulation of cdk5, hyperphosphorylation, and cytoskeletal pathology in the Niemann-Pick type C murine model, *J Neurosci* **22**: 6515–6525 (2002).
74. D. C. German, E. M. Quintero, C. L. Liang, B. Ng, S. Punia, C. Xie and J. M. Dietschy, Selective neurodegeneration, without neurofibrillary tangles, in a mouse model of Niemann-Pick C disease, *J Comp Neurol* **433**: 415–425 (2001).
75. C. D. Pacheco, M. J. Elrick and A. P. Lieberman, Tau deletion exacerbates the phenotype of Niemann-Pick type C mice and implicates autophagy in pathogenesis, *Hum Mol Genet* **18**: 956–965 (2009).
76. D. C. Ko, L. Milenkovic, S. M. Beier, H. Manuel, J. Buchanan and M. P. Scott, Cell-autonomous death of cerebellar Purkinje neurons with autophagy in Niemann-Pick type C disease, *PLoS Genet* **1**: 81–95 (2005).

77. G. Liao, Y. Yao, J. Liu, Z. Yu, S. Cheung, A. Xie, X. Liang and X. Bi, Cholesterol accumulation is associated with lysosomal dysfunction and autophagic stress in Npcl–/– mouse brain, *Am J Pathol* **171**: 962–975 (2007).
78. A. Nunes, S. N. Pressey, J. D. Cooper and S. Soriano, Loss of amyloid precursor protein in a mouse model of Niemann-Pick type C disease exacerbates its phenotype and disrupts tau homeostasis, *Neurobiol Dis* **42**: 349–359 (2011).
79. N. Mattsson, H. Zetterberg, S. Bianconi, N. M. Yanjanin, R. Fu, J. E. Mansson, F. D. Porter and K. Blennow, Gamma-secretase-dependent amyloid-beta is increased in Niemann-Pick type C: a cross-sectional study, *Neurology* **76**: 366–372 (2011).
80. Y. Saito, K. Suzuki, E. Nanba, T. Yamamoto, K. Ohno and S. Murayama, Niemann-Pick type C disease: accelerated neurofibrillary tangle formation and amyloid beta deposition associated with apolipoprotein E epsilon 4 homozygosity, *Ann Neurol* **52**: 351–355 (2002).
81. C. R. Scherzer, K. Offe, M. Gearing, H. D. Rees, G. Fang, C. J. Heilman, C. Schaller, H. Bujo, A. I. Levey and J. J. Lah, Loss of apolipoprotein E receptor LR11 in Alzheimer disease, *Arch Neurol* **61**: 1200–1205 (2004).
82. M. P. Marzolo and G. Bu, Lipoprotein receptors and cholesterol in APP trafficking and proteolytic processing, implications for Alzheimer's disease, *Semin Cell Dev Biol* **20**: 191–200 (2009).
83. L. W. Jin, F. S. Shie, I. Maezawa, I. Vincent and T. Bird, Intracellular accumulation of amyloidogenic fragments of amyloid-beta precursor protein in neurons with Niemann-Pick type C defects is associated with endosomal abnormalities, *Am J Pathol* **164**: 975–985 (2004).
84. Z. S. Ji, K. Mullendorff, I. H. Cheng, R. D. Miranda, Y. Huang and R. W. Mahley, Reactivity of apolipoprotein E4 and amyloid beta peptide: lysosomal stability and neurodegeneration, *J Biol Chem* **281**: 2683–2692 (2006).
85. J. Ezaki and E. Kominami, The intracellular location and function of proteins of neuronal ceroid lipofuscinoses, *Brain Pathol* **14**: 77–85 (2004).
86. H. H. Goebel, The neuronal ceroid-lipofuscinoses, *J Child Neurol* **10**: 424–437 (1995).
87. H. M. Mitchison and S. E. Mole, Neurodegenerative disease: the neuronal ceroid lipofuscinoses (Batten disease), *Curr Opin Neurol* **14**: 795–803 (2001).
88. J. M. Holopainen, J. Saarikoski, P. K. Kinnunen and I. Jarvela, Elevated lysosomal pH in neuronal ceroid lipofuscinoses (NCLs), *Eur J Biochem* **268**: 5851–5856 (2001).
89. J. M. Weimer, J. W. Benedict, A. L. Getty, C. C. Pontikis, M. J. Lim, J. D. Cooper and D. A. Pearce, Cerebellar defects in a mouse model of juvenile neuronal ceroid lipofuscinosis, *Brain Res* **1266**: 93–107 (2009).
90. D. N. Palmer, I. M. Fearnley, J. E. Walker, N. A. Hall, B. D. Lake, L. S. Wolfe, M. Haltia, R. D. Martinus and R. D. Jolly, Mitochondrial ATP synthase subunit c storage in the ceroid-lipofuscinoses (Batten disease), *Am J Med Genet* **42**: 561–567 (1992).

91. H. H. Goebel and K. E. Wisniewski, Current state of clinical and morphological features in human NCL, *Brain Pathol* **14**: 61–69 (2004).
92. E. Fossale, P. Wolf, J. A. Espinola, T. Lubicz-Nawrocka, A. M. Teed, H. Gao, D. Rigamonti, E. Cattaneo, M. E. MacDonald and S. L. Cotman, Membrane trafficking and mitochondrial abnormalities precede subunit c deposition in a cerebellar cell model of juvenile neuronal ceroid lipofuscinosis, *BMC Neurosci* **5**: 57 (2004).
93. D. A. Pearce, S. A. Nosel and F. Sherman, Studies of pH regulation by Btn1p, the yeast homolog of human Cln3p, *Mol Genet Metab* **66**: 320–323 (1999).
94. S. Vergarajauregui, P. S. Connelly, M. P. Daniels and R. Puertollano, Autophagic dysfunction in mucolipidosis type IV patients, *Hum Mol Genet* **17**: 2723–2737 (2008).
95. T. Otomo, K. Higaki, E. Nanba, K. Ozono and N. Sakai, Inhibition of autophagosome formation restores mitochondrial function in mucolipidosis II and III skin fibroblasts, *Mol Genet Metab* **98**: 393–399 (2009).
96. A. Takamura, K. Higaki, K. Kajimaki, S. Otsuka, H. Ninomiya, J. Matsuda, K. Ohno, Y. Suzuki and E. Nanba, Enhanced autophagy and mitochondrial aberrations in murine G(M1)-gangliosidosis, *Biochem Biophys Res Commun* **367**: 616–622 (2008).
97. M. Manzoni, E. Monti, R. Bresciani, A. Bozzato, S. Barlati, M. T. Bassi and G. Borsani, Overexpression of wild-type and mutant mucolipin proteins in mammalian cells: effects on the late endocytic compartment organization, *FEBS Lett* **567**: 219–224 (2004).
98. P. R. Pryor, F. Reimann, F. M. Gribble and J. P. Luzio, Mucolipin-1 is a lysosomal membrane protein required for intracellular lactosylceramide traffic, *Traffic* **7**: 1388–1398 (2006).
99. S. Vergarajauregui and R. Puertollano, Two di-leucine motifs regulate trafficking of mucolipin-1 to lysosomes, *Traffic* **7**: 337–353 (2006).
100. G. Bach, Mucolipidosis type IV, *Mol Genet Metab* **73**: 197–203 (2001).
101. J. M. LaPlante, C. P. Ye, S. J. Quinn, E. Goldin, E. M. Brown, S. A. Slaugenhaupt and P. M. Vassilev, Functional links between mucolipin-1 and $Ca2^{+}$-dependent membrane trafficking in mucolipidosis IV, *Biochem Biophys Res Commun* **322**: 1384–1391 (2004).
102. H. F. Cantiello, N. Montalbetti, W. H. Goldmann, M. K. Raychowdhury, S. Gonzalez-Perrett, G. A. Timpanaro and B. Chasan, Cation channel activity of mucolipin-1: the effect of calcium, *Pflugers Archiv* **451**: 304–312 (2005).
103. A. A. Soyombo, S. Tjon-Kon-Sang, Y. Rbaibi, E. Bashllari, J. Bisceglia, S. Muallem and K. Kiselyov, TRP-ML1 regulates lysosomal pH and acidic lysosomal lipid hydrolytic activity, *J Biol Chem* **281**: 7294–7301 (2006).
104. G. Bach, M. M. Cohen and G. Kohn, Abnormal ganglioside accumulation in cultured fibroblasts from patients with mucolipidosis IV, *Biochem Biophys Res Commun* **66**: 1483–1490 (1975).

105. B. F. Crandall, M. Philippart, W. J. Brown and D. A. Bluestone, Review article: mucolipidosis IV, *Am J Med Genet* **12**: 301–308 (1982).
106. K. Venkatachalam, A. A. Long, R. Elsaesser, D. Nikolaeva, K. Broadie and C. Montell, Motor deficit in a Drosophila model of mucolipidosis type IV due to defective clearance of apoptotic cells, *Cell* **135**: 838–851 (2008).
107. Y. Suzuki, E. Nanba, J. Matsuda and A. Oshima, β-galactosidase deficiency (β-galactosidosis): G_{M1}-gangliosidosis and Morquio B disease. In: *The Online Metabolic and Molecular Bases of Inherited Disease*, eds. D.Valle, A. L. Beaudet, B. Vogelstein, K. W. Kinzler, *et al.* (McGraw-Hill, New York, 2007).
108. K. Suzuki and G. C. Chen, Morphological, histochemical and biochemical studies on a case of systemic late infantile lipidosis (generalized gangliosidosis), *J Neuropathol Exp Neurol* **27**: 15–38 (1968).
109. J. P. van der Voorn, W. Kamphorst, M. S. van der Knaap and J. M. Powers, The leukoencephalopathy of infantile GM1 gangliosidosis: oligodendrocytic loss and axonal dysfunction, *Acta Neuropathol* **107**: 539–545 (2004).
110. L. Svennerholm, Ganglioside loss is a primary event in Alzheimer disease type I, *Progr Brain Res* **101**: 391–404 (1994).
111. I. Kracun, S. Kalanj, C. Cosovic and J. Talan-Hranilovic, Brain gangliosides in Alzheimer's disease, *J Hirnforsch* **31**: 789–793 (1990).
112. S. Kalanj, I. Kracun, H. Rosner and C. Cosovic, Regional distribution of brain gangliosides in Alzheimer's disease, *Neurol Croat* **40**: 269–281 (1991).
113. I. Y. Tamboli, H. Hampel, N. T. Tien, K. Tolksdorf, B. Breiden, P. M. Mathews, P. Saftig, K. Sandhoff and J. Walter, Sphingolipid storage affects autophagic metabolism of the amyloid precursor protein and promotes Abeta generation, *J Neurosci* **31**: 1837–1849 (2011).
114. S. U. Walkley and K. Suzuki, Consequences of NPC1 and NPC2 loss of function in mammalian neurons, *Biochim Biophys Acta* **1685**: 48–62 (2004).
115. S. U. Walkley, H. J. Baker, M. C. Rattazzi, M. E. Haskins and J. Y. Wu, Neuroaxonal dystrophy in neuronal storage disorders: evidence for major GABAergic neuron involvement, *J Neurol Sci* **104**: 1–8 (1991).
116. S. D. Ginsberg, J. E. Galvin, V. M. Lee, L. B. Rorke, D. W. Dickson, J. H. Wolfe, M. Z. Jones and J. Q. Trojanowski, Accumulation of intracellular amyloid-beta peptide (A beta 1-40) in mucopolysaccharidosis brains, *J Neuropathol Exp Neurol* **58**: 815–824 (1999).

CHAPTER 11

AUTOPHAGY IN HUNTINGTON'S AND PARKINSON'S DISEASES: PATHOGENIC MECHANISMS AND THERAPEUTIC POTENTIALS

Junghyun Lim, Lauren G. Friedman, Nicole C. McKnight and Zhenyu Yue

ABSTRACT

Emerging evidence demonstrates the neuroprotective role of autophagy in protein quality and cellular organelle damage control. The interference of genetic mutations and environmental toxins with basal autophagic activity in neurons is detrimental and restoring or enhancing neuronal autophagy in the context of some neurodegenerative diseases is believed to offer protection. This chapter will highlight recent research progress in understanding how autophagy is involved in the pathogenesis of Huntington's and Parkinson's diseases as well as the increasing evidence for autophagy as an attractive drug target for these two neurodegenerative disorders.

INTRODUCTION

Huntington's disease (HD) and Parkinson's disease (PD) are two devastating neurodegenerative movement disorders that are characterized by the degeneration of selective neurons and the abnormal accumulation of protein aggregates or inclusions in affected neuronal populations. Despite decades of research, the pathogenic pathways underlying the onset of these diseases are poorly understood. Recent studies point to dysfunctional protein clearance systems, which may partially contribute to pathogenesis, and suggests that restoration and/or enhancement of protein aggregate degradation may alleviate the progression of the disease.

Macroautophagy (autophagy) is a mostly non-specific, bulk lysosomal degradation pathway that occurs in almost all cells and tissues. Emerging evidence has increasingly shown that autophagy is a catabolic mechanism for the turnover of aggregate-prone proteins, macromolecular complexes and organelles. Many human disease-related proteins form oligomers/aggregates that are neurotoxic; examples include polyQ-containing proteins such as huntingtin in HD,[1,2] ataxin-1 in spinocerebellar ataxia (SCA1),[3] and androgen receptor in spinobulbar muscular atrophy.[4] Intact intracellular catabolism is therefore critical in regulating harmful protein levels in neurons. In addition, pathogenic mutants of α-synuclein (α-syn) in Parkinson's disease, SOD1 in amyotrophic lateral sclerosis (ALS), and tau in Alzheimer's disease (AD) and other tauopathies are believed to be autophagy substrates.[5–8] It has become clear that basal autophagy plays a vital neuroprotective role in central nervous system (CNS) neurons.[9–11] Accordingly, therapeutic strategies for enhancing autophagic activity have become increasingly attractive as potential treatments for neurodegenerative disorders associated with protein aggregation. This chapter will highlight the current progress in autophagy research that is focused on HD and PD.

PART 1. AUTOPHAGY AND HUNTINGTON'S DISEASE

Huntington's Disease

Huntington's disease is a dominantly-inherited, neurodegenerative disorder. It belongs to the polyglutamine disease family; the CAG repeat expansion in the *huntingtin* gene leads to long polyglutamine tracts within the mutant proteins.[12,13] The typical symptoms of HD are involuntary muscle jerks and twitches and these generally accompany motor, cognitive, and psychological dysfunction.[14] The onset of HD usually occurs between 35–50 years, though rare cases before the age of 20 or as late as 80 are reported. The pathology of HD is brain specific, showing massive neuronal loss in the caudate and putamen.[15] Pioneering work identified the genetic inheritance of an expanded trinucleotide (CAG) tract in exon 1 of the *huntingtin* (*Htt*) gene as the cause of the disease.[13] The resulting mutant Htt protein product carrying polyQ tracts in its N-terminal region forms insoluble aggregates. CAG repeats beyond 35 in number accelerates the disease risk and the severity of the disease and protein aggregation is correlated with the length of polyQ tracts as well as *Htt* fragment length and amount.[16] In addition, N-terminal Htt fragments carrying expanded polyQ tracts are more prone to forming aggregates and are more toxic than full-length mutant Htt.[17,18] While the presence of aggregates is a hallmark of ongoing molecular pathology of HD, it is believed that the toxicity is mediated by the monomers and/or oligomers with specific conformations.

Early Evidence Implicating Autophagy in HD

Using electron microscopy analysis, Kegel and DiFiglia examined postmortem HD brains and provided early evidence linking autophagy and HD. They observed Htt-labeled vacuoles with morphological features of autophagosomes in HD neurons. In addition, they found multivesicular bodies and a proliferation of endosomal and lysosomal bodies in affected cortical neurons. These authors also showed that overexpression of mutant Htt in mouse striatal culture cells stimulated lysosomal activity and accumulation of autophagic vacuoles.[19] Using another mouse model expressing full-length Htt mutant protein, Li *et al.* found accumulation of Htt aggregates concomitant with formation of vacuoles resembling late autophagosomes (autolysosomes) in lateral globus pallidus (LGP) axons.[20] These intriguing observations stirred further investigations into the potential role of autophagic function in the pathogenesis of HD.

Degradation of Huntingtin

Htt is a large and complex protein consisting of an N-terminal region (where polyQ repeats occur), HEAT (Htt, elongation factor 3, protein phosphatase 2A, and TOR 1) repeats, a functional nuclear export signal (NES), and a nuclear localization signal (NLS)[14] (Figure 1A). It is ubiquitously expressed in CNS neurons[21,22] and it is distributed among multiple subcellular compartments, including endoplasmic reticulum, the Golgi complex, the nucleus, and mitochondria.[14,23,24] Even though the physiological function of Htt has yet to be fully understood, available evidence suggests that it plays a role in the regulation of a wide range of processes including embryonic development, apoptosis, BDNF production, axonal and vesicle transport, and synaptic activity.[14] In contrast to normally-functioning Htt, Htt carrying a polyQ tract greater than 35 becomes toxic, as this mutant Htt is insoluble and forms aggregates or inclusion bodies. Because neurons cannot dissipate the toxic materials by cell division, the clearance of mutant Htt and its aggregates becomes critical for their viability. Using an inducible mouse model expressing mutated Htt fragments, Yamamoto *et al.* demonstrated that there is an intrinsic degradation system of neurons to clear polyQ-containing aggregates after the continuous synthesis of the mutant fragment is inhibited.[25] This result also suggests that, despite the presence of such a clearance system, its efficiency is insufficient to handle the influx of mutant proteins.

The ubiquitin proteasome system (UPS) and autophagy are the main routes for Htt degradation and cross-talk and functional complementation between the UPS

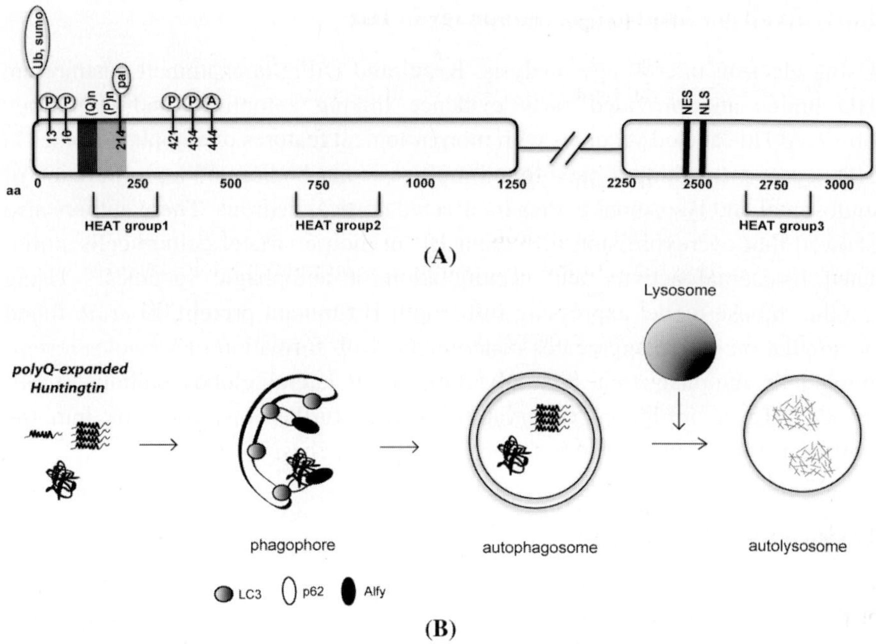

Figure 1. Schematic diagrams of Huntingtin structure and possible degradation of polyQ-expanded Huntingtin by autophagy. (A) Sequence domains of Huntingtin and residues that are regulated through post-translational modification. (Q)n indicates the polyglutamine tract, which is followed by the polyproline sequence, (P)n. Huntingtin has three HEAT groups, and each consists of clusters of HEAT repeats. Huntingtin is ubiquitinated and sumoylated (Ub, sumo) at the end of N-terminus. Phosphorylations at serine 13 and serine 16 (p) mediate its own clearance perhaps through autophagy. Palmitoylation (Pal) at cysteine 214 regulates vesicle trafficking and acetylation (A) at lysine 444 is important for the Htt degradation through autophagy. Phosphorylations (p) at serine 421 and 434 regulate the cleavage of itself and aggregation formation. NES and NLS indicate nuclear export signal and nuclear localization signal, respectively. (**B**) Degradation of polyQ-expanded Htt by selective autophagy. PolyQ-expanded Htt forms oligomers and these oligomers can form protein aggregates. Autophagic adaptor proteins, such as p62 and Alfy, recruit polyQ-expanded Huntingtin aggregates or oligomers to the phagophores. Once the formation of autophagosomes completes, autophagsomes fuse with lysosomes and then they become autolysosomes, wherein the Htt protein aggregates are degraded.

and autophagy has been suggested. However, the relative efficiency of these two systems to remove toxic forms of mutant Htt has not been fully examined. A recent study of cellular and mouse HD models showed that inhibition of the UPS led to a greater accumulation of soluble N-terminal mutant Htt than did inhibition of autophagy.[26] This study suggests the UPS and autophagy may have different capacities to remove soluble mutant Htt in cells. Further studies investigating specific Htt species that are prone to regulation by the UPS versus autophagy in the cytoplasm or the nucleus will be very useful for the therapeutic goal of accelerating the clearance of toxic Htt species.

Degradation by the ubiquitin-proteasome system

The UPS regulates turnover of proteins by tagging them with ubiquitin and subsequently targeting them to the proteasome for degradation. Wildtype Htt, some of the toxic fragments, and even the soluble form of mutant Htt are known substrates for the UPS, and specific proteasome inhibitors increase Htt aggregates in cellular and animal models of HD.[27–29] In contrast, overexpression of heat shock proteins or chaperones reduces Htt aggregates and increases viability in HD animal models.[30,31] The presence of the UPS in the nucleus enables nuclear inclusions containing mutant Htt to be cleared.[32] UPS activity, however, declines with age, and such impairment is more prevalent in neurons than in glial cells. This notion may in part explain why neurons are particularly vulnerable and why misfolded proteins are preferentially accumulated in neurons.[14,33] Co-localization of polyQ inclusions and ubiquitin was observed in a variety of HD models and postmortem HD brains.[18,34] Additional evidence revealed that Htt aggregates also co-localize with proteasome subunits and chaperones,[35] suggesting that UPS dysfunction causes insufficient clearance of mutant Htt. While the proteasome can process soluble Htt (mostly wildtype), oligomeric and aggregated mutant Htt cannot penetrate the narrow pore of the proteasome.[36] Furthermore, the proteasome may degrade polyQ-flanking sequences, but not polyQ stretches, leaving polyQ stretches without flanking sequences.[28,37] These "bare" polyQ's may be more aggregate-prone and toxic.

Degradation by the autophagy-lysosomal system

Early reports first described increased numbers of autophagosome-like structures in brains of HD patients.[19,38–40] These first observations suggested that the autophagic pathway might be involved in Htt degradation, although it was unclear how autophagy was involved. Current evidence suggests that autophagy activation can ameliorate HD pathology in HD animal models through the clearance of disease protein aggregates. Administration of autophagy activators such as rapamycin, or by increasing *beclin 1*, *LC3*, or *Atg5* gene expression levels, promotes the clearance of the mutant Htt and reduces the aggregate levels, leading to the improvement of behavioral phenotypes in a variety of HD animal models.[1,41,42] In contrast, inhibition of autophagy using 3-methyadenine (3-MA) or inhibition of autophagosome turnover using Bafilomycin A1 attenuates the clearance of mutant Htt, leading to increased levels of soluble mutant Htt and aggregates. Furthermore, full-length or truncated wild-type or mutant Htt are observed at various stages of autophagic vacuole formation and maturation.[43,44] Other autophagy proteins including LC3, Atg5, Atg12 or Atg16L are also recruited to protein aggregates.[45] Interestingly, a recent study revealed that overexpression of the transcription factor EB (TFEB), a master regulator gene of most lysosomal genes, increases lysosomal activity and degradation of the mutant protein in an inducible HD cell model.[46] Thus, it is likely that autophagy ameliorates toxicity by clearing both soluble monomeric and aggregated Htt.[47] Alternatively, clearance of aggregates might be due to degradation of the aggregate precursors (including both the soluble and oligomeric species), rather than large aggregates.

Regulation of Autophagy by Huntingtin

Htt protein itself may have a regulatory role in its own degradation through autophagy. It was initially proposed that there may be a protective role of protein aggregates in HD which results from stimulation of the autophagic process and subsequent degradation of mutant Htt.[48] The early observation linking Htt and autophagy was that overexpression of mutant Htt in mouse striatal cell culture stimulated lysosomal activity and formation of autophagic structures.[19] Excessive endosome/lysosome-like organelles were observed in brains of HD patients and animal models.[40,49] One study suggested that increased autophagy induction was due to the sequestration of mTOR by polyQ aggregates.[42] mTOR is a negative regulator of autophagy and its sequestration by aggregates would be expected to activate autophagy and subsequent clearance of protein aggregates. In line with these findings, a knockin mouse model expressing Htt with 200 glutamines showed an early activation of the autophagic responses, along with HD-related phenotypes and accumulation of inclusions.[50]

In contrast, an inhibitory effect of mutant Htt on autophagy efficiency has also been reported. Martinez-Vicente *et al.* showed a defect in the ability of autophagosomes to sequester cytosolic cargo in HD cells. The authors proposed that the reduced capacity of autophagosomes to take up cytosolic components may contribute decreased efficiency of autophagic degradation and subsequent accumulation of protein aggregates.[51] Additional evidence suggests that Htt aggregates may recruit and sequester the autophagy protein Beclin 1, potentially interfering with the normal function of Beclin 1 in autophagy.[52] A previous study has also provided genetic evidence that deletion of the polyQ stretch in Htt causes an increase in autophagic activity and in turn, increased longevity in mice.[53]

The effect of polyQ-expanded Htt may represent a gain-of-function that is associated with increased autophagosome synthesis and alleviation of HD neuropathology. Future studies should investigate how autophagy is altered in HD animal models and whether there is disease stage-specific change in autophagic activity.

Potential Role of p62/SQSTM1 and Alfy in Degradation of Htt via Selective Autophagy

Until recently, autophagy was considered a bulk degradation system with little selectivity for specific cargos. Recent studies, however, provide evidence that autophagy can selectively remove certain proteins, protein aggregates, and misfolded proteins using a specific form of autophagy called aggrephagy[51] (see Chapter 2). This selective form of autophagy is perhaps mediated by autophagic adaptor proteins, such as p62/SQSTM1, NBR1, and Alfy.[54] Among them, p62/SQSTM1 has been extensively investigated because of an early observation that p62 is involved in the autophagic degradation of protein aggregates.[55] Several distinctive domains of p62 are required for its action as an autophagic adaptor. p62 interacts with ubiquitinated proteins via its UBA domain and oligomerizes via its PB1 domain.[56] By interacting with LC3 through an LC3-interacting region (LIR) motif, p62 recruits polyubiquitinated proteins or aggregates to autophagic compartments, wherein ubiquitinated proteins or protein aggregates are ultimately degraded.[57] During this process, because p62 itself is degraded by autophagy,[55] the level of p62 may be used to monitor autophagic activity.[58–60] The involvement of p62 in mutant Htt aggregate formation and degradation supports the emerging role of p62 in Huntington's disease.[55,61,62] The above evidence, therefore, suggests a model where the adaptor protein p62 recruits ubiquitinated proteins or protein aggregates to autophagosomes for degradation. However, it is not clear how p62-mediated autophagy achieves the selective degradation of disease aggregate proteins (e.g. polyQ-Htt). Recent work has suggested that PI3P-binding Autophagy-linked FYVE domain protein (Alfy) is also an autophagic adaptor, regulating the degradation of mutant Htt with p62 in a coordinated manner. Alfy, acting as a scaffolding adaptor protein, connects p62-bound protein aggregates to the core autophagic machinery and ensures autophagic degradation of aggregates[63,64] (Figure 1B). Future study will be needed to understand the exact mechanisms whereby the autophagic adaptor proteins function in selective autophagy for the degradation of aggregate-prone proteins including mutant Htt in HD.

The Effect of Post-Translational Modifications in Htt Clearance

While autophagic adaptor-mediated degradation of protein aggregates requires ubiquitination before the aggregates meet the autophagic machinery, it has been recently suggested that other post-translational modifications of Htt also play a critical role in Htt degradation. For example, mutant Htt can be acetylated at K444 by CREB binding protein (CBP), which increases the trafficking of mutant Htt to autophagosomes and subsequent degradation by autophagy in a p62-dependent manner.[65] An Htt mutant resistant to acetylation was accumulated in neurons, leading to neurodegeneration. Thus, acetylation of Htt renders Htt prone to autophagic degradation. In another line of study, IKK (IκB kinase) was shown to mediate phosphorylation of Htt at Serine 13, and this phosphorylation has a regulatory role in Htt degradation through autophagy.[66] This modification may regulate additional post-translational modifications of Htt, such as ubiquitination, sumoylation, and acetylation, which in turn lead to normal clearance of Htt by the UPS and autophagy. Lysosome-associated membrane protein LAMP-2 and heat shock protein Hsc70 are involved in this process. While phosphorylation of Ser13 and Ser16 seems to affect the degradation of Htt, the presence of a polyQ tract reduces the efficiency of the phosphorylation and further degradation by the UPS and autophagy[67] (Figure 1A). It is possible that the phosphorylation of Htt may activate Htt recognition by a chaperone complex required for selective protein degradation. Further study of the complex network of various post-translational modifications and their regulatory roles in the turnover of mutant Htt proteins will yield important information for the development of therapeutic strategies targeting the process of autophagic degradation of Htt.

Therapeutic Approaches for Huntington's Disease: Enhancing Autophagy for Mutant Huntingtin Clearance

Currently there are no therapeutic treatments available for HD. Considering the monogenic nature of HD, therapeutic approaches targeting polyQ-expanded Htt to enhance its degradation through autophagy could be promising[68,69] (see Chapter 13). Emerging studies have provided proof-of-principle evidence that Htt mutant aggregates can be degraded by autophagy via genetic and pharmacological intervention.[70] A series of investigations by Rubinsztein and colleagues has tested various chemicals in boosting autophagic degradation and mutant Htt clearance. mTOR-dependent as well as mTOR-independent autophagy modulators were found to affect the levels of Htt aggregates and alleviate the toxicity. Rapamycin and its analog CCI-779 exerted a protective role in

Drosophila and mouse HD models by inhibiting mTOR activity and enhancing autophagy.[1,42] Lithium or trehalose treatment were shown to upregulate autophagy and in turn enhance degradation of mutant Htt in mTOR-independent pathways.[71–74] Through a small molecule screen, three novel small-molecule enhancers (SMERs) were identified that are capable of increasing the clearance of mutant Htt fragments, reducing mutant Htt aggregation and offering protection in cells and fly models of HD.[75] A recent report showed that rilmenidine, an anti-hypertensive drug, induced autophagy via an mTOR-independent pathway, and the drug attenuated the disease process in an HD mouse model and reduced levels of the mutant huntingtin fragment.[76] The current challenge is to understand the detailed mechanism of action for each potential "autophagy drug" and to develop assays to monitor autophagic activity in tissues. Future work should also focus on identifying highly specific and potent autophagy modulators.

PART 2. PARKINSON'S DISEASE AND AUTOPHAGY

Parkinson's Disease

Parkinson's disease (PD) is the most prevalent neurodegenerative movement disorder. PD is clinically characterized by tremor, bradykinesia, rigidity, and postural instability. PD patients often exhibit concomitant psychopathological symptoms including cognitive deterioration and mood disorders.[77] The progressive loss of midbrain dopaminergic (DA) neurons of the nigrostriatal pathway and the presence of intracellular neuronal inclusions (Lewy bodies) that contain ubiquitin and α-synuclein (α-syn) are the pathological hallmarks of the disease. While DA neuron degeneration in the substantia nigra is the predominant feature, neuronal cell loss and α-syn aggregates are also observed in the locus coeruleus, raphe nucleus, cerebral cortex, anterior thalamus, hypothalamus, and brainstem of postmortem PD brains.[78]

Both genetic and environmental factors contribute to PD etiology, however little is known about the precise pathogenic mechanisms. The vast majority of PD cases occur sporadically, while 5%–10% of cases are attributable to familial mutations.[79] Autosomal dominant mutations in LRRK2 account for the most common genetic forms of PD and some sporadic cases of PD.[80,81] Pathogenic mutations in α-syn, the primary component of Lewy bodies,[82] are associated with autosomal dominant inherited forms of the disease.[83–85] Autosomal recessive mutations in PINK-1, Parkin, and DJ-1 cause early-onset PD.[86–88] The identification of inherited genetic mutations has not only shed light on the underlying causes of the genetic forms of PD, but may also provide clues to the underlying causes of idiopathic PD.

Autophagy in Toxin-Based PD Models: Early Observations

The selective degeneration of dopamine (DA) neurons in PD patients has been linked to oxidative stress. At physiological levels, DA is transported into synaptic vesicles and is protected from auto-oxidation. However, excess DA and DA metabolites can cause oxidative injury.[89] In sporadic PD, it was hypothesized that various environmental neurotoxins can produce oxidative damage. A large number of studies have used toxin-based models to study the cellular mechanisms involved in PD pathogenesis. Interestingly, autophagosome proliferation in degenerating DA neurons has been observed in response to acute treatment with a variety of toxins including 6-hydroxydopamine (6-OHDA),[90] rotenone,[91] paraquat,[92] methamphetamine (METH),[93] and 1-methyl-4-phenylpyridinium (MPP^+).[94] Knockdown of genes required for autophagosome formation rescued MPP^+-induced neurotoxicity.[94] Taken together, these studies indicate that the pharmacological toxins cause dysfunctional autophagy, which could be, in part, responsible for the neurotoxicity. While autophagy is well known as a neuroprotective mechanism, aberrant hyperactivity of autophagy can be destructive and cause neurite degeneration in certain toxin models of PD (see Chapter 17).

Protein Degradation Pathways in PD

Degradation by the UPS

Intact intracellular catabolism is vital for the prevention of late-onset neurodegenerative diseases. While recent evidence reveals a link between autophagy and the pathogenesis of several chronic neurodegenerative diseases, it is clear that autophagy also acts in concert with the UPS to maintain proper protein turnover (see Chapter 3). The UPS mediates the breakdown of short-lived proteins whereby ubiquitin-tagged substrates are degraded by proteases in the catalytic core of the proteasome.[95] Several reports have identified a potential role for UPS dysfunction in PD pathogenesis. Aberrant accumulations of α-syn are caused by genomic expansion and decreased UPS activity.[96,97] Additionally, genetic studies revealed that two forms of familial PD are caused by mutations in Parkin and UCH-L1 (ubiquitin carboxy-terminal hydrolase-L1), which are both involved in proteasomal degradation.[98,99] Sporadic PD has also been associated with a dysfunctional UPS. Impaired proteasome function was observed in the substantia nigra in postmortem PD brains[100] and treatment with proteasomal inhibitors in neuronal culture and rat models produced PD-like pathology and behavioral abnormalities.[101,102] However, these findings remain controversial due to conflicting evidence.[103,104] Interestingly, a recent study showed that deletion of

26S proteasomes in CNS neurons caused neurodegeneration and Lewy body-like inclusions,[105] providing genetic evidence that impairment of proteasomal degradation may, in part, contribute to PD-like pathology.

Degradation by chaperone-mediated autophagy

Chaperone-mediated autophagy (CMA) is a highly selective form of lysosomal degradation. Soluble cytosolic proteins containing a consensus motif are recognized by the Hsc70 chaperone and are then targeted to the lysosome for degradation by lysosomal surface receptor LAMP2. Wild-type α-syn contains a CMA motif and is degraded by CMA in ventral midbrain neuronal cultures. On the other hand, while pathogenic forms of α-syn bind strongly to the lysosome, they cannot be internalized.[106] Therefore mutant α-syn may disrupt CMA degradation of other cytosolic proteins, including wild-type α-syn, leading to their accumulation. Recent studies have revealed that impaired CMA mediated by exogenous wild-type or mutant α-syn leads to enhanced cytoplasmic levels of pro-survival protein, MEF2D (myocyte enhancer factor 2D) which provides a potential mechanism for the loss of neuronal viability in PD.[107]

Analysis in postmortem brains has shown that both Hsc70 and LAMP2 expression levels were reduced in the substantia nigra, suggesting the possibility of decreased CMA function.[108] Several studies have shown that macroautophagy is induced when CMA is compromised, while selective blockage of LAMP2 in mouse fibroblasts leads to upregulation of macroautophagy.[109] These findings suggest that when CMA is impaired due to the effects of aging[110] or the overexpression of wild-type or mutant forms of α-syn,[106,111] macroautophagy may serve as a compensatory clearance pathway. Whether changes in CMA activity are a primary cause or a secondary effect of the disease remains unclear.

Degradation by macroautophagy

Accumulating evidence has revealed that basal levels of macroautophagy (autophagy) are neuroprotective. The genetic ablation of two essential autophagy genes, *Atg5* and *Atg7,* in neurons leads to accumulation of ubiquitinated aggregates and subsequent neuronal loss in mice.[9,10,112] These studies highlight the beneficial role of autophagy in the prevention of ubiquitinated protein accumulation. Several reports have now demonstrated a functional link between α-syn clearance and autophagy. Earlier studies in cultured cells showed that α-syn was degraded by the UPS as well as autophagy.[5,113] Autophagy induction was observed in mice overexpressing both human wild-type and pathogenic A53T mutant α-syn.[114] Interestingly, the A53T mutation in α-syn produced a more robust activation of autophagy in mice. This might be explained by the observation that mutated α-syn variants can block CMA and subsequently promote autophagic clearance in a compensatory manner.[106]

Insufficient autophagosome formation may predispose to neurodegeneration associated with cytotoxic α-syn accumulation, and enhancing autophagy may reverse these pathological effects.[74] While blocking autophagy initiation promoted α-syn oligomer accumulation,[114] stimulating autophagy with rapamycin ameliorated the accumulation of exogenous wild-type and mutant forms of α-syn in neuronal-like cells.[5] Similarly, viral transfer of autophagy gene *beclin 1* reduced α-syn aggregates and rescued neurite shortening or synaptic morphology in neuronal-like cells and α-syn transgenic mice.[115] The above studies provide important evidence for the role of autophagy in the clearance of α-syn, particularly its oligomeric form. A more recent study has suggested that the UPS is the main degradation pathway for endogenous α-syn under normal conditions *in vivo*, whereas autophagy is recruited when neurons are overburdened with increased α-syn levels.[116]

Autosomal Dominant PD: Deregulation of Autophagy

Mutations in *SNCA* (α-syn) and *LRRK2* cause autosomal dominant PD. Whether or not the two genes are in involved in the same pathogenic pathway leading to PD is unclear. Limited evidence in animal models has shown that α-syn-associated PD-like pathology is regulated by LRRK2.[117,118] Recent studies, however, have suggested a role for the two genes in autophagy regulation. Whereas mutant α-syn was believed to affect CMA as described above, overexpression of wildtype α-syn, which can be caused by *SNCA* multiplication, may inhibit macroautophagy.[119]

Mutations of *LRRK2* are the most common genetic cause of PD. Expression of pathogenic *LRRK2* mutations produces alterations in autophagic activity.[120–122] The common PD mutation G2019S in LRRK2 leads to neurite shortening and concomitant accumulation of autophagosomes in the somata and neurites. Reducing the levels of core autophagic proteins rescues neuritic degeneration,[123] indicating a role for autophagy in neurite degeneration that is mediated by a LRRK2 mutation. In *LRRK2* knockout mice, autophagy is apparently impaired in the kidney, and is accompanied by increased levels of α-syn and inflammation.[117] However, it is unclear why autophagy-associated pathology is restricted to certain tissues such as the kidney but not the brain. Taken together, the above evidence demonstrates a link between the two autosomal dominant genes (*SNCA* and *LRRK2)* and autophagy deregulation.

Autosomal Recessive PD: Impairment of Mitophagy

Organelles and proteins of aging neurons are particularly vulnerable to cumulative oxidative damage and genetic and environmental toxins;[124] in particular, mitochondria are susceptible to oxidative stress. Damaged or old mitochondria are selectively removed by autophagy in a process called mitophagy.[125] Oxidative stress is implicated in the pathogenesis of PD and the failure of autophagy to clear damaged mitochondria may be one of the underlying causes of the disease.

Some forms of familial PD have been linked to autosomal recessive mutations in two proteins involved in mitophagy. Recent emerging evidence has demonstrated a surprising role for Parkin (encoded by *PARK2*), a component of an E3 ubiquitin ligase complex, in the removal of damaged mitochondria mediated by autophagy. Parkin is selectively recruited to depolarized (damaged) mitochondrial membranes following treatment with mitochondrial uncoupler, carbonyl cyanide m-chlorophenylhydrazone (CCCP).[126] The CCCP treatment also causes co-localization of Parkin with LC3, suggesting that Parkin recruits autophagic machinery to the damaged mitochondria.[126] Parkin-mediated mitophagy was reversed when cells were treated with autophagy inhibitors. In heteroplasmic cybrid cells, which contain a stable mixture of wild-type and mutant mitochondrial DNA, Parkin was selectively translocated to uncoupled mitochondria but not to intact mitochondria,[127] demonstrating its specificity for injured mitochondria.

PINK1 (PTEN induced putative kinase 1) is a mitochondrial kinase that is localized to the mitochondrial membrane. Mutations in the kinase domain of PINK1 have been shown to cause early-onset Parkinson's disease through increased susceptibility to cellular stress.[86] Recent studies have suggested that PINK1 may functionally link Parkin to dysfunctional mitochondria. PINK1 expression is markedly increased following treatment with CCCP and is preferentially found on depolarized mitochondria.[128] PINK1 overexpression induces mitochondrial aggregation and elimination. Additionally, when PINK1 mutants were exogenously expressed, mitochondria were diffusely distributed in cells and mitochondrial removal was impaired.[129] Therefore, PINK1 is required for Parkin recruitment in the CCCP model of mitophagy.

Several PD-pathogenic mutations of *Parkin* caused defects in mitophagy. Notably, Parkin translocation was affected to varying degrees by different mutations in *Parkin*.[127,128,130] In addition, PD-linked *PINK1* mutations led to defective Parkin recruitment and Parkin-induced mitophagy.[129–131] These studies provide important evidence for the potential role of mutant Parkin and PINK1 in dysfunctional mitophagy. Furthermore, it is noted that the proteasome also plays a role in mitochondrial turnover. For example, a recent report found that the inner mitochondrial membrane and matrix proteins are degraded by mitophagy in cultured fibroblasts, whereas the outer mitochondrial proteins were degraded primarily by the proteasome.[131] Another study showed that the proteasome and p97 mediate Parkin-related degradation of mitofusins, which regulate mitochondria fusion reactions, and therefore indirectly affects mitophagy.[132] It is likely that the UPS and autophagy coordinate in the mitophagy process and both contribute to mitochondria turnover. Thus, mitophagy is a critical mechanism that maintains mitochondria homeostasis to meet metabolic demand and/or perform quality control to remove damaged mitochondria. The inherited mutations of Parkin and PINK1 disrupt the clearance of depolarized mitochondria (Figure 2), resulting in oxidative stress, which may explain why these mutations cause early-onset disease.

Implication of Dysfunctional Autophagy in PD Pathogenesis and Therapeutic Development

As described above, the investigation of the normal functions of PD genes and the pathophysiology related to their mutations has implicated impaired autophagy in the pathogenesis of PD. Since the vast majority of PD cases occur sporadically, it is critical to understand the neuroprotective mechanisms that may deteriorate

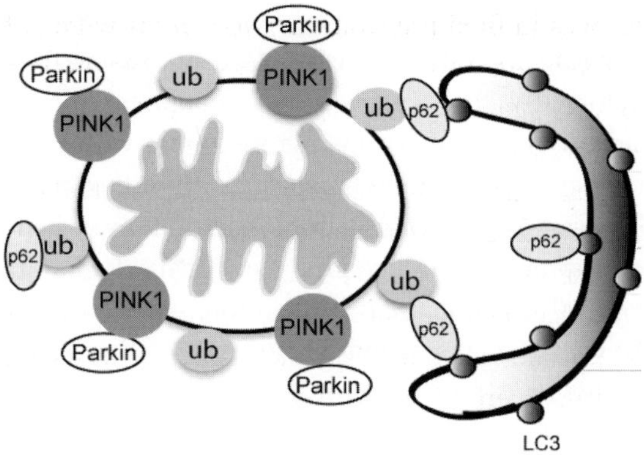

Figure 2. Regulation of mitophagy by PD-related genes, *PINK1* and *parkin*. When PINK1 is stabilized at depolarized mitochondria, Parkin is recruited to damaged mitochondria through binding to PINK1. Parkin mediates ubiquitinations of unknown substrates at the mitochondria, which in turn helps recruit autophagic adaptor protein p62 and other autophagy machinery proteins. This process might be important for selective degradation of mitochondria by autophagy. PD recessive mutations in *PINK1* and *parkin* lead to dysregulation of mitophagy

with age and lead to idiopathic PD. Since autophagosome formation and autophagosome-lysosome fusion progressively decline in normal aging,[133] it raises a possibility that genetic or environmental impairment of autophagy, facilitated by aging, is one of the underlying causes to idiopathic PD. Despite evidence from cell culture experiments, the role of autophagy in the onset and development of PD has not been elucidated in mammalian models. Data from our laboratory suggests that genetic suppression of autophagy in DA neurons contributes to PD-related pathologies, including progressive degeneration of DA terminals, loss of cell bodies, motor function deficits, and accumulation of intraneuronal ubiquitinated inclusions and presynaptic α-syn. In addition, we found that inactivation of autophagy leads to upregulation of LRRK2 expression in fibroblast cells and accumulation of LRRK2 protein in certain neural cell types.[134] While genetic PD mutations may compromise neuronal autophagy at discrete steps, insufficient autophagy may represent a risk for the development of idiopathic PD. Future experiments should continue to investigate the potential for autophagy as a means of preventing the accumulation of neurotoxic α-syn aggregates and damaged mitochondria. Towards this goal, the identification of specific autophagy enhancers and their subsequent testing in various animal models will be critical in determining beneficial strategies for the treatment of PD.

Additional Note

While this chapter was in final preparation, two studies were published back-to-back in *Human Molecular Genetics*. They provide intriguing genetic evidence that α-syn levels modulate Huntington's disease in mice.[135,136] The evidence suggests cross-talk between the two distinct pathogenic pathways mediated by the two different disease genes. For instance, it is possible that the two pathogenic pathways merge at autophagy deregulation. Importantly, the two studies highlight the possibility that the therapeutic strategy of antagonizing α-syn toxicity might alleviate HD. Given that increased autophagy can enhance the clearance of both α-syn and polyQ-expanded Htt, targeting autophagy might serve as a stone to kill two birds in the development of drugs to treat HD.

ACKNOWLEDGMENTS

This work was supported by the United States National Institutes of Health–National Institute of Neurological Disorders and Stroke Grants R01NS060123, NS060809, RNS055683A; the Michael J. Fox Foundation; the Bachmann-Strauss

Dystonia and Parkinson Foundation; and CHDI Foundation (to Z. Yue). We thank members of Dr. Yue's lab for critical reading and comments.

REFERENCES

1. B. Ravikumar, R. Duden and D. C. Rubinsztein, Aggregate-prone proteins with polyglutamine and polyalanine expansions are degraded by autophagy, *Hum Mol Genet* **11**: 1107–1117 (2002).
2. P. O. Bauer, A. Goswami, H. K. Wong, M. Okuno, M. Kurosawa, M. Yamada, H. Miyazaki, G. Matsumoto, Y. Kino, Y. Nagai and N. Nukina, Harnessing chaperone-mediated autophagy for the selective degradation of mutant huntingtin protein, *Nat Biotechnol* **28**: 256–263 (2010).
3. H. Y. Zoghbi and H. T. Orr, Pathogenic mechanisms of a polyglutamine-mediated neurodegenerative disease, spinocerebellar ataxia type 1, *J Biol Chem* **284**: 7425–7429 (2009).
4. U. B. Pandey, Z. Nie, Y. Batlevi, B. A. McCray, G. P. Ritson, N. B. Nedelsky, S. L. Schwartz, N. A. DiProspero, M. A. Knight, O. Schuldiner, R. Padmanabhan, M. Hild, D. L. Berry, D. Garza, C. C. Hubbert, T. P. Yao, E. H. Baehrecke and J. P. Taylor, HDAC6 rescues neurodegeneration and provides an essential link between autophagy and the UPS, *Nature* **447**: 859–863 (2007).
5. J. L. Webb, B. Ravikumar, J. Atkins, J. N. Skepper and D. C. Rubinsztein, Alpha-synuclein is degraded by both autophagy and the proteasome, *J Biol Chem* **278**: 25009–25013 (2003).
6. T. Kabuta, Y. Suzuki and K. Wada, Degradation of amyotrophic lateral sclerosis-linked mutant Cu, Zn-superoxide dismutase proteins by macroautophagy and the proteasome, *J Biol Chem* **281**: 30524–30533 (2006).
7. Z. Berger, B. Ravikumar, F. M. Menzies, L. G. Oroz, B. R. Underwood, M. N. Pangalos, I. Schmitt, U. Wullner, B. O. Evert, C. J. O'Kane and D. C. Rubinsztein, Rapamycin alleviates toxicity of different aggregate-prone proteins, *Hum Mol Genet* **15**: 433–442 (2006).
8. W. H. Yu, A. M. Cuervo, A. Kumar, C. M. Peterhoff, S. D. Schmidt, J. H. Lee, P. S. Mohan, M. Mercken, M. R. Farmery, L. O. Tjernberg, Y. Jiang, K. Duff, Y. Uchiyama, J. Naslund, P. M. Mathews, A. M. Cataldo and R. A. Nixon, Macroautophagy — a novel beta-amyloid peptide-generating pathway activated in Alzheimer's disease, *J Cell Biol* **171**: 87–98 (2005).
9. T. Hara, K. Nakamura, M. Matsui, A. Yamamoto, Y. Nakahara, R. Suzuki-Migishima, M. Yokoyama, K. Mishima, I. Saito, H. Okano and N. Mizushima, Suppression of basal autophagy in neural cells causes neurodegenerative disease in mice, *Nature* **441**: 885–889 (2006).
10. M. Komatsu, S. Waguri, T. Chiba, S. Murata, J. Iwata, I. Tanida, T. Ueno, M. Koike, Y. Uchiyama, E. Kominami and K. Tanaka, Loss of autophagy in the central nervous system causes neurodegeneration in mice, *Nature* **441**: 880–884 (2006).

11. C. B. Lucking, N. Abbas, A. Durr, V. Bonifati, A. M. Bonnet, T. de Broucker, G. De Michele, N. W. Wood, Y. Agid and A. Brice, Homozygous deletions in parkin gene in European and North African families with autosomal recessive juvenile parkinsonism. The European Consortium on Genetic Susceptibility in Parkinson's Disease and the French Parkinson's Disease Genetics Study Group, *Lancet* **352**: 1355–1356 (1998).
12. H. T. Orr and H. Y. Zoghbi, Trinucleotide repeat disorders, *Annu Rev Neurosci* **30**: 575–621 (2007).
13. Anonymous, A novel gene containing a trinucleotide repeat that is expanded and unstable on Huntington's disease chromosomes. The Huntington's Disease Collaborative Research Group, *Cell* **72**: 971–983 (1993).
14. C. Zuccato, M. Valenza and E. Cattaneo, Molecular mechanisms and potential therapeutical targets in Huntington's disease, *Physiol Rev* **90**: 905–981 (2010).
15. A. Reiner, R. L. Albin, K. D. Anderson, C. J. D'Amato, J. B. Penney and A. B. Young, Differential loss of striatal projection neurons in Huntington disease, *Proc Natl Acad Sci USA* **85**: 5733–5737 (1988).
16. Z. R. Crook and D. Housman, Huntington's disease: can mice lead the way to treatment?, *Neuron* **69**: 423–435 (2011).
17. A. S. Hackam, R. Singaraja, C. L. Wellington, M. Metzler, K. McCutcheon, T. Zhang, M. Kalchman and M. R. Hayden, The influence of huntingtin protein size on nuclear localization and cellular toxicity, *J Cell Biol* **141**: 1097–1105 (1998).
18. C. A. Gutekunst, S. H. Li, H. Yi, J. S. Mulroy, S. Kuemmerle, R. Jones, D. Rye, R. J. Ferrante, S. M. Hersch and X. J. Li, Nuclear and neuropil aggregates in Huntington's disease: relationship to neuropathology, *J Neurosci* **19**: 2522–2534 (1999).
19. K. B. Kegel, M. Kim, E. Sapp, C. McIntyre, J. G. Castano, N. Aronin and M. DiFiglia, Huntingtin expression stimulates endosomal-lysosomal activity, endosome tubulation, and autophagy, *J Neurosci* **20**: 7268–7278 (2000).
20. H. Li, S. H. Li, Z. X. Yu, P. Shelbourne and X. J. Li, Huntingtin aggregate-associated axonal degeneration is an early pathological event in Huntington's disease mice, *J Neurosci* **21**: 8473–8481 (2001).
21. M. DiFiglia, E. Sapp, K. Chase, C. Schwarz, A. Meloni, C. Young, E. Martin, J. P. Vonsattel, R. Carraway, S. A. Reeves, *et al.*, Huntingtin is a cytoplasmic protein associated with vesicles in human and rat brain neurons, *Neuron* **14**: 1075–1081 (1995).
22. F. R. Fusco, Q. Chen, W. J. Lamoreaux, G. Figueredo-Cardenas, Y. Jiao, J. A. Coffman, D. J. Surmeier, M. G. Honig, L. R. Carlock and A. Reiner, Cellular localization of huntingtin in striatal and cortical neurons in rats: lack of correlation with neuronal vulnerability in Huntington's disease, *J Neurosci* **19**: 1189–1202 (1999).
23. Y. Trottier, D. Devys, G. Imbert, F. Saudou, I. An, Y. Lutz, C. Weber, Y. Agid, E. C. Hirsch and J. L. Mandel, Cellular localization of the Huntington's disease protein and discrimination of the normal and mutated form, *Nat Genet* **10**: 104–110 (1995).

24. A. N. Strehlow, J. Z. Li and R. M. Myers, Wild-type huntingtin participates in protein trafficking between the Golgi and the extracellular space, *Hum Mol Genet* **16**: 391–409 (2007).
25. A. Yamamoto, J. J. Lucas and R. Hen, Reversal of neuropathology and motor dysfunction in a conditional model of Huntington's disease, *Cell* **101**: 57–66 (2000).
26. X. Li, C. E. Wang, S. Huang, X. Xu, X. J. Li, H. Li and S. Li, Inhibiting the ubiquitin-proteasome system leads to preferential accumulation of toxic N-terminal mutant huntingtin fragments, *Hum Mol Genet* **19**: 2445–2455 (2010).
27. N. R. Jana, M. Tanaka, G. Wang and N. Nukina, Polyglutamine length-dependent interaction of Hsp40 and Hsp70 family chaperones with truncated N-terminal huntingtin: their role in suppression of aggregation and cellular toxicity, *Hum Mol Genet* **9**: 2009–2018 (2000).
28. C. I. Holmberg, K. E. Staniszewski, K. N. Mensah, A. Matouschek and R. I. Morimoto, Inefficient degradation of truncated polyglutamine proteins by the proteasome, *EMBO J* **23**: 4307–4318 (2004).
29. N. F. Bence, R. M. Sampat and R. R. Kopito, Impairment of the ubiquitin-proteasome system by protein aggregation, *Science* **292**: 1552–1555 (2001).
30. A. Wyttenbach, J. Carmichael, J. Swartz, R. A. Furlong, Y. Narain, J. Rankin and D. C. Rubinsztein, Effects of heat shock, heat shock protein 40 (HDJ-2), and proteasome inhibition on protein aggregation in cellular models of Huntington's disease, *Proc Natl Acad Sci USA* **97**: 2898–2903 (2000).
31. A. Klettner, The induction of heat shock proteins as a potential strategy to treat neurodegenerative disorders, *Drug News Perspect* **17**: 299–306 (2004).
32. A. Iwata, Y. Nagashima, L. Matsumoto, T. Suzuki, T. Yamanaka, H. Date, K. Deoka, N. Nukina and S. Tsuji, Intranuclear degradation of polyglutamine aggregates by the ubiquitin-proteasome system, *J Biol Chem* **284**: 9796–9803 (2009).
33. S. Tydlacka, C. E. Wang, X. Wang, S. Li and X. J. Li, Differential activities of the ubiquitin-proteasome system in neurons versus glia may account for the preferential accumulation of misfolded proteins in neurons, *J Neurosci* **28**: 13285–13295 (2008).
34. M. DiFiglia, E. Sapp, K. O. Chase, S. W. Davies, G. P. Bates, J. P. Vonsattel and N. Aronin, Aggregation of huntingtin in neuronal intranuclear inclusions and dystrophic neurites in brain, *Science* **277**: 1990–1993 (1997).
35. J. A. Johnston, C. L. Ward and R. R. Kopito, Aggresomes: a cellular response to misfolded proteins, *J Cell Biol* **143**: 1883–1898 (1998).
36. M. Renna, M. Jimenez-Sanchez, S. Sarkar and D. C. Rubinsztein, Chemical inducers of autophagy that enhance the clearance of mutant proteins in neurodegenerative diseases, *J Biol Chem* **285**: 11061–11067 (2010).
37. P. Venkatraman, R. Wetzel, M. Tanaka, N. Nukina and A. L. Goldberg, Eukaryotic proteasomes cannot digest polyglutamine sequences and release them during degradation of polyglutamine-containing proteins, *Mol Cell* **14**: 95–104 (2004).

38. A. Petersen, K. E. Larsen, G. G. Behr, N. Romero, S. Przedborski, P. Brundin and D. Sulzer, Expanded CAG repeats in exon 1 of the Huntington's disease gene stimulate dopamine-mediated striatal neuron autophagy and degeneration, *Hum Mol Genet* **10**: 1243–1254 (2001).
39. Z. H. Qin, Y. Wang, K. B. Kegel, A. Kazantsev, B. L. Apostol, L. M. Thompson, J. Yoder, N. Aronin and M. DiFiglia, Autophagy regulates the processing of amino terminal huntingtin fragments, *Hum Mol Genet* **12**: 3231–3244 (2003).
40. E. Sapp, C. Schwarz, K. Chase, P. G. Bhide, A. B. Young, J. Penney, J. P. Vonsattel, N. Aronin and M. DiFiglia, Huntingtin localization in brains of normal and Huntington's disease patients, *Ann Neurol* **42**: 604–612 (1997).
41. K. Jia, A. C. Hart and B. Levine, Autophagy genes protect against disease caused by polyglutamine expansion proteins in *Caenorhabditis elegans*, *Autophagy* **3**: 21–25 (2007).
42. B. Ravikumar, C. Vacher, Z. Berger, J. E. Davies, S. Luo, L. G. Oroz, F. Scaravilli, D. F. Easton, R. Duden, C. J. O'Kane and D. C. Rubinsztein, Inhibition of mTOR induces autophagy and reduces toxicity of polyglutamine expansions in fly and mouse models of Huntington disease, *Nat Genet* **36**: 585–595 (2004).
43. P. O. Seglen and P. B. Gordon, 3-Methyladenine: specific inhibitor of autophagic/lysosomal protein degradation in isolated rat hepatocytes, *Proc Natl Acad Sci USA* **79**: 1889–1892 (1982).
44. A. Yamamoto, Y. Tagawa, T. Yoshimori, Y. Moriyama, R. Masaki and Y. Tashiro, Bafilomycin A1 prevents maturation of autophagic vacuoles by inhibiting fusion between autophagosomes and lysosomes in rat hepatoma cell line, H-4-II-E cells, *Cell Struct Funct* **23**: 33–42 (1998).
45. A. Iwata, J. C. Christianson, M. Bucci, L. M. Ellerby, N. Nukina, L. S. Forno and R. R. Kopito, Increased susceptibility of cytoplasmic over nuclear polyglutamine aggregates to autophagic degradation, *Proc Natl Acad Sci USA* **102**: 13135–13140 (2005).
46. M. Sardiello, M. Palmieri, A. di Ronza, D. L. Medina, M. Valenza, V. A. Gennarino, C. Di Malta, F. Donaudy, V. Embrione, R. S. Polishchuk, S. Banfi, G. Parenti, E. Cattaneo and A. Ballabio, A gene network regulating lysosomal biogenesis and function, *Science* **325**: 473–477 (2009).
47. S. Sarkar and D. C. Rubinsztein, Huntington's disease: degradation of mutant huntingtin by autophagy, *FEBS J* **275**: 4263–4270 (2008).
48. A. Ventruti and A. M. Cuervo, Autophagy and neurodegeneration, *Curr Neurol Neurosci Rep* **7**: 443–451 (2007).
49. S. W. Davies, M. Turmaine, B. A. Cozens, M. DiFiglia, A. H. Sharp, C. A. Ross, E. Scherzinger, E. E. Wanker, L. Mangiarini and G. P. Bates, Formation of neuronal intranuclear inclusions underlies the neurological dysfunction in mice transgenic for the HD mutation, *Cell* **90**: 537–548 (1997).
50. M. Y. Heng, D. K. Duong, R. L. Albin, S. J. Tallaksen-Greene, J. M. Hunter, M. J. Lesort, A. Osmand, H. L. Paulson and P. J. Detloff, Early autophagic response in a

novel knock-in model of Huntington disease, *Hum Mol Genet* **19**: 3702–3720 (2010).

51. M. Martinez-Vicente, Z. Talloczy, E. Wong, G. Tang, H. Koga, S. Kaushik, R. de Vries, E. Arias, S. Harris, D. Sulzer and A. M. Cuervo, Cargo recognition failure is responsible for inefficient autophagy in Huntington's disease, *Nat Neurosci* **13**: 567–576 (2010).
52. M. Shibata, T. Lu, T. Furuya, A. Degterev, N. Mizushima, T. Yoshimori, M. MacDonald, B. Yankner and J. Yuan, Regulation of intracellular accumulation of mutant Huntingtin by Beclin 1, *J Biol Chem* **281**: 14474–14485 (2006).
53. S. Zheng, E. B. Clabough, S. Sarkar, M. Futter, D. C. Rubinsztein and S. O. Zeitlin, Deletion of the huntingtin polyglutamine stretch enhances neuronal autophagy and longevity in mice, *PLoS Genet* **6**: e1000838 (2010).
54. V. Kirkin, D. G. McEwan, I. Novak and I. Dikic, A role for ubiquitin in selective autophagy, *Mol Cell* **34**: 259–269 (2009).
55. G. Bjørkøy, T. Lamark, A. Brech, H. Outzen, M. Perander, A. Øvervatn, H. Stenmark and T. Johansen, p62/SQSTM1 forms protein aggregates degraded by autophagy and has a protective effect on huntingtin-induced cell death, *J Cell Biol* **171**: 603–614 (2005).
56. M. L. Seibenhener, T. Geetha and M. W. Wooten, Sequestosome 1/p62 — more than just a scaffold, *FEBS Lett* **581**: 175-179 (2007).
57. S. Pankiv, T. H. Clausen, T. Lamark, A. Brech, J. A. Bruun, H. Outzen, A. Øvervatn, G. Bjørkøy and T. Johansen, p62/SQSTM1 binds directly to Atg8/LC3 to facilitate degradation of ubiquitinated protein aggregates by autophagy, *J Biol Chem* **282**: 24131–24145 (2007).
58. A. BenYounes, N. Tajeddine, M. Tailler, S. A. Malik, S. Shen, D. Metivier, O. Kepp, I. Vitale, M. C. Maiuri and G. Kroemer, A fluorescence-microscopic and cytofluorometric system for monitoring the turnover of the autophagic substrate p62/SQSTM1, *Autophagy* **7**: 883–891 (2011).
59. Q. J. Wang, Y. Ding, D. S. Kohtz, N. Mizushima, I. M. Cristea, M. P. Rout, B. T. Chait, Y. Zhong, N. Heintz and Z. Yue, Induction of autophagy in axonal dystrophy and degeneration, *J Neurosci* **26**: 8057–8068 (2006).
60. Z. Yue, Regulation of neuronal autophagy in axon: implication of autophagy in axonal function and dysfunction/degeneration, *Autophagy* **3**: 139–141 (2007).
61. K. Zatloukal, C. Stumptner, A. Fuchsbichler, H. Heid, M. Schnoelzer, L. Kenner, R. Kleinert, M. Prinz, A. Aguzzi and H. Denk, p62 is a common component of cytoplasmic inclusions in protein aggregation diseases, *Am J Pathol* **160**: 255–263 (2002).
62. K. M. Donaldson, W. Li, K. A. Ching, S. Batalov, C. C. Tsai and C. A. Joazeiro, Ubiquitin-mediated sequestration of normal cellular proteins into polyglutamine aggregates, *Proc Natl Acad Sci USA* **100**: 8892–8897 (2003).
63. M. Filimonenko, P. Isakson, K. D. Finley, M. Anderson, H. Jeong, T. J. Melia, B. J. Bartlett, K. M. Myers, H. C. Birkeland, T. Lamark, D. Krainc, A. Brech,

H. Stenmark, A. Simonsen and A. Yamamoto, The selective macroautophagic degradation of aggregated proteins requires the PI3P-binding protein Alfy, *Mol Cell* **38**: 265–279 (2010).

64. A. Simonsen, H. C. Birkeland, D. J. Gillooly, N. Mizushima, A. Kuma, T. Yoshimori, T. Slagsvold, A. Brech and H. Stenmark, Alfy, a novel FYVE-domain-containing protein associated with protein granules and autophagic membranes, *J Cell Sci* **117**: 4239–4251 (2004).
65. H. Jeong, F. Then, T. J. Melia, Jr., J. R. Mazzulli, L. Cui, J. N. Savas, C. Voisine, P. Paganetti, N. Tanese, A. C. Hart, A. Yamamoto and D. Krainc, Acetylation targets mutant huntingtin to autophagosomes for degradation, *Cell* **137**: 60–72 (2009).
66. L. M. Thompson, C. T. Aiken, L. S. Kaltenbach, N. Agrawal, K. Illes, A. Khoshnan, M. Martinez-Vincente, M. Arrasate, J. G. O'Rourke, H. Khashwji, T. Lukacsovich, Y. Z. Zhu, A. L. Lau, A. Massey, M. R. Hayden, S. O. Zeitlin, S. Finkbeiner, K. N. Green, F. M. LaFerla, G. Bates, L. Huang, P. H. Patterson, D. C. Lo, A. M. Cuervo, J. L. Marsh and J. S. Steffan, IKK phosphorylates Huntingtin and targets it for degradation by the proteasome and lysosome, *J Cell Biol* **187**: 1083–1099 (2009).
67. X. Gu, E. R. Greiner, R. Mishra, R. Kodali, A. Osmand, S. Finkbeiner, J. S. Steffan, L. M. Thompson, R. Wetzel and X. W. Yang, Serines 13 and 16 are critical determinants of full-length human mutant huntingtin induced disease pathogenesis in HD mice, *Neuron* **64**: 828–840 (2009).
68. M. Jimenez-Sanchez, F. Thompson, E. Zavodsky and D. C. Rubinsztein, Autophagy and polyglutamine diseases, *Prog Neurobiol*, Sep 10 [Epub ahead of print] (2011).
69. I. Munoz-Sanjuan and G. P. Bates, The importance of integrating basic and clinical research toward the development of new therapies for Huntington disease, *J Clin Invest* **121**: 476–483 (2011).
70. S. Imarisio, J. Carmichael, V. Korolchuk, C. W. Chen, S. Saiki, C. Rose, G. Krishna, J. E. Davies, E. Ttofi, B. R. Underwood and D. C. Rubinsztein, Huntington's disease: from pathology and genetics to potential therapies, *Biochem J* **412**: 191–209 (2008).
71. S. Sarkar, R. A. Floto, Z. Berger, S. Imarisio, A. Cordenier, M. Pasco, L. J. Cook and D. C. Rubinsztein, Lithium induces autophagy by inhibiting inositol monophosphatase, *J Cell Biol* **170**: 1101–1111 (2005).
72. S. Sarkar and D. C. Rubinsztein, Inositol and IP3 levels regulate autophagy: biology and therapeutic speculations, *Autophagy* **2**: 132–134 (2006).
73. M. Tanaka, Y. Machida, S. Niu, T. Ikeda, N. R. Jana, H. Doi, M. Kurosawa, M. Nekooki and N. Nukina, Trehalose alleviates polyglutamine-mediated pathology in a mouse model of Huntington disease, *Nat Med* **10**: 148–154 (2004).
74. S. Sarkar, J. E. Davies, Z. Huang, A. Tunnacliffe and D. C. Rubinsztein, Trehalose, a novel mTOR-independent autophagy enhancer, accelerates the clearance of mutant huntingtin and alpha-synuclein, *J Biol Chem* **282**: 5641–5652 (2007).
75. S. Sarkar, E. O. Perlstein, S. Imarisio, S. Pineau, A. Cordenier, R. L. Maglathlin, J. A. Webster, T. A. Lewis, C. J. O'Kane, S. L. Schreiber and D. C. Rubinsztein, Small

molecules enhance autophagy and reduce toxicity in Huntington's disease models, *Nat Chem Biol* **3**: 331–338 (2007).
76. C. Rose, F. M. Menzies, M. Renna, A. Acevedo-Arozena, S. Corrochano, O. Sadiq, S. D. Brown and D. C. Rubinsztein, Rilmenidine attenuates toxicity of polyglutamine expansions in a mouse model of Huntington's disease, *Hum Mol Genet* **19**: 2144–2153 (2010).
77. J. S. Reijnders, U. Ehrt, W. E. Weber, D. Aarsland and A. F. Leentjens, A systematic review of prevalence studies of depression in Parkinson's disease, *Mov Disord* **23**: 183–189; quiz 313 (2008).
78. Y. Zhang, V. L. Dawson and T. M. Dawson, Oxidative stress and genetics in the pathogenesis of Parkinson's disease, *Neurobiol Dis* **7**: 240–250 (2000).
79. W. Dauer and S. Przedborski, Parkinson's disease: mechanisms and models, *Neuron* **39**: 889–909 (2003).
80. A. Zimprich, S. Biskup, P. Leitner, P. Lichtner, M. Farrer, S. Lincoln, J. Kachergus, M. Hulihan, R. J. Uitti, D. B. Calne, A. J. Stoessl, R. F. Pfeiffer, N. Patenge, I. C. Carbajal, P. Vieregge, F. Asmus, B. Muller-Myhsok, D. W. Dickson, T. Meitinger, T. M. Strom, Z. K. Wszolek and T. Gasser, Mutations in LRRK2 cause autosomal-dominant parkinsonism with pleomorphic pathology, *Neuron* **44**: 601–607 (2004).
81. C. Paisan-Ruiz, S. Jain, E. W. Evans, W. P. Gilks, J. Simon, M. van der Brug, A. Lopez de Munain, S. Aparicio, A. M. Gil, N. Khan, J. Johnson, J. R. Martinez, D. Nicholl, I. M. Carrera, A. S. Pena, R. de Silva, A. Lees, J. F. Marti-Masso, J. Perez-Tur, N. W. Wood and A. B. Singleton, Cloning of the gene containing mutations that cause PARK8-linked Parkinson's disease, *Neuron* **44**: 595–600 (2004).
82. M. G. Spillantini, M. L. Schmidt, V. M. Lee, J. Q. Trojanowski, R. Jakes and M. Goedert, Alpha-synuclein in Lewy bodies, *Nature* **388**: 839–840 (1997).
83. M. H. Polymeropoulos, C. Lavedan, E. Leroy, S. E. Ide, A. Dehejia, A. Dutra, B. Pike, H. Root, J. Rubenstein, R. Boyer, E. S. Stenroos, S. Chandrasekharappa, A. Athanassiadou, T. Papapetropoulos, W. G. Johnson, A. M. Lazzarini, R. C. Duvoisin, G. Di Iorio, L. I. Golbe and R. L. Nussbaum, Mutation in the alpha-synuclein gene identified in families with Parkinson's disease, *Science* **276**: 2045–2047 (1997).
84. R. Kruger, W. Kuhn, T. Muller, D. Woitalla, M. Graeber, S. Kosel, H. Przuntek, J. T. Epplen, L. Schols and O. Riess, Ala30Pro mutation in the gene encoding alpha-synuclein in Parkinson's disease, *Nat Genet* **18**: 106–108 (1998).
85. J. J. Zarranz, J. Alegre, J. C. Gomez-Esteban, E. Lezcano, R. Ros, I. Ampuero, L. Vidal, J. Hoenicka, O. Rodriguez, B. Atares, V. Llorens, E. Gomez Tortosa, T. del Ser, D. G. Munoz and J. G. de Yebenes, The new mutation, E46K, of alpha-synuclein causes Parkinson and Lewy body dementia, *Ann Neurol* **55**: 164–173 (2004).
86. E. M. Valente, P. M. Abou-Sleiman, V. Caputo, M. M. Muqit, K. Harvey, S. Gispert, Z. Ali, D. Del Turco, A. R. Bentivoglio, D. G. Healy, A. Albanese, R. Nussbaum, R. Gonzalez-Maldonado, T. Deller, S. Salvi, P. Cortelli, W. P. Gilks, D. S. Latchman, R. J. Harvey, B. Dallapiccola, G. Auburger and N. W. Wood, Hereditary

early-onset Parkinson's disease caused by mutations in PINK1, *Science* **304**: 1158–1160 (2004).

87. T. M. Dawson, Parkin and defective ubiquitination in Parkinson's disease, *J Neural Transm Suppl*: 209–213 (2006).
88. V. Bonifati, P. Rizzu, M. J. van Baren, O. Schaap, G. J. Breedveld, E. Krieger, M. C. Dekker, F. Squitieri, P. Ibanez, M. Joosse, J. W. van Dongen, N. Vanacore, J. C. van Swieten, A. Brice, G. Meco, C. M. van Duijn, B. A. Oostra and P. Heutink, Mutations in the DJ-1 gene associated with autosomal recessive early-onset parkinsonism, *Science* **299**: 256–259 (2003).
89. K. A. Maguire-Zeiss, D. W. Short and H. J. Federoff, Synuclein, dopamine and oxidative stress: co-conspirators in Parkinson's disease?, *Brain Res Mol Brain Res* **134**: 18–23 (2005).
90. R. K. Dagda, J. Zhu, S. M. Kulich and C. T. Chu, Mitochondrially localized ERK2 regulates mitophagy and autophagic cell stress: implications for Parkinson's disease, *Autophagy* **4**: 770–782 (2008).
91. Y. Chen, E. McMillan-Ward, J. Kong, S. J. Israels and S. B. Gibson, Mitochondrial electron-transport-chain inhibitors of complexes I and II induce autophagic cell death mediated by reactive oxygen species, *J Cell Sci* **120**: 4155–4166 (2007).
92. R. A. Gonzalez-Polo, M. Niso-Santano, M. A. Ortiz-Ortiz, A. Gomez-Martin, J. M. Moran, L. Garcia-Rubio, J. Francisco-Morcillo, C. Zaragoza, G. Soler and J. M. Fuentes, Inhibition of paraquat-induced autophagy accelerates the apoptotic cell death in neuroblastoma SH-SY5Y cells, *Toxicol Sci* **97**: 448–458 (2007).
93. K. E. Larsen, E. A. Fon, T. G. Hastings, R. H. Edwards and D. Sulzer, Methamphetamine-induced degeneration of dopaminergic neurons involves autophagy and upregulation of dopamine synthesis, *J Neurosci* **22**: 8951–8960 (2002).
94. J. H. Zhu, C. Horbinski, F. Guo, S. Watkins, Y. Uchiyama and C. T. Chu, Regulation of autophagy by extracellular signal-regulated protein kinases during 1-methyl-4-phenylpyridinium-induced cell death, *Am J Pathol* **170**: 75–86 (2007).
95. A. L. Goldberg, Protein degradation and protection against misfolded or damaged proteins, *Nature* **426**: 895–899 (2003).
96. A. B. Singleton, M. Farrer, J. Johnson, A. Singleton, S. Hague, J. Kachergus, M. Hulihan, T. Peuralinna, A. Dutra, R. Nussbaum, S. Lincoln, A. Crawley, M. Hanson, D. Maraganore, C. Adler, M. R. Cookson, M. Muenter, M. Baptista, D. Miller, J. Blancato, J. Hardy and K. Gwinn-Hardy, alpha-Synuclein locus triplication causes Parkinson's disease, *Science* **302**: 841 (2003).
97. Y. Tanaka, S. Engelender, S. Igarashi, R. K. Rao, T. Wanner, R. E. Tanzi, A. Sawa, V. Dawson L, T. M. Dawson and C. A. Ross, Inducible expression of mutant alpha-synuclein decreases proteasome activity and increases sensitivity to mitochondria-dependent apoptosis, *Hum Mol Genet* **10**: 919–926 (2001).
98. B. I. Giasson and V. M. Lee, Parkin and the molecular pathways of Parkinson's disease, *Neuron* **31**: 885–888 (2001).

99. E. Leroy, R. Boyer, G. Auburger, B. Leube, G. Ulm, E. Mezey, G. Harta, M. J. Brownstein, S. Jonnalagada, T. Chernova, A. Dehejia, C. Lavedan, T. Gasser, P. J. Steinbach, K. D. Wilkinson and M. H. Polymeropoulos, The ubiquitin pathway in Parkinson's disease, *Nature* **395**: 451–452 (1998).
100. K. S. McNaught and P. Jenner, Proteasomal function is impaired in substantia nigra in Parkinson's disease, *Neurosci Lett* **297**: 191–194 (2001).
101. K. S. McNaught, C. Mytilineou, R. Jnobaptiste, J. Yabut, P. Shashidharan, P. Jennert and C. W. Olanow, Impairment of the ubiquitin-proteasome system causes dopaminergic cell death and inclusion body formation in ventral mesencephalic cultures, *J Neurochem* **81**: 301–306 (2002).
102. K. S. McNaught, D. P. Perl, A. L. Brownell and C. W. Olanow, Systemic exposure to proteasome inhibitors causes a progressive model of Parkinson's disease, *Ann Neurol* **56**: 149–162 (2004).
103. J. H. Kordower, N. M. Kanaan, Y. Chu, R. Suresh Babu, J. Stansell, 3rd, B. T. Terpstra, C. E. Sortwell, K. Steece-Collier and T. J. Collier, Failure of proteasome inhibitor administration to provide a model of Parkinson's disease in rats and monkeys, *Ann Neurol* **60**: 264–268 (2006).
104. C. Cook and L. Petrucelli, A critical evaluation of the ubiquitin-proteasome system in Parkinson's disease, *Biochim Biophys Acta* **1792**: 664–675 (2009).
105. L. Bedford, D. Hay, A. Devoy, S. Paine, D. G. Powe, R. Seth, T. Gray, I. Topham, K. Fone, N. Rezvani, M. Mee, T. Soane, R. Layfield, P. W. Sheppard, T. Ebendal, D. Usoskin, J. Lowe and R. J. Mayer, Depletion of 26S proteasomes in mouse brain neurons causes neurodegeneration and Lewy-like inclusions resembling human pale bodies, *J Neurosci* **28**: 8189–8198 (2008).
106. A. M. Cuervo, L. Stefanis, R. Fredenburg, P. T. Lansbury and D. Sulzer, Impaired degradation of mutant alpha-synuclein by chaperone-mediated autophagy, *Science* **305**: 1292–1295 (2004).
107. Q. Yang, H. She, M. Gearing, E. Colla, M. Lee, J. J. Shacka and Z. Mao, Regulation of neuronal survival factor MEF2D by chaperone-mediated autophagy, *Science* **323**: 124–127 (2009).
108. L. Alvarez-Erviti, M. C. Rodriguez-Oroz, J. M. Cooper, C. Caballero, I. Ferrer, J. A. Obeso and A. H. Schapira, Chaperone-mediated autophagy markers in Parkinson disease brains, *Arch Neurol* **67**: 1464–1472 (2010).
109. A. C. Massey, S. Kaushik, G. Sovak, R. Kiffin and A. M. Cuervo, Consequences of the selective blockage of chaperone-mediated autophagy, *Proc Natl Acad Sci USA* **103**: 5805–5810 (2006).
110. A. M. Cuervo and J. F. Dice, Age-related decline in chaperone-mediated autophagy, *J Biol Chem* **275**: 31505–31513 (2000).
111. M. Xilouri, T. Vogiatzi, K. Vekrellis, D. Park and L. Stefanis, Abberant alpha-synuclein confers toxicity to neurons in part through inhibition of chaperone-mediated autophagy, *PLoS One* **4**: e5515 (2009).
112. M. Komatsu, Q. J. Wang, G. R. Holstein, V. L. Friedrich, Jr., J. Iwata, E. Kominami, B. T. Chait, K. Tanaka and Z. Yue, Essential role for autophagy protein Atg7 in the

maintenance of axonal homeostasis and the prevention of axonal degeneration, *Proc Natl Acad Sci USA* **104**: 14489–14494 (2007).

113. T. Vogiatzi, M. Xilouri, K. Vekrellis and L. Stefanis, Wild type alpha-synuclein is degraded by chaperone-mediated autophagy and macroautophagy in neuronal cells, *J Biol Chem* **283**: 23542–23556 (2008).
114. W. H. Yu, B. Dorado, H. Y. Figueroa, L. Wang, E. Planel, M. R. Cookson, L. N. Clark and K. E. Duff, Metabolic activity determines efficacy of macroautophagic clearance of pathological oligomeric alpha-synuclein, *Am J Pathol* **175**: 736–747 (2009).
115. B. Spencer, R. Potkar, M. Trejo, E. Rockenstein, C. Patrick, R. Gindi, A. Adame, T. Wyss-Coray and E. Masliah, Beclin 1 gene transfer activates autophagy and ameliorates the neurodegenerative pathology in alpha-synuclein models of Parkinson's and Lewy body diseases, *J Neurosci* **29**: 13578–13588 (2009).
116. D. Ebrahimi-Fakhari, I. Cantuti-Castelvetri, Z. Fan, E. Rockenstein, E. Masliah, B. T. Hyman, P. J. McLean and V. K. Unni, Distinct roles *in vivo* for the ubiquitin-proteasome system and the autophagy-lysosomal pathway in the degradation of {alpha}-synuclein, *J Neurosci* **31**: 14508–14520 (2011).
117. Y. Tong, H. Yamaguchi, E. Giaime, S. Boyle, R. Kopan, R. J. Kelleher, 3rd and J. Shen, Loss of leucine-rich repeat kinase 2 causes impairment of protein degradation pathways, accumulation of alpha-synuclein, and apoptotic cell death in aged mice, *Proc Natl Acad Sci USA* **107**: 9879–9884 (2010).
118. X. Lin, L. Parisiadou, X. L. Gu, L. Wang, H. Shim, L. Sun, C. Xie, C. X. Long, W. J. Yang, J. Ding, Z. Z. Chen, P. E. Gallant, J. H. Tao-Cheng, G. Rudow, J. C. Troncoso, Z. Liu, Z. Li and H. Cai, Leucine-rich repeat kinase 2 regulates the progression of neuropathology induced by Parkinson's-disease-related mutant alpha-synuclein, *Neuron* **64**: 807–827 (2009).
119. A. R. Winslow, C. W. Chen, S. Corrochano, A. Acevedo-Arozena, D. E. Gordon, A. A. Peden, M. Lichtenberg, F. M. Menzies, B. Ravikumar, S. Imarisio, S. Brown, C. J. O'Kane and D. C. Rubinsztein, alpha-Synuclein impairs macroautophagy: implications for Parkinson's disease, *J Cell Biol* **190**: 1023–1037 (2010).
120. J. Alegre-Abarrategui, H. Christian, M. M. Lufino, R. Mutihac, L. L. Venda, O. Ansorge and R. Wade-Martins, LRRK2 regulates autophagic activity and localizes to specific membrane microdomains in a novel human genomic reporter cellular model, *Hum Mol Genet* **18**: 4022–4034 (2009).
121. D. Ramonet, J. P. Daher, B. M. Lin, K. Stafa, J. Kim, R. Banerjee, M. Westerlund, O. Pletnikova, L. Glauser, L. Yang, Y. Liu, D. A. Swing, M. F. Beal, J. C. Troncoso, J. M. McCaffery, N. A. Jenkins, N. G. Copeland, D. Galter, B. Thomas, M. K. Lee, T. M. Dawson, V. L. Dawson and D. J. Moore, Dopaminergic neuronal loss, reduced neurite complexity and autophagic abnormalities in transgenic mice expressing G2019S mutant LRRK2, *PLoS One* **6**: e18568 (2011).
122. P. Gomez-Suaga, B. Luzon-Toro, D. Churamani, L. Zhang, D. Bloor-Young, S. Patel, P. G. Woodman, G. C. Churchill and S. Hilfiker, Leucine-rich repeat kinase 2 regulates autophagy through a calcium-dependent pathway involving NAADP, *Hum Mol Genet* **21**: 511–525 (2011).

123. E. D. Plowey, S. J. Cherra, 3rd, Y. J. Liu and C. T. Chu, Role of autophagy in G2019S-LRRK2-associated neurite shortening in differentiated SH-SY5Y cells, *J Neurochem* **105**: 1048–1056 (2008).
124. R. A. Nixon, Autophagy in neurodegenerative disease: friend, foe or turncoat?, *Trends Neurosci* **29**: 528–535 (2006).
125. J. J. Lemasters, Selective mitochondrial autophagy, or mitophagy, as a targeted defense against oxidative stress, mitochondrial dysfunction, and aging, *Rejuvenation Res* **8**: 3–5 (2005).
126. D. Narendra, A. Tanaka, D. F. Suen and R. J. Youle, Parkin is recruited selectively to impaired mitochondria and promotes their autophagy, *J Cell Biol* **183**: 795–803 (2008).
127. D. F. Suen, D. P. Narendra, A. Tanaka, G. Manfredi and R. J. Youle, Parkin overexpression selects against a deleterious mtDNA mutation in heteroplasmic cybrid cells, *Proc Natl Acad Sci USA* **107**: 11835–11840 (2010).
128. D. P. Narendra, S. M. Jin, A. Tanaka, D. F. Suen, C. A. Gautier, J. Shen, M. R. Cookson and R. J. Youle, PINK1 is selectively stabilized on impaired mitochondria to activate Parkin, *PLoS Biol* **8**: e1000298 (2010).
129. S. Kawajiri, S. Saiki, S. Sato, F. Sato, T. Hatano, H. Eguchi and N. Hattori, PINK1 is recruited to mitochondria with parkin and associates with LC3 in mitophagy, *FEBS Lett* **584**: 1073–1079 (2010).
130. R. J. Youle and D. P. Narendra, Mechanisms of mitophagy, *Nat Rev Mol Cell Biol* **12**: 9–14 (2011).
131. S. R. Yoshii, C. Kishi, N. Ishihara and N. Mizushima, Parkin mediates proteasome-dependent protein degradation and rupture of the outer mitochondrial membrane, *J Biol Chem* **286**: 19630–19640 (2011).
132. A. Tanaka, M. M. Cleland, S. Xu, D. P. Narendra, D. F. Suen, M. Karbowski and R. J. Youle, Proteasome and p97 mediate mitophagy and degradation of mitofusins induced by Parkin, *J Cell Biol* **191**: 1367–1380 (2010).
133. B. A. McCray and J. P. Taylor, The role of autophagy in age-related neurodegeneration, *Neurosignals* **16**: 75–84 (2008).
134 L. G. Friedman, M. L. Lachenmayer, J. Wang, L. He, S.M. Poulose, M. Komatsu, G. R. Holstein and Z. Yue, Disrupted autophagy leads to dopaminergic axon and dendrite degeneration and promotes presynaptic accummulation of α-synuclein and LRRK2 in the brain, *J Neurosci* **32**: 7585–7593 (2012).
135. S. Corrochano, M. Renna, S. Carter, N. Chrobot, R. Kent, M. Stewart, J. Cooper, S. D. Brown, D. C. Rubinsztein and A. Acevedo-Arozena, alpha-Synuclein levels modulate Huntington's disease in mice, *Hum Mol Genet* **21**: 485–494 (2012).
136. C. Tomas-Zapico, M. Diez-Zaera, I. Ferrer, P. Gomez-Ramos, M. A. Moran, M. T. Miras-Portugal, M. Diaz-Hernandez and J. J. Lucas, alpha-Synuclein accumulates in huntingtin inclusions but forms independent filaments and its deficiency attenuates early phenotype in a mouse model of Huntington's disease, *Hum Mol Genet* **21**: 495–510 (2012).

CHAPTER 12

METABOLISM, AUTOPHAGY AND NEURODEGENERATION

W. Haung Yu and Karen E. Duff

ABSTRACT

An alteration in macroautophagic activity is a central tenant of some of the more recent theories as to why neurodegenerative disorders and proteinopathies may occur. Macroautophagy is an energy rich process, which may be susceptible to perturbations in mitochondrial activity. Mitochondria are essential for all metabolic processes and without this organelle, there would be no cellular respiration to drive cellular activity. Borne of putative xenobiotic origin, the mitochondrion uses its own DNA in addition to nuclear genetic material to generate essential proteins critical for respiration (ATP generation). In support of this link, extensive archival data has implicated mitochondrial defects as a causal effector of neurodegeneration, either due to insufficiency in energy generation, or increased apoptosis due to calcium release and activation of caspases. This chapter will review how neurodegenerative diseases are impacted by mitochondrial dysfunction and macroautophagy impairment, and provide evidence that links the two pathways together. A reduction in mitochondrial function may result in lower macroautophagy and contribute to neurodegeneration, with important therapeutic implications for Parkinson's disease, Alzheimer's disease and Huntington's disease.

MITOCHONDRIA FUNCTION AND DYSFUNCTION

The core of all metabolic activity in the cell lies within the mitochondria. It produces adenosine triphosphate (ATP) and is responsible for generating the energy that drives all cellular processes. While some cellular activity can be powered by anaerobic glycolysis, it is generally inefficient relative to ATP production which is needed to drive high-energy requirements. Perturbations in ATP generation can lead to deleterious downstream events. A second energy source, guanosine triphosphate (GTP), is used for mRNA expression and some GTP-ase driven pathways.

The typical mitochondria are approximately 0.5 μm in size, equivalent to the nascent autophagosome, which can degrade defective mitochondria in a process called mitophagy.[1–3] Mitochondria are double-membraned organelles, where the inner membrane has folds (cristae) that are essential for the production of adenosine triphosphate (ATP) as part of the electron transfer system (and citric acid/Krebs cycle). Nobel Prizes were awarded in 1977 (Peter Mitchell) and 1997 (P. D. Boyer and J. E. Walker, shared with J. C. Skou for the discovery of NA^+/K^+ ATPase), for work in defining the electron transfer system (chemiosmosis) and the ATP synthase, respectively.[4–10]

ATP drives most biochemical processes making it indispensible to life. ATP production results from several chemical reactions involving proton donation during aerobic respiration (via electron transfer from one system to another) from four complexes (complex I — nicotinamide adenine dinucleotide dehydrogenase (NADH); II — succinate dehydrogenase; III — cytochrome b/c oxidase; IV — cytochrome c oxidase). ATP production occurs in the final complex.[2,11] ATP can also be regenerated from ADP (adenosine diphosphate) via phosphorylation mediated by ATP synthase. Chemical inhibitors have been identified that can specifically inhibit aerobic respiration, including rotenone, a broad-spectrum pesticide linked to Parkinsonism, which specifically inhibits the electron transfer from complex I to a complex intermediary, ubiquinone.[12] The downstream effect is a loss of electron transfer to complexes III and IV and the production of ATP. Other chemicals associated with Parkinsonism, including paraquat and MPTP (1-methyl-4-phenyl-1,2,3,6-tetrahydropyridine), are both associated with inhibition of complex I activity.[13–15] ATP production and aerobic respiration is more efficient than glycolysis — the metabolism of glucose to produce energy — which can help generate ATP. It is for this reason that multicellular organisms rely on ATP generation to drive cellular activity.

Metabolism and Macroautophagy

If there is a loss in mitochondrial activity and cellular respiration, it is likely that energy-dependent activities such as protein and organelle degradation will also decline.[16–19] Protein degradation is critical for cellular homeostasis as multiple labs have shown that a decline in proteasomal, macroautophagic activity or chaperone-mediated autophagy (CMA) results in proteinopathy and reduced longevity.[20–25] All three degradation pathways are affected by aging, with proteasomal and chaperone-mediated autophagy being significantly reduced with age. The decline in protein quality control makes macroautophagy — the only inducible catabolic pathway — essential for maintaining cellular homeostasis.

Moreover, insufficient macroautophagy has been shown to evoke dramatic effects in the brain. Conditional knockout of essential macroautophagy genes (Atg5 and Atg7) in mice leads to neuropathology as indicated by accumulation of polyubiquitinated proteins and aggregates.[26,27]

One hypothesis is that neuronal macroautophagy is generally highly efficient. Boland *et al.* (2008) demonstrated that upregulation of macroautophagy using rapamycin does not normally lead to a greater number of autophagic vacuoles (AVs) or an increase in LC3-II due to a highly efficient mechanism that effectively clears AVs. It is only under conditions where transport of AVs has been blocked (vinblastine destabilization of microtubules) or where degradation is impaired (by preventing acidification of lysosomes with bafilomycin A1 or inhibiting protease activity) that AVs accumulate. Consequently, blocking autophagic induction, or inhibiting autophagic flux[28] could lead to neuronal disruption and neurodegeneration.

While it is well known that macroautophagy is essential to generate the building blocks of metabolism including amino acids, sugars, simple fatty acids and energy, it places a significant energetic burden on the cell. In general, when mTOR (mammalian target of rapamycin) is inactive, protein translation is inhibited, but the *de novo* generation of Atg proteins may be required for the activation and elongation of the autophagosome membrane.[29] While protein translation is predominantly GTP-driven, ATP is necessary for the annealing of nucleic acids by helicases and various other (deoxyribonucleic acid/ribonucleic acid)/RNA synthesis steps. In addition, ATP is essential for protein synthesis of macroautophagic components, along with all other proteins.

The generation of newly synthesized proteins and the formation of an autophagic vacuole requires post-translational conversion of some proteins (phospolipidation of LC3/Atg8; covalent bonding of Atg5-Atg12) and recruitment of lipids and elongation factors that regulate the AV's size. These processes are all reliant on ATP to drive the enzymes and fusion events, and they may also be coupled with GTP-driven GTPases. The motor proteins kinesin and dynein that are responsible for transport are all driven by ATP hydrolysis.[30–32] Moreover, microtubule assembly is dependent on ATP, making this energy source a critical element in macroautophagy. Finally, to complete autophagic flux, the nascent autophagosome which contains cytoplasmic material fated for degradation must be transported alongside acidic organelles such as endosomes and lysosomes to acquire the ambient pH and degradation enzymes (proteases, glucosidases and lipases) to degrade the vacuolar content. While successful autophagic flux will produce sufficient energy to warrant this costly endeavor, alterations in ATP generation may severely impact the cells capacity to successfully generate the required energy to complete the process. In a recent study,[33] we identified that rotenone, a specific NADH oxidase (Class I) inhibitor, initially stimulated macroautophagy to turn over defective mitochondria. With acute rotenone exposure, there was still a significant level of ATP in the cell (presumably from ADP:ATP recycling). Chronic treatment of rotenone, however, resulted in a significant reduction in ATP levels as mitochondrial generation of ATP was ablated with probable reduction in ADP conversion. As a consequence, macroautophagy was impaired and AV production was halted (indicated by lack of LC3-II).

Paradoxically, macroautophagy is stimulated when the cell senses low levels of ATP or nutrients. AMPK can inhibit mTOR when it senses low ATP levels in the cell,[34] and when the cell has low nutrients it can activate Gcn2 (general control non-repressed 2) and eIF2α-P (eukaryotic initiation factor) suggesting that there is a feedback mechanism that recognizes a low metabolic state of the cell, resulting in enhanced cellular content turnover through macroautophagy. Under glucose deprivation, AMPK directly phosphorylates Ulk1 at positions Ser 317 and Ser 777 reported by the Guan group and S467, S555, S574, and S637 from Shaw's group[35] to activate its kinase function.[36] While breakdown of glucose produces 13 molecules of ATP, 1 molecule of the fatty acid palmitoleic acid conversion can generate 131 ATP molecules. In general, successful completion of macroautophagy theoretically yields more energy than it utilizes. However, when this process fails or is disrupted it may be detrimental to the cell. If a cell is at low energy (e.g. when ATP or nutrient levels are low), then it must extend its energy resources to induce macroautophagy. Should that process fail along the way, it can impair the cell even more, so it is imperative that it surpasses this cellular energy deficit to generate more energy. This may be the case for MPP^+, where blocking autophagy prevents neuronal death.[37]

ATP regulated activity is also critical for lysosomal processes as it has been demonstrated that macroautophagy can be disrupted by the reduced localization of the V-ATPase (vacuolar-type H^+-ATPase) pump required for lysosomal acidification.[38] This acidification pump is reliant on ATP to drive its activity. In presenilin1 (PS1)-deficient cells, the V-ATPase pump subunit (v-ATPase V0a1) does not properly relocalize from the ER to the lysosome due to PS1-dependent maturation issues. It is conceivable that other events may also limit its relocalization to the lysosomal membrane or impair the enzyme's proton pump capacity necessary for organelle acidification and catabolic competency. Indeed, the treatment with Bafilomycin A1, which inhibits this pump, is a classic method to block lysosomal degradation, leading to the buildup of AVs and undegraded proteins. Finally proteases, lipases and glucosidases that are responsible for the catabolic events in lysosomes are all dependent on ATP to drive their enzymatic activity.

Cumulatively, macroautophagy is dependent on ATP levels to drive its activity from transcription to proteolysis/lipolysis/glycolysis. In elegant studies by Meijer *et al.* it was shown that a small reduction in cellular respiration negatively impacted macroautophagic proteolysis. As little as 15% loss in NADH oxidase activity resulted in 50% loss in macroautophagic protein degradation.[34,39–42]

In general, neurons are considered to be in a nutrient-rich environment as they benefit from continual sources of energy due to its greater than average blood supply on a per weight basis. Reduction in this energy supply may have a significant impact on neuronal function. PET (positron emission tomography) and MRI (magnetic resonance imaging) studies have shown that with aging and with neurodegeneration (including Huntington's, Alzheimer's and Parkinson's diseases), there are significant reductions in the levels of glucose utilization, suggesting that both events correlate with reduced nutrient delivery.[43–48] If these events become extensive, the need for cellular derived energy production through macroautophagy may become imperative for neuronal viability.

Aging and Mitochondrial Function

As all multicellular organisms are reliant on mitochondrial production of energy, perturbations in this system will have negative effects on the organism. One potential defect in mitochondrial activity is the loss of energy "leaking" out of the mitochondria. Electrons from the respiratory chain are lost in the form of reactive oxygen species. In addition to reduced ATP production, the leaked free radicals can propagate oxidative stress, which in turn can destabilize mitochondrial DNA and activate caspases for apoptosis.

The axiom of mitochondrial damage and aging has been theorized for a good part of a century, with several studies showing that aged individuals (>50 years) have lower mitochondrial respiration than their younger counterparts.[19,49,50] Moreover, oxidative stress is a hallmark of aging, and of neurodegeneration (in particular Alzheimer's disease (AD) and Parkinson's disease (PD)). Cumulatively, this data would link mitochondrial oxidative stress with aging. As a consequence of mitochondrial damage, there is lower respiration that leads to loss of cellular activity. This in turn may increase the number of dysfunctional mitochondria in aging, presenting a vicious circle of reduced metabolic activity and reduced macroautophagic proteolysis as there is evidence that CMA declines with aging. There is similar, but limited, evidence that macroautophagy may be stunted with aging.[51] To date, however, there is no *in vivo* evidence directly linking metabolic dysfunction in aging with decline in macroautophagy.

DISEASES OF THE MITOCHONDRIA AND NEURODEGENERATION

Mitochondrial imbalances are seen in an array of rare genetic disorders. Diseases are caused by many mitochondrial DNA mutations including syndromes such as Kearns-Sayre, Pearson's, MELAS and Leber's optic neuropathy.[52] Kearns-Sayre and Pearson's syndrome are a consequence of rearrangement of the mitochondrial DNA, while others like MELAS (mitochondrial encephalomyopathy, lactic acidosis, and stroke like) syndrome and Leber's optic neuropathy result from point mutations. Still other diseases are a consequence of autosomal dominant defects in genes located in the nucleus but with mitochondrial function. This list includes Friedrich's ataxia, Wilson's disease, and spastic paraplegia. Finally, defects in genes regulating phosphorylation of enzyme function can produce chronic disease such as Barth syndrome and genetic forms of coenzyme Q10 deficiency.[53]

Although these diseases manifest with a range of clinical symptoms such as myopathy, diabetes and disruption of the endocrine pathway, they often present with neurological deficits reflecting the susceptibility of the nervous system to energy depletion resulting from mitochondrial dysfunction. The brain consumes an inordinate amount of energy relative to its weight. It is estimated that the brain is approximately 5% of the mass of a human body, but requires 20% of all energy produced.

It is not surprising therefore that there has been a lengthy body of research examining mitochondrial dysfunction in an array of neurological disorders and dementias, including epilepsy, schizophrenia, bipolar disorders, Alzheimer's, Parkinson's and Huntington's diseases. Peripheral disorders that include cardiovascular disease and diabetes manifest with increased oxidative stress beginning in the mitochondria. Both of these diseases have also been linked to the onset and progression of AD, suggesting that peripheral changes may impact neurodegeneration.

Mitochondrial Dysfunction in Parkinson's Disease

Affecting roughly five million people worldwide, PD and its associated spectrum of dementias is a devastating disease condition that destroys the dopamine neurons in the substantia nigra (and other areas) many years before the onset of hallmark symptoms such as tremors, muscle rigidity and slow movements. Current therapies attempt to reduce the side effects of dopaminergic cell loss and bolster depleted levels of dopamine through L-dopa supplementation, but supplementation strategies are of limited efficacy as the diseases progresses and dopaminergic neurons continue to die.

One explanation for why neurons die in neurodegenerative disease is related to loss of mitochondrial function, and potentially free radical toxicity generated from the mitochondria as it becomes destabilized. The reactive oxygen species that are created from dying mitochondria can promote a cascade of free radical generation and apoptosis. The double-hit of reduced electron-transfer system generation of ATP and pro-apoptotic conditions are critical elements of mitochondrial dysfunction in Parkinson's disease. Glucose and oxygen delivery/consumption in PD brains is dramatically impaired as PET images of PD patients show a substantial loss of glucose utilization in the dopaminergic pathways, suggesting that there is substantial neuronal loss and significantly reduced glucose use.[54] These studies have been supported by data from functional magnetic resonance imaging (fMRI) studies that measure blood-oxygen consumption, or delivery to the brain, which suggest that affected regions suffer from low metabolic activity.[55] Loss of dopaminergic neurons has also been seen with PET labeling using [18F]-Dopa and SPECT (single-photon emission computerized tomography) imaging of dopamine transporters which has confirmed the regional vulnerability of the brain.[56]

There is considerable evidence that genetic and environmental factors play a role in Parkinsonism. The genetic risk factors are centered around the stabilization of mitochondria and promotion of lysosomal function. Some forms of familial Parkinson's disease are linked to loss of function of genes called Parkin and PINK1 and recent evidence suggests that these genes are important for normal functioning of the mitochondria.[57] It has been shown that PINK1 and Parkin functionally interact to maintain mitochondrial function as loss of Parkin or PINK1 function impairs the morphology and activity of mitochondria, which then produce less ATP. In turn, mitophagy (macroautophagic turnover of mitochondria) is affected.[57,58] As previously mentioned, lysosomal activity is crucial to macroautophagic function, and lysosomal-resident proteins such as glucocerebrosidase and ATP13a2 have been genetically linked to Parkinsonism.[59–61] The former is a glucosidase, most commonly associated with Gaucher's disease, but more recently identified as a genetic risk factor for the development of PD and other Parkinsonisms.

Environmental and Exogenous Influences on Mitochondrial Function and Parkinsonism

MPTP

The neuropathological development of Parkinson's disease has long been associated with mitochondrial dysfunction. Neurotoxins like MPTP and rotenone interfere with mitochondrial function and have been associated with Parkinsonian-like phenotype and loss of dopaminergic neurons since 1947 in primates. It was not until a similar compound, MPPP (1-methyl-4-phenyl-4-propionoxypiperidine) was found to be contaminated with MPTP and the link to Parkinsonism was made. MPPP (Desmethylprodine; an analog of meperidine, commonly known as Demerol) is an opioid analgesic drug that is no longer used clinically,[62] but it is synthesized illegally for recreational drug use. In 1977, a 23-year-old graduate student (Barry Kidston) attempted to reproduce the synthesis of MPPP from a 1947 paper published by Dr. Albert Ziering.[63,64] Kidston made an error in the process of esterifying the meperidine backbone (to give it its analgesic effect). Within a few days of trying the home-made drug, Kidson developed a severe Parkinsonian phenotype. To relieve the motor deficits, he was placed on Sinemet, but he committed suicide a couple of years later. At autopsy, there was clear evidence of significant cell loss in the substantia nigra, similar to that seen in PD patients. It has subsequently been shown that Kidston synthesized MPTP which undergoes a rapid conversion by monoamine oxidase B to MPP^+ (1-methyl-4-phenylpyridinium).[65] MPP^+ is selectively taken up by dopamine transporters at the synaptic terminal and specifically inhibits complex I of the mitochondria, selectively killing dopaminergic neurons of the substantia nigra and causing Parkinsonian symptoms. However, further studies involving MPTP have failed to show development of Lewy bodies, a key component of PD pathology.[65–67] Therefore, the mechanism behind MPTP toxicity as it relates to PD is not fully understood. Interestingly, MPTP toxicity induces macroautophagy *in vivo*, based on evidence of increased AV-like structures, but the accumulation of AVs may also be a result of less than optimal autophagic flux.[37,68]

Rotenone

Emmanuel Geoffroy first isolated rotenone from a specimen of *Robinia nicou* (*Lonchocarpus nicou*), a plant found in French Guyana. Researchers later determined that the substance that Geoffroy termed *nicouline* was identical to rotenone. Rotenone is used in solution as a pesticide and insecticide, and in emulsion as a piscicide. Several indigenous tribes in French Guyana extract rotenone from plants, release it into the water and catch the poisoned fish that float to the surface. Rotenone is lethal to fish (and insects), as the lipophilic compound is easily taken up through the gills or trachea, but fish caught in this way are not poisonous to humans as rotenone is very poorly absorbed by the gastrointestinal tract of mammals.[69] The lowest lethal dose for a child is 143 mg/kg but human deaths attributed to rotenone are rare because it is an irritant that causes vomiting.

In the US and elsewhere, rotenone has been used in the environment for several purposes as it has minor and transient environmental side-effects. It is used by researchers studying the biodiversity of marine fish as a piscicide to collect "hidden" fish that represent an important component of shoreline fish communities. It is also used in powdered form to reduce parasitic mites on chickens and other fowl, and it is used as an organic pesticide dust for the garden. It was previously classified as a non-synthetic and was used in organic farming until 2005 when it was added to the list of prohibited substances due to concerns about its safety. These concerns followed studies in 2000 that showed that injecting rotenone into rats led to the development of symptoms similar to those of PD. Rotenone was continuously applied over a period of five weeks, mixed with DMSO dimethylsulfoxide) and PEG (polyethylene glycol) to enhance tissue penetration, and injected into the jugular vein. Further studies with primary cultures of rat neurons and microglia or acute rat brain slices have shown that low doses of rotenone (below 10 nM) induce oxidative damage and death of dopaminergic neurons.[70] Similar to MPTP, rotenone was shown to inhibit complex I transfer of electrons from iron-sulfur centers to ubiquinone in the electron transport chain preventing NADH conversion to ATP, thereby killing dopaminergic neurons in the substantia nigra.

It is also important for the cell that macroautophagy does not exacerbate cellular energy deficits by excessive clearance of mitochondria, such as may occur following chronic exposure to organic pesticides like rotenone that targets the mitochondrial complex I (NADH-ubiquinone oxidoreductase), or agents that promote energy depletion. Our lab was able to show that rotenone could inhibit ATP production and that with chronic exposure,[33] it significantly reduced a cell's capacity to generate autophagic vacuoles which were necessary to clear the dysfunctional mitochondria. This resulted in an increase in α-synuclein aggregation and oligomerization, indicating protein quality control had shifted to increased aberrant forms of the toxic protein that are associated with Parkinsonism neuropathology. Our data show that the levels of macroautophagy induced, and the efficiency of clearance of aggregated α-synuclein was dependent on the metabolic health of the cells as chronic, but not acute, exposure to rotenone led to a drop in metabolic activity that correlated with a steep reduction in LC3-II levels and a dramatic increase in oligomers. Overall, this implies that metabolic deficits significantly reduce the ability of macroautophagy to degrade α-synuclein.[33]

Although these studies do not directly suggest that rotenone exposure is responsible for PD in humans, lipophilic rotenone (and MPTP) can cross the blood-brain barrier and it is thus a potential causative agent for Parkinsonian intoxications. In the United States and in Canada, all uses of rotenone, except as a piscicide, are being phased out.

Parkinson's Disease and Macroautophagy

Failure of macroautophagy has been proposed as a mechanism for neuropathology in PD as it is implicated in reduced clearance of defective mitochondria and removal of oligomeric/aggregated α-synuclein. Spencer *et al.* propose that beclin 1 levels are reduced in PD and in an animal model of α-synucleinopathy.[71] A recent paper suggested that beclin 1 may interact with Parkin to remove redundant mitochondria.[72] Our lab reported the presence of α-synuclein oligomers upon reduction of Atg5 or beclin 1 using siRNA targeting of those genes,[33] while the Stefanis group have indicated the importance of macroautophagy in α-synuclein homeostasis (along with CMA, which was first described by Cuervo, Stefanis, Sulzer and their colleagues).[73,74] Cumulatively, macroautophagy is essential for protein quality control and the removal of redundant, or aberrant protein confirmations, but the pathway is also intertwined with mitochondrial control. Our studies have indicated that while acute mitochondrial damage may support macroautophagic induction, prolonged exposure to rotenone resulted in the loss of macroautophagy and increased synucleinopathy, which is in agreement with other studies linking macroautophagy with synucleinopathy.

Alzheimer's Disease and Macroautophagy

Alzheimer's disease is the most common dementia, affecting five million Americans and almost 30 million worldwide, with approximately one in eight adults over the age of 65 suffering from the disease. It is believed that most cases develop due to various environmental or non-genetic events. In 2005, Nixon and colleagues reported the accumulation of undegraded autophagic vacuoles in neuronal projections of patients who were subsequently diagnosed with Alzheimer's disease.[75] This was the first indication that macroautophagy may not be working efficiently in AD. Moreover, Pickford *et al.* reported the reduction in beclin 1, suggesting that macroautophagy may not be induced sufficiently to maintain protein quality control.[76]

Alzheimer's disease is associated with three major neuropathological events — the accumulation of extracellular plaques composed principally of the beta-amyloid (Aβ) peptide, accumulation of intracellular aggregates of misfolded tau, a microtubule binding protein; and neuronal loss. Aβ has been shown to disrupt mitochondrial function, and more recently it has been shown to be produced by the AV.[77–79] Aβ is also thought to stimulate autophagy, thought it may not be beneficial in that respect as it can also disrupt membrane constituency.[80,81] While the main purpose of autophagy is protein clearance, Aβ may disrupt the AV by destabilizing it, resulting in Aβ release from the organellar/membrane and leaving undegraded content behind, as seen in the neurites of AD patients.

The loss of autophagic clearance is coupled with the deleterious destabilization of mitochondria by Aβ. Beta-amyloid can interact with several mitochondrial components, with all interactions contributing to mitochondrial dysfunction. Aβ is translocated into the mitochondria by TOM-40 (translocase of the outer membrane complex) to interact with the Aβ-binding alcohol dehydrogenase (ABAD) where it generates reactive oxygen species (ROS) by impairing subunit 1 cytochrome oxidase activity and promoting cytochrome c release.[82] This in turn activates caspase 3 and apoptosis. It may also directly impact mitochondria and inhibit complex IV activity. The net effect is a loss of ATP generation and dysfunctional mitochondria, which need to be degraded through macroautophagy. A loss of energy metabolism has been linked to AD and aging for many years, with newer PET and MRI evidence showing a functional decline in glucose and oxygen utilization indicating vulnerable regions of mitochondrial disruption. Aβ may contribute to failure of macroautophagic flux (reduced induction and formation of AVs), and this may be coupled with the accumulation of hyperphosphorylated tau, which misfolds and aggregates.

Tau participates in cellular transport (based on its microtubule binding domains), but misfolded and aggregated tau may impede axonal microtubule transport of proteins and vacuoles. It has been shown that the P301L mutant variant of tau (which aggregates faster than its wild-type isoform) has slower mitochondrial transport dynamics.[83] At present, there is no data on autophagic vacuole transport with tau hyperphosphorylation, but if the data replicates the mitochondrial events, the stunted transport of AVs to lysosomes (which are abundant in the cell body) may be strong biochemical evidence linking tau pathology with autophagosome defects seen in the AD biopsy material.[75] As previously mentioned, ATP is required for microtubule/transport dynamics, and given the evidence of slowed mitochondrial dynamics, it is feasible that a loss of energy delivery may result in a snowball effect leading to transport failure and accumulation of proteins, AVs, and mitochondria in dystrophic neurites.

Huntington's Disease and Macroautophagy

Polyglutamine repeat extensions of the huntingtin gene and expression of the protein (htt) are associated with preferential loss of medium spiny neurons, such as striatal neurons. A correlation was made between the polyglutamine repeat regions and their neuropathological effect has significantly advanced research of this neurodegenerative disease. Functional MRI and PET scans of a typical Huntington's disease patient show significant atrophy of the cerebral and striatal regions, and ventricular enlargement, indicative of neuronal loss and reduced energy utilization. These functional changes can also be seen long before clinical manifestations arise suggesting that there may be protracted cellular changes that occur. Htt is involved in several cellular pathways that impair neuronal function; for example, it has been shown to inhibit PGC-1α (peroxisome proliferator-activated receptor gamma co-activator 1-alpha) activity.[84] PGC-1α is a nuclear transcription factor that regulates the transcription of many mitochondrial proteins derived in the nucleus. The inhibition of mitochondrial biogenesis may be a critical factor in neuronal loss as the cell becomes highly dependent on exogenous energy sources and recycling of ADP:ATP.

Mutant htt itself can promote aggregation that can impair several processes including transport of mitochondria along microtubules to local sites where ATP generation is needed. It can specifically inhibit mitochondrial complex II (succinate dehydrogenase) of the ATP generation pathway, directly inhibiting energy production.[85] Finally, it can block NMDA receptors that help regulate intracellular calcium, a process that is highly dependent on mitochondrial stores of calcium.[86] Htt can therefore effectively impair mitochondrial function and potentially increase apoptotic activity.

Not surprisingly, catabolic conversion of high-poly-glutamine copy proteins is stunted. Proteasomes and CMA mechanisms can be physically blocked by the aggregated protein at the proteasomal pore, or at the mouth of the LAMP2a receptor, effectively blocking its activity. This may add additional stress on macroautophagy to maintain protein quality control in the cell. Protein refolding would be required to linearize the protein for proteasomal processing, otherwise, macroautophagy is a primary mechanism for htt aggregate degradation. This is likely the case for all aggregated proteins in neurodegeneration that undergo amyloid-like misfolding. Macroautophagy, however, is not spared by htt pathogenesis, as it has been reported that htt may reduce the efficiency of macroautophagic degradation. Cuervo and colleagues have reported that the htt protein may inhibit vacuolar cargo recognition resulting in nascent AVs that are devoid of aggregates (mostly cytoplasm) or delimiting membranes that do not fuse properly, rendering macroautophagy inefficient. It is also possible that these events are slowed by the htt effects on mitochondria that reduce the metabolic capacity of the cell.[87] Yamamoto and colleagues, and the Rubinsztein group have independently shown that in cell and animal models,[88–90] it is possible to reduce the aggregate burden through stimulation of macroautophagy, physically shifting the kinetics to promote more protein turnover and energy production, compensating for lost energy potential due to reduced mitochondrial number and activity.

CONCLUSION

Aging and neurodegeneration are both associated with reduced mitochondrial function and reduced protein quality control. This may be a consequence of reduced efficiency in maintaining mitochondria; health (fusion/fission events, as well as ROS and clearance of dysfunctional mitochondria). It may also be a product of a vicious circle where poor protein homeostasis leads to the accumulation of aggregated or aberrant proteins which negatively impact mitochondria function. Downstream, the clearance of redundant or dysfunctional mitochondria by macroautophagy may be inhibited by energy deficits that impair the induction/development of the autophagosome, or downstream transport of AVs to lysosomes, as well as lysosomal competency (acidification or protease activity). Relative to other catabolic machinery, macroautophagy has multiple steps reliant on ATP and general metabolic health of the cell. The interplay between mitochondrial activity and autophagy occurs on multiple levels, as the function, or dysfunction of mitochondria impacts or impairs macroautophagic activity. This is even more critical in neurodegeneration where events such as proteinopathy can affect neuronal function and viability. The two are so interlinked that it is conceivable that dysfunction of one can lead to impairment of the other.

REFERENCES

1. J. J. Lemasters, Selective mitochondrial autophagy, or mitophagy, as a targeted defense against oxidative stress, mitochondrial dysfunction, and aging, *Rejuvenation Res* **8**: 3–5 (2005).
2. S. Ferguson-Miller, D. L. Brautigan and E. Margoliash, Correlation of the kinetics of electron transfer activity of various eukaryotic cytochromes c with binding to mitochondrial cytochrome c oxidase, *J Biol Chem* **251**: 1104–1115 (1976).
3. S. P. Bagchi, Trace dosages of the neurotoxins MPTP and MPP^+ may affect brain dopamine *in vivo*, *Life Sci* **51**: 389–396 (1992).
4. H. A. Krebs, The citric acid cycle and the Szent-Gyorgyi cycle in pigeon breast muscle, *Biochem J* **34**: 775–779 (1940).
5. H. A. Krebs, E. Salvin and W. A. Johnson, The formation of citric and alpha-ketoglutaric acids in the mammalian body, *Biochem J* **32**: 113–117 (1938).

6. P. Mitchell and J. Moyle, Stoichiometry of proton translocation through the respiratory chain and adenosine triphosphatase systems of rat liver mitochondria, *Nature* **208**: 147–151 (1965).
7. J. C. Skou, Studies on the Na ion and K ion activated ATP hydrolysing enzyme system. The role of SH groups, *Biochem Biophys Res Commun* **10**: 79–84 (1963).
8. M. Gresser, J. Cardon, G. Rosen and P. D. Boyer, Demonstration and quantitation of catalytic and noncatalytic bound ATP in submitochondrial particles during oxidative phosphorylation, *J Biol Chem* **254**: 10649–10653 (1979).
9. N. J. Gay and J. E. Walker, The atp operon: nucleotide sequence of the region encoding the alpha-subunit of *Escherichia coli* ATP-synthase, *Nucleic Acids Res* **9**: 2187–2194 (1981).
10. P. R. Walker, The regulation of enzyme synthesis in animal cells, *Essays Biochem* **13**: 39–69 (1977).
11. H. H. Herman, K. Wimalasena, L. C. Fowler, C. A. Beard and S. W. May, Demonstration of the ascorbate dependence of membrane-bound dopamine beta-monooxygenase in adrenal chromaffin granule ghosts, *J Biol Chem* **263**: 666–672 (1988).
12. A. Hartley, J. M. Stone, C. Heron, J. M. Cooper and A. H. Schapira, Complex I inhibitors induce dose-dependent apoptosis in PC12 cells: relevance to Parkinson's disease, *J Neurochem* **63**: 1987–1990 (1994).
13. T. Fukushima, T. Tawara, A. Isobe, N. Hojo, K. Shiwaku and Y. Yamane, Radical formation site of cerebral complex I and Parkinson's disease, *J Neurosci Res* **42**: 385–390 (1995).
14. S. Ramachandiran, J. M. Hansen, D. P. Jones, J. R. Richardson and G. W. Miller, Divergent mechanisms of paraquat, MPP^+, and rotenone toxicity: oxidation of thioredoxin and caspase-3 activation, *Toxicol Sci* **95**: 163–171 (2007).
15. Y. Mizuno, N. Sone, K. Suzuki and T. Saitoh, Studies on the toxicity of 1-methyl-4-phenylpyridinium ion (MPP^+) against mitochondria of mouse brain, *J Neurol Sci* **86**: 97–110 (1988).
16. S. Goodell and G. Cortopassi, Analysis of oxygen consumption and mitochondrial permeability with age in mice, *Mech Ageing Dev* **101**: 245–256 (1998).
17. T. M. Hagen, C. M. Wehr and B. N. Ames, Mitochondrial decay in aging. Reversal through supplementation of acetyl-L-carnitine and N-tert-butyl-alpha-phenyl-nitrone, *Ann N Y Acad Sci* **854**: 214–223 (1998).
18. G. Lenaz, C. Bovina, M. D'Aurelio, R. Fato, G. Formiggini, M. L. Genova, G. Giuliano, M. Merlo Pich, U. Paolucci, G. Parenti Castelli and B. Ventura, Role of mitochondria in oxidative stress and aging, *Ann N Y Acad Sci* **959**: 199–213 (2002).
19. E. Maklashina and B. A. Ackrell, Is defective electron transport at the hub of aging? *Aging Cell* **3**: 21–27 (2004).
20. S. Kaushik and A. M. Cuervo, Chaperone-mediated autophagy, *Methods Mol Biol* **445**: 227–244 (2008).
21. R. Kiffin, S. Kaushik, M. Zeng, U. Bandyopadhyay, C. Zhang, A. C. Massey, M. Martinez-Vicente and A. M. Cuervo, Altered dynamics of the lysosomal receptor for chaperone-mediated autophagy with age, *J Cell Sci* **120**: 782–791 (2007).

22. A. Massey, R. Kiffin and A. M. Cuervo, Pathophysiology of chaperone-mediated autophagy, *Int J Biochem Cell Biol* **36**: 2420–2434 (2004).
23. A. Melendez, Z. Talloczy, M. Seaman, E. L. Eskelinen, D. H. Hall and B. Levine, Autophagy genes are essential for dauer development and life-span extension in *C. elegans*, *Science* **301**: 1387–1391 (2003).
24. J. N. Keller, J. Gee and Q. Ding, The proteasome in brain aging, *Ageing Res Rev* **1**: 279–293 (2002).
25. J. N. Keller, K. B. Hanni and W. R. Markesbery, Possible involvement of proteasome inhibition in aging: implications for oxidative stress, *Mech Ageing Dev* **113**: 61–70 (2000).
26. M. Komatsu, S. Waguri, T. Chiba, S. Murata, J. Iwata, I. Tanida, T. Ueno, M. Koike, Y. Uchiyama, E. Kominami and K. Tanaka, Loss of autophagy in the central nervous system causes neurodegeneration in mice, *Nature* **441**: 880–884 (2006).
27. T. Hara, K. Nakamura, M. Matsui, A. Yamamoto, Y. Nakahara, R. Suzuki-Migishima, M. Yokoyama, K. Mishima, I. Saito, H. Okano and N. Mizushima, Suppression of basal autophagy in neural cells causes neurodegenerative disease in mice, *Nature* **441**: 885–889 (2006).
28. B. Boland, A. Kumar, S. Lee, F. M. Platt, J. Wegiel, W. H. Yu and R. A. Nixon, Autophagy induction and autophagosome clearance in neurons: relationship to autophagic pathology in Alzheimer's disease, *J Neurosci* **28**: 6926–6937 (2008).
29. D. J. Klionsky and S. D. Emr, Autophagy as a regulated pathway of cellular degradation, *Science* **290**: 1717–1721 (2000).
30. S. Kumar, I. H. Lee and M. Plamann, Cytoplasmic dynein ATPase activity is regulated by dynactin-dependent phosphorylation, *J Biol Chem* **275**: 31798–31804 (2000).
31. S. T. Brady, A novel brain ATPase with properties expected for the fast axonal transport motor, *Nature* **317**: 73–75 (1985).
32. S. A. Cohn, A. L. Ingold and J. M. Scholey, Correlation between the ATPase and microtubule translocating activities of sea urchin egg kinesin, *Nature* **328**: 160–163 (1987).
33. W. H. Yu, B. Dorado, H. Y. Figueroa, L. Wang, E. Planel, M. R. Cookson, L. N. Clark and K. E. Duff, Metabolic activity determines efficacy of macroautophagic clearance of pathological oligomeric alpha-synuclein, *Am J Pathol* **175**: 736–747 (2009).
34. A. J. Meijer and P. Codogno, Autophagy: regulation by energy sensing, *Curr Biol* **21**: R227–R229 (2011).
35. D. F. Egan, D. B. Shcackelford, M. M. Mihaylova, S. Gelino, R. A. Kohn, W. Mair, D. S. Vasquez, *et al.*, Phosphorylation of ULK1 (hATG1) by AMP-activated protein kinase connects energy sening to mitophagy, *Science* 331: 456–461 (2011).
36. J. Kim, M. Kundu, B. Viollet and K. L. Guan, AMPK and mTOR regulate autophagy through direct phosphorylation of Ulk1, *Nat Cell Biol* **13**: 132–141 (2011).
37. J. H. Zhu, C. Horbinski, F. Guo, S. Watkins, Y. Uchiyama and C. T. Chu, Regulation of autophagy by extracellular signal-regulated protein kinases during 1-methyl-4-phenylpyridinium-induced cell death, *Am J Pathol* **170**: 75–86 (2007).

38. J. H. Lee, W. H. Yu, A. Kumar, S. Lee, P. S. Mohan, C. M. Peterhoff, D. M. Wolfe, M. Martinez-Vicente, A. C. Massey, G. Sovak, Y. Uchiyama, D. Westaway, A. M. Cuervo and R. A. Nixon, Lysosomal proteolysis and autophagy require presenilin 1 and are disrupted by Alzheimer-related PS1 mutations, *Cell* **141**: 1146–1158 (2010).
39. J. P. Schellens, H. Vreeling-Sindelarova, P. J. Plomp and A. J. Meijer, Cytochemical and morphometric analysis of autophagy in energy depleted rat hepatocytes, *Cell Biol Int Rep* **14**: 805–814 (1990).
40. P. J. Plomp, E. J. Wolvetang, A. K. Groen, A. J. Meijer, P. B. Gordon and P. O. Seglen, Energy dependence of autophagic protein degradation in isolated rat hepatocytes, *Eur J Biochem* **164**: 197–203 (1987).
41. P. J. Plomp, P. B. Gordon, A. J. Meijer, H. Hoyvik and P. O. Seglen, Energy dependence of different steps in the autophagic-lysosomal pathway, *J Biol Chem* **264**: 6699–6704 (1989).
42. J. P. Schellens and A. J. Meijer, Energy depletion and autophagy. Cytochemical and biochemical studies in isolated rat hepatocytes, *Histochem J* **23**: 460–466 (1991).
43. P. Pietrini, S. J. Teipel, P. Bartenstein, S. I. Rapoport, H. J. Moller and H. Hampel, PET and the effects of aging and neurodegeneration on brain function: basic principles, *Drug News Perspect* **11**: 161–165 (1998).
44. K. Asanuma, V. Dhawan, M. Carbon and D. Eidelberg, Assessment of disease progression in parkinsonism, *J Neurol* **251**(Suppl 7): vII4–8 (2004).
45. S. Thobois, S. Guillouet and E. Broussolle, Contributions of PET and SPECT to the understanding of the pathophysiology of Parkinson's disease, *Neurophysiol Clin* **31**: 321–340 (2001).
46. H. Nagasawa, H. Tanji, H. Nomura, H. Saito, Y. Itoyama, I. Kimura, S. Tuji, T. Fujiwara, R. Iwata, M. Itoh and T. Ido, PET study of cerebral glucose metabolism and fluorodopa uptake in patients with corticobasal degeneration, *J Neurol Sci* **139**: 210–217 (1996).
47. F. Chretien, J. Servan, J. Mikol, M. Trierweiller, D. Elghozi and F. Gray, A 70-year-old man with extrapyramidal symptoms, dementia and hemosiderosis, *Brain Pathol* **16**: 235–236 (2006).
48. M. Zarei, B. Patenaude, J. Damoiseaux, C. Morgese, S. Smith, P. M. Matthews, F. Barkhof, S. A. Rombouts, E. Sanz-Arigita and M. Jenkinson, Combining shape and connectivity analysis: an MRI study of thalamic degeneration in Alzheimer's disease, *Neuroimage* **49**: 1–8 (2010).
49. K. E. Conley, D. J. Marcinek and J. Villarin, Mitochondrial dysfunction and age, *Curr Opin Clin Nutr Metab Care* **10**: 688–692 (2007).
50. R. H. Weindruch, M. K. Cheung, M. A. Verity and R. L. Walford, Modification of mitochondrial respiration by aging and dietary restriction, *Mech Ageing Dev* **12**: 375–392 (1980).
51. A. Del Roso, S. Vittorini, G. Cavallini, A. Donati, Z. Gori, M. Masini, M. Pollera and E. Bergamini, Ageing-related changes in the *in vivo* function of rat liver macroautophagy and proteolysis, *Exp Gerontol* **38**: 519–527 (2003).

52. P. Lestienne and N. Bataille, Mitochondrial DNA alterations and genetic diseases: a review, *Biomed Pharmacother* **48**: 199–214 (1994).
53. S. DiMauro, Mitochondrial myopathies, *Curr Opin Rheumatol* **18**: 636–641 (2006).
54. T. Eckert, A. Barnes, V. Dhawan, S. Frucht, M. F. Gordon, A. S. Feigin and D. Eidelberg, FDG PET in the differential diagnosis of parkinsonian disorders, *Neuroimage* **26**: 912–921 (2005).
55. A. Berardelli and A. Suppa, Recent advances in the pathophysiology of Parkinson's disease: evidence from fMRI and TMS studies, *Exp Neurol* **227**: 10–12 (2011).
56. D. J. Brooks, V. Ibanez, G. V. Sawle, N. Quinn, A. J. Lees, C. J. Mathias, R. Bannister, C. D. Marsden and R. S. Frackowiak, Differing patterns of striatal 18F-dopa uptake in Parkinson's disease, multiple system atrophy, and progressive supranuclear palsy, *Ann Neurol* **28**: 547–555 (1990).
57. D. P. Narendra, S. M. Jin, A. Tanaka, D. F. Suen, C. A. Gautier, J. Shen, M. R. Cookson and R. J. Youle, PINK1 is selectively stabilized on impaired mitochondria to activate Parkin, *PLoS Biol* **8**: e1000298 (2010).
58. D. Narendra, A. Tanaka, D. F. Suen and R. J. Youle, Parkin is recruited selectively to impaired mitochondria and promotes their autophagy, *J Cell Biol* **183**: 795–803 (2008).
59. A. Velayati, W. H. Yu and E. Sidransky, The role of glucocerebrosidase mutations in Parkinson disease and Lewy body disorders, *Curr Neurol Neurosci Rep* **10**: 190–198 (2010).
60. A. J. Lees and A. B. Singleton, Clinical heterogeneity of ATP13A2 linked disease (Kufor-Rakeb) justifies a PARK designation, *Neurology* **68**: 1553–1554 (2007).
61. A. Ramirez, A. Heimbach, J. Grundemann, B. Stiller, D. Hampshire, L. P. Cid, I. Goebel, A. F. Mubaidin, A. L. Wriekat, J. Roeper, A. Al-Din, A. M. Hillmer, M. Karsak, B. Liss, C. G. Woods, M. I. Behrens and C. Kubisch, Hereditary parkinsonism with dementia is caused by mutations in ATP13A2, encoding a lysosomal type 5 P-type ATPase, *Nat Genet* **38**: 1184–1191 (2006).
62. J. W. Langston, P. Ballard, J. W. Tetrud and I. Irwin, Chronic Parkinsonism in humans due to a product of meperidine-analog synthesis, *Science* **219**: 979–980 (1983).
63. A. Ziering, L. Berger, *et al.*, Piperidine derivatives; 4-arylpiperidines, *J Org Chem* **12**: 894–903 (1947).
64. A. Ziering and J. Lee, Piperidine derivatives; 1,3-dialkyl-4-aryl-4-acyloxypiperidines, *J Org Chem* **12**: 911–914 (1947).
65. R. R. Ramsay, A. T. Kowal, M. K. Johnson, J. I. Salach and T. P. Singer, The inhibition site of MPP^+, the neurotoxic bioactivation product of 1-methyl-4-phenyl-1,2,3,6-tetrahydropyridine is near the Q-binding site of NADH dehydrogenase, *Arch Biochem Biophys* **259**: 645–649 (1987).
66. F. R. Martin, J. Sanchez-Ramos and M. Rosenthal, Selective and nonselective effects of 1-methyl-4-phenylpyridinium on oxygen consumption in rat striatal and hippocampal slices, *J Neurochem* **57**: 1340–1346 (1991).
67. D. A. Di Monte, Mitochondrial DNA and Parkinson's disease, *Neurology* **41**: 38–42; discussion 42–33 (1991).

68. G. E. Meredith, S. Totterdell, M. Beales and C. K. Meshul, Impaired glutamate homeostasis and programmed cell death in a chronic MPTP mouse model of Parkinson's disease, *Exp Neurol* **219**: 334–340 (2009).
69. R. Gosselin, R. Smih and H. Hodge, *Clinical Toxicology of Commercial Products* (Williams & Wilkins, Baltimore, 1984), p. 2.
70. R. Betarbet, T. B. Sherer, G. MacKenzie, M. Garcia-Osuna, A. V. Panov and J. T. Greenamyre, Chronic systemic pesticide exposure reproduces features of Parkinson's disease, *Nat Neurosci* **3**: 1301–1306 (2000).
71. B. Spencer, R. Potkar, M. Trejo, E. Rockenstein, C. Patrick, R. Gindi, A. Adame, T. Wyss-Coray and E. Masliah, Beclin 1 gene transfer activates autophagy and ameliorates the neurodegenerative pathology in alpha-synuclein models of Parkinson's and Lewy body diseases, *J Neurosci* **29**: 13578–13588 (2009).
72. S. Michiorri, V. Gelmetti, E. Giarda, F. Lombardi, F. Romano, R. Marongiu, S. Nerini-Molteni, P. Sale, R. Vago, G. Arena, L. Torosantucci, L. Cassina, M. A. Russo, B. Dallapiccola, E. M. Valente and G. Casari, The Parkinson-associated protein PINK1 interacts with Beclin1 and promotes autophagy, *Cell Death Differ* **17**: 962–974 (2010).
73. M. Xilouri and L. Stefanis, Autophagic pathways in Parkinson disease and related disorders, *Expert Rev Mol Med* **13**: e8 (2011).
74. A. M. Cuervo, L. Stefanis, R. Fredenburg, P. T. Lansbury and D. Sulzer, Impaired degradation of mutant alpha-synuclein by chaperone-mediated autophagy, *Science* **305**: 1292–1295 (2004).
75. R. A. Nixon, J. Wegiel, A. Kumar, W. H. Yu, C. Peterhoff, A. Cataldo and A. M. Cuervo, Extensive involvement of autophagy in Alzheimer disease: an immuno-electron microscopy study, *J Neuropathol Exp Neurol* **64**: 113–122 (2005).
76. F. Pickford, E. Masliah, M. Britschgi, K. Lucin, R. Narasimhan, P. A. Jaeger, S. Small, B. Spencer, E. Rockenstein, B. Levine and T. Wyss-Coray, The autophagy-related protein beclin 1 shows reduced expression in early Alzheimer disease and regulates amyloid beta accumulation in mice, *J Clin Invest* **118**: 2190–2199 (2008).
77. S. M. Cardoso, C. F. Pereira, P. I. Moreira, D. M. Arduino, A. R. Esteves and C. R. Oliveira, Mitochondrial control of autophagic lysosomal pathway in Alzheimer's disease, *Exp Neurol* **223**: 294–298 (2010).
78. B. Pajak, M. Songin, J. B. Strosznajder and B. Gajkowska, Alzheimer's disease genetic mutation evokes ultrastructural alterations: correlation to an intracellular Abeta deposition and the level of GSK-3beta-P(Y216) phosphorylated form, *Neurotoxicology* **30**: 581–588 (2009).
79. D. Ling, H. J. Song, D. Garza, T. P. Neufeld and P. M. Salvaterra, Abeta42-induced neurodegeneration via an age-dependent autophagic-lysosomal injury in Drosophila, *PLoS One* **4**: e4201 (2009).
80. W. H. Yu, A. M. Cuervo, A. Kumar, C. M. Peterhoff, S. D. Schmidt, J. H. Lee, P. S. Mohan, M. Mercken, M. R. Farmery, L. O. Tjernberg, Y. Jiang, K. Duff, Y. Uchiyama, J. Naslund, P. M. Mathews, A. M. Cataldo and R. A. Nixon, Macroautophagy — a

novel beta-amyloid peptide-generating pathway activated in Alzheimer's disease, *J Cell Biol* **171**: 87–98 (2005).

81. W. H. Yu, A. Kumar, C. Peterhoff, L. Shapiro Kulnane, Y. Uchiyama, B. T. Lamb, A. M. Cuervo and R. A. Nixon, Autophagic vacuoles are enriched in amyloid precursor protein-secretase activities: implications for beta-amyloid peptide over-production and localization in Alzheimer's disease, *Int J Biochem Cell Biol* **36**: 2531–2540 (2004).
82. L. Devi, B. M. Prabhu, D. F. Galati, N. G. Avadhani and H. K. Anandatheerthavarada, Accumulation of amyloid precursor protein in the mitochondrial import channels of human Alzheimer's disease brain is associated with mitochondrial dysfunction, *J Neurosci* **26**: 9057–9068 (2006).
83. W. Stoothoff, P. B. Jones, T. L. Spires-Jones, D. Joyner, E. Chhabra, K. Bercury, Z. Fan, H. Xie, B. Bacskai, J. Edd, D. Irimia and B. T. Hyman, Differential effect of three-repeat and four-repeat tau on mitochondrial axonal transport, *J Neurochem* **111**: 417–427 (2009).
84. H. V. Che, S. Metzger, E. Portal, C. Deyle, O. Riess and H. P. Nguyen, Localization of sequence variations in PGC-1alpha influence their modifying effect in Huntington disease, *Mol Neurodegen* **6**: 1 (2011).
85. M. Napolitano, D. Centonze, P. Gubellini, S. Rossi, S. Spiezia, G. Bernardi, A. Gulino and P. Calabresi, Inhibition of mitochondrial complex II alters striatal expression of genes involved in glutamatergic and dopaminergic signaling: possible implications for Huntington's disease, *Neurobiol Dis* **15**: 407–414 (2004).
86. M. M. Zeron, O. Hansson, N. Chen, C. L. Wellington, B. R. Leavitt, P. Brundin, M. R. Hayden and L. A. Raymond, Increased sensitivity to N-methyl-D-aspartate receptor-mediated excitotoxicity in a mouse model of Huntington's disease, *Neuron* **33**: 849–860 (2002).
87. M. Martinez-Vicente, Z. Talloczy, E. Wong, G. Tang, H. Koga, S. Kaushik, R. de Vries, E. Arias, S. Harris, D. Sulzer and A. M. Cuervo, Cargo recognition failure is responsible for inefficient autophagy in Huntington's disease, *Nat Neurosci* **13**: 567–576 (2010).
88. B. Ravikumar, C. Vacher, Z. Berger, J. E. Davies, S. Luo, L. G. Oroz, F. Scaravilli, D. F. Easton, R. Duden, C. J. O'Kane and D. C. Rubinsztein, Inhibition of mTOR induces autophagy and reduces toxicity of polyglutamine expansions in fly and mouse models of Huntington disease, *Nat Genet* **36**: 585–595 (2004).
89. B. Ravikumar, R. Duden and D. C. Rubinsztein, Aggregate-prone proteins with polyglutamine and polyalanine expansions are degraded by autophagy, *Hum Mol Genet* **11**: 1107–1117 (2002).
90. A. Yamamoto, M. L. Cremona and J. E. Rothman, Autophagy-mediated clearance of huntingtin aggregates triggered by the insulin-signaling pathway, *J Cell Biol* **172**: 719–731 (2006).

CHAPTER 13

THE POTENTIAL OF AUTOPHAGY REGULATION IN THE TREATMENT OF NEURODEGENERATIVE DISEASES

Ashley R. Winslow, Zeyn W. Green-Thompson and David C. Rubinsztein

ABSTRACT

This chapter will provide an overview of the possible uses of autophagy modulation as a therapeutic for neurodegenerative disorders. Due to the diversity of neurodegenerative diseases, the chapter is subdivided into a discussion of the role of autophagy inhibition and modulation in the context of each disorder. Understanding the molecular pathogenesis of each disease will help consideration of how autophagy modulation may be potentially useful as a therapeutic.

INTRODUCTION

Neurodegeneration

Neurodegeneration generally refers to the structural or functional loss of brain tissue. This loss usually begins at the molecular level and progresses to a more systemic degeneration of entire regions of the brain. Neurodegeneration at the molecular level is usually characterised by mitochondrial dysfunction, axonal swelling which is thought to represent disrupted axonal transport, and cell death. A major risk factor for developing many neurodegenerative disorders is aging.

Some researchers have begun to focus on groups of diseases defined by similar pathogenesis. Through this approach, researchers hope to identify common pathways relevant to maintaining health in numerous neurodegenerative diseases. This approach is advantageous for both patients, through therapeutic drug discovery, and researchers, by extending general understanding of disease and health. One such group of diseases are the proteinopathies. Proteinopathies are characterised by the accumulation of intracellular or extracellular toxic proteins within the brain and include diseases such as Parkinson's disease (PD), Alzheimer's disease (AD), prion diseases, tauopathies, Huntington's disease (HD) and spinocerebellar ataxia type 3 (SCA3).

Proteinopathies

Common to some proteinopathies, such as HD, SCA3, and some forms of PD, is a dominant, toxic gain-of-function mutation. These mutations confer an altered tertiary protein structure that results from misfolding of the protein, leading to the eventual formation of aggregate bodies or inclusions. Other proteinopathies are not as simple. Diseases like AD, amyotrophic lateral sclerosis (ALS), and other forms of PD are often associated with the accumulation of a wildtype protein. The change in conformation of the protein alters the thermodynamic stability of the protein in its aqueous environment, leading to a decrease in its solubility. These unstable aggregate-prone proteins usually exist as soluble monomeric or oligomeric forms or insoluble aggregates. Aggregation is one means through which hydrophobic proteins can stabilise themselves in an aqueous environment. Although the aggregate is insoluble, it is thought to exist in a constant state of flux, whereby soluble monomers and oligomers are constantly added and removed from the aggregate.

The extent to which monomers, oligomers, and aggregates contribute to disease progression and at which stage remain controversial.[1] While some argue that the aggregate is a protective form of the mutated protein that sequesters the toxic misfolded protein into an insoluble aggregate,[2] other data supports the theory that the aggregate can act as a toxic entity. A number of elegant studies utilized aggregate-prone protein mutants that favoured either fibril or oligomeric formation to show that the oligomer-forming species were more toxic *in vivo*.[2,3] Although these studies strongly support the theory of a "soluble oligomeric-induced toxicity", the aggregate forming species of these proteins are not necessarily benign. There exists substantial evidence for the inhibition of key pathways by the presence of large intracellular and nuclear aggregates. It is thought that aggregation blocks these processes both by its physical presence and by the sequestration of important proteins into the aggregate itself, making them unavailable for normal functions. In the case of Huntington's disease, the mutant form of huntingtin is known to co-localise with a number of cellular proteins that mediate various cellular processes, such as vesicle trafficking, protein degradation, and proteolysis.[4,5] Thus, it is possible that both oligomers and larger aggregates may contribute to pathogenesis and the relative toxicities may depend on factors such as subcellular localisation.

Protein misfolding is a common occurrence in both diseased and healthy cells. An understanding of how these misfolded proteins are degraded may provide clues to possible mechanisms for clearing aggregates or preventing aggregation. Induction of normal cellular protein degradation processes can increase clearance of misfolded proteins, prevent aggregation and ameliorate symptoms of neurodegeneration.[6] This may provide a rational therapeutic strategy for these diseases. The ubiquitin-proteasome system (UPS) and autophagy-lysosome machinery are the two main pathways available to the cell for protein degradation.

Protein Degradation Machinery

UPS

The proteasome is responsible for the selective degradation and recycling of short-lived cytosolic and nuclear proteins. While the half-life of proteins degraded by the proteasome is variable, the proteasome generally degrades short-lived proteins through a highly specialised process.[7]

Attachment of a polyubiquitin chain to a protein targets the protein for degradation by the catalytic core of the proteasome. The 26S proteasome, a 2000 kDa, multisubunit protein complex, consists of a 20S subunit flanked on the top and bottom by two 19S subunits. These subunits assume a specific hollow barrel conformation that allows for degradation of proteins that generally require unfolding prior to degradation by the uppermost 19S subunit. The catalytic activity occurs within the barrel of the proteasome after entry of the cargo into the structure. If the cargo is unable to be unfolded and enter the proteasome barrel it will not be degraded and may potentially restrict the movement of other cargo into the proteasome pore. Due to the size of the narrow barrel of the proteasome and specificity of the process, most oligomerized or aggregated proteins are thought to be precluded from entering the structure.

Autophagy

Proteins that are precluded from the proteasome are usually degraded by the less selective cellular process, macroautophagy. This process relies on the acidic hydrolases of the lysosomes to degrade substrates. Since the lysosome is an expandable vesicle, lysosomal-dependent degradation is generally less specific and allows for the degradation of large proteins and organelles, without the need for unfolding of the protein.

Macroautophagy, more commonly referred to as autophagy, is responsible for the bulk degradation of cytoplasmic proteins, organelles, and long-lived proteins. In most cell types autophagy operates constitutively, but it can be upregulated in response to signalling pathways. Autophagy requires the formation of autophagosomes, double-membrane vesicles that engulf cytoplasmic contents, and the subsequent fusion of autophagosomes with lysosomes. This fusion produces an autophagolysosome (also called autolysosome) within which the targeted cargo is degraded by acidic hydrolases stored in the lysosome. Importantly, due to the ability to of autophagosomes to engulf large cargo, autophagy has been shown to clear toxic and aggregate-prone proteins that are thought to be precluded from clearance by the UPS. While autophagy is classically referred to as a "non-selective" degradative system, recent evidence has shown potential targeting of substrates to the autophagosome, possibly through the involvement of ubiquitin.[8] The mechanistic control of autophagy and its role in health and disease will be discussed in more detail in later sections.

Autophagy induction

The mammalian target of rapamycin (mTOR) is a 289 kDa serine-threonine kinase that participates in two complexes that regulate metabolism and cell survival.[9,10] The first of the two complexes is the mTOR complex 1 (mTORC1) which positively regulates cell proliferation and cell growth. mTORC1 acts on anabolic processes by phosporylating downstream targets, 4E-binding protein 1 (4EBP1) and ribosomal protein S6 kinase polypeptide 1 (S6K).[11] Functions of the second complex, mTORC2, are still unclear due to a lack of known specific inhibitors and its lethality in mice when deleted. Despite these difficulties, scientists have discovered roles for mTORC2 in cell survival, proliferation, metabolism, AKT phosphorylation, and cytosketeletal organisation. Very little is known about the mechanisms by which mTORC2 acts in these roles.[12]

mTORC1 and mTORC2 are also known as the rapamycin-sensitive and insensitive complexes, respectively. Rapamycin is an antifungal, bacterially-produced compound that is used as an immunosuppressant and anticancer agent. Treatment of cells with rapamycin inhibits the activity of mTORC1 but not mTORC2.[9] Inhibition of mTOR is known to induce autophagy through inhibition of the mTORC1 complex.[13] mTOR activity can be monitored by assaying changes in the phosphorylation states of downstream kinase substrates 4EBP1 and S6K (although these proteins are not thought to be direct regulators of mammalian autophagy). The phosphorylation state of these proteins often correlates inversely with autophagy induction.

Autophagy can be induced independently of the mTOR pathway through treatment with the fungal disaccharide trehalose,[14,15] via an unknown mechanism, or by the lowering of intracellular inositol or inositol-1,4,5,-triphosphate (IP3).[16] Genetic inhibition of the IP3 receptor[17] or depletion of inositol through the use of compounds has been shown to induce autophagy without altering mTOR activity.[16,18] Conversely, activation of autophagy through mTOR inhibition has no affect on inositol levels, indicating that these two pathways act on autophagy independently of one another.

Therapeutic upregulation of autophagy by decreasing inositol is an appealing alternative to manipulation of mTOR. The commonly used mood stabilisers, lithium, carbamezapine, and sodium valporate lower inositol levels and are likely to have less deleterious side effects on cell growth when compared to rapamycin.[16,19] It may also be possible to use a combination of mTOR-dependent and -independent pathways to induce autophagy, with the added advantage of using less of each compound yet achieving greater upregulation of autophagy.[16]

HUNTINGTON'S DISEASE

Huntington's disease (HD) is an autosomal dominant, monogenic disorder that causes neuronal degeneration of cells in the frontal lobes, basal ganglia, and striatum. HD pathogenesis involves cognitive decline, psychiatric disturbances, motor dysfunction, and eventual dementia and death.[20] Death follows about 15 to 20 years after onset of HD symptoms, with a median age-of-onset at age 37. Prevalence is 5–10 per 100,000 in most populations.

The normal huntingtin protein is ubiquitously expressed and found at the highest levels within the brain and testes. Normally, the wildtype htt gene contains a stretch of polyglutamine repeats encoded by a series of 35 or fewer CAG codons. Expansion of this CAG codon repeat above the 35-repeat threshold causes disease; as this repeat number increases the age of onset decreases. At a molecular level, the translation of the CAG expansion to a pathogenic polyglutamine repeat confers an alternative conformation on the huntingtin protein that is thought to be responsible for its toxicity through a gain-of-function mechanism. This altered tertiary structure causes the HD protein to form intranuclear inclusions and cytoplasmic aggregates.[21–23]

Autophagy and HD

Recent research has shown that mutant htt may impair autophagy degradation through inefficient cargo recognition and sequestration. Overexpression of the mutant (polyQ expanded htt) in comparison to wildtype htt does not affect autophagosome synthesis or lysosomal fusion, but upon autophagy induction in the presence of mutant htt overexpression, the vacuolar autophagosome structures that are formed contain less cargo than those formed under normal, wildtype conditions (Figure 1).[24] This study has important implications for both the understanding of HD pathogenesis and the possibility of autophagy-related therapeutic strategies.

Upregulation of Autophagy and Its Therapeutic Potential in HD

Since the exact mode of molecular pathogenesis of mutant huntingtin is still not understood, current research has focussed on reducing the amount of intracellular mutant huntingtin in the hope of decreasing cytotoxicity. Enhancing degradation and decreasing expression of mutant huntingtin are promising therapeutic strategies. Autophagy is integral in efforts to clear the mutant huntingtin protein.[25] Treatment with the autophagy inducer rapamycin has been shown to enhance clearance of mutant huntingtin aggregates and soluble huntingtin, thus reducing HD-associated toxicity.[26,27] The rapamycin analogue CCI-779 has been shown to both decrease aggregate formation and enhance behavioural performance in an HD mouse model.[4] Conversely, blocking autophagy by preventing autophagosome-lysosome fusion or autophagosome formation increases soluble and aggregated mutant huntingtin and enhances toxicity in cell and mouse models.[26,28]

Due to the immunosuppressant qualities of rapamycin, long-term treatment of HD by mTOR inhibition with rapamycin may not be an ideal method for treatment, particularly if one is aiming to delay disease onset in presymptomatic mutation carriers.[29] Through a screen of FDA-approved drugs conducted to identify alternative autophagy-inducing pathways, an mTOR-independent pathway has been elucidated which upregulates autophagy through cAMP modulation of IP_3 and subsequent regulation of calpain activity. The pathway is potentially circular, as Gs_α, which regulates IP3 levels, is then cleaved by calpains leading to constitutive activation. Fly and zebrafish models of disease have been used to provide proof-of-principle for the use of therapeutic upregulation of autophagy, via this mTOR-independent pathway, to ameliorate disease phenotypes and decrease mutant huntingtin levels.[18]

In terms of future therapeutic approaches in humans, rilmenidine and clonidine (α_2 adrenoceptor and noradrenergic imidazoline I_1 receptor agonists), prescribed as anti-hypertensive agents, emerged from the above-mentioned screen as inducers of autophagy. Rilmenidine may prove more clinically tolerable, as it has a 30-fold greater selectivity for imidazoline I_1 receptors compared to α_2 receptors, and therefore has a lesser tendency to evoke side effects such as sedation.[30] In the N171-82Q Huntington's disease mice, rilmenidine reduced mutant huntingtin levels and ameliorated disease severity.[31]

At present there is no therapeutic treatment for Huntington's disease. Current approaches are limited to symptomatic treatment and palliative care. Data from the studies outlined above give clear indication of the potential for autophagy-upregulating agents to reduce levels of mutant huntingtin, diminish pathological hallmarks of the disease, and improve neurological function.

AMYOTROPHIC LATERAL SCLEROSIS

Amyotrophic lateral sclerosis (ALS) is a form of motor neuron disease characterised by muscular atrophy. Neurons of ALS patients exhibit progressive degeneration of the upper and lower motor neurons in the cerebral cortex, the brain stem, and spinal cord.[32] Damage to both glial and Renshaw cells are seen, with the latter occurring first.[33,34] The most common known causes of familial ALS include mutations in SOD1 (CuZn superoxide dismutase-1), TDP43 (TAR DNA-binding protein 43), FUS (fused in sarcoma), and C90RF72, which are associated with the marked accumulation of inclusions in patient tissues.[34,35]

Autophagy, Dynein, ESCRT, and ALS

A number of mutations affecting microtubule transport have also been implicated in the development of motor neuron disease (MND) in mouse models and humans, including the p150Glued mutation of the dynactin complex,[36–38,39] which regulates dynein binding or cargo. Reduced dynein activity impairs autophagy flux by compromising autophagosome-lysosome fusion which leads to accumulation of autophagy substrates (Figure 1).[28] As disruption in dynein-dependent transport is implicated in the cause of this group of diseases, it is likely that the aggregation of substrates seen in mouse models could, at least in part, result from the disruption of the microtubule-dependent movement of autophagosomes, autophagosome maturation, and autophagosome-lysosome fusion, resulting in an overall accumulation of autophagosomes. Indeed, an increase in autophagosome number and LC3-II levels can be observed in dynein-deficient transgenic mice, consistent with this theory.[28,40]

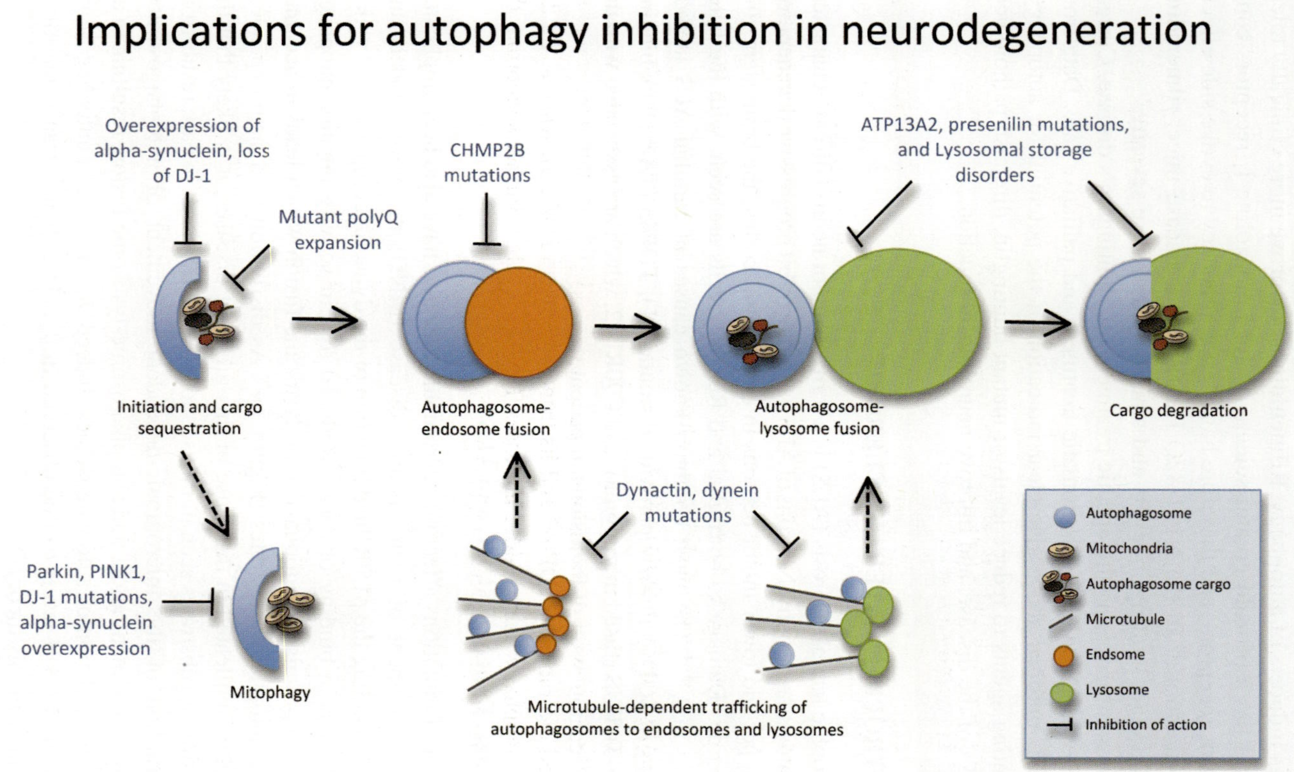

Figure 1. Does autophagy dysfunction cause neurodegeneration? Example showing autophagy dysfunction at a number of stages in the autophagosome-lysosome pathway, implications for Huntington's disease, Parkinson's Alzheimer's disease, motor neiron disease, and lysosomal storage disorders.

Autophagy inhibition may also be relevant to ESCRT (endosomal sorting complexes required for transport) mutation-associated forms of ALS. Endosomal incorporation of proteins and endosomal-lysosomal fusion is maintained by the sequential interaction of ESCRT complexes. In a subset of the familial form of ALS, mutations of the CHMP2B subunit of ESCRT-III have been shown to associate with large TDP-43 aggregates and the accumulation of autophagosomes. Indeed, CHMP2B mutations cause autophagosome accumulation due to impaired autophagosome-endosome fusion.[41]

Autophagy Modulation: Implications from a Lithium Clinical Trial and Animal Models

Lithium has been used successfully in a drug trial for treatment of an amyotrophic lateral sclerosis (ALS) mouse model expressing a SOD1 mutation. Tissue collected from these mice showed an increase in autophagic activity and a general decrease in disease phenotypes such as ubiquitin and alpha-synuclein accumulation.[33] Administration of lithium carbonate (in combination with riluzole) to human patients in a small drug trial significantly delayed the onset of disability and death in comparison to those administered riluzole alone.[33] These findings were further supported by a study in SOD1 G93A mice showing that treatment with lithium and valproic acid delayed onset of symptoms, decreased disease severity, and prolonged life-span.[42] Similarly, autophagy induction by treatment with trehalose enhanced clearance of TDP-43 and its truncated form (TDP-25), whereas inhibition of autophagy using 3-MA increases TDP-43 levels,[43] further supporting the potential benefit of upregulating autophagy for the treatment of ALS.

However, these findings will need to be further validated in human ALS patients with larger cohorts. A recent clinical trial found no beneficial effects of combinatorial treatment with LiCl and riluzole over riluzole alone in patients with ALS.[44] Although the cohort in this study was larger than the previous lithium combinatorial study, the cohort consisted of a heterogenous group of both sporadic and familial forms of ALS and was stopped after only six months of drug administration. The trial ended early due to a lack of beneficial effects (not worsening of disease symptoms). Autophagy upregulation may prove more useful as a preventative treatment as the onset of symptoms generally corresponds to a high degree of neuronal degeneration, which will not be recoverable through autophagy induction. It is also important to consider that while lithium treatment has been shown to upregulate autophagy in a number of models and systems, lithium is capable of slowing the clearance of aggregate-prone proteins through its actions on GSK-3β.[45]

Understanding the different mechanisms of ALS pathogenesis may be vital to effective treatment by autophagy regulation in motor neuron diseases. While upregulation of autophagy by mTOR-dependent and -independent treatment appears to be a promising therapeutic strategy for reducing the accumulation of proteins and dysfunctional organelles, it is likely that this will not be beneficial in forms caused by ESCRT mutations. The blockage in later autophagosome fusion events required for degradation of autophagy substrates may not be ameliorated by inducing autophagosome formation. At present, it is generally believed that if autophagy clearance can be upregulated within a certain disease background, then therapeutic upregulation of autophagy may provide some beneficial effects through clearance of damaged mitochondria, aggregate-prone species, and toxic oligomers. Consequently, those diseases characterised by a nearly complete block in autophagy clearance, usually through lysosomal fusion dysfunction, may not be good candidates for autophagy induction therapies. For these diseases, future studies will need to focus on the lysosome for therapeutic intervention, while autophagy kinetic regulation may play a secondary role in treatment.

ALZHEIMER'S DISEASE

Alzheimer's disease (AD) is the most common neurodegenerative disease and the principal cause of dementia in the elderly. Although many of the genetic and environmental factors determining AD risk are still largely unknown, it is thought that the extracellular beta-amyloid deposits play a causative role in the development of the disease, along with intraneuronal tau aggregates.

Lysosomal Dysfunction and AD: A Complicated Role for Autophagy in AD Pathogenesis

Autophagosomes and autophagolysosomes accumulate in the brains of patients suffering from Alzheimer's disease and in a number of mouse models of AD (reviewed in Ref. 46). Immunohistochemical staining in AD patient brains with autophagy marker antibodies, such as Atg12 and LC3, indicate a possible upregulation of autophagy in early stages of disease. Interestingly, immunostaining for these markers was not uniform in every cell, but rather, immunoreactivity seemed to correlate with severity of neuritic degeneration and proximity to plaques,[47] suggesting a specific induction of autophagy in response to toxic species.

Presenilin 1 is a transmembrane protein that is mutated in rare Mendelian forms of AD, leading to the production of neurotoxic beta-amyloid. Presenilin mutations have also been recently implicated in lysosomal dysfunction through inefficient lysosomal acidification by vATPase both *in vitro* and *in vivo*. This has important implications for autophagy, as autophagosome-lysosome fusion will be deficient.[48] Additionally, blocking later steps in the autophagic process by inhibiting cathepsin-mediated proteolysis within autolysosomes in mice results in a strikingly similar phenotype to AD pathogenesis in humans.[49] These data are particularly compelling in light of the theory of lysosomal dysfunction in AD.

However, lysosomal dysfunction and late-stage autophagy defects may not be the only contributors to AD pathogenesis. Affected areas of AD patient brains may have reduced Beclin 1 mRNA and protein levels, which would be predicted to impair autophagosome formation. Also, the neuronal tissues where the decreases in Beclin-1 are seen, are specific to areas of the brain normally affected by AD pathology. Overexpression of Beclin-1 in AD mice models reduced intracellular accumulation of beta-amyloid and extracellular deposition of beta-amyloid plaques, while beta-amyloid deposition and neurodegeneration were exacerbated by reduced Beclin 1 expression, thus providing support for its causal role in disease.[50] These Beclin 1 experiments suggest that, while there may be some block in autophagosome-lysosome fusion in some forms of AD, there may still be beneficial effects of autophagy induction in some forms of AD. However, some care needs to be taken when comparing effects in different mouse models of AD with different genetic lesions, as well as in comparing genetic forms of AD in mice to sporadic AD in humans, as the latter is determined by a host of largely unknown genetic and environmental factors. Beclin-1 may have roles distinct from autophagy.[51] Additionally, it seems that a number of variables may transcriptionally regulate Beclin-1. Studies have suggested that decreased levels of Beclin-1 may correlate with both progression of AD and age, as younger AD patients and younger AD mouse models did not have lower levels of the protein.[50]

Autophagy Modulation and AD

These data raise the possibility that autophagy upregulation may be beneficial in late stage AD by decreasing the levels of beta-amyloid. Treatment of transgenic mice expressing a familial Alzheimer's disease mutation of the amyloid protein precursor (FAD-F717V mutant APP, or PDAPP for short) with rapamycin improved learning and spatial tasks and decreased beta-amyloid[42] levels, the main toxic species.[52] These findings were corroborated in a study showing that learning deficits, increased beta-amyloid levels, and tau pathology were rescued with rapamycin treatment in a triple-transgenic (3-tg: tau, beta-amyloid, and presenilin mutations) AD mouse model. This study suggests that there may also be a feedback loop between mTOR signalling and beta-amyloid levels, where triple transgenic mice accumulate beta-amyloid and have a concomitant increase in mTOR signalling specifically in the hippocampus and cortex (areas of AD-associated pathogenesis in humans).[53] This finding has important implications for the understanding of both AD molecular pathogenesis and the molecular actions of rapamycin treatment in an AD *in vivo* model.

The presence of tau dense neurofibrillary tangles and neurophil threads in patients with AD suggests a role for tau in AD pathogenesis. Humans normally express six different tau isoforms as a result of alternative splicing of three different exons within the tau gene. The alternative splicing of exon 10 results in the formation of either three (3R) or four repeat (4R) tau, whereby the repeat region is thought to confer microtubule-binding properties. Differences in the ratio of 3R to 4R, calpain and caspase cleavages of tau, and hyperphosphorylation of tau have all been implicated in tau pathogenesis in AD and other tauopathies.[54–57]

Treatment of Drosophila expressing wildtype or mutant tau within the fly eye (a model for neurodegeneration) with rapamycin alleviated toxicity and increased survival. Cell-based models suggest that autophagy induction decreased primarily insoluble forms of tau while treatment of cells expressing tau with lysosomotropic agents delays clearance of tau.[58,59] Additionally, treatment of neuronal cultures, wildtype, and beta-amyloid expressing rats with lithium reduced tau phosphorylation likely through GSK-3.[60,61] Interestingly, treatment of mutant tau transgenic mice with lithium decreases tau accumulation and neurofibrillary-tangle formation, yet the treatment had little to no benefits for memory function.[62] Some of the beneficial effects of lithium on NFT and tau load clearance may be attributed to lithium's autophagy-activating activity.

PARKINSON'S DISEASE

Parkinson's disease is characterised by the degeneration of the substantia nigra and the development of proteinaceous inclusions termed Lewy bodies. Alpha-synuclein, a protein known to play a causal role in PD, is a major constituent of Lewy bodies.[63] The neurons found in the substantia nigra are responsible for the secretion of dopamine and it is thought that cell death within this region decreases dopamine secretion, ultimately leading to decreased stimulation of the motor cortex by the basal ganglia. This degeneration is thought to be responsible for muscle rigidity, tremors, and bradykinesia that are characteristic of PD patients.[64] Interestingly, the common development of cognitive and behavioural impairments in PD patients suggests degeneration of brain areas outside of the substantia nigra.

Although the vast majority of PD cases are sporadic, about 5% to 10% of all cases are rare hereditary forms of the disease. These cases present as autosomal dominant or autosomal recessive forms of the disease. Family studies have identified about 13 loci, named PARK genes, with mutations that strongly associate with the development of PD (dominant forms: alpha-synuclein, LRRK2 recessive forms: PINK1, parkin, DJ-1, ATP13A2). Alpha-synuclein was the first PARK locus identified. Rare point mutations (A53T, A30P, E46K) and chromosomal multiplications of the wildtype alpha-synuclein gene cause familial PD.[65,66] In these multiplication cases, age-of-onset and severity of the disease correlate closely with the expression level of alpha-synuclein suggesting a dose-response relative to the amount of alpha-synuclein protein present.[67] Thus, increased levels of wildtype alpha-synuclein increase the severity of PD and decrease the age-of-onset. The accumulation of alpha-synuclein and the dominant mode of inheritance of most alpha-synuclein-associated forms of PD suggest a gain-of-function of synuclein in the development of PD. Therefore, accelerated clearance of synuclein, whether mutant or excess wildtype, may decrease the toxic species within a cell and ameliorate disease pathology.

Mitophagy and Autophagy Dysfunction, Significant Contributors to PD Pathogenesis?

Research in a range of models and organisms involving various PARK genes have identified common pathways that are likely to be defective in PD. Two of the most significantly recurring themes are the identification of mitochondrial disruption and deficiencies in the protein degradation pathways. Wildtype PINK1 regulates localisation of parkin (an E3 ligase) to mitochondria, whereby both proteins regulate autophagy-dependent clearance of mitochondria via mitophagy. Loss of PINK1, parkin, and DJ-1 and overexpression of wildtype alpha-synuclein cause increased mitochondrial dysfunction including mitochondria fragmentation, abnormal membrane potential, and increased ROS production.[68–74]

Since autophagy-dependent degradation of dysfunctional mitochondria through mitophagy is important for maintaining intracellular homeostasis, it is important to understand how autophagy, and protein degradation in general, regulates these deficiencies. The accumulation of ubiquitinated proteinaceious inclusions in PD brains hints at inefficient or inhibited protein degradation. Alpha-synuclein, parkin, and UCH-L1[75–77] are thought to participate in UPS function either through direct or indirect interaction of the proteins with the proteasome. Overexpression of alpha-synuclein or expression of loss-of-function mutations of parkin have been shown to increase ubiquitination dysfunction and substrate accumulation.[78–80]

While these studies provide compelling evidence for UPS dysfunction, more recent research has found that inhibition of the autophagosome-lyosome pathway may contribute to PD and alpha-synuclein associated dementia with Lewy bodies (DLB). Levels of mTOR were increased and Atg7 levels decreased in the brains of patients wih DLB. Similar changes in protein levels were seen in alpha-synuclein transgenic mice.[81] Overexpression of alpha-synuclein, mimicking gene duplication effects in PD, in cells and transgenic mice decreased autophagosome synthesis by inhibiting the ER-to-Golgi traffic regulator, Rab1a;[82] similar findings were reported for loss of DJ-1 (Figure 1).[83] The effect of wildtype and mutant forms of LRRK2 on autophagy is controversial. Some studies suggest that LRRK2 may have an active role in autophagy modulation in response to stress conditions and neurite outgrowth, but the exact effect of LRRK2 remains unclear.[84,85]

Autophagy Induction Can Clear Mutant and Wildtype Alpha-Synuclein

Both wildtype alpha-synuclein and the A53T point mutant are degraded by the proteasome and the autophagy-lysosome pathway. One of the first studies which looked at the kinetics and degradative route of wildtype and mutant alpha-synuclein showed that proteasome inhibition by beta-lactone increased accumulation of overexpressed wildtype and mutant alpha-synuclein. Interestingly, the kinetic data from this study showed that the mutant A53T alpha-synuclein had a slowed rate of degradation, indicating a possible block in the degradation of this protein compared to its wildtype counterpart.[86] Since this early study, the A53T point mutation has been shown to be preferentially cleared by autophagy, possibly due to a conformational change in the protein that precludes its entrance into the narrow barrel of the proteasome.[87] These data are consistent with the early findings of alpha-synuclein degradation by the UPS, as treatment with proteasome inhibitors slowed degradation of the wildtype and mutant protein but did not completely block degradation. Indeed, autophagy appears to be responsible for a portion of alpha-synuclein degradation at least for point mutants.[87,88] It is also important to note that wildtype alpha-synuclein can be degraded by direct translocation of the protein into the lysosome by CMA.[89] CMA-mediated degradation of A53T and A30P is thought to be less efficient as both mutant proteins aberrantly bind CMA receptors and can block direct translocation, thus decreasing their own degradation.[89]

Therapeutically, enhanced clearance of alpha-synuclein provides a compelling option for treatment of synuclein-associated forms of PD through both chemical and genetic autophagy enhancers. Overexpression of Beclin 1 reduced alpha-synuclein accumulation and associated neurodegenerative phenotypes. These beneficial effects of Beclin 1 on alpha-synuclein toxicity are also seen with rapamycin.[90]

Autophagy Modulation May Provide General Health Benefits in Dying Cells

In addition to enhanced clearance of synuclein, upregulation of autophagy may have a number of general benefits for treatment of PD by increasing clearance of dysfunctional mitochondria, aggregate-prone proteins, and decreasing susceptibility to apoptotic insults. Indeed, general upregulation of autophagy has also been shown to protect against mitochondrial inhibitors paraquat, and rotenone, which are often used to generate PD-like phenotypes in animal models.[91,92] While the effects of these compounds in animal models may not mimic the pathogenic pathways in PD, these data provide further evidence for the potential benefits of upregulation of autophagy, particularly through the enhanced clearance of dysfunctional mitochondria.

Although accelerated degradation of alpha-synuclein through regulation of autophagy appears to be a promising form of treatment for PD, it is important to keep in mind what is currently known about PD pathogenesis and how different genetic factors may complicate therapeutic strategies. A number of studies have shown various points in the autophagy-lysosome pathway to be dysfunctional in cellular and animal models of PD. Mutations in the lysosomal ATPase ATP13A2 are associated with early-onset juvenile forms of PD.[93] Mutations in this gene have been shown to inhibit autophagosome-lysosome fusion causing accumulation of autophagosomes and undegraded cargo.[93,94] Similar to ESCRT-associated mutations in ALS, decreased autophagosome clearance in ATP13A2 mutated forms of PD may exacerbate neurodegeneration through increased accumulation of dysfunctional organelles and proteins.

LYSOSOMAL STORAGE DISORDER

Lysosomal storage disorders (LSD) are a group of neurodegenerative and myopathic diseases which are pathologically defined by the accumulation of undegraded intracellular cargo. This cargo accumulation is a result of complete or partial lysosomal dysfunction, due to the insufficient delivery or mutation of proteins necessary for lysosomal function. Since the lysosome is vital for degradation of cargo from a number of organelles including endosomes, MVBs, and autophagosomes, some LSDs can affect many pathways.

Lysosomal Dysfunction in LSDs

Since effective autophagy degradation depends on proper lysosomal function, autophagy has been found to be dysfunctional in many LSDs. Many of these studies observe an accumulation of autophagy substrates and an accumulation of LC3-II due to a block in autophagosome-lysosome fusion or incomplete degradation of LC3-II by lyosomal hydrolases [Mucolipidosis type IV (MLIV)]. Due to possible feedback loops, a number of LSDs including Pompe disease, have shown an additional upregulation of autophagosome synthesis in addition to the autophagosome-lysosome fusion block.[95]

In addition to autophagy defects, mitochondrial accumulation and dysfunction and increased apoptosis have been observed in some LSDs. Since autophagy regulates mitochondrial health through degradation of dysfunctional mitochondria via mitophagy and chemical and genetic inhibition of autophagy are able to recapitulate these mitochondrial phenotypes,[96] it is plausible that a significant portion of LSD pathology results from improper degradation of mitochondria by autophagy.

Autophagy Downregulation as a Therapeutic for LSDs?

To date, enzyme replacement therapy has provided the most promising option for future treatment of LSDs with the hope of replacing dysfunctional proteins and restoring lysosomal function. Unlike a number of the previously discussed disorders, many LSDs may not be good candidate diseases for primary treatment with autophagy-inducing drugs due to the downstream block in autophagosome-lysosome fusion.[97] As the blocks in autophagy flux in these diseases are not complete, it is unlikely that blocking autophagosome formation would be beneficial as it would lead to even greater flux impairment.

AGING AND AUTOPHAGY

Intracellular homeostasis is dependent upon a balance between catabolism and anabolism. The loss of normal function and regulation of these processes during aging can result in the build-up and or deficiency of vital proteins and organelles. Mounting evidence shows that both autophagy and the UPS become impaired during a normal aging process. Understanding how the kinetics of these pathways naturally change over time may be critical to our understanding of disease, the treatment of disease, and quality of life for the elderly.

Recent studies have focussed on the role of nutrient sensing pathways in life-span extension. Knockdown of key receptors or chemical inhibition of PKA,[98] PKB,[99] and TOR[100,101] in yeast, and the IGF-1[102] pathway in *C. elegans*, or caloric restriction in mice[103] all increase life-span and increase autophagy. The daf-2 gene in *C. elegans* encodes an insulin receptor-like protein that when mutated can double the life-span of the organism and increase their resistance to stress. A reduction in BEC-1 (Beclin 1 orthologue) in *C. elegans* inhibits the life-span extension affects of daf-2 mutants, and decreases the life-span of wildtype animals.[104]

Changes in autophagy flux affect the life-span of model organisms such as *C. elegans*[104] and *Drosophila.*[105] Knockout of autophagy genes in *C. elegans* and *Drosophila* accelerate aging and decrease life-span while induction of autophagy increases life-span and decreases mitochondrial accumulation in *Drosophila.*[106] Aging studies done in Atg8a (autophagy-related 8a, LC3 ortholog) mutant flies show that a reduction in autophagy reduces life-span while an increase in Atg8a levels extends life-span by 56%, increases resistance to oxidative stress, and increases polyubiquitinated protein formation.[106] Treatment of flies with rapamycin extends life-span and that this extension can increase longevity beyond the extension achieved with dietary restriction. These effects were shown to be mediated by both autophagy and translation-dependent effects of TORC1 downregulation,[107] providing further support for the role for autophagy in longevity and life-span extension. Despite the important findings in the above genetic model system, the validation of such a role for autophagy in mammalian models has not been reported thus far.

CONCLUSION

The chemical and genetic control of autophagy flux holds promise for the treatment of a range of neurodegenerative diseases. Future research will need to focus on identifying more specific regulators of autophagy and how to modulate this flux appropriately within those diseases associated with dysfunctional autophagy. In order to solve these issues we need to understand the disease mechanisms better.

ACKNOWLEDGEMENTS

The authors of this chapter are funded by the following organizations: Wellcome Trust (Principal Fellowship to DCR), the European Community's Seventh Framework Programme (FP7/2007–2013) under grant agreement number 241791 (MEFOPA), Wellcome Trust/MRC Strategic Award for Neurodegeneration (DCR), the NIHR Biomedical Research Centre at Addenbrooke's Hospital.

REFERENCES

1. R. R. Kopito and D. Ron, Conformational disease, *Nat Cell Biol* **2**: E207–E209 (2000).
2. M. Arrasate, S. Mitra, E. S. Schweitzer, M. R. Segal and S. Finkbeiner, Inclusion body formation reduces levels of mutant huntingtin and the risk of neuronal death, *Nature* **431**: 805–810 (2004).

3. B. Winner, R. Jappelli, S. K. Maji, P. A. Desplats, L. Boyer, S. Aigner, C. Hetzer, T. Loher, M. Vilar, S. Campioni, C. Tzitzilonis, A. Soragni, S. Jessberger, H. Mira, A. Consiglio, E. Pham, E. Masliah, F. H. Gage and R. Riek, *In vivo* demonstration that alpha-synuclein oligomers are toxic, *Proc Natl Acad Sci USA* **108**: 4194–4199 (2011).
4. B. Ravikumar, C. Vacher, Z. Berger, J. E. Davies, S. Luo, L. G. Oroz, F. Scaravilli, D. F. Easton, R. Duden, C. J. O'Kane and D. C. Rubinsztein, Inhibition of mTOR induces autophagy and reduces toxicity of polyglutamine expansions in fly and mouse models of Huntington disease, *Nat Genet* **36**: 585–595 (2004).
5. A. W. Dunah, H. Jeong, A. Griffin, Y. M. Kim, D. G. Standaert, S. M. Hersch, M. M. Mouradian, A. B. Young, N. Tanese and D. Krainc, Sp1 and TAFII130 transcriptional activity disrupted in early Huntington's disease, *Science* **296**: 2238–2243 (2002).
6. Z. Berger, B. Ravikumar, F. M. Menzies, L. G. Oroz, B. R. Underwood, M. N. Pangalos, I. Schmitt, U. Wullner, B. O. Evert, C. J. O'Kane and D. C. Rubinsztein, Rapamycin alleviates toxicity of different aggregate-prone proteins, *Hum Mol Genet* **15**: 433–442 (2006).
7. A. L. Goldberg, Protein degradation and protection against misfolded or damaged proteins, *Nature* **426**: 895–899 (2003).
8. P. K. Kim, D. W. Hailey, R. T. Mullen and J. Lippincott-Schwartz, Ubiquitin signals autophagic degradation of cytosolic proteins and peroxisomes, *Proc Natl Acad Sci USA* **105**: 20567–20574 (2008).
9. J. Heitman, N. R. Movva and M. N. Hall, Targets for cell cycle arrest by the immunosuppressant rapamycin in yeast, *Science* **253**: 905–909 (1991).
10. E. Jacinto, R. Loewith, A. Schmidt, S. Lin, M. A. Ruegg, A. Hall and M. N. Hall, Mammalian TOR complex 2 controls the actin cytoskeleton and is rapamycin insensitive, *Nat Cell Biol* **6**: 1122–1128 (2004).
11. D. A. Guertin and D. M. Sabatini, Defining the role of mTOR in cancer, *Cancer Cell* **12**: 9–22 (2007).
12. M. Laplante and D. M. Sabatini, mTOR signaling at a glance, *J Cell Sci* **122**: 3589–3594 (2009).
13. T. Noda and Y. Ohsumi, Tor, a phosphatidylinositol kinase homologue, controls autophagy in yeast, *J Biol Chem* **273**: 3963–3966 (1998).
14. W. J. Welch and C. R. Brown, Influence of molecular and chemical chaperones on protein folding, *Cell Stress Chaperones* **1**: 109–115 (1996).
15. S. Sarkar, J. E. Davies, Z. Huang, A. Tunnacliffe and D. C. Rubinsztein, Trehalose, a novel mTOR-independent autophagy enhancer, accelerates the clearance of mutant huntingtin and alpha-synuclein, *J Biol Chem* **282**: 5641–5652 (2007).
16. S. Sarkar, R. A. Floto, Z. Berger, S. Imarisio, A. Cordenier, M. Pasco, L. J. Cook and D. C. Rubinsztein, Lithium induces autophagy by inhibiting inositol monophosphatase, *J Cell Biol* **170**: 1101–1111 (2005).
17. A. Criollo, M. C. Maiuri, E. Tasdemir, I. Vitale, A. A. Fiebig, D. Andrews, J. Molgo, J. Diaz, S. Lavandero, F. Harper, G. Pierron, D. di Stefano, R. Rizzuto, G. Szabadkai

and G. Kroemer, Regulation of autophagy by the inositol trisphosphate receptor, *Cell Death Differ* **14**: 1029–1039 (2007).
18. A. Williams, S. Sarkar, P. Cuddon, E. K. Ttofi, S. Saiki, F. H. Siddiqi, L. Jahreiss, A. Fleming, D. Pask, P. Goldsmith, C. J. O'Kane, R. A. Floto and D. C. Rubinsztein, Novel targets for Huntington's disease in an mTOR-independent autophagy pathway, *Nat Chem Biol* **4**: 295–305 (2008).
19. S. Sarkar, E. O. Perlstein, S. Imarisio, S. Pineau, A. Cordenier, R. L. Maglathlin, J. A. Webster, T. A. Lewis, C. J. O'Kane, S. L. Schreiber and D. C. Rubinsztein, Small molecules enhance autophagy and reduce toxicity in Huntington's disease models, *Nat Chem Biol* **3**: 331–338 (2007).
20. R. H. Myers, Huntington's disease genetics, *NeuroRx* **1**: 255–262 (2004).
21. Anonymous, A novel gene containing a trinucleotide repeat that is expanded and unstable on Huntington's disease chromosomes. The Huntington's Disease Collaborative Research Group, *Cell* **72**: 971–983 (1993).
22. K. L. Sugars and D. C. Rubinsztein, Transcriptional abnormalities in Huntington disease, *Trends Genet* **19**: 233–238 (2003).
23. C. A. Ross, When more is less: pathogenesis of glutamine repeat neurodegenerative diseases, *Neuron* **15**: 493–496 (1995).
24. M. Martinez-Vicente, Z. Talloczy, E. Wong, G. Tang, H. Koga, S. Kaushik, R. de Vries, E. Arias, S. Harris, D. Sulzer and A. M. Cuervo, Cargo recognition failure is responsible for inefficient autophagy in Huntington's disease, *Nat Neurosci* **13**: 567–576 (2010).
25. D. C. Rubinsztein, The roles of intracellular protein-degradation pathways in neurodegeneration, *Nature* **443**: 780–786 (2006).
26. B. Ravikumar, R. Duden and D. C. Rubinsztein, Aggregate-prone proteins with polyglutamine and polyalanine expansions are degraded by autophagy, *Hum Mol Genet* **11**: 1107–1117 (2002).
27. S. Sarkar and D. C. Rubinsztein, Huntington's disease: degradation of mutant huntingtin by autophagy, *FEBS J* **275**: 4263–4270 (2008).
28. B. Ravikumar, A. Acevedo-Arozena, S. Imarisio, Z. Berger, C. Vacher, C. J. O'Kane, S. D. Brown and D. C. Rubinsztein, Dynein mutations impair autophagic clearance of aggregate-prone proteins, *Nat Genet* **37**: 771–776 (2005).
29. J. J. Bissler, F. X. McCormack, L. R. Young, J. M. Elwing, G. Chuck, J. M. Leonard, V. J. Schmithorst, T. Laor, A. S. Brody, J. Bean, S. Salisbury and D. N. Franz, Sirolimus for angiomyolipoma in tuberous sclerosis complex or lymphangioleiomyomatosis, *N Engl J Med* **358**: 140–151 (2008).
30. A. Yu and W. H. Frishman, Imidazoline receptor agonist drugs: a new approach to the treatment of systemic hypertension, *J Clin Pharmacol* **36**: 98–111 (1996).
31. C. Rose, F. M. Menzies, M. Renna, A. Acevedo-Arozena, S. Corrochano, O. Sadiq, S. D. Brown and D. C. Rubinsztein, Rilmenidine attenuates toxicity of polyglutamine expansions in a mouse model of Huntington's disease, *Hum Mol Genet* **19**: 2144–2153 (2010).

32. L. P. Rowland, Amyotrophic lateral sclerosis, *Curr Opin Neurol* **7**: 310–315 (1994).
33. F. Fornai, P. Longone, L. Cafaro, O. Kastsiuchenka, M. Ferrucci, M. L. Manca, G. Lazzeri, A. Spalloni, N. Bellio, P. Lenzi, N. Modugno, G. Siciliano, C. Isidoro, L. Murri, S. Ruggieri and A. Paparelli, Lithium delays progression of amyotrophic lateral sclerosis, *Proc Natl Acad Sci USA* **105**: 2052–2057 (2008).
34. P. G. Ince, J. R. Highley, J. Kirby, S. B. Wharton, H. Takahashi, M. J. Strong and P. J. Shaw, Molecular pathology and genetic advances in amyotrophic lateral sclerosis: an emerging molecular pathway and the significance of glial pathology, *Acta Neuropathol* **122**: 657–671 (2011).
35. L. Pasquali, P. Longone, C. Isidoro, S. Ruggieri, A. Paparelli and F. Fornai, Autophagy, lithium, and amyotrophic lateral sclerosis, *Muscle Nerve* **40**: 173–194 (2009).
36. M. Hafezparast, R. Klocke, C. Ruhrberg, A. Marquardt, A. Ahmad-Annuar, S. Bowen, G. Lalli, A. S. Witherden, H. Hummerich, S. Nicholson, P. J. Morgan, R. Oozageer, J. V. Priestley, S. Averill, V. R. King, S. Ball, J. Peters, T. Toda, A. Yamamoto, Y. Hiraoka, M. Augustin, D. Korthaus, S. Wattler, P. Wabnitz, C. Dickneite, S. Lampel, F. Boehme, G. Peraus, A. Popp, M. Rudelius, J. Schlegel, H. Fuchs, M. Hrabe de Angelis, G. Schiavo, D. T. Shima, A. P. Russ, G. Stumm, J.E. Martin and E. M. Fisher, Mutations in dynein link motor neuron degeneration to defects in retrograde transport, *Science* **300**: 808–812 (2003).
37. S. J. King and T. A. Schroer, Dynactin increases the processivity of the cytoplasmic dynein motor, *Nat Cell Biol* **2**: 20–24 (2000).
38. B. H. LaMonte, K. E. Wallace, B. A. Holloway, S. S. Shelly, J. Ascano, M. Tokito, T. Van Winkle, D. S. Howland and E. L. Holzbaur, Disruption of dynein/dynactin inhibits axonal transport in motor neurons causing late-onset progressive degeneration, *Neuron* **34**: 715–727 (2002).
39. I. Puls, C. Jonnakuty, B. H. LaMonte, E. L. Holzbaur, M. Tokito, E. Mann, M. K. Floeter, K. Bidus, D. Drayna, S. J. Oh, R. H. Brown, Jr., C. L. Ludlow and K. H. Fischbeck, Mutant dynactin in motor neuron disease, *Nat Genet* **33**: 455–456 (2003).
40. F. M. Laird, M. H. Farah, S. Ackerley, A. Hoke, N. Maragakis, J. D. Rothstein, J. Griffin, D. L. Price, L. J. Martin and P. C. Wong, Motor neuron disease occurring in a mutant dynactin mouse model is characterized by defects in vesicular trafficking, *J Neurosci* **28**: 1997–2005 (2008).
41. M. Filimonenko, S. Stuffers, C. Raiborg, A. Yamamoto, L. Malerod, E. M. Fisher, A. Isaacs, A. Brech, H. Stenmark and A. Simonsen, Functional multivesicular bodies are required for autophagic clearance of protein aggregates associated with neurodegenerative disease, *J Cell Biol* **179**: 485–500 (2007).
42. H. L. Feng, Y. Leng, C. H. Ma, J. Zhang, M. Ren and D. M. Chuang, Combined lithium and valproate treatment delays disease onset, reduces neurological deficits and prolongs survival in an amyotrophic lateral sclerosis mouse model, *Neuroscience* **155**: 567–572 (2008).

43. X. Wang, H. Fan, Z. Ying, B. Li, H. Wang and G. Wang, Degradation of TDP-43 and its pathogenic form by autophagy and the ubiquitin-proteasome system, *Neurosci Lett* **469**: 112–116 (2010).
44. S. P. Aggarwal, L. Zinman, E. Simpson, J. McKinley, K. E. Jackson, H. Pinto, P. Kaufman, R. A. Conwit, D. Schoenfeld, J. Shefner and M. Cudkowicz, Safety and efficacy of lithium in combination with riluzole for treatment of amyotrophic lateral sclerosis: a randomised, double-blind, placebo-controlled trial, *Lancet Neurol* **9**: 481–488 (2010).
45. S. Sarkar, R. A. Floto, Z. Berger, S. Imarisio, A. Cordenier, M. Pasco, L. J. Cook and D. C. Rubinsztein, Lithium induces autophagy by inhibiting inositol monophosphatase, *J Cell Biol* **170**: 1101–1111 (2005).
46. R. A. Nixon and A. M. Cataldo, Lysosomal system pathways: genes to neurodegeneration in Alzheimer's disease, *J Alzheimers Dis* **9**: 277–289 (2006).
47. J. F. Ma, Y. Huang, S. D. Chen and G. Halliday, Immunohistochemical evidence for macroautophagy in neurones and endothelial cells in Alzheimer's disease, *Neuropathol Appl Neurobiol* **36**: 312–319 (2010).
48. J. H. Lee, W. H. Yu, A. Kumar, S. Lee, P. S. Mohan, C. M. Peterhoff, D. M. Wolfe, M. Martinez-Vicente, A. C. Massey, G. Sovak, Y. Uchiyama, D. Westaway, A. M. Cuervo and R. A. Nixon, Lysosomal proteolysis and autophagy require presenilin 1 and are disrupted by Alzheimer-related PS1 mutations, *Cell* **141**: 1146–1158 (2010).
49. B. Boland, A. Kumar, S. Lee, F. M. Platt, J. Wegiel, W. H. Yu and R. A. Nixon, Autophagy induction and autophagosome clearance in neurons: relationship to autophagic pathology in Alzheimer's disease, *J Neurosci* **28**: 6926–6937 (2008).
50. F. Pickford, E. Masliah, M. Britschgi, K. Lucin, R. Narasimhan, P. A. Jaeger, S. Small, B. Spencer, E. Rockenstein, B. Levine and T. Wyss-Coray, The autophagy-related protein beclin 1 shows reduced expression in early Alzheimer disease and regulates amyloid beta accumulation in mice, *J Clin Invest* **118**: 2190–2199 (2008).
51. S. F. Funderburk, Q. J. Wang and Z. Yue, The Beclin 1-VPS34 complex — at the crossroads of autophagy and beyond, *Trends Cell Biol* **20**: 355–362 (2010).
52. P. Spilman, N. Podlutskaya, M. J. Hart, J. Debnath, O. Gorostiza, D. Bredesen, A. Richardson, R. Strong and V. Galvan, Inhibition of mTOR by rapamycin abolishes cognitive deficits and reduces amyloid-beta levels in a mouse model of Alzheimer's disease, *PLoS One* **5**: e9979 (2010).
53. A. Caccamo, S. Majumder, A. Richardson, R. Strong and S. Oddo, Molecular interplay between mammalian target of rapamycin (mTOR), amyloid-beta, and Tau: effects on cognitive impairments, *J Biol Chem* **285**: 13107–13120 (2010).
54. T. C. Gamblin, F. Chen, A. Zambrano, A. Abraha, S. Lagalwar, A. L. Guillozet, M. Lu, Y. Fu, F. Garcia-Sierra, N. LaPointe, R. Miller, R. W. Berry, L. I. Binder and V. L. Cryns, Caspase cleavage of tau: linking amyloid and neurofibrillary tangles in Alzheimer's disease, *Proc Natl Acad Sci USA* **100**: 10032–10037 (2003).

55. L. S. Yang and H. Ksiezak-Reding, Calpain-induced proteolysis of normal human tau and tau associated with paired helical filaments, *Eur J Biochem* **233**: 9–17 (1995).
56. J. P. Brion, D. P. Hanger, A. M. Couck and B. H. Anderton, A68 proteins in Alzheimer's disease are composed of several tau isoforms in a phosphorylated state which affects their electrophoretic mobilities, *Biochem J* **279**(Pt 3): 831–836 (1991).
57. J. P. Brion, D. P. Hanger, M. T. Bruce, A. M. Couck, J. Flament-Durand and B. H. Anderton, Tau in Alzheimer neurofibrillary tangles. N- and C-terminal regions are differentially associated with paired helical filaments and the location of a putative abnormal phosphorylation site, *Biochem J* **273**(Pt 1): 127–133 (1991).
58. T. Hamano, T. F. Gendron, E. Causevic, S. H. Yen, W. L. Lin, C. Isidoro, M. Deture and L. W. Ko, Autophagic-lysosomal perturbation enhances tau aggregation in transfectants with induced wild-type tau expression, *Eur J Neurosci* **27**: 1119–1130 (2008).
59. Z. Berger, E. K. Ttofi, C. H. Michel, M. Y. Pasco, S. Tenant, D. C. Rubinsztein and C. J. O'Kane, Lithium rescues toxicity of aggregate-prone proteins in Drosophila by perturbing Wnt pathway, *Hum Mol Genet* **14**: 3003–3011 (2005).
60. J. R. Munoz-Montano, F. J. Moreno, J. Avila and J. Diaz-Nido, Lithium inhibits Alzheimer's disease-like tau protein phosphorylation in neurons, *FEBS Lett* **411**: 183–188 (1997).
61. P. S. Klein and D. A. Melton, A molecular mechanism for the effect of lithium on development, *Proc Natl Acad Sci USA* **93**: 8455–8459 (1996).
62. K. Leroy, K. Ando, C. Heraud, Z. Yilmaz, M. Authelet, J. M. Boeynaems, L. Buee, R. De Decker and J. P. Brion, Lithium treatment arrests the development of neurofibrillary tangles in mutant tau transgenic mice with advanced neurofibrillary pathology, *J Alzheimers Dis* **19**: 705–719 (2010).
63. M. G. Spillantini, M. L. Schmidt, V. M. Lee, J. Q. Trojanowski, R. Jakes and M. Goedert, Alpha-synuclein in Lewy bodies, *Nature* **388**: 839–840 (1997).
64. W. R. Gibb and A. J. Lees, The relevance of the Lewy body to the pathogenesis of idiopathic Parkinson's disease, *J Neurol Neurosurg Psychiatry* **51**: 745–752 (1988).
65. A. B. Singleton, M. Farrer, J. Johnson, A. Singleton, S. Hague, J. Kachergus, M. Hulihan, T. Peuralinna, A. Dutra, R. Nussbaum, S. Lincoln, A. Crawley, M. Hanson, D. Maraganore, C. Adler, M. R. Cookson, M. Muenter, M. Baptista, D. Miller, J. Blancato, J. Hardy and K. Gwinn-Hardy, alpha-Synuclein locus triplication causes Parkinson's disease, *Science* **302**: 841 (2003).
66. M. C. Chartier-Harlin, J. Kachergus, C. Roumier, V. Mouroux, X. Douay, S. Lincoln, C. Levecque, L. Larvor, J. Andrieux, M. Hulihan, N. Waucquier, L. Defebvre, P. Amouyel, M. Farrer and A. Destee, Alpha-synuclein locus duplication as a cause of familial Parkinson's disease, *Lancet* **364**: 1167–1169 (2004).
67. O. A. Ross, A. T. Braithwaite, L. M. Skipper, J. Kachergus, M. M. Hulihan, F. A. Middleton, K. Nishioka, J. Fuchs, T. Gasser, D. M. Maraganore, C. H. Adler, L. Larvor, M. C. Chartier-Harlin, C. Nilsson, J. W. Langston, K. Gwinn, N. Hattori

and M. J. Farrer, Genomic investigation of alpha-synuclein multiplication and parkinsonism, *Ann Neurol* **63**: 743–750 (2008).
68. I. E. Clark, M. W. Dodson, C. Jiang, J. H. Cao, J. R. Huh, J. H. Seol, S. J. Yoo, B. A. Hay and M. Guo, Drosophila pink1 is required for mitochondrial function and interacts genetically with parkin, *Nature* **441**: 1162–1166 (2006).
69. F. Kamp, N. Exner, A. K. Lutz, N. Wender, J. Hegermann, B. Brunner, B. Nuscher, T. Bartels, A. Giese, K. Beyer, S. Eimer, K. F. Winklhofer and C. Haass, Inhibition of mitochondrial fusion by alpha-synuclein is rescued by PINK1, Parkin and DJ-1, *EMBO J* **29**: 3571–3589 (2010).
70. J. Park, S. B. Lee, S. Lee, Y. Kim, S. Song, S. Kim, E. Bae, J. Kim, M. Shong, J. M. Kim and J. Chung, Mitochondrial dysfunction in Drosophila PINK1 mutants is complemented by parkin, *Nature* **441**: 1157–1161 (2006).
71. S. Vasseur, S. Afzal, J. Tardivel-Lacombe, D. S. Park, J. L. Iovanna and T. W. Mak, DJ-1/PARK7 is an important mediator of hypoxia-induced cellular responses, *Proc Natl Acad Sci USA* **106**: 1111–1116 (2009).
72. I. Irrcher, H. Aleyasin, E. L. Seifert, S. J. Hewitt, S. Chhabra, M. Phillips, A. K. Lutz, M. W. Rousseaux, L. Bevilacqua, A. Jahani-Asl, S. Callaghan, J. G. MacLaurin, K. F. Winklhofer, P. Rizzu, P. Rippstein, R. H. Kim, C. X. Chen, E. A. Fon, R. S. Slack, M. E. Harper, H. M. McBride, T. W. Mak and D. S. Park, Loss of the Parkinson's disease-linked gene DJ-1 perturbs mitochondrial dynamics, *Hum Mol Genet* **19**: 3734–3746 (2010).
73. L. Zhang, M. Shimoji, B. Thomas, D. J. Moore, S. W. Yu, N. I. Marupudi, R. Torp, I. A. Torgner, O. P. Ottersen, T. M. Dawson and V. L. Dawson, Mitochondrial localization of the Parkinson's disease related protein DJ-1: implications for pathogenesis, *Hum Mol Genet* **14**: 2063–2073 (2005).
74. V. Loeb, E. Yakunin, A. Saada and R. Sharon, The transgenic overexpression of alpha-synuclein and not its related pathology associates with complex I inhibition, *J Biol Chem* **285**: 7334–7343 (2010).
75. H. Shimura, N. Hattori, S. Kubo, Y. Mizuno, S. Asakawa, S. Minoshima, N. Shimizu, K. Iwai, T. Chiba, K. Tanaka and T. Suzuki, Familial Parkinson disease gene product, parkin, is a ubiquitin-protein ligase, *Nat Genet* **25**: 302–305 (2000).
76. A. D'Andrea and D. Pellman, Deubiquitinating enzymes: a new class of biological regulators, *Crit Rev Biochem Mol Biol* **33**: 337–352 (1998).
77. E. Leroy, R. Boyer, G. Auburger, B. Leube, G. Ulm, E. Mezey, G. Harta, M. J. Brownstein, S. Jonnalagada, T. Chernova, A. Dehejia, C. Lavedan, T. Gasser, P. J. Steinbach, K. D. Wilkinson and M. H. Polymeropoulos, The ubiquitin pathway in Parkinson's disease, *Nature* **395**: 451–452 (1998).
78. H. Snyder, K. Mensah, C. Theisler, J. Lee, A. Matouschek and B. Wolozin, Aggregated and monomeric alpha-synuclein bind to the S6' proteasomal protein and inhibit proteasomal function, *J Biol Chem* **278**: 11753–11759 (2003).
79. Y. Zhang, J. Gao, K. K. Chung, H. Huang, V. L. Dawson and T. M. Dawson, Parkin functions as an E2-dependent ubiquitin-protein ligase and promotes the degradation

of the synaptic vesicle-associated protein, CDCrel-1, *Proc Natl Acad Sci USA* **97**: 13354–13359 (2000).
80. K. K. Chung, Y. Zhang, K. L. Lim, Y. Tanaka, H. Huang, J. Gao, C. A. Ross, V. L. Dawson and T. M. Dawson, Parkin ubiquitinates the alpha-synuclein-interacting protein, synphilin-1: implications for Lewy-body formation in Parkinson disease, *Nat Med* **7**: 1144–1150 (2001).
81. L. Crews, B. Spencer, P. Desplats, C. Patrick, A. Paulino, E. Rockenstein, L. Hansen, A. Adame, D. Galasko and E. Masliah, Selective molecular alterations in the autophagy pathway in patients with Lewy body disease and in models of alpha-synucleinopathy, *PLoS One* **5**: e9313 (2010).
82. A. R. Winslow, C. W. Chen, S. Corrochano, A. Acevedo-Arozena, D. E. Gordon, A. A. Peden, M. Lichtenberg, F. M. Menzies, B. Ravikumar, S. Imarisio, S. Brown, C. J. O'Kane and D. C. Rubinsztein, alpha-Synuclein impairs macroautophagy: implications for Parkinson's disease, *J Cell Biol* **190**: 1023–1037 (2010).
83. G. Krebiehl, S. Ruckerbauer, L. F. Burbulla, N. Kieper, B. Maurer, J. Waak, H. Wolburg, Z. Gizatullina, F. N. Gellerich, D. Woitalla, O. Riess, P. J. Kahle, T. Proikas-Cezanne and R. Kruger, Reduced basal autophagy and impaired mitochondrial dynamics due to loss of Parkinson's disease-associated protein DJ-1, *PLoS One* **5**: e9367 (2010).
84. J. Alegre-Abarrategui, H. Christian, M. M. Lufino, R. Mutihac, L. L. Venda, O. Ansorge and R. Wade-Martins, LRRK2 regulates autophagic activity and localizes to specific membrane microdomains in a novel human genomic reporter cellular model, *Hum Mol Genet* **18**: 4022–4034 (2009).
85. E. D. Plowey, S. J. Cherra, 3rd, Y. J. Liu and C. T. Chu, Role of autophagy in G2019S-LRRK2-associated neurite shortening in differentiated SH-SY5Y cells, *J Neurochem* **105**: 1048–1056 (2008).
86. M. C. Bennett, J. F. Bishop, Y. Leng, P. B. Chock, T. N. Chase and M. M. Mouradian, Degradation of alpha-synuclein by proteasome, *J Biol Chem* **274**: 33855–33858 (1999).
87. J. L. Webb, B. Ravikumar, J. Atkins, J. N. Skepper and D. C. Rubinsztein, alpha-Synuclein is degraded by both autophagy and the proteasome, *J Biol Chem* **278**: 25009–25013 (2003).
88. H. J. Lee, F. Khoshaghideh, S. Patel and S. J. Lee, Clearance of alpha-synuclein oligomeric intermediates via the lysosomal degradation pathway, *J Neurosci* **24**: 1888–1896 (2004).
89. A. M. Cuervo, L. Stefanis, R. Fredenburg, P. T. Lansbury and D. Sulzer, Impaired degradation of mutant alpha-synuclein by chaperone-mediated autophagy, *Science* **305**: 1292–1295 (2004).
90. B. Spencer, R. Potkar, M. Trejo, E. Rockenstein, C. Patrick, R. Gindi, A. Adame, T. Wyss-Coray and E. Masliah, Beclin 1 gene transfer activates autophagy and ameliorates the neurodegenerative pathology in alpha-synuclein models of Parkinson's and Lewy body diseases, *J Neurosci* **29**: 13578–13588 (2009).

91. T. Pan, P. Rawal, Y. Wu, W. Xie, J. Jankovic and W. Le, Rapamycin protects against rotenone-induced apoptosis through autophagy induction, *Neuroscience* **164**: 541–551 (2009).
92. B. Ravikumar, Z. Berger, C. Vacher, C. J. O'Kane and D. C. Rubinsztein, Rapamycin pre-treatment protects against apoptosis, *Hum Mol Genet* **15**: 1209–1216 (2006).
93. A. Ramirez, A. Heimbach, J. Grundemann, B. Stiller, D. Hampshire, L. P. Cid, I. Goebel, A. F. Mubaidin, A. L. Wriekat, J. Roeper, A. Al-Din, A. M. Hillmer, M. Karsak, B. Liss, C. G. Woods, M. I. Behrens and C. Kubisch, Hereditary parkinsonism with dementia is caused by mutations in ATP13A2, encoding a lysosomal type 5 P-type ATPase, *Nat Genet* **38**: 1184–1191 (2006).
94. A. Di Fonzo, H. F. Chien, M. Socal, S. Giraudo, C. Tassorelli, G. Iliceto, G. Fabbrini, R. Marconi, E. Fincati, G. Abbruzzese, P. Marini, F. Squitieri, M. W. Horstink, P. Montagna, A. D. Libera, F. Stocchi, S. Goldwurm, J. J. Ferreira, G. Meco, E. Martignoni, L. Lopiano, L. B. Jardim, B. A. Oostra, E. R. Barbosa and V. Bonifati, ATP13A2 missense mutations in juvenile parkinsonism and young onset Parkinson disease, *Neurology* **68**: 1557–1562 (2007).
95. N. Raben, V. Hill, L. Shea, S. Takikita, R. Baum, N. Mizushima, E. Ralston and P. Plotz, Suppression of autophagy in skeletal muscle uncovers the accumulation of ubiquitinated proteins and their potential role in muscle damage in Pompe disease, *Hum Mol Genet* **17**: 3897–3908 (2008).
96. J. J. Jennings, Jr., J. H. Zhu, Y. Rbaibi, X. Luo, C. T. Chu and K. Kiselyov, Mitochondrial aberrations in mucolipidosis Type IV, *J Biol Chem* **281**: 39041–39050 (2006).
97. N. Raben, C. Schreiner, R. Baum, S. Takikita, S. Xu, T. Xie, R. Myerowitz, M. Komatsu, J. H. Van Der Meulen, K. Nagaraju, E. Ralston and P. H. Plotz, Suppression of autophagy permits successful enzyme replacement therapy in a lysosomal storage disorder-murine Pompe disease, *Autophagy* **6**: 1078–1089 (2010).
98. J. M. Thevelein, L. Cauwenberg, S. Colombo, J. H. De Winde, M. Donation, F. Dumortier, L. Kraakman, K. Lemaire, P. Ma, D. Nauwelaers, F. Rolland, A. Teunissen, P. Van Dijck, M. Versele, S. Wera and J. Winderickx, Nutrient-induced signal transduction through the protein kinase A pathway and its role in the control of metabolism, stress resistance, and growth in yeast, *Enzyme Microb Technol* **26**: 819–825 (2000).
99. M. Wei, P. Fabrizio, J. Hu, H. Ge, C. Cheng, L. Li and V. D. Longo, Life span extension by calorie restriction depends on Rim15 and transcription factors downstream of Ras/PKA, Tor, and Sch9, *PLoS Genet* **4**: e13 (2008).
100. N. D. Bonawitz, M. Chatenay-Lapointe, Y. Pan and G. S. Shadel, Reduced TOR signaling extends chronological life span via increased respiration and upregulation of mitochondrial gene expression, *Cell Metab* **5**: 265–277 (2007).
101. M. Kaeberlein, R. W. Powers, 3rd, K. K. Steffen, E. A. Westman, D. Hu, N. Dang, E. O. Kerr, K. T. Kirkland, S. Fields and B. K. Kennedy, Regulation of yeast replicative life span by TOR and Sch9 in response to nutrients, *Science* **310**: 1193–1196 (2005).

102. R. Y. Lee, J. Hench and G. Ruvkun, Regulation of *C. elegans* DAF-16 and its human ortholog FKHRL1 by the daf-2 insulin-like signaling pathway, *Curr Biol* **11**: 1950–1957 (2001).
103. R. Weindruch, P. H. Naylor, A. L. Goldstein and R. L. Walford, Influences of aging and dietary restriction on serum thymosin alpha 1 levels in mice, *J Gerontol* **43**: B40–B42 (1988).
104. A. Melendez, Z. Talloczy, M. Seaman, E. L. Eskelinen, D. H. Hall and B. Levine, Autophagy genes are essential for dauer development and life-span extension in *C. elegans*, *Science* **301**: 1387–1391 (2003).
105. G. Juhasz, B. Erdi, M. Sass and T. P. Neufeld, Atg7-dependent autophagy promotes neuronal health, stress tolerance, and longevity but is dispensable for metamorphosis in Drosophila, *Genes Dev* **21**: 3061–3066 (2007).
106. A. Simonsen, R. C. Cumming, A. Brech, P. Isakson, D. R. Schubert and K. D. Finley, Promoting basal levels of autophagy in the nervous system enhances longevity and oxidant resistance in adult Drosophila, *Autophagy* **4**: 176–184 (2008).
107. I. Bjedov, J. M. Toivonen, F. Kerr, C. Slack, J. Jacobson, A. Foley and L. Partridge, Mechanisms of life span extension by rapamycin in the fruit fly *Drosophila melanogaster*, *Cell Metab* **11**: 35–46 (2010).

CHAPTER 14

LYSOSOME STORAGE DISORDERS ON THE BRAIN: THE AUTOPHAGY LYSOSOME PATHWAY CONTRIBUTES TO DISEASE PATHOPHYSIOLOGY AND MAY BE UTILIZED FOR THERAPEUTIC BENEFIT

John J. Shacka

ABSTRACT

Lysosome storage disorders (LSDs) are defined by genetic mutations that either directly or indirectly compromise lysosome function, many of which exhibit documented CNS pathophysiology. As lysosome dysfunction is a principal contributing factor to the onset and progression of LSDs, it is not surprising to learn of several reports documenting inherent abnormalities in the autophagy-lysosome pathway (ALP), as the ALP requires itntact lysosome function for its proper clearance of substrates and homeostatic maintenance of energy balance. In addition, accumulating evidence suggests that alterations in ALP signaling contribute to the onset and progression of adult onset neurodegenerative disease. Thus the study of LSD mutations and their effects on the ALP has aided in our understanding of ALP-associated contributions to diseases such as Alzheimer's and Parkinson's. Herein we will review LSDs with known CNS pathophysiology and will discuss the often heterogeneous ALP abnormalities amongst these diseases. LSD-specific therapies will also be discussed and how the ALP may be a direct or indirect target of such therapies. Finally, the function of the ALP in LSDs will be discussed relative to its putative role in adult onset neurodegenerative disease.

INTRODUCTION

Nowhere in human disease is the importance of intact lysosome function more apparent than in a group of over 40 diseases known collectively as lysosome storage disorders (LSDs). LSDs by definition involve a gene mutation (typically a single gene but not always) that results in the primary dysfunction (partial or complete) of a lysosomal protein, or may result in secondary lysosome dysfunction if, for instance the mutation alters the proper delivery of substrate to the lysosome. Mutations may affect enzymes (e.g. lysosomal hydrolases) or non-enzymatic proteins (e.g. membrane-associated proteins without known hydrolase activity). While their pathology is quite heterogeneous, all LSDs exhibit a pronounced accumulation of undigested storage material that is secondary to the mutation-derived dysfunction. In addition, many LSDs exhibit dramatic neuropathology, which highlights the exquisite sensitivity of neurons to targeted defects in lysosome function and the continued need to develop therapies for LSDs that adequately and effectively target the CNS.

The autophagy-lysosome pathway (ALP) is a principal intracellular substrate degradation pathway and studies using human tissue from LSD patients or experimental models of LSDs highlight the need for intact lysosome function to maintain proper ALP function. Together, defects in lysosome function along with resultant induction of autophagic stress may very well contribute to the induction of cell death, and models of ALP-associated cell death in LSDs have been postulated. LSDs typically manifest during childhood and are often fatal. However, real progress has been made in the development of therapeutics for the treatment of LSDs, many of which are a direct or indirect result of our knowledge of proper ALP function and its dysfunction in LSDs. Furthermore, the study of the ALP in LSDs has also enhanced our understanding of age-related neurodegenerative diseases including Alzheimer's, Parkinson's and Huntington's, where ALP dysfunction is implicated in familial onset disease (via mutations that directly or indirectly affect ALP function) as well as in idiopathic disease.

The goals of this review are to provide an overview of LSDs and their known or hypothesized effects on the ALP, with a specific focus on LSDs with documented CNS pathology. LSD-specific therapies will also be discussed and how the ALP may be a direct or indirect target of such therapies. Finally, the function of the ALP in LSDs will be discussed relative to its putative role in adult onset neurodegenerative disease.

LSDs WITH KNOWN CNS PATHOLOGY

Many LSDs are considered primary CNS diseases and exhibit dramatic neurological and neuropathological features. However there also exist several LSDs with profound effects on peripheral organs (e.g. kidney, heart, muscle) that also have reported disruption of CNS function or associated neuropathology. Listed in Table 1 are LSDs with known effects on the CNS, grouped according to the general class of protein that is mutated, the known mutations and resulting generated storage product.

Lipidoses result from mutations that affect lipid metabolism and are a predominant type of LSDs affecting the CNS. Of the lipidoses, ***neuronal ceroid lipofuscinoses (NCL)*** are the most prevalent pediatric neurodegenerative disease with an incidence rate of one in 12,500 children (reviewed in Refs. 1 to 3). NCL mutations are autosomal recessive and produce either partial or total loss of function of lysosomal hydrolases or lysosomal membrane proteins (see Table 1). Such mutations result in the lysosomal accumulation of undigested, autofluorescent lipoprotein. NCL are further classified by their relative age of onset (congenital, infantile, late infantile, juvenile and the rare adult form) and in general, the earlier onset forms (e.g. congenital CD deficiency and infantile forms) have the most severe clinical presentation and often the poorest prognosis.

Gaucher disease (GD) is an inherited, autosomal recessive disorder that results from mutations in the lysosomal enzyme glucocerebrocidase (GCase) and results in the abnormal storage of glucosylceramide (GC) and glucosylsphingosine (GS) (reviewed in Refs. 4 and 5). Saposin C is an enzyme that has been shown to enhance the normal activity of GCase, and mutations in the saposin C domain of the prosaposin gene have also been identified in a small variant subset of GD cases.[4] GD is classified into two categories, non-neuronopathic and neuronopathic.[4] Of the neuronopathic types, type II (acute) is the rarer form with onset in the first two to three months of life and is characterized by extensive and progressive brain damage, eye movement problems, motor dysfunction, seizures and a poor ability to swallow, with death usually occurring by two years of age. Type III (subacute) has similar features but exhibits a much more chronic progression, with affected patients often surviving into their teenage years or into early adulthood.

Fabry disease is a rare, X-linked LSD caused by a deficiency in the lysosomal enzyme α-galactosidase A, and results in the accumulation of globotriaosylceramide (Gb_3) (reviewed in Ref. 6). While Fabry patients exhibit prevalent renal dysfunction and cardiomyopathy, neurological dysfunction has been documented

Table 1. Lysosomal storage disorders with known mutations and CNS pathophysiology.

Classification	Disease	Gene mutation	Localization	Function	Storage product(s)	Ref(s).
Lipidoses	NCL/Batten	*Congenital: cathepsin D*	Lysosome/ endosome	Bulk substrate hydrolysis	Subunit c of mitochondrial ATPase	1–3
		Infantile (CLN1): palmitoyl protein thioesterase-1 (PPT-1)	Lysosome	Hydrolysis of thioester-linked fatty acyl groups from cysteine residues	Sphingolipid activator proteins (saposins) A&D	1, 2
		Late infantile (CLN2): tripeptidyl peptidase-1 (TPP-1)	Lysosome	Hydrolysis of N-terminal tripeptides from substrates	Subunit c of mitochondrial ATPase	1, 2
		Late infantile variant: CLN6	ER	Transmembrane protein; function unclear	Subunit c of mitochondrial ATPase	1, 2
		Juvenile: CLN3 protein (CLN3P)	Lysosome/ endosome membrane	Unclear	Subunit c of mitochondrial ATPase	1, 2
	Gaucher	*Glucocerebrosidase (β-glucosidase)*	Lysosome	Hydrolysis of glucocerebrosides	Glucosylceramide; glucosylsphingosine	1, 2
	Fabry	*α-galactosidase A*	Lysosome	Hydrolysis of terminal α-galactosyl moieties from glycolipids and glycoproteins	Globotrialosylceramide (Gb_3)	4, 5

(*Continued*)

Table 1. *(Continued)*

Classification	Disease	Gene mutation	Localization	Function	Storage product(s)	Ref(s).
	Niemann-Pick C	*Npc1*	Endo-lysosomal	Binds intracellular cholesterol and regulates its transport to lysosomes	Unesterified cholesterol and glycolipids	6, 7
		Npc2	Endo-lysosomal	Secreted intracellular cholesterol binding protein	Unesterified cholesterol and glycolipids	8, 9
	GM1 gangliosidosis	*β-galactosidase*	Lysosome	Hydrolysis of β-galactosides to monosaccharides	GM1 gangliosides	8, 9
	GM2 gangliosidosis (Sandhoff disease)	*Hexosaminidase A&B*	Lysosome	Hydrolysis of terminal N-acetyl-D-hexosamine residues in N-acetyl-β-D-hexosaminides	GM2 gangliosides	10
	Multiple sulfatase deficiency (Austin disease)	*Sulfatase modifying factor 1 (SUMF)*	ER	Regulates post-translational modification and in turn activation of sulfatases	Glycosaminoglycans (GAGs)	11
Mucopolysaccharidosis (MPS)	MPS III (Sanfilippo syndrome)	*MPS IIIA: sulfamidase*	Lysosome	Hydrolysis of sulfates from glycosaminoglycans (GAGs)	Glycosaminoglycans (GAGs)	12

(Continued)

Table 1. (*Continued*)

Classification	Disease	Gene mutation	Localization	Function	Storage product(s)	Ref(s).
		MPS IIIB: N-acetylglucosaminidase	Lysosome	Hydrolysis of sulfates from glycosaminoglycans (GAGs)	Glycosaminoglycans (GAGs)	13, 14
Glycoproteinosis	Mucolipidosis (ML)	*ML-II: UDP-N-acetylglucosamine-lysosomal-enzyme N-acetylglucosamine-phosphotransferase (GlcNAc phosphotransferase)*	Golgi	Phosphorylates mannose to mannose-6-phosphate	Lipoprotein	16
		ML-IV: mucolipin 1 (TRP-ML1)	Lysosome membrane	Ion channel important for extrusion of divalent cations	Lipoprotein	16, 20
Glycogen storage	Pompe disease	*Acid α-glucosidase (GAA)*	Lysosome	Hydrolysis of glycogen to glucose	Glycogen	22, 23
	Danon disease	*LAMP-2*	Lysosome/autophagic vacuole membrane	Lysosome-autophagic vacuole fusion	Glycogen	24–26

Note: When possible, review articles were cited as references due to compliance with limits on reference number.

and includes an increased risk for stroke (due most likely to cerebrovascular complications), neuropathic pain, vertigo and tinnitus.[7]

Niemann-Pick C types 1 and 2 (Npc1/2) are autosomal recessive diseases for cholesterol binding proteins; Npc1 is a transmembrane-spanning late endosomal/lysosomal protein and Npc2 is a secreted protein (reviewed in Ref. 8). Mutations in both Npc1 and Npc2 are considered "secondary" lysosomal disorders, in that their resultant increase in unesterified cholesterol is predicted to alter endosomal/lysosomal membrane integrity as well as trafficking defects that result in the downstream accumulation of undegraded sphingolipids. Npc1 and Npc2 mutations induce progressive ataxia with dystonia and dementia that typically present in early childhood, with death likely by the early teens.[9]

GM1 gangliosidosis is an autosomal recessive disorder caused by a mutation in the lysosomal enzyme β-galactosidase, which leads to the accumulation of GM1 gangliosides primarily in neurons (reviewed in Ref. 10). There are three types of GM1 gangliosidosis that produce progressive neurodegeneration and are defined by age of onset and severity (early infantile vs. late infantile vs. adult) and produce progressive neurodegeneration with pathophysiology including motor dysfunction, mental retardation, seizures, blindness and deafness (reviewed in Ref. 11). ***Sandhoff disease*** is a rare autosomal recessive lipid storage disorder caused by mutations and resultant deficiency in the lysosomal hydrolases hexosaminidase A and B. Sandhoff disease is a representative ***GM2 gangliosidosis*** and results in the progressive accumulation of GM2 gangliosides in neuronal cells with concomitant progressive degeneration of the CNS. As with GM1 gangliosidosis, there are three types (infantile vs. juvenile vs. adult) with the earlier onset forms exhibiting the most severe pathophysiology, including motor dysfunction/paralysis, mental retardation, seizures and vision/hearing loss (reviewed in Ref. 11).

Multiple sulfatase deficiency (MSD), also known as Austin disease, is a very rare autosomal recessive disease with a mutation in sulfatase modifying factor 1 (SUMF1), which regulates the post-translational modification responsible for activating lysosomal sulfatases, and thus its deficiency results in the downstream inactivation of multiple sulfatases (reviewed in Ref. 12). Sulfatases are normally responsible for the breakdown of lipids and mucopolycaccharides (otherwise known as glycosaminoglycans; GAGs) within lysosomes. Accumulation of GAGs and lipids induces profound neurodegeneration including mental retardation, motor dysfunction and hearing/vision loss (reviewed in Ref. 12). Because MSD induces GAG storage, it is considered similar to that of mucopolysaccharodosis (MPS), which cause defects in enzymes that break down GAGs.

Mucopolysaccharidosis (MPS) are autosomal recessive disorders and represent the subclass of LSDs caused by defects in lysosomal enzymes required to hydrolyze GAGs. There are several types of MPS, many exhibiting varying degrees of CNS pathophysiology including mental retardation, motor dysfunction, and loss of sensory organ (ears/eyes) function (reviewed in Ref. 13). Of the known types of MPS, MPS III and MPS VI have documented alterations in the ALP.[14,15] MPS III (Sanfilippo syndrome) is characterized by the aberrant accumulation of heparan sulfate as a result of deficiency in the lysosomal enzymes sulfamidase (***MPS IIIA***) or lysosomal N-acetylglucosaminidase (***MPS IIIB***).[14] ***MPS VI*** (Maroteaux-Lamy syndrome) results from a deficiency in the lysosomal enzyme N-acetylgalactosamine-4-sulfatase (also known as arylsulfatase B) with subsequent storage of dermatan sulfate, but does not exhibit any known CNS pathophysiology.[15]

Glycoproteinosis are a subclass of LSDs that affect glycoprotein-associated post-translational modification of lysosomal enzymes or glycoprotein hydrolysis. ***Mucolipidosis (ML)*** comprise autosomal recessive diseases characterized by the abnormal storage of lipids and lipoproteins. At least one form (MLIV) is associated with profound CNS deficits including mental and motor retardation as well as retinal degeneration (reviewed in Ref. 16) whereas both ML-II and ML-IV exhibit defects in the ALP.[17–19] ***ML-II*** is caused by a deficiency in the enzyme UDP-N-acetylglucosamine-lysosomal-enzyme N-acetylglucosamine-phosphotransferase (GlcNAc phosphotransferase), which resides in the Golgi and ordinarily phosphorylates mannose to mannose 6-phosphate. Without mannose 6-phosphate, glycoprotein hydrolases are not properly targeted to the lysosome but rather are transported from the Golgi to the extracellular space. Intracellular lipoproteins ordinarily hydrolyzed by functional lysosomes thus form inclusions. ***ML-IV*** is caused by a mutation in the lysosomal/endosomal ion channel mucolipin 1 protein (TRP-ML1), which is important for the extrusion of divalent cations including iron[20] and possibly zinc[21] from endosomes and lysosomes, and contributes directly to the intralysosomal accumulation of undigested lipoprotein.

Glycogen storage diseases result from the inefficient breakdown of glycogen that is most prevalent in the muscle of affected patients, but CNS pathophysiology has also been reported.

Pompe disease is an autosomal recessive LSD caused by an accumulation of glycogen in the lysosome resulting from deficiency of the lysosomal hydrolase acid α-glucosidase (GAA).[22] The most severe form of Pompe is infantile onset with complete GAA deficiency, where affected children usually die by age 2 from cardiac failure but also exhibit profound neuronal and skeletal muscle damage.[23]

Danon disease results from mutations in the lysosomal membrane protein Lamp II and resultant loss of its transmembrane domain, which impairs lysosome trafficking and promotes vacuolar accumulation of glycogen (reviewed in

Ref. 24). Cardiomyopathy, vacuolar myopathy and mental retardation are common pathophysiological findings in Danon patients although CNS deficits are typically not found in female patients.[25,26]

HOW IS THE AUTOPHAGY-LYSOSOME PATHWAY ALTERED IN LSDs?

Over the last ten years great progress has been made in not only confirming alterations in the ALP in LSDs, but also in delineating the manner of its disruption, as well as determining that ALP dysfunction contributes to the regulation of cell death in LSDs. As all LSDs by definition are intimately associated with some manner of lysosome dysfunction, it is easy to imagine a direct functional consequence of LSDs on the ALP. However, it is still quite impressive how such heterogeneous mutations and resultant storage products can induce very similar effects on ALP function.

Although all of the LSDs reviewed in this chapter have documented CNS pathology, much of what we know about ALP dysfunction in these LSDs has resulted from the study of peripheral tissues and/or fibroblasts derived from human patients. As such the relative character and importance of ALP alterations may not be the same as in post-mitotic neurons affected by persistent lysosome dysfunction. Regardless, important information relating to ALP dysfunction has been gleaned from studying non-neuronal aspects of LSDs, and as such these findings will be discussed as part of this review, in addition to any documented, neuronal-specific disruptions in ALP function. In this section we will discuss evidence supporting effects of LSD mutations on ALP completion vs. induction and how this may be in part mutation specific.

Inhibition of Autophagy Completion is a Common Feature of LSDs

Disruption of the ALP in LSDs is in large part a consequence of lysosome dysfunction and is characterized by the accumulation of a heterogeneous population of autophagic vacuoles (AVs) including autophagosomes (immature AVs that have not yet fused with lysosomes) and autolysosomes (mature, "degradative" AVs that have already fused with lysosomes).[27] Lysosome dysfunction is believed to induce accumulation of AVs in one of two ways: (1) a decrease in effective fusion of autophagosomes with lysosomes, which may lead to an increase in autophagosomes; or (2) a block in the degradative capacity of lysosomes leading to an increase in autolysosomes.[27,28] Biochemically, AV accumulation is easily identified by an increase in levels of microtubule-associated protein 1 light chain

3-II (LC3-II), a 14 kDa microtubule-associated protein that upon its cleavage and lipidation is inserted into the membrane of autophagosomes.[29] Both autophagosomes and autolysosomes are positive for LC3-II, although typically to a lesser extent in autolysosomes as they exhibit a more acidic pH responsible for effective LC3-II turnover.[30]

Assays have also been developed to measure the manner in which AVs accumulate, including co-localization of LC3-positive autophagosomes with lysosomes (either via Lamp or Lysotracker Red; LTR) using fluorescence microscopy, a simple assessment of autophagosome-lysosome fusion, or assessment of autophagic flux via Western blot analysis of LC3-II levels following treatment with lysosomal inhibitors.[31] The majority of experimental models for LSDs have used a combination of the above methods to report alterations in autophagosome-lysosome fusion or autophagic flux.[6,14,15,18,24,32–37] Non-specific consequences of autophagy compromise include the accumulation of ubiquitin and p62/SQSTM1 (p62) as their levels may also be affected by proteasomal and transcriptional activation; although such markers are useful complements to assays that directly assess autophagy completion. An increase in ubiquitin is associated with the accumulation of insoluble toxic protein species that are ordinarily cleared by the ALP.[38] The protein p62 is a scaffold protein that binds both LC3 and polyubiquitin, is itself an autophagy substrate and has been shown to increase following the inhibition of autophagy completion.[39] Increases in ubiquitin, ubiquitin-positive inclusions and p62 are reported in many experimental models of LSDs and in fibroblasts obtained from LSD patients.[5,6,14,15,18,19,40–44] Together, these published findings indicate an impairment of autophagy completion that is a common component of most LSDs.

Some studies also report heterogeneous effects of LSD mutations on autophagy completion. For instance, fibroblasts cultured from MPS VI patients deficient in lysosomal arylsulfatase do not exhibit alterations in LAMP2-LC3 co-localization although their lysosomes exhibit a decreased ability to clear exogenous fluorescent-tagged EGF along with increased p62.[15] In addition, autophagosome-lysosome fusion (as monitored by LC3-Lamp2 and LC3-LTR co-localization) is similar in ML II/III fibroblasts deficient in the non-lysosomal GlcNac phosphotransferase, although its deficiency appears to compromise the proper targeting of cathepsins to lysosomes that may in turn decrease the degradation of p62 and ubiquitin aggregates.[19] Fibroblasts cultured from humans harboring mutations in the extralysosomal *NPC1* gene, which lead to increased levels of intracellular cholesterol and indirectly affect lysosome function, also exhibit intact autophagosome-lysosome fusion[45] yet exhibit increased ubiquitin[42] and a decrease in the clearance of APP C-terminal fragments[35] in brains of Npc1-deficient mice, again suggesting that certain LSD mutations may not result in a noticeable disruption in autophagy completion.

Autophagy Induction in LSDs

In addition to their ability to inhibit AV completion, many LSD mutations result in an enhanced ability of affected cells to induce macroautophagy. Autophagy induction is commonly measured by an increase in LC3-II following serum starvation or pharmacological induction (e.g. rapamycin). Stress-induced autophagy induction is considered to be greater than control if there is a further increase in LC3-II following the pharmacological inhibition of lysosomal enzymes or autophagosome-lysosome fusion that is greater than that observed with these treatments alone.[31,46] Another direct method of measuring autophagy induction is through assessment of long-lived protein degradation.[31] Studies of LSD mutations utilizing the above methods indicate that autophagy induction remains intact in LSD mutant cells, and in some cases the degree of autophagy induction in mutant cells is greater than that of non-mutant controls, despite a mutation-specific partial impairment in autophagosome-lysosome fusion or degradative flux.[17,18,40,45,47] Indirect evidence of autophagy induction has also been observed in many studies of LSD mutations in association with increased levels of LC3-II and includes upregulation of Beclin1.[15,18,48,49] important for regulating lipid kinase signaling of autophagy induction through its interaction with Class III PI3K (Vps34), and upregulation of Atg7 as shown in Pompe (GAA-deficient) mice,[49] which is important for autophagosome maturation. A reduction in phosphorylation of p70S6K has also been reported with Cln3 deficiency,[50] and represents indirect evidence implicating mTOR as a negative regulator of autophagy induction. However, other NCL mutations do not exhibit alterations in mTOR-dependent phosphorylation, such as those conducted in Npc1-deficient vs. wild-type fibroblasts,[45] or rather exhibit enhancement of mTOR and p70S6K phosphorylation as in GM1-gangliosidosis-deficient vs. wild-type brain and cultured astrocytes,[10] suggesting in these cases that autophagy induction in LSDs may also occur in a "Beclin-dependent, mTOR-independent" manner.

It is not entirely clear why autophagy is induced in LSDs but several possibilities have been proposed. Based on results obtained using experimental models of Npc,[45] it was suggested that autophagy induction is a compensatory mechanism for alterations in lipid trafficking, serving a role in promoting survival early in disease but inducing cell death later. Along these lines, it has been shown that lipid storage products, such as ceramide and sphingosine-1, directly induce autophagy, and have also been shown themselves to regulate cell death.[51,52]

Induction of autophagy may also be a compensatory response to a partial block in autophagosome-lysosome fusion as has been shown in experimental models of ML-IV[18,20] and Gaucher disease.[40] It has also been shown recently that lysosome dysfunction disrupts an mTOR negative feedback loop that in turn prolongs the activation of constitutive autophagy.[53] It has been hypothesized that this combined effect of persistent autophagy induction and inhibition of autophagy completion would induce a massive accumulation of AVs and in turn more severe neurotoxicity.[54] Evidence supporting a deleterious role for autophagy induction in the setting of lysosome inhibition is supported by evidence that Atg7 knockdown (and resultant inhibition of autophagy induction) attenuates apoptosis induced by the disruption of lysosome function.[55] Specific to LSDs, experimental Atg7 deficiency in a mouse model of Pompe disease was also shown to decrease the accumulation of autophagosomes in muscle and glycogen storage.[33,49] Further studies are necessary to determine if the induction of neuronal cell death due to LSD mutations is regulated by such combined imbalances in autophagy completion and induction and if inhibiting autophagy induction (to be discussed below) is a viable therapeutic option.

ALP-TARGETED THERAPIES FOR LSDs

There are several therapeutic approaches for LSDs that utilize direct or indirect manipulation of the ALP (see Table 2). Both enzyme replacement therapy (ERT) or substrate reduction therapy (SRT) are viable treatment options for several LSDs. Results of ERT have been published recently that indicate an attenuation of autophagic stress (and in some cases the attenuation of cell death) associated with LSDs. Renal biopsies obtained from Fabry patients subjected to ERT via treatment with agalsidase alfa exhibited a decrease in vacuoles, p62 and p62-ubiquitin co-localization.[6] In a mouse model of MPS VI (which as a disease exhibits neither signs of autophagic stress nor accumulation of dermatan sulfate in the brain), AAV-mediated overexpression of lysosomal arylsulfatase B resulted in a decrease in LC3-II, p62 and ubiquitin in mouse liver, spleen and kidney combined with a decreased storage of dermatan sulfate.[15] In an extension of studies focused on cathepsin D (CD) deficiency (a model of NCL/Batten Disease), overexpression of wild-type CD attenuated the death of neuroblastoma cells and *C. elegans* dopamine neurons that also overexpressed human alpha synuclein (ASYN), which is known to increase following CD deficiency,[43] although the protective effects of CD overexpression were not exactly a form of ERT as they were performed in wild-type and not CD-deficient neuronal cells.

Table 2. Putative ALP-associated therapies for LSDs with known CNS pathology or for closely-related age-related neurodegenerative disease.

Classification	Disease	Therapy	Therapeutic benefit	Comments	Ref(s).
Enzyme replacement therapy (ERT)	Fabry	Agalsidase alfa	Decrease in AV accumulation Decrease in p62 and ubiquitin-p62 co-localization	Not yet validated in neuronal tissue	6
	NCL/Parkinson's disease	Cathepsin D overexpression in neuroblastoma cells and *C. elegans*	Decreased neuronal cell death following overexpression of wild-type ASYN	Need to validate benefit from replacement of CD function in CD-deficient cells	32
	NCL	CLN3P overexpression	Decrease in stimulus-induced neuronal cell death	Need to validate benefit from replacement of CLN3P function in CLN3P-deficient cells	51
Substrate reduction therapy (SRT)	Pompe disease	Conditional Atg7 deficiency in skeletal muscle of *Gaa*-deficient mice	Potent decrease in AV accumulation along with 50%–60% decrease in glycogen storage	Not yet validated in neuronal tissue	42
	Npc1	Allopregnanolone	Reduces esterified cholesterol and accumulation of AVs and lysosomal cathepsins in Npc1-deficient mice		50

(*Continued*)

Table 2. (*Continued*)

Classification	Disease	Therapy	Therapeutic benefit	Comments	Ref(s).
	MLII/III	Inhibition of autophagy in GlcNac-phosphotransferase-deficient fibroblasts with 3MA	Temporary restoration of mitochondrial function	Requires validation in neuronal cells; long-term inhibition of autophagy may also be detrimental to mitochondrial function	49
ERT + SRT	Pompe disease	Human acid α-glucosidase (GAA) plus Atg7 deficiency in skeletal muscle of *Gaa*-deficient mice	Decrease in AV accumulation and glycogen storage that approached WT control levels more so than SRT alone	Not yet validated in neuronal tissue	42
Lysosome biogenesis	LSDs/age-related neurodegenerative disease	TFEB-regulated transactivation of targets that promote lysosome biogenesis and/or function	Maintenance of lysosome and autophagy function	May only work for LSD mutations with indirect effects on lysosome function	52, 53

The constitutive production of human GAA in the *Gaa*-deficient mouse model of Pompe disease allowed for clearance of glycogen from cardiac/type I (slow-twitch) skeletal muscle fibers, but not type II fibers that exhibit more extensive accumulation of AVs.[56] Perhaps since type II fibers have documented, preferential defects in trafficking and processing of enzymes along the endocytic pathway[56] combined with AV accumulation, this may make ERT more problematic in getting recombinant GAA to the lysosome to aid in the clearance of glycogen in type II skeletal muscle. While ERT alone has proven ineffective in decreasing AV accumulation and glycogen storage in Pompe (*Gaa*-deficient) mice, SRT with a focus on manipulating autophagy induction has been shown to provide therapeutic benefit. Conditional Atg7 deficiency in skeletal muscle of *Gaa*-deficient mice exhibited a potent decrease in AV accumulation and a resultant 50%–60% decrease in decreased glycogen storage.[49] When skeletal muscle Atg7 deficiency was combined with ERT of human GAA, glycogen levels in *Gaa*-deficient mice approached levels seen in wild-type controls.[49]

Other studies have manipulated autophagy induction for therapeutic benefit in LSDs. For instance, in a mouse model of gangliosidosis, α-gal-deficient astrocytes were more sensitive to oxidative stress-induced death compared to wild-type controls, an effect that was attenuated upon treatment with the class III PI3K inhibitor 3-methyladenine (3-MA), a potent inhibitor of autophagy induction.[10] This report suggests that autophagy induction leads to greater mitochondrial dysfunction as a result of α-gal deficiency, but however does not explain the exact cause of this phenomenon. In a study of ML II/III, where mitochondrial function was shown to be severely affected by deficiency of lysosomal GlcNac-phosphotransferase, treatment of affected fibroblasts with 3-MA resulted in a temporary restoration of mitochondrial function.[57] Together these findings indicate that autophagy induction may deplete pools of functional mitochondria as the mutation precludes their effective recycling. It has also been shown that the targeted removal of an upstream storage product that induces downstream lysosomal dysfunction may have therapeutic benefit in association with a decrease in autophagic stress, as brains of Npc1-deficient mice treated with allopregnanolone, which reduces unesterified cholesterol, exhibited treatment-specific decreases in AV accumulation and levels of lysosomal cathepsins.[58]

However, evidence also exists supporting a "protective" role of autophagy induction in LSDs. For instance, 3-MA was found to be more effective at reducing viability of *Cln3*-deficient vs. wild-type cerebellar granule neurons and [50,51] which may be related to an accumulation of ceramide specific to *Cln3*-deficient cells,[51] clearance of which is mediated in part via autophagy induction.[52] However it is also known that in addition to its ability to inhibit class III PI3-K, 3-MA can effectively inhibit the class I (pro-survival) PI3-K-Akt pathway.[59] Thus further investigation would be needed to confirm the effects of 3-MA specific for blocking autophagy induction under conditions of Cln3 deficiency, as well as investigating a more direct inhibition of autophagy induction via Atg gene knockdown studies. The importance of the Cln3 protein (CLN3P) in regulating cell survival in LSDs was also indicated recently by a significant attenuation of stimulus-induced cell death in neuroblastoma cells following its overexpression.[60] While the ability of ERT- or SRT-driven therapies to attenuate CNS pathophysiology has not yet been assessed for many LSDs, it is reasonable to predict that such study may further reveal therapeutic benefit for neuronal populations in a manner that involves manipulation of the ALP.

Recent evidence has identified the "CLEAR" (Coordinated Lysosomal Expression and Regulation) network of genes, which involves the coordinated transactivation of several gene targets directly related to lysosome composition and function. Transcription factor EB (TFEB) acts as a master regulator of transcriptional activation for these lysosome-associated gene targets, and the nuclear translocation of TFEB has been shown to be induced following an enhanced lysosomal storage of undegraded material.[61,62] Thus it is possible that the therapy of certain LSD mutations, in particular those that indirectly compromise lysosome function may one day benefit from a "CLEAR"-targeted enhancement of lysosome biogenesis that in turn enhances lysosome function. It was also shown recently that overexpression of TFEB regulates neuronal cell death following treatment with the dopaminergic neurotoxin MPP+,[63] suggesting that age-related neurodegenerative diseases such as Parkinson's may also benefit from a therapeutic strategy that promotes lysosome biogenesis and function.

LSD MUTATIONS AND THEIR RELEVANCE TO ADULT ONSET NEURODEGENERATIVE DISEASE

The study of LSD-specific mutations has generated a wealth of important information relating to the role of the lysosome and autophagy in the etiology and progression of many adult onset neurodegenerative diseases. Reported age-related declines in autophagy and lysosome function (reviewed in Refs. 64 and 65) may produce effects similar to those observed in LSDs, but in a less accelerated manner due to their sporadic, idiopathic nature with no definitive genetic link. The CNS pathophysiology of LSDs is also similar in many ways to that of adult onset neurodegenerative disease. For instance, NCL mutations are reported to generate axonal dystrophy,[37,66,67] altered synaptic transmission,[36,68] a robust brain inflammatory response,[12,69] alterations in brain cytoskeletal and trafficking proteins,[70,71] ubiquitin-rich inclusions,[18,43] dysfunctional mitochondria[14] and progressive neuron loss and death.[8,12,43,60,69,72–74]

Pathological hallmarks of neurodegenerative disease include protein species that under normal circumstances are cleared by the ALP, but with persistent compromise of the ALP accumulate as insoluble aggregates or inclusions that are hypothesized to contribute to disease onset and/or progression. Recent publications of cathepsin D deficiency (a model of congenital NCL/Batten disease) are examples of how mutation of a specific lysosomal hydrolase can contribute to the accumulation of a putative toxic protein species, in this case ASYN, a pathological hallmark of Parkinson's disease and other Lewy Body diseases.[43,44] Cathepsin D was identified recently as the principal enzyme mediating ASYN degradation[75] and its overexpression attenuates ASYN aggregation and DA neuronal cell death.[43] Other recent studies of cathepsin D deficiency indicate its relative importance for neuronal survival in age-related neurodegenerative disease. For instance, cathepsin D deficiency leads to an increase in neurotoxic forms of the cytoskeletal protein tau that are relevant to Alzheimer's disease pathogenesis.[71] Such c-terminal tau cleavage products are formed putatively via activated caspase present in cathepsin D-deficient neuronal cells.[71,72,74] Cathepsin D has also been shown to play a pivotal role in the ability of chaperone-mediated autophagy to effectively clear the transcription factor myocyte enhancing factor 2D (MEF2D), which is ordinarily survival promoting but when its degradation is inhibited it enhances accumulation of ASYN and neuronal cell death relevant to Parkinson's disease.[76]

Recent evidence has also indicated a strong link between Parkinson's disease and rare pathogenic variants of GBA, the gene that is mutated in Gaucher disease,[77] even though the overwhelming majority of Gaucher disease carriers and patients do not develop parkinsonism.[78] Gaucher disease mouse models have been shown recently to exhibit progressive, ASYN- and ubiquitin-positive aggregates in the CNS in association with accumulation of glucosylceramide,[79] and glucosylceramide has been shown to be localized to ASYN-positive inclusions in Parkinson's disease.[80] As the ALP is known to be important for ASYN degradation,[81,82] it is hopeful that the study of Gaucher disease mutations will shed further light on the mechanisms of ASYN accumulation and toxicity in Parkinson's disease.

Studies of other LSD mutations have further revealed the importance of intact lysosomal function for the clearance of proteins associated with age-related neurodegenerative disease. The effective metabolism of APP was shown recently to be critically regulated by intact lysosome function, as brain tissue and neurons derived from mouse models of Npc1, GM1 gangliosidosis and Sandhoff's disease all exhibit an enhanced accumulation of endogenous amyloid precursor protein C-terminal fragments.[35] However, deficiency of cathepsin D was shown to have little effect on the processing of APP, thus precluding the importance of lysosomal aspartic acid proteolytic activity in regulating secretase-induced cleavage of APP,[83] although steady-state lysosomal turnover of APP was not directly assessed in this study. In a study of MSD, the degradation of mutant ASYN or mutant huntingtin (leading to expanded CAG repeats characteristic of Huntington's disease) was further compromised upon their overexpression in cells obtained from MSD-deficient mice.[14]

CONCLUSIONS

LSD mutations invoke a striking heterogeneity of effects on the ALP which may be related in part to the heterogeneous manner in which they disrupt lysosome function. It is clear that a disruption in the ALP, either as a result of lysosome dysfunction or as a result of aberrant autophagy induction, contributes in a disease-specific manner to the pathophysiology of LSDs. In addition, evidence suggests that targeting the ALP may be a viable therapeutic approach for LSDs, and may actually provide additive protection when used in combination with other therapies such as ERT, again further demonstrating the intimate relationship of the ALP to LSD mutations. Future studies are certainly warranted that more specifically address the role of the ALP in nervous system tissue affected by LSD mutations, as well as the utility of targeting the ALP as a therapeutic approach to promote neuronal survival.

ACKNOWLEDGMENTS

I wish to thank Rhonda Carr for expert assistance in preparation of this manuscript. Dr. Shacka is supported by a VA Merit Award 1 I01 BX000957-01.

REFERENCES

1. J. Ezaki and E. Kominami, The intracellular location and function of proteins of neuronal ceroid lipofuscinoses, *Brain Pathol* **14**: 77–85 (2004).
2. J. J. Shacka and K. A. Roth, Cathepsin deficiency as a model for neuronal ceroid lipofuscinoses, *Am J Pathol* **167**: 1473–1476 (2005).
3. J. J. Shacka and K. A. Roth, Cathepsin D deficiency and NCL/Batten disease: there's more to death than apoptosis, *Autophagy* **3**: 474–476 (2007).

4. Y. Sun, B. Liou, H. Ran, M. R. Skelton, M. T. Williams, C. V. Vorhees, K. Kitatani, Y. A. Hannun, D. P. Witte, Y. H. Xu and G. A. Grabowski, Neuronopathic Gaucher disease in the mouse: viable combined selective saposin C deficiency and mutant glucocerebrosidase (V394L) mice with glucosylsphingosine and glucosylceramide accumulation and progressive neurological deficits, *Hum Mol Genet* **19**: 1088–1097 (2010).
5. Y. Sun and G. A. Grabowski, Impaired autophagosomes and lysosomes in neuronopathic Gaucher disease, *Autophagy* **6**: 648–649 (2010).
6. M. Chevrier, N. Brakch, C. Lesueur, D. Genty, Y. Ramdani, S. Moll, M. Djavaheri-Mergny, C. Brasse-Lagnel, A. Laquerriere, F. Barbey and S. Bekri, Autophagosome maturation is impaired in Fabry disease, *Autophagy* **6**: 589–599 (2010).
7. J. R. Berger, D. A. Dillon, B. A. Young, S. J. Goldstein and P. Nelson, Cystinosis of the brain and spinal cord with associated vasculopathy, *J Neurol Sci* **284**: 182–185 (2009).
8. D. C. Ko, L. Milenkovic, S. M. Beier, H. Manuel, J. Buchanan and M. P. Scott, Cell-autonomous death of cerebellar purkinje neurons with autophagy in Niemann-Pick type C disease, *PLoS Genet* **1**: 81–95 (2005).
9. J. J. Higgins, M. C. Patterson, J. M. Dambrosia, A. T. Pikus, P. G. Pentchev, S. Sato, R. O. Brady and N. W. Barton, A clinical staging classification for type C Niemann-Pick disease, *Neurology* **42**: 2286–2290 (1992).
10. A. Takamura, K. Higaki, K. Kajimaki, S. Otsuka, H. Ninomiya, J. Matsuda, K. Ohno, Y. Suzuki and E. Nanba, Enhanced autophagy and mitochondrial aberrations in murine G(M1)-gangliosidosis, *Biochem Biophys Res Commun* **367**: 616–622 (2008).
11. T. Kolter and K. Sandhoff, Sphingolipid metabolism diseases, *Biochim Biophys Acta* **1758**: 2057–2079 (2006).
12. C. Settembre, I. Annunziata, C. Spampanato, D. Zarcone, G. Cobellis, E. Nusco, E. Zito, C. Tacchetti, M. P. Cosma and A. Ballabio, Systemic inflammation and neurodegeneration in a mouse model of multiple sulfatase deficiency, *Proc Natl Acad Sci USA* **104**: 4506–4511 (2007).
13. M. J. Valstar, H. T. Bruggenwirth, R. Olmer, R. A. Wevers, F. W. Verheijen, B. J. Poorthuis, D. J. Halley and F. A. Wijburg, Mucopolysaccharidosis type IIIB may predominantly present with an attenuated clinical phenotype, *J Inherit Metab Dis* **33**: 759–767 (2010).
14. C. Settembre, A. Fraldi, L. Jahreiss, C. Spampanato, C. Venturi, D. Medina, R. de Pablo, C. Tacchetti, D. C. Rubinsztein and A. Ballabio, A block of autophagy in lysosomal storage disorders, *Hum Mol Genet* **17**: 119–129 (2008).
15. A. Tessitore, M. Pirozzi and A. Auricchio, Abnormal autophagy, ubiquitination, inflammation and apoptosis are dependent upon lysosomal storage and are useful biomarkers of mucopolysaccharidosis VI, *Pathogenetics* **2**: 4 (2009).
16. G. Altarescu, M. Sun, D. F. Moore, J. A. Smith, E. A. Wiggs, B. I. Solomon, N. J. Patronas, K. P. Frei, S. Gupta, C. R. Kaneski, O. W. Quarrell, S. A. Slaugenhaupt, E. Goldin and R. Schiffmann, The neurogenetics of mucolipidosis type IV, *Neurology* **59**: 306–313 (2002).

17. J. J. Jennings, Jr., J. H. Zhu, Y. Rbaibi, X. Luo, C. T. Chu and K. Kiselyov, Mitochondrial aberrations in mucolipidosis Type IV, *J Biol Chem* **281**: 39041–39050 (2006).
18. S. Vergarajauregui, P. S. Connelly, M. P. Daniels and R. Puertollano, Autophagic dysfunction in mucolipidosis type IV patients, *Hum Mol Genet* **17**: 2723–2737 (2008).
19. T. Otomo, K. Higaki, E. Nanba, K. Ozono and N. Sakai, Inhibition of autophagosome formation restores mitochondrial function in mucolipidosis II and III skin fibroblasts, *Mol Genet Metab* **98**: 393–399 (2009).
20. X. P. Dong, X. Cheng, E. Mills, M. Delling, F. Wang, T. Kurz and H. Xu, The type IV mucolipidosis-associated protein TRPML1 is an endolysosomal iron release channel, *Nature* **455**: 992–996 (2008).
21. J. L. Eichelsdoerfer, J. A. Evans, S. A. Slaugenhaupt and M. P. Cuajungco, Zinc dyshomeostasis is linked with the loss of mucolipidosis IV-associated TRPML1 ion channel, *J Biol Chem* **285**: 34304–34308 (2010).
22. T. Fukuda, A. Roberts, M. Ahearn, K. Zaal, E. Ralston, P. H. Plotz and N. Raben, Autophagy and lysosomes in Pompe disease, *Autophagy* **2**: 318–320 (2006).
23. L. R. DeRuisseau, D. D. Fuller, K. Qiu, K. C. DeRuisseau, W. H. Donnelly, Jr., C. Mah, P. J. Reier and B. J. Byrne, Neural deficits contribute to respiratory insufficiency in Pompe disease, *Proc Natl Acad Sci USA* **106**: 9419–9424 (2009).
24. P. Saftig, W. Beertsen and E. L. Eskelinen, LAMP-2: a control step for phagosome and autophagosome maturation, *Autophagy* **4**: 510–512 (2008).
25. Y. K. Bui, P. Renella, J. A. Martinez-Agosto, A. Verity, A. Madikians and J. C. Alejos, Danon disease with typical early-onset cardiomyopathy in a male: focus on a novel LAMP-2 mutation, *Pediatr Transplant* **12**: 246–250 (2008).
26. H. Kim, A. Cho, B. C. Lim, M. J. Kim, K. J. Kim, I. Nishino, Y. S. Hwang and J. H. Chae, A 13-year-old girl with proximal weakness and hypertrophic cardiomyopathy with Danon disease, *Muscle Nerve* **41**: 879–882 (2010).
27. E. L. Eskelinen and P. Saftig, Autophagy: a lysosomal degradation pathway with a central role in health and disease, *Biochim Biophys Acta* **1793**: 664–673 (2009).
28. R. A. Nixon, D. S. Yang and J. H. Lee, Neurodegenerative lysosomal disorders — a continuum from development to late age, *Autophagy* **4**: 590–599 (2008).
29. Y. Kabeya, N. Mizushima, T. Ueno, A. Yamamoto, T. Kirisako, T. Noda, E. Kominami, Y. Ohsumi and T. Yoshimori, LC3, a mammalian homologue of yeast Apg8p, is localized in autophagosome membranes after processing, *EMBO J* **19**: 5720–5728 (2000).
30. H. M. Ni, A. Bockus, A. L. Wozniak, K. Jones, S. Weinman, X. M. Yin and W. X. Ding, Dissecting the dynamic turnover of GFP-LC3 in the autolysosome, *Autophagy* **7**: 188–204 (2011).
31. D. J. Klionsky, H. Abeliovich, P. Agostinis, D. K. Agrawal, *et al.*, Guidelines for the use and interpretation of assays for monitoring autophagy in higher eukaryotes, *Autophagy* **4**: 151–175 (2007).

32. T. Fukuda, L. Ewan, M. Bauer, R. J. Mattaliano, K. Zaal, E. Ralston, P. H. Plotz and N. Raben, Dysfunction of endocytic and autophagic pathways in a lysosomal storage disease, *Ann Neurol* **59**: 700–708 (2006).
33. N. Raben, V. Hill, L. Shea, S. Takikita, R. Baum, N. Mizushima, E. Ralston and P. Plotz, Suppression of autophagy in skeletal muscle uncovers the accumulation of ubiquitinated proteins and their potential role in muscle damage in Pompe disease, *Hum Mol Genet* **17**: 3897–3908 (2008).
34. A. Fidzianska, E. Walczak and M. Walski, Abnormal chaperone-mediated autophagy (CMA) in cardiomyocytes of a boy with Danon disease, *Folia Neuropathol* **45**: 133–139 (2007).
35. B. Boland, D. A. Smith, D. Mooney, S. S. Jung, D. M. Walsh and F. M. Platt, Macroautophagy is not directly involved in the metabolism of amyloid precursor protein, *J Biol Chem* **285**: 37415–37428 (2010).
36. M. Koike, H. Nakanishi, P. Saftig, J. Ezaki, K. Isahara, Y. Ohsawa, W. Schulz-Schaeffer, T. Watanabe, S. Waguri, S. Kametaka, M. Shibata, K. Yamamoto, E. Kominami, C. Peters, K. von Figura and Y. Uchiyama, Cathepsin D deficiency induces lysosomal storage with ceroid lipofuscin in mouse CNS neurons, *J Neurosci* **20**: 6898–6906 (2000).
37. M. Koike, M. Shibata, S. Waguri, K. Yoshimura, I. Tanida, E. Kominami, T. Gotow, C. Peters, K. von Figura, N. Mizushima, P. Saftig and Y. Uchiyama, Participation of autophagy in storage of lysosomes in neurons from mouse models of neuronal ceroid-lipofuscinoses (Batten disease), *Am J Pathol* **167**: 1713–1728 (2005).
38. T. Hara, K. Nakamura, M. Matsui, A. Yamamoto, Y. Nakahara, R. Suzuki-Migishima, M. Yokoyama, K. Mishima, I. Saito, H. Okano and N. Mizushima, Suppression of basal autophagy in neural cells causes neurodegenerative disease in mice, *Nature* **441**: 885–889 (2006).
39. G. Bjørkøy, T. Lamark, A. Brech, H. Outzen, M. Perander, A. Øvervatn, H. Stenmark and T. Johansen, p62/SQSTM1 forms protein aggregates degraded by autophagy and has a protective effect on huntingtin-induced cell death, *J Cell Biol* **171**: 603–614 (2005).
40. A. M. Vaccaro, M. Motta, M. Tatti, S. Scarpa, L. Masuelli, M. Bhat, M. T. Vanier, A. Tylki-Szymanska and R. Salvioli, Saposin C mutations in Gaucher disease patients resulting in lysosomal lipid accumulation, saposin C deficiency, but normal prosaposin processing and sorting, *Hum Mol Genet* **19**: 2987–2997 (2010).
41. S. Vergarajauregui and R. Puertollano, Mucolipidosis type IV: the importance of functional lysosomes for efficient autophagy, *Autophagy* **4**: 832–834 (2008).
42. G. Liao, Y. Yao, J. Liu, Z. Yu, S. Cheung, A. Xie, X. Liang and X. Bi, Cholesterol accumulation is associated with lysosomal dysfunction and autophagic stress in Npc1 -/- mouse brain, *Am J Pathol* **171**: 962–975 (2007).
43. L. Qiao, S. Hamamichi, K. A. Caldwell, G. A. Caldwell, T. A. Yacoubian, S. Wilson, Z. L. Xie, L. D. Speake, R. Parks, D. Crabtree, Q. Liang, S. Crimmins, L. Schneider, Y. Uchiyama, T. Iwatsubo, Y. Zhou, L. Peng, Y. Lu, D. G. Standaert, K. C. Walls,

J. J. Shacka, K. A. Roth and J. Zhang, Lysosomal enzyme cathepsin D protects against alpha-synuclein aggregation and toxicity, *Mol Brain* **1**: 1–18 (2008).

44. V. Cullen, M. Lindfors, J. Ng, A. Paetau, E. Swinton, P. Kolodziej, H. Boston, P. Saftig, J. Woulfe, M. B. Feany, L. Myllykangas, M. G. Schlossmacher and J. Tyynela, Cathepsin D expression level affects alpha-synuclein processing, aggregation, and toxicity *in vivo*, *Mol Brain* **2**: 1–17 (2009).
45. C. D. Pacheco, R. Kunkel and A. P. Lieberman, Autophagy in Niemann-Pick C disease is dependent upon Beclin-1 and responsive to lipid trafficking defects, *Hum Mol Genet* **16**: 1495–1503 (2007).
46. N. Mizushima and T. Yoshimori, How to interpret LC3 immunoblotting, *Autophagy* **3**: 542–545 (2007).
47. C. Settembre, A. Fraldi, L. Jahreiss, C. Spampanato, C. Venturi, D. Medina, R. de Pablo, C. Tacchetti, D. C. Rubinsztein and A. Ballabio, A block of autophagy in lysosomal storage disorders, *Hum Mol Genet* **17**: 119–129 (2007).
48. X. S. Xie, D. Padron, X. Liao, J. Wang, M. G. Roth and J. K. de Brabander, Salicylihalamide A inhibits the V0 sector of the V-ATPase through a mechanism distinct from bafilomycin A1, *J Biol Chem* **279**: 19755–19763 (2004).
49. N. Raben, C. Schreiner, R. Baum, S. Takikita, S. Xu, T. Xie, R. Myerowitz, M. Komatsu, J. H. Van Der Meulen, K. Nagaraju, E. Ralston and P. H. Plotz, Suppression of autophagy permits successful enzyme replacement therapy in a lysosomal storage disorder-murine Pompe disease, *Autophagy* **6**: 1078–1089 (2010).
50. Y. Cao, J. A. Espinola, E. Fossale, A. C. Massey, A. M. Cuervo, M. E. MacDonald and S. L. Cotman, Autophagy is disrupted in a knock-in mouse model of juvenile neuronal ceroid lipofuscinosis, *J Biol Chem* **281**: 20483–20493 (2006).
51. D. A. Persaud-Sawin and R. M. Boustany, Cell death pathways in juvenile Batten disease, *Apoptosis* **10**: 973–985 (2005).
52. S. Lepine, J. C. Allegood, M. Park, P. Dent, S. Milstien and S. Spiegel, Sphingosine-1-phosphate phosphohydrolase-1 regulates ER stress-induced autophagy, *Cell Death Differ* **18**: 350–361 (2010).
53. L. Yu, C. K. McPhee, L. Zheng, G. A. Mardones, Y. Rong, J. Peng, N. Mi, Y. Zhao, Z. Liu, F. Wan, D. W. Hailey, V. Oorschot, J. Klumperman, E. H. Baehrecke and M. J. Lenardo, Termination of autophagy and reformation of lysosomes regulated by mTOR, *Nature* **465**: 942–946 (2010).
54. Z. Yue, L. Friedman, M. Komatsu and K. Tanaka, The cellular pathways of neuronal autophagy and their implication in neurodegenerative diseases, *Biochim Biophys Acta* **1793**: 1496–1507 (2009).
55. K. C. Walls, A. P. Ghosh, A. V. Franklin, B. J. Klocke, M. Ballestas, J. J. Shacka, J. Zhang and K. A. Roth, Lysosome dysfunction triggers Atg7-dependent neural apoptosis, *J Biol Chem* **285**: 10497–10507 (2010).
56. T. Fukuda, M. Ahearn, A. Roberts, R. J. Mattaliano, K. Zaal, E. Ralston, P. H. Plotz and N. Raben, Autophagy and mistargeting of therapeutic enzyme in skeletal muscle in Pompe disease, *Mol Ther* **14**: 831–839 (2006).

57. H. J. Gundersen, P. Bagger, T. F. Bendtsen, S. M. Evans, *et al.*, The new stereological tools: disector, fractionator, nucleator and point sampled intercepts and their use in pathological research and diagnosis, *APMIS* **96**: 857–881 (1988).
58. G. Liao, S. Cheung, J. Galeano, A. X. Ji, Q. Qin and X. Bi, Allopregnanolone treatment delays cholesterol accumulation and reduces autophagic/lysosomal dysfunction and inflammation in Npc1-/- mouse brain, *Brain Res* **1270**: 140–151 (2009).
59. Y. T. Wu, H. L. Tan, G. Shui, C. Bauvy, Q. Huang, M. R. Wenk, C. N. Ong, P. Codogno and H. M. Shen, Dual role of 3-methyladenine in modulation of autophagy via different temporal patterns of inhibition on class I and III phosphoinositide 3-kinase, *J Biol Chem* **285**: 10850–10861 (2010).
60. S. B. Narayan, D. Rakheja, J. V. Pastor, K. Rosenblatt, S. R. Greene, J. Yang, B. A. Wolf and M. J. Bennett, Over-expression of CLN3P, the Batten disease protein, inhibits PANDER-induced apoptosis in neuroblastoma cells: Further evidence that CLN3P has anti-apoptotic properties, *Mol Genet Metab* **88**: 178–183 (2006).
61. M. Sardiello, M. Palmieri, A. di Ronza, D. L. Medina, M. Valenza, V. A. Gennarino, C. Di Malta, F. Donaudy, V. Embrione, R. S. Polishchuk, S. Banfi, G. Parenti, E. Cattaneo and A. Ballabio, A gene network regulating lysosomal biogenesis and function, *Science* **325**: 473–477 (2009).
62. M. Sardiello and A. Ballabio, Lysosomal enhancement: a CLEAR answer to cellular degradative needs, *Cell Cycle* **8**: 4021–4022 (2009).
63. B. Dehay, J. Bove, N. Rodriguez-Muela, C. Perier, A. Recasens, P. Boya and M. Vila, Pathogenic lysosomal depletion in Parkinson's disease, *J Neurosci* **30**: 12535–12544 (2010).
64. J. E. Slemmer, J. J. Shacka, M. I. Sweeney and J. T. Weber, Antioxidants and free radical scavengers for the treatment of stroke, traumatic brain injury and aging, *Curr Med Chem* **15**: 404–414 (2008).
65. V. N. Pivtoraiko, S. L. Stone, K. A. Roth and J. J. Shacka, Oxidative stress and autophagy in the regulation of lysosome-dependent neuron death, *Antioxid Redox Signal* **11**: 481–496 (2009).
66. S. Singh and M. Dikshit, Apoptotic neuronal death in Parkinson's disease: involvement of nitric oxide, *Brain Res Rev* **54**: 233–250 (2007).
67. S. Love, L. R. Bridges and C. P. Case, Neurofibrillary tangles in Niemann-Pick disease type C, *Brain* **118**(Pt 1): 119–129 (1995).
68. K. Venkatachalam, A. A. Long, R. Elsaesser, D. Nikolaeva, K. Broadie and C. Montell, Motor deficit in a Drosophila model of mucolipidosis type IV due to defective clearance of apoptotic cells, *Cell* **135**: 838–851 (2008).
69. H. Nakanishi, J. Zhang, M. Koike, T. Nishioku, Y. Okamoto, E. Kominami, K. von Figura, C. Peters, K. Yamamoto, P. Saftig and Y. Uchiyama, Involvement of nitric oxide released from microglia-macrophages in pathological changes of cathepsin D-deficient mice, *J Neurosci* **21**: 7526–7533 (2001).
70. C. D. Pacheco, M. J. Elrick and A. P. Lieberman, Tau deletion exacerbates the phenotype of Niemann-Pick type C mice and implicates autophagy in pathogenesis, *Hum Mol Genet* **18**: 956–965 (2009).

71. V. Khurana, I. Elson-Schwab, T. A. Fulga, K. A. Sharp, C. A. Loewen, E. Mulkearns, J. Tyynela, C. R. Scherzer and M. B. Feany, Lysosomal dysfunction promotes cleavage and neurotoxicity of tau *in vivo*, *PLoS Genet* **6**: e1001026 (2010).
72. M. Koike, M. Shibata, Y. Ohsawa, H. Nakanishi, T. Koga, S. Kametaka, S. Waguri, T. Momoi, E. Kominami, C. Peters, K. von Figura, P. Saftig and Y. Uchiyama, Involvement of two different cell death pathways in retinal atrophy of cathepsin D-deficient mice, *Mol Cell Neurosci* **22**: 146–161 (2003).
73. K. C. Walls, B. J. Klocke, P. Saftig, M. Shibata, Y. Uchiyama, K. A. Roth and J. J. Shacka, Altered regulation of phosphatidylinositol 3-kinase signaling in cathepsin D-deficient brain, *Autophagy* **3**: 222–229 (2007).
74. J. J. Shacka, B. J. Klocke, C. Young, M. Shibata, J. W. Olney, Y. Uchiyama, P. Saftig and K. A. Roth, Cathepsin D deficiency induces persistent neurodegeneration in the absence of Bax-dependent apoptosis, *J Neurosci* **27**: 2081–2090 (2007).
75. D. Sevlever, P. Jiang and S. H. Yen, Cathepsin D is the main lysosomal enzyme involved in the degradation of alpha-synuclein and generation of its carboxy-terminally truncated species, *Biochemistry* **47**: 9678–9687 (2008).
76. Q. Yang, H. She, M. Gearing, E. Colla, M. Lee, J. J. Shacka and Z. Mao, Regulation of neuronal survival factor MEF2D by chaperone-mediated autophagy, *Science* **323**: 124–127 (2009).
77. J. Mitsui, I. Mizuta, A. Toyoda, R. Ashida, Y. Takahashi, J. Goto, Y. Fukuda, H. Date, A. Iwata, M. Yamamoto, N. Hattori, M. Murata, T. Toda and S. Tsuji, Mutations for Gaucher disease confer high susceptibility to Parkinson disease, *Arch Neurol* **66**: 571–576 (2009).
78. K. S. Hruska, O. Goker-Alpan and E. Sidransky, Gaucher disease and the synucleinopathies, *J Biomed Biotechnol* **78549**: 1–6 (2006).
79. Y. H. Xu, Y. Sun, H. Ran, B. Quinn, D. Witte and G. A. Grabowski, Accumulation and distribution of alpha-synuclein and ubiquitin in the CNS of Gaucher disease mouse models, *Mol Genet Metab* **102**: 436–447 (2011).
80. O. Goker-Alpan, B. K. Stubblefield, B. I. Giasson and E. Sidransky, Glucocerebrosidase is present in alpha-synuclein inclusions in Lewy body disorders, *Acta Neuropathol* **120**: 641–649 (2010).
81. J. J. Shacka, K. A. Roth and J. Zhang, The autophagy-lysosomal degradation pathway: role in neurodegenerative disease and therapy, *Front Biosci* **13**: 718–736 (2008).
82. V. N. Pivtoraiko, A. J. Harrington, B. J. Mader, A. M. Luker, G. A. Caldwell, K. A. Caldwell, K. A. Roth and J. J. Shacka, Low-dose bafilomycin attenuates neuronal cell death associated with autophagy-lysosome pathway dysfunction, *J Neurochem* **114**: 1193–1204 (2010).
83. P. Saftig, C. Peters, K. von Figura, K. Craessaerts, F. Van Leuven and B. De Strooper, Amyloidogenic processing of human amyloid precursor protein in hippocampal neurons devoid of cathepsin D, *J Biol Chem* **271**: 27241–27244 (1996).

SECTION 3

SPECIALIZED AUTOPHAGY: THE NEW FRONTIER

CHAPTER 15

AUTOPHAGY — ROLES IN SYNAPTIC STRUCTURE AND FUNCTION

Daniela Hernandez and David Sulzer

ABSTRACT

Macroautophagy is widely studied as a stress-induced response, but the nervous system appears to have adapted this process to modulate synapses under conditions of normal development and plasticity.

INTRODUCTION

Given recent widespread interest in neuronal macroautophagy, it may seem surprising that until recently many biologists believed that this process was absent in neurons[1] and consequently, studies carried out on this topic a decade ago could be substantially delayed in publication. There were nevertheless early reports of neuronal autophagy, including the prescient electron microscopy studies of Leon Roizin in the 1970s that showed high levels of autophagic vacuoles (AVs) in the neurons of patients with Huntington's disease.[2] Studies by Herb Barden indicated that lipofuscin and neuromelanin, the pigments found in neurons of aging brains, were luminal components of autophagic lysosomes,[3] a claim supported in recent years.[4]

As the field was primarily focused on induction of autophagy by amino acid or serum deprivation, it is easy to understand why the existence of neuronal autophagy was often discarded. Neurons are quite resistant under starvation conditions. For example, postnatal neurons can be cultured for weeks in serum-free medium.[5] The lack of readily observed CNS autophagy extended to *in vivo* studies as in food-deprived animals, glycogen, fat, and muscle are rapidly broken down peripherally to supply the brain with nutrients. The hypothalamus, moreover, tightly regulates plasma glucose levels by stimulating release of insulin from the pancreas, which activates mTORC1: this process inhibits autophagy while enhancing glucose uptake, decreasing the need for autophagy-derived nutrients.[1,6] It is further possible that turnover of neuronal AVs is rapid, and so observation of AVs is rare under many conditions.[7] Consistent with these observations, AV induction was rarely reported in neuronal cultures or CNS tissue.

We are still at an early stage in understanding the roles of autophagy in the turnover of synaptic components. As of this writing, we are not aware of any studies detailing synaptic functions affected by chaperone-mediated autophagy (CMA), although CMA is involved in the turnover of at least one protein that regulates synaptic function, alpha-synuclein.[8,9] Roles for lysosomal multi-vesicular bodies at synapses are also relatively unclear. This chapter is thus mostly limited to studies on macroautophagy and associated pathways.

STRESS-INDUCED AUTOPHAGY: IMPLICATIONS FOR NEURONAL MORPHOLOGY

A marked induction of AVs under many forms of neuronal stress is now widely reported.[7,10,11] Neurotoxicity can result from oxidative stress following exposure to L-DOPA, increased cytosolic dopamine (DA), and glutamate-triggered excitoxocity, each of which induce autophagy.[12–14] Neurodegenerative disorders such as Parkinson's disease (PD),[15] Huntington's disease,[16–18] prion diseases,[19,20] and Alzheimer's disease[21] are also associated with increased autophagy (see also Section 2 and Chapter 5). For example, leucine-rich repeat kinase 2 (LRRK2) mutations implicated in Parkinson's disease appear to decrease neurite outgrowth via effects on AVs, in part via effects on LC3 phosphorylation (see below).[22,23]

In some instances of neurodegeneration and neurotoxicity, stress-induced autophagy is related to altered/compromised neurites and synapses. For example, the presence of AVs at presynaptic terminals and dendritic spines has been documented in patients with transmissible spongiform encephalopathies, implicating macroautophagy in the loss of synapses,[19,20] and the authors speculate that entire synapses damaged by disease could be degraded via autophagy. AVs accumulate in early stages of neurodegeneration in some models, and inhibition of their formation can suppress degeneration.[24]

Classic axonal *Wallerian* or anterograde degeneration, in which the axon distal to scission from the cell body degenerates, may involve autophagy:[25] it is indeed difficult to fathom how else large amounts of neuritic material are selectively degraded in the absence of an alternate means to mark neurites for phagocytosis. Recent work on the nigrostriatal dopaminergic projection indicates a role for macroautophagy in retrograde axonal degeneration as well, in that a protective effect of Akt following axotomy may be due to enhanced mTOR activity, which suppresses AV formation.[26]

A striking example of neurite alteration associated with stress-induced autophagy occurs with methamphetamine, a drug long known to produce neurotoxic loss of dopaminergic axons. Axonal loss appears to be caused by a redistribution of DA from synaptic vesicles in the terminals to the cytoplasm,[27,28] where it produces reactive oxygen species that trigger macroautophagy as a stress-induced response. This induction of autophagy in cultured neurons by methamphetamine is associated with a nearly complete loss of neurites in the absence of cell body loss.[13] The collapse of lysosomal pH gradients by the "weak base" effects of methamphetamine[29] disrupts the function of lysosomal proteases and further contributes to a buildup of AVs, so that cultured DA neurons without neurites exhibit massive levels of AVs in the cell body.

In the case of excitotoxicity, Purkinje cells and hippocampal neurons of Lurcher mice, which express constitutively active delta2 glutamate receptors (GluRδ2), exhibit dystrophic axons with focal swellings that contain AVs. Autophagy is induced due to an increase in Na^+ influx, which also leads to neuronal depolarization and increased Ca^{2+} influx through voltage-gated calcium channels. Increased cytosolic Ca^{2+} ultimately results in cell death via activation of calpains.[30–32] As genetic deletion of Atg5, an essential macroautophagy-related gene, exacerbates cell death, macroautophagy appears to provide a protective function,[31] although this issue is not yet clear, in part due to the age at which the protein becomes deficient.[33] The balance between excitatory stress and ongoing autophagy as a stress-induced response could be particularly important for maintaining normal neuronal function under conditions of high cytosolic Ca^{2+}.

Accumulation of AVs in axons may also be predicted in lysosomal storage disorders as both a stress induced response and decreased AV turnover.[4] Indeed, axonal AV accumulation is reported in mouse models of Batten disease expressing mutant lysosomal cathepsin enzymes[34] and is implicated in dysfunction of hippocampal GABAergic interneurons in cathepsin D-deficient mice.[35]

ROLES FOR MACROAUTOPHAGY IN NEURITE REMODELING AND RETRACTION

In contrast to its role as a response to neurodegenerative or toxic stress, what role does autophagy play under normal conditions in the normal turnover of synaptic components and/or neurites? mTOR signaling, which regulates protein synthesis and autophagy induction, is implicated in synaptic plasticity, including the maintenance of hippocampal long term potentiation.[36] Synapse disassembly or "pruning" has also been implicated in normal development and during learning and memory.[37] Thus, alterations in synaptic structure involving clearance of synaptic components, including secretory vesicles, receptors and transporters, and mitochondria, with attendant changes in synaptic function, might rely on autophagy.[38]

Early evidence that neuronal autophagy sculpts neurites was provided in Mary Bartlett Bunge's 1973 study on the presence of AVs in retracting neuronal processes of rat sympathetic motor neurons, indicating that neuronal remodeling could occur through autophagic clearance.[39] She described the local formation of lysosomal-like structures at the growth cone and the presence of AVs at various stages of formation along nerve fibers.

Peter Hollenbeck (1993) expanded on these findings by demonstrating that AVs formed in the distal regions of axons in cultured sympathetic neurons.[40] AVs accumulated both cytosolic and endocytic products. Following trituration, localization of Texas red-dextran label in the cytosol changed from diffuse to punctuate within 72 hours, indicating sequestration by AVs. At the axonal growth cone, Lucifer yellow in the medium was sequestered by neurons into AVs, which then traveled along the axon back to the cell body, corroborating the notion that endocytic structures and their contents can be degraded through amphisomes/AVs in neurons.

Hollenbeck reported both retrograde and anterograde movement of AVs. The net direction of movement of these organelles was tightly coupled to the energetic state of the neuron. Hollenbeck demonstrated this concept using forskolin, which enhances cAMP levels and has more recently been found to promote phosphorylation of LC3, which inhibits autophagy,[23] and the actin polymerization inhibitor, cytochalasin. Forskolin caused AVs to move anterogradely, thus supplying the growth cone with nutrients and membrane, while cytocholasin favored retrograde transport and inhibited axonal growth. These findings suggest that neurons modulate autophagy in a dynamic manner to fit their physiological requirements, and Hollenbeck's study introduced the idea that the rate and vector of AV movement allows neurons to redistribute membrane and alter neuronal structure.

More recent studies have identified roles for Ulk1/Atg1, a protein kinase involved in AV formation and axonal transport, in neurite remodeling. (The term Ulk1 is most commonly used for a mammalian Atg1 homologue.) In Drosophila, disruption of AV formation or AV-lysosomal fusion decreased the size of the neuromuscular junction (NMJ), while overexpression of Atg1 enhanced autophagy and increased the number of synaptic boutons and neuritic branches. Consistently, exposure of wild type *Drosophila* to the mTOR inhibitor, rapamycin,[7] resulted in synaptic overgrowth.[41] Deletion of *unc-51*, a homologue of *atg1*, in Drosophila resulted in glutamate receptor clusters unapposed to Bruchpilot, an active zone protein, the development of aberrant synapses with fewer active zones, a decrease in the density of synapses in general, and a decrease in evoked neurotransmitter release at the Drosophila NMJ.[42] In another *Drosophila* model, deletion of *unc-51* resulted in accumulation of synaptic vesicles in axon shafts, a decrease in the number of synaptic boutons, and the clustering of dense core vesicles and multivesicular bodies.[43] In *C. elegans*, mutations in *unc-51* and *unc-14* resulted in abnormal localization of the axon guidance netrin receptor in neuronal cell bodies and abnormal axonal growth, further consistent with a role for Atg1 activity in the regulation of axonal morphology.[44] In mammals, mutations in *unc-51*/Ulk1/Atg1 homologues disrupted neurite formation/extension in mouse cerebral granule cells, and increased branching in cultured dorsal root ganglion neurons.[45] Although these phenotypes could be due to abnormal localization of axon guidance receptors or deficits in axonal transport, changes in lysosomal degradation may also play a role.

Neurite alteration via autophagy could also be mediated by changes in AV-lysosomal fusion. For example, overexpression of the N-terminal region of TI-VAMP/VAMP7, a v-SNARE found on late endosomes necessary for amphisome fusion with the lysosome,[46] disrupts axonal and dendritic outgrowth in mouse hippocampal neurons.[45] While AV induction in these animals may be normal, fusion and consequently degradation is disturbed, leading to defective neuritic morphology.

Multiple means of autophagy regulation converge on the covalent lipidation of microtubule-associated protein 1 light chain 3 (LC3), which converts the protein to the AV-associated membrane component LC3-II. Such a step has been recently implicated in neurite outgrowth and retraction,[23] so that protein kinase A-mediated LC3 phosphorylation suppressed AV formation and blocks neurite retraction, while rapamycin or neuronal stress enhances LC3 phosphorylation, AV formation, and neurite retraction.

Characterizing a role for constitutive autophagy in regulating CNS synaptic transmission and plasticity has become more straightforward with experiments using conditional autophagy-deficient mice. The use of "floxed" genetic technology allows particular sets of neurons to be autophagy deficient. For example, autophagy deficiency selectively in mouse Purkinje cells leads to axonal, but not dendritic, dystrophy and accumulation of aberrant membrane structures,[30] suggesting that constitutive autophagy is necessary for normal axonal morphology. In our recent studies, we find that DA neurons deficient in autophagy display larger presynaptic terminals and enhanced evoked DA release. Rapamycin induced AV formation in wild type, but not autophagy-deficient, DA synaptic terminals, resulting in a decrease in presynaptic terminal volume.[47]

Together, current data suggest that autophagy at synaptic terminals can degrade synaptic components. It may be that low levels or lack of autophagy initially results in synaptic or neuronal overgrowth, but that with continued autophagy deficiency, there may be an accumulation of damaged mitochondria and other cellular constituents that lead to synaptic loss.

TURNOVER OF PRESYNAPTIC COMPONENTS BY MACROAUTOPHAGY

In 1971, Eric Holtzman and colleagues reported that recycling synaptic vesicles at the lobster NMJ can fuse with multivesicular bodies and lysosomes, suggesting a means for normal synaptic vesicle degradation.[48] Thus, the neuron can control the fate of synaptic vesicle membrane following fusion so that, following reaccumulation into coated vesicles or endosomes, vesicle membrane can either be used to reform synaptic vesicles or be degraded. These studies also introduced an important point about neuronal autophagy: in many cases, neuronal AVs appear to function as so-called "amphisomes", meaning they are derived from both autophagic organelles and components of endosomes, in this case containing both endocytosed synaptic vesicle membrane and associated proteins.

The means by which recycling synaptic vesicle membrane can be differentially targeted to the lysosome are unknown. The adaptor proteins AP2 and AP3 can be selectively used for the recovery of fused synaptic vesicle membrane from the plasma membrane depending on neuronal activity, with AP3 providing bulk endocytosis with high activity levels.[49,50] These AP3-dependent endosomes may provide a structure that can donate the membrane to AVs or lysosomes. This suggests the presence of an early step that determines whether synaptic vesicle membrane is recycled or degraded (Figure 1). Another potential regulatory step for synaptic vesicle degradation is provided by Rab5, which both regulates endosomal function and vesicle recycling[51] and inhibits mTORC1 function,[52] which enhances macroautophagy.[53–55]

There are already thus several candidates that may determine how neuronal activity could provide selective degradation of presynaptic elements and organelles, including synaptic vesicles, mitochondria, endoplasmic reticulum, cytoskeletal elements, presynaptic receptors and channels, and endosomes. An important role for the appropriate localization of presynaptic mitochondria is particularly demonstrated in the modulation of synaptic strength due to effects on the availability of synaptic vesicles for fusion under periods of intense stimulus.[56]

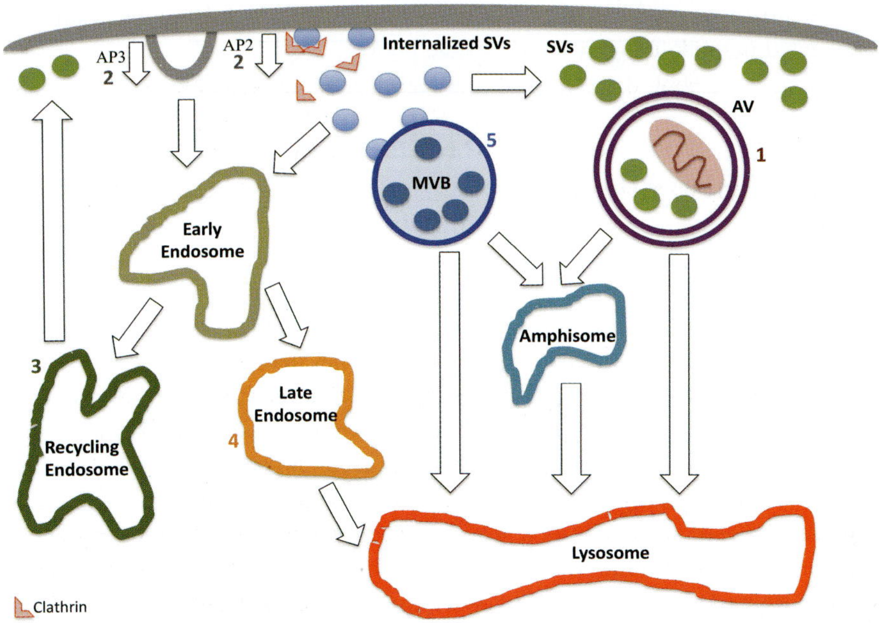

Figure 1. Proposed synaptic vesicle recycling and degradation pathways. AVs may directly sequester cytosolic synaptic vesicles and provide a system to deliver these organelles to lysosomes for degradation (1). In addition, for those synaptic vesicles that undergo full fusion with the plasma membrane, synaptic vesicle membrane can be internalized and targeted either for recycling or degradation. The adaptor proteins AP2 or AP3 mediate recovery of vesicle membrane, depending on the level of neuronal activity. AP2-mediated SV internalization involves clathrin-dependent endocytosis, for presumed uncaging and recycling (2), while AP3-mediated internalization involves an endosomal intermediate, from which membrane can either enter the recycling endosome to reform SVs (3), or a later stage endosome (4), Later stage endosomes may fuse with local AVs/amphisomes for transport to mature lysosomes where vesicular membranes are degraded. Following internalization, apparent synaptic vesicles are also observed within multivesicular bodies (MVBs) (5), which can subsequently fuse with an AV/amphisome or directly with the lysosome.

Recent evidence indicating autophagic degradation of synaptic vesicles is observed in dopaminergic neurons, for which the probability of neurotransmitter release appears to be related in part to the size of the readily releasable pool.[57] We observed a decrease in the number of synaptic vesicles following macroautophagy induction by rapamycin in striatum.[47] Cyclic voltammetry recordings[58] show a decrease in evoked dopamine release from striatal terminals following macroautophagy induction. This response was absent in conditional autophagy-deficient dopaminergic neurons, consistent with a relatively rapid induction of autophagic synaptic vesicle degradation. The data also suggest that rapamycin may decrease the number of presynaptic mitochondria, which as mentioned is implicated in modulating synaptic vesicle pools.

TURNOVER OF POSTSYNAPTIC RECEPTORS BY MACROAUTOPHAGY

GABA Receptors

In a study of the *C. elegans* NMJ, Bruce Bamber and colleagues identified a role for selective macroautophagy in the degradation of $GABA_A$ receptors (GABAR).[38] They disrupted expression of the netrin receptor, which is required for axon pathfinding, in GABAergic or cholinergic neurons, or both. Loss of GABAergic innervation decreased GABAR clustering. When muscle cells were completely denervated, GABARs accumulated in AVs via endocytosis. When autophagy was blocked by a UNC-51 mutation, the decrease in GABA-mediated currents was rescued. These data suggest that receptor accumulation in AVs is inhibited by contact, rather than function, of presynaptic inputs. Remarkably, these results were specific for GABARs, as when the netrin receptor was selectively deleted from cholinergic inputs, neither accumulation of acetylcholine receptors in AVs nor a decrease in currents evoked by cholinergic agonists was observed. The authors suggest that innervation provides an intercellular signal independent of neurotransmitter release that regulates receptor localization and blocks degradation of GABARs through macroautophagy.[38] This is consistent with Hollenbeck's earlier suggestion that neurons regulate autophagy to degrade synaptic components depending on their physiological state.[48]

Additional studies suggest further crosstalk between the biochemical systems that regulate autophagy and GABAergic signaling. The $GABA_A$ receptor-associated protein (GABARAP), a homolog of atg8/LC3, an autophagy-related protein found on the outer and inner AV membranes, is involved in GABAR trafficking from the Golgi apparatus to the plasma membrane,[59,60] autophagy induction, AV maturation, and clearance of mitochondria.[61,62] Like atg8 and its mammalian homologue, LC3, unlipidated GABARAP is cytosolic, but its localization changes to AV membranes following lipidation.[59,60] During extinction of fear conditioning, GABARAP activity is necessary for NMDA-dependent insertion of GABARs into the postsynaptic membrane in hippocampal CA1 neurons, indicating a function in strengthening inhibitory synapses during learning and memory.[63]

Glutamate receptor activation has also been reported to target $GABA_B$ receptors to lysosomes in cortical neurons in a calcium-dependent manner,[64] reminiscent of calcium-regulated stress response induction of autophagy as discussed above.

More broadly, the state of phosphorylation and ubiquitination, beta-arrestins, and interaction with adaptor proteins can each regulate endocytosis[65] and subsequent trafficking of neurotransmitter receptors for recycling or degradation. Rab5[50,52,66] or VAMP7[46] activity may be required for receptor targeting to amphisomes/lysosomes for degradation. Additional candidates have been introduced in experiments in yeast, including the "endosomal sorting complex required for transport" (ESCRT) signals introduced by Scott Emr and collaborators.[67,68]

Glutamate Receptors

There is also evidence for autophagic turnover of glutamate receptors. While to our knowledge, AMPAR accumulation in AVs has not yet been shown to affect synaptic transmission, it is well established that a net decrease in the number of surface postsynaptic AMPARs results in LTD expression.[32] Moreover, NMDAR activation can induce AMPAR internalization followed by degradation in the lysosome, with lysosomal protease but not proteasome inhibitors inhibiting the loss of AMPARs in hippocampal neurons.

Hippocampal neurons and Purkinje cells of mice carrying a null mutation for the beta subunit of the adaptor protein AP4, which is involved in basolateral sorting in epithelial cells and in neurons and associates with lysosomal associated protein 2 (LAMP2),[69] exhibit axonal swellings and an accumulation of axonal AVs. To study the effect of this AP4 mutation on protein sorting, Matsuda and colleagues focused on Purkinje cells because their dendrites, which project to the molecular layer of the cerebellum, are spatially separated from their axon, which extends into the deep cerebellar nuclei (DCN).[70] AMPARs are normally localized to dendritic spines, but Purkinje cell axons from mice lacking the beta subunit of AP4 were immunolabeled for AMPARs both *in vivo* and *in vitro*, indicating receptor mislocalization. Improper targeting of AMPARs to axons was mediated by a lack of interaction between AP4 and transmembrane AMPA receptor regulatory proteins (TARPs). Electron microscopy indicated that the bulging regions of axons from AP4 null mutant mice contained AV-like organelles. Occasionally these structures were found at axon terminals that appeared normal and formed synapses with DCN neurons. AP4 mutant mice exhibited an increase in the lipidated, AV-membrane bound form of LC3, LC3-II, and AMPARs co-localized with GFP-LC3 puncta in axons of Purkinje cells *in vivo* and in hippocampal cells *in vitro*,[70] indicating AMPAR accumulation in axonal and terminal AVs.

Reminiscent of autophagy of GABARs discussed above, the effects were specific to AMPARs, as the metabotrophic glutamate receptor 1 (mGLUR1) was not found in axons in Purkinje cells axons *in vivo* nor in hippocampal neurons *in vitro*,[70] and the lack of AP4 did not affect localization of vesicle-associated membrane protein 2/synaptobrevin II (VAMP2), part of the SNARE complex,[71] nor the NR1 subunit of the NMDAR.

This delivery of AMPAR to lysosomes may be dependent on subunit composition, as GluR1 preferentially enters the recycling pathway following activation with AMPA, whereas GluR2 and GluR1/GluR2 heterodimers are trafficked to the late endosome/lysosome for degradation following activation with NMDA.[72] Sorting of AMPAR subunits between the recycling and lysosomal pathways is driven by the interaction of subunit cytoplasmic tails with NSF, suggesting that such interactions may provide a mechanism for the selective degradation of synaptic receptors through autophagy (Figure 2).

This fate depends on the mode of activation and AMPAR subunit composition. After AMPA-mediated internalization, GluR2 subunits enter the recycling pathway, but following activation with NMDA, they are targeted for lysosomal degradation. GluR1 enters the recycling pathway and GluR3 is trafficked to the lysosome, regardless of NMDAR activation. NSF interaction with the GluR2 cytoplasmic tail is implicated in determining whether GluR2 receptors enter the recycling or lysosomal degradation pathway, while GluR3, which lacks an NSF-binding domain, is preferentially trafficked to the lysosome. Thus, subunit-specific interactions with adaptor proteins have profound effects on receptor degradation. These interactions may also differ depending on whether delivery to the lysosome is via macroautophagy or multivesicular bodies.

While it is likely that receptors, transporters, and channels can be degraded by autophagy, the mechanisms controlling degradation may be different for specific components. This could entail early regulatory steps: for example, AP3 and its adaptor complex regulator AGAP1 enable the recycling of endocytosed m5 muscarinic receptors in DA neurons,[73] which in turn regulates the level of evoked DA release. Thus, AP3/AGAP1 activity could target m5 muscarinic receptors in DA neurons for degradation without loss of other synaptic receptors or channels. Uncovering these processes is complicated by the strong possibility that degradation requires accumulation of membrane-associated

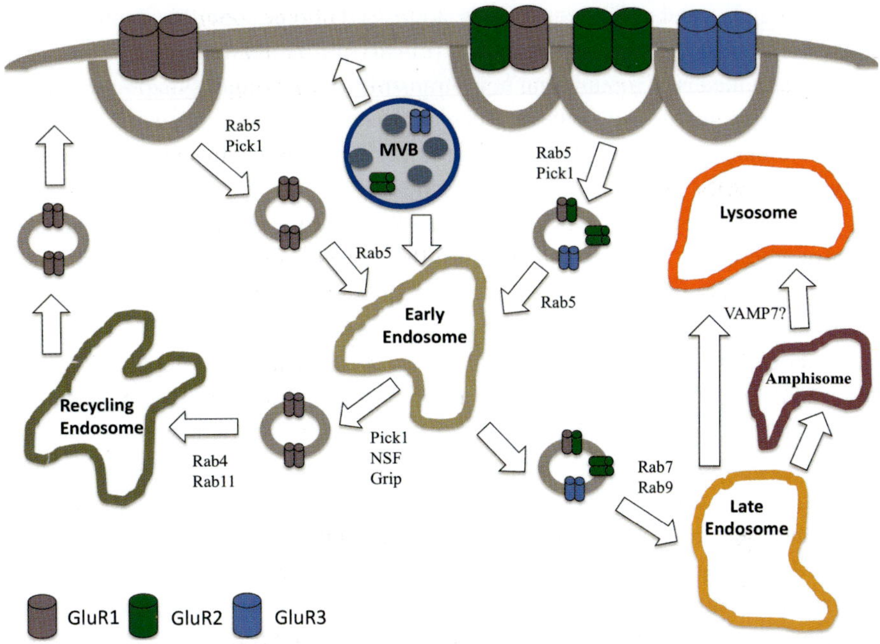

Figure 2. Model of AMPA receptor recycling and degradation. The fate of endocytosed receptors, either to be recycled to the cell surface or degraded, impacts synaptic function. Following LTP induction, AMPARs are inserted into the postsynaptic membrane, while following LTD induction, AMPARs are internalized via clathrin-dependent endocytosis. Upon internalization, receptors are first trafficked to the early endosome, which requires PICK1 and Rab5 activity. In the early endosome, internalized receptors are associated with Rab4 and Rab5. PICK1 and Grip are though to facilitate AMPAR recycling. If trafficked to the recycling endosome, receptors become associated with Rab4 and Rab11; if trafficked to the late endosome, they become associated with Rab7 and Rab9.

proteins in multivesicular bodies rather than AVs, a topic little explored in the CNS to date.

SUMMARY

After decades in obscurity, the field of neuroscience has revisited Hollenbeck's and Bunge's pioneering work indicating that autophagy plays roles in trafficking membrane to and from the synapse, with consequences that regulate structure, excitation/inhibition balance at synapses, synaptic morphology, and synaptic transmission. While we are now in preliminary stages of characterizing these phenomena, they appear to be related to selective autophagy of synaptic constituents. Thus, in addition to macroautophagy's well-established roles in stress response and cellular homeostasis as addressed throughout this volume, data presented here suggest that synaptic terminals utilize this anciently conserved process as a means of modulating neurotransmitter release and synaptic plasticity. From this perspective, autophagic processes resemble other fundamental biological phenomena that were adopted by the nervous system to provide means to provide synapses with a wide range of states.

The observations on autophagic targeting of GABARs and AMPARs have important implications for the roles of autophagy in synaptic function under nominally unstressed conditions.[38,70] First, they indicate that presynaptic GABAergic contact provides an intercellular signal postsynaptically that may be independent of neurotransmitter release to regulate receptor localization and blocked degradation of GABARs through autophagy.[38] Second, they suggest that particular types of synaptic receptors may be selectively degraded through autophagy within the same cellular compartment, apparently in part due to particular interactions with adaptor proteins.[38,70,72]

Perhaps most importantly, these findings point towards a general role for autophagy in regulating synaptic plasticity. For example, GABA and acetylcholine receptors are inhibitory and excitatory at the *C. elegans* NMJ, and so selective degradation of inhibitory GABARs can alter the balance between excitatory and inhibitory signals at these synapses. Such regulation is important in circuits where the balance between excitation and inhibition alters plasticity, such as in the visual system, which relies on GABAergic signaling during development of ocular dominance columns[74,75] or the corticostriatal synapse, where either long term depression or potentiation can be induced depending on spike timing.[76] Thus, crosstalk between trafficking molecules and the autophagy pathway might result in changes in synaptic plasticity[52,72] and a major future issue will be to determine how synaptic components are differentially targeted for recycling or lysosomal degradation.

While the mechanisms of synaptic autophagy are incompletely defined at present, many other signaling pathways and mechanisms identified in other cell types have been subsequently found to be adapted in complex ways in the nervous system, and there is no reason to presume that lysosomal degradative pathways will provide an exception to this observation.

ACKNOWLEDGMENTS

We thank NIDA, NINDS, and the Picower, Simons, and Parkinson's Disease Foundations for support of our research and lab members and our collaborators for conversation about these issues, including Ana Maria Cuervo and Robert Burke.

REFERENCES

1. B. Boland and R. A. Nixon, Neuronal macroautophagy: from development to degeneration, *Mol Aspects Med* **27**: 503–519 (2006).
2. L. Roizin, S. Stellar, N. Willson, J. Whittier and J. C. Liu, Electron microscope and enzyme studies in cerebral biopsies of Huntington's chorea, *Trans Am Neurol Assoc* **99**: 240–243 (1974).
3. H. Barden, Relationship of Golgi thiaminepyrosphosphatase and lysosomal acid phosphatase to neuromelanin and lipofuscin in cerebral neurons of the aging rhesus monkey, *J Neuropathol Exp Neurol* **29**: 225–240 (1970).
4. D. Sulzer, E. Mosharov, Z. Talloczy, F. A. Zucca, J. D. Simon and L. Zecca, Neuronal pigmented autophagic vacuoles: lipofuscin, neuromelanin, and ceroid as macroautophagic responses during aging and disease, *J Neurochem* **106**: 24–36 (2008).
5. G. J. Brewer, Serum-free B27/neurobasal medium supports differentiated growth of neurons from the striatum, substantia nigra, septum, cerebral cortex, cerebellum, and dentate gyrus, *J Neurosci Res* **42**: 674–683 (1995).
6. W. L. Yen and D. J. Klionsky, How to live long and prosper: autophagy, mitochondria, and aging, *Physiology (Bethesda)* **23**: 248–262 (2008).
7. S. A. Tooze and G. Schiavo, Liaisons dangereuses: autophagy, neuronal survival and neurodegeneration, *Curr Opin Neurobiol* **18**: 504–515 (2008).
8. M. Martinez-Vicente, Z. Talloczy, S. Kaushik, A. C. Massey, J. Mazzulli, E. V. Mosharov, R. Hodara, R. Fredenburg, D. C. Wu, A. Follenzi, W. Dauer, S. Przedborski, H. Ischiropoulos, P. T. Lansbury, D. Sulzer and A. M. Cuervo, Dopamine-modified alpha-synuclein blocks chaperone-mediated autophagy, *J Clin Invest* **118**: 777–788 (2008).
9. A. M. Cuervo, L. Stefanis, R. Fredenburg, P. T. Lansbury and D. Sulzer, Impaired degradation of mutant alpha-synuclein by chaperone-mediated autophagy, *Science* **305**: 1292–1295 (2004).
10. K. E. Larsen and D. Sulzer, Autophagy in neurons: a review, *Histol Histopathol* **17**: 897–908 (2002).
11. D. C. Rubinsztein, M. DiFiglia, N. Heintz, R. A. Nixon, Z. H. Qin, B. Ravikumar, L. Stefanis and A. Tolkovsky, Autophagy and its possible roles in nervous system diseases, damage and repair, *Autophagy* **1**: 11–22 (2005).
12. D. Sulzer, J. Bogulavsky, K. E. Larsen, G. Behr, E. Karatekin, M. H. Kleinman, N. Turro, D. Krantz, R. H. Edwards, L. A. Greene and L. Zecca, Neuromelanin biosynthesis is driven by excess cytosolic catecholamines not accumulated by synaptic vesicles, *Proc Natl Acad Sci USA* **97**: 11869–11874 (2000).
13. K. E. Larsen, E. A. Fon, T. G. Hastings, R. H. Edwards and D. Sulzer, Methamphetamine-induced degeneration of dopaminergic neurons involves autophagy and upregulation of dopamine synthesis, *J Neurosci* **22**: 8951–8960 (2002).
14. Z. Yue, Regulation of neuronal autophagy in axon: implication of autophagy in axonal function and dysfunction/degeneration, *Autophagy* **3**: 139–141 (2007).

15. P. Anglade, S. Vyas, F. Javoy-Agid, M. T. Herrero, P. P. Michel, J. Marquez, A. Mouatt-Prigent, M. Ruberg, E. C. Hirsch and Y. Agid, Apoptosis and autophagy in nigral neurons of patients with Parkinson's disease, *Histol Histopathol* **12**: 25–31 (1997).
16. K. B. Kegel, M. Kim, E. Sapp, C. McIntyre, J. G. Castano, N. Aronin and M. DiFiglia, Huntingtin expression stimulates endosomal-lysosomal activity, endosome tubulation, and autophagy, *J Neurosci* **20**: 7268–7278 (2000).
17. Y. J. Kim, E. Sapp, B. G. Cuiffo, L. Sobin, J. Yoder, K. B. Kegel, Z. H. Qin, P. Detloff, N. Aronin and M. DiFiglia, Lysosomal proteases are involved in generation of N-terminal huntingtin fragments, *Neurobiol Dis* **22**: 346–356 (2006).
18. A. Petersen, K. E. Larsen, G. G. Behr, N. Romero, S. Przedborski, P. Brundin and D. Sulzer, Expanded CAG repeats in exon 1 of the Huntington's disease gene stimulate dopamine-mediated striatal neuron autophagy and degeneration, *Hum Mol Genet* **10**: 1243–1254. (2001).
19. P. P. Liberski, B. Sikorska, J. Bratosiewicz-Wasik, D. C. Gajdusek and P. Brown, Neuronal cell death in transmissible spongiform encephalopathies (prion diseases) revisited: from apoptosis to autophagy, *Int J Biochem Cell Biol* **36**: 2473–2490 (2004).
20. B. Sikorska, P. P. Liberski, P. Giraud, N. Kopp and P. Brown, Autophagy is a part of ultrastructural synaptic pathology in Creutzfeldt-Jakob disease: a brain biopsy study, *Int J Biochem Cell Biol* **36**: 2563–2573 (2004).
21. J. A. Rodriguez-Navarro, L. Rodriguez, M. J. Casarejos, R. M. Solano, A. Gomez, J. Perucho, A. M. Cuervo, J. Garcia de Yebenes and M. A. Mena, Trehalose ameliorates dopaminergic and tau pathology in parkin deleted/tau overexpressing mice through autophagy activation, *Neurobiol Dis* **39**: 423–438 (2010).
22. E. D. Plowey, S. J. Cherra, 3rd, Y. J. Liu and C. T. Chu, Role of autophagy in G2019S-LRRK2-associated neurite shortening in differentiated SH-SY5Y cells, *J Neurochem* **105**: 1048–1056 (2008).
23. S. J. Cherra, 3rd, S. M. Kulich, G. Uechi, M. Balasubramani, J. Mountzouris, B. W. Day and C. T. Chu, Regulation of the autophagy protein LC3 by phosphorylation, *J Cell Biol* **190**: 533–539 (2010).
24. Y. Yang, K. Fukui, T. Koike and X. Zheng, Induction of autophagy in neurite degeneration of mouse superior cervical ganglion neurons, *Eur J Neurosci* **26**: 2979–2988 (2007).
25. Z. X. Piao, W. S. Wang, X. J. Xu, Q. W. Wang, X. Huo, M. H. Han and Y. J. Piao, Autophagy of neuron axon during regeneration of rat sciatic nerves, *Di Yi Jun Yi Da Xue Xue Bao* **24**: 361–364 (2004).
26. H. C. Cheng, S. R. Kim, T. F. Oo, T. Kareva, O. Yarygina, M. Rzhetskaya, C. Wang, M. During, Z. Talloczy, K. Tanaka, M. Komatsu, K. Kobayashi, H. Okano, N. Kholodilov and R. E. Burke, Akt suppresses retrograde degeneration of dopaminergic axons by inhibition of macroautophagy, *J Neurosci* **31**: 2125–2135 (2011).

27. E. V. Mosharov, K. E. Larsen, E. Kanter, K. A. Phillips, K. Wilson, Y. Schmitz, D. E. Krantz, K. Kobayashi, R. H. Edwards and D. Sulzer, Interplay between cytosolic dopamine, calcium, and alpha-synuclein causes selective death of substantia nigra neurons, *Neuron* **62**: 218–229 (2009).
28. F. Fornai, P. Lenzi, L. Capobianco, L. Iacovelli, P. Scarselli, G. Lazzeri and A. De Blasi, Involvement of dopamine receptors and beta-arrestin in metamphetamine-induced inclusions formation in PC12 cells, *J Neurochem* **105**: 1939–1947 (2008).
29. D. Sulzer and S. Rayport, Amphetamine and other psychostimulants reduce pH gradients in midbrain dopaminergic neurons and chromaffin granules: a mechanism of action, *Neuron* **5**: 797–808 (1990).
30. Z. Yue, A. Horton, M. Bravin, P. L. DeJager, F. Selimi and N. Heintz, A novel protein complex linking the delta 2 glutamate receptor and autophagy: implications for neurodegeneration in lurcher mice, *Neuron* **35**: 921–933 (2002).
31. J. Nishiyama, K. Matsuda, W. Kakegawa, N. Yamada, J. Motohashi, N. Mizushima and M. Yuzaki, Reevaluation of neurodegeneration in lurcher mice: constitutive ion fluxes cause cell death with, not by, autophagy, *J Neurosci* **30**: 2177–2187 (2010).
32. Y. T. Wang, Probing the role of AMPAR endocytosis and long-term depression in behavioural sensitization: relevance to treatment of brain disorders, including drug addiction, *Br J Pharmacol* **153**(Suppl 1): S389–S395 (2008).
33. Z. Yue, Autophagy in lurcher mice: indicted but yet to be acquitted for the death of Purkinje cells, *Autophagy* **6**: 571–572 (2010).
34. M. Koike, M. Shibata, S. Waguri, K. Yoshimura, I. Tanida, E. Kominami, T. Gotow, C. Peters, K. von Figura, N. Mizushima, P. Saftig and Y. Uchiyama, Participation of autophagy in storage of lysosomes in neurons from mouse models of neuronal ceroid-lipofuscinoses (Batten disease), *Am J Pathol* **167**: 1713–1728 (2005).
35. T. Shimizu, Y. Hayashi, R. Yamasaki, J. Yamada, J. Zhang, K. Ukai, M. Koike, K. Mine, K. von Figura, C. Peters, P. Saftig, T. Fukuda, Y. Uchiyama and H. Nakanishi, Proteolytic degradation of glutamate decarboxylase mediates disinhibition of hippocampal CA3 pyramidal cells in cathepsin D-deficient mice, *J Neurochem* **94**: 680–690 (2005).
36. S. J. Tang, G. Reis, H. Kang, A. C. Gingras, N. Sonenberg and E. M. Schuman, A rapamycin-sensitive signaling pathway contributes to long-term synaptic plasticity in the hippocampus, *Proc Natl Acad Sci USA* **99**: 467–472 (2002).
37. Y. Goda and G. W. Davis, Mechanisms of synapse assembly and disassembly, *Neuron* **40**: 243–264 (2003).
38. A. M. Rowland, J. E. Richmond, J. G. Olsen, D. H. Hall and B. A. Bamber, Presynaptic terminals independently regulate synaptic clustering and autophagy of GABAA receptors in *Caenorhabditis elegans*, *J Neurosci* **26**: 1711–1720 (2006).
39. M. B. Bunge, Fine structure of nerve fibers and growth cones of isolated sympathetic neurons in culture, *J Cell Biol* **56**: 713–735 (1973).

40. P. Hollenbeck, Products of endocytosis and autophagy are retrieved from axons by regulated retrograde organelle transport, *J Cell Biol* **121**: 305–315 (1993).
41. W. Shen and B. Ganetzky, Autophagy promotes synapse development in Drosophila, *J Cell Biol* **187**: 71–79 (2009).
42. Y. P. Wairkar, H. Toda, H. Mochizuki, K. Furukubo-Tokunaga, T. Tomoda and A. Diantonio, Unc-51 controls active zone density and protein composition by downregulating ERK signaling, *J Neurosci* **29**: 517–528 (2009).
43. H. Toda, H. Mochizuki, R. Flores, 3rd, R. Josowitz, T. B. Krasieva, V. J. Lamorte, E. Suzuki, J. G. Gindhart, K. Furukubo-Tokunaga and T. Tomoda, UNC-51/ATG1 kinase regulates axonal transport by mediating motor-cargo assembly, *Genes Dev* **22**: 3292–3307 (2008).
44. K. Ogura and Y. Goshima, The autophagy-related kinase UNC-51 and its binding partner UNC-14 regulate the subcellular localization of the Netrin receptor UNC-5 in *Caenorhabditis elegans*, *Development* **133**: 3441–3450 (2006).
45. S. Sann, Z. Wang, H. Brown and Y. Jin, Roles of endosomal trafficking in neurite outgrowth and guidance, *Trends Cell Biol* **19**: 317–324 (2009).
46. C. M. Fader, D. G. Sanchez, M. B. Mestre and M. I. Colombo, TI-VAMP/VAMP7 and VAMP3/cellubrevin: two v-SNARE proteins involved in specific steps of the autophagy/multivesicular body pathways, *Biochim Biophys Acta* **1793**: 1901–1916 (2009).
47. D. Hernandez, C. A. Torres, W. Setlik, C. Cebrián, E. V. Mosharov, G. Tang, H. C. Cheng, N. Kholodilov, O. Yarygina, R. E. Burke, M. Gershon and D. Sulzer, *Neuron* **74**: 277–284 (2012).
48. E. Holtzman, A. R. Freeman and L. A. Kashner, Stimulation-dependent alterations in peroxidase uptake at lobster neuromuscular junctions, *Science* **173**: 733–736 (1971).
49. S. M. Voglmaier and R. H. Edwards, Do different endocytic pathways make different synaptic vesicles?, *Curr Opin Neurobiol* **17**: 374–380 (2007).
50. S. M. Voglmaier, K. Kam, H. Yang, D. L. Fortin, Z. Hua, R. A. Nicoll and R. H. Edwards, Distinct endocytic pathways control the rate and extent of synaptic vesicle protein recycling, *Neuron* **51**: 71–84 (2006).
51. V. N. Murthy and P. De Camilli, Cell biology of the presynaptic terminal, *Annu Rev Neurosci* **26**: 701–728 (2003).
52. J. G. Hanley, Endosomal sorting of AMPA receptors in hippocampal neurons, *Biochem Soc Trans* **38**: 460–465 (2010).
53. R. J. Flinn, Y. Yan, S. Goswami, P. J. Parker and J. M. Backer, The late endosome is essential for mTORC1 signaling, *Mol Biol Cell* **21**: 833–841 (2010).
54. L. Li, E. Kim, H. Yuan, K. Inoki, P. Goraksha-Hicks, R. L. Schiesher, T. P. Neufeld and K. L. Guan, Regulation of mTORC1 by the Rab and Arf GTPases, *J Biol Chem* **285**: 19705–19709 (2010).
55. B. Ravikumar, S. Imarisio, S. Sarkar, C. J. O'Kane and D. C. Rubinsztein, Rab5 modulates aggregation and toxicity of mutant huntingtin through macroautophagy in cell and fly models of Huntington disease, *J Cell Sci* **121**: 1649–1660 (2008).

56. P. Verstreken, C. V. Ly, K. J. Venken, T. W. Koh, Y. Zhou and H. J. Bellen, Synaptic mitochondria are critical for mobilization of reserve pool vesicles at Drosophila neuromuscular junctions, *Neuron* **47**: 365–378 (2005).
57. J. A. Daniel, S. Galbraith, L. Iacovitti, A. Abdipranoto and B. Vissel, Functional heterogeneity at dopamine release sites, *J Neurosci* **29**: 14670–14680 (2009).
58. D. L. Robinson, B. J. Venton, M. L. Heien and R. M. Wightman, Detecting subsecond dopamine release with fast-scan cyclic voltammetry *in vivo*, *Clin Chem* **49**: 1763–1773 (2003).
59. Y. Kabeya, N. Mizushima, A. Yamamoto, S. Oshitani-Okamoto, Y. Ohsumi and T. Yoshimori, LC3, GABARAP and GATE16 localize to autophagosomal membrane depending on form-II formation, *J Cell Sci* **117**: 2805–2812 (2004).
60. T. A. Leil, Z. W. Chen, C. S. Chang and R. W. Olsen, GABAA receptor-associated protein traffics GABAA receptors to the plasma membrane in neurons, *J Neurosci* **24**: 11429–11438 (2004).
61. H. Weidberg, E. Shvets, T. Shpilka, F. Shimron, V. Shinder and Z. Elazar, LC3 and GATE-16/GABARAP subfamilies are both essential yet act differently in autophagosome biogenesis, *EMBO J* **29**: 1792–1802 (2010).
62. I. Novak, V. Kirkin, D. G. McEwan, J. Zhang, P. Wild, A. Rozenknop, V. Rogov, F. Lohr, D. Popovic, A. Occhipinti, A. S. Reichert, J. Terzic, V. Dotsch, P. A. Ney and I. Dikic, Nix is a selective autophagy receptor for mitochondrial clearance, *EMBO Rep* **11**: 45–51 (2010).
63. K. C. Marsden, J. B. Beattie, J. Friedenthal and R. C. Carroll, NMDA receptor activation potentiates inhibitory transmission through GABA receptor-associated protein-dependent exocytosis of GABA(A) receptors, *J Neurosci* **27**: 14326–14337 (2007).
64. P. J. Maier, I. Marin, T. Grampp, A. Sommer and D. Benke, Sustained glutamate receptor activation down-regulates GABAB receptors by shifting the balance from recycling to lysosomal degradation, *J Biol Chem* **285**: 35606–35614 (2010).
65. J. N. Hislop and M. Von Zastrow, Role of ubiquitination in endocytic trafficking of G-protein-coupled receptors, *Traffic* **12**: 137–148 (2010).
66. S. K. Kaasinen, L. Harvey, A. J. Reynolds and I. A. Hendry, Autophagy generates retrogradely transported organelles: a hypothesis, *Int J Dev Neurosci* **26**: 625–634 (2008).
67. D. J. Katzmann, M. Babst and S. D. Emr, Ubiquitin-dependent sorting into the multivesicular body pathway requires the function of a conserved endosomal protein sorting complex, ESCRT-I, *Cell* **106**: 145–155 (2001).
68. D. J. Katzmann, G. Odorizzi and S. D. Emr, Receptor downregulation and multivesicular-body sorting, *Nat Rev Mol Cell Biol* **3**: 893–905 (2002).
69. R. C. Aguilar, M. Boehm, I. Gorshkova, R. J. Crouch, K. Tomita, T. Saito, H. Ohno and J. S. Bonifacino, Signal-binding specificity of the mu4 subunit of the adaptor protein complex AP-4, *J Biol Chem* **276**: 13145–13152 (2001).
70. S. Matsuda, E. Miura, K. Matsuda, W. Kakegawa, K. Kohda, M. Watanabe and M. Yuzaki, Accumulation of AMPA receptors in autophagosomes in neuronal axons lacking adaptor protein AP-4, *Neuron* **57**: 730–745 (2008).

71. A. N. Ngatchou, K. Kisler, Q. Fang, A. M. Walter, Y. Zhao, D. Bruns, J. B. Sorensen and M. Lindau, Role of the synaptobrevin C terminus in fusion pore formation, *Proc Natl Acad Sci USA* **107**: 18463–18468 (2010).
72. S. H. Lee, A. Simonetta and M. Sheng, Subunit rules governing the sorting of internalized AMPA receptors in hippocampal neurons, *Neuron* **43**: 221–236 (2004).
73. J. Bendor, J. E. Lizardi-Ortiz, R. I. Westphalen, M. Brandstetter, H. C. Hemmings, Jr., D. Sulzer, M. Flajolet and P. Greengard, AGAP1/AP-3-dependent endocytic recycling of M5 muscarinic receptors promotes dopamine release, *EMBO J* **29**: 2813–2826 (2010).
74. J. G. Pinto, K. R. Hornby, D. G. Jones and K. M. Murphy, Developmental changes in GABAergic mechanisms in human visual cortex across the lifespan, *Front Cell Neurosci* **4**: 16 (2010).
75. T. K. Hensch, M. Fagiolini, N. Mataga, M. P. Stryker, S. Baekkeskov and S. F. Kash, Local GABA circuit control of experience-dependent plasticity in developing visual cortex, *Science* **282**: 1504–1508 (1998).
76. E. Fino, V. Paille, Y. Cui, T. Morera-Herreras, J. M. Deniau and L. Venance, Distinct coincidence detectors govern the corticostriatal spike timing-dependent plasticity, *J Physiol* **588**: 3045–3062 (2010).

CHAPTER 16

NEURONAL MITOCHONDRIAL TRANSPORT AND TURNOVER VIA MITOPHAGY

Zu-Hang Sheng and Charleen T. Chu

ABSTRACT

Mitochondria are essential organelles for neuronal survival and function due to their key roles in ATP generation, calcium buffering, and apoptotic signalling. As neurons are highly polarized cells, the complex mechanisms of mitochondrial transport and docking are particularly important for neuronal development and synaptic function. Mitochondria are transported to and dock in areas with high demand for ATP and calcium homeostasis. Conversely, effete, or damaged mitochondria are transported for lysosomal degradation through the process of macroautophagy. Axonal mitochondria display complex mobility patterns and undergo saltatory and bidirectional movement. At any given time, approximately one-third of axonal mitochondria are mobile in cultured neurons. The balance between mobile and stationary mitochondria, however, is dynamic and responds quickly to changes in axonal and synaptic physiology. Defects in mitochondrial transport and turnover have been implicated in the pathologic processes of amyotrophic lateral sclerosis, Huntington's disease and Parkinson's disease. Further investigation into the mechanisms that regulate axonal mitochondrial transport, docking, and targeting for autophagy will provide important insight into these disease processes. Here, we highlight recent advances in our understanding of the motor-adaptor complexes and docking machinery that mediate mitochondrial transport and distribution along axons. We will also discuss new mechanisms underlying cargo recognition and autophagic turnover of damaged mitochondria, with particular reference to Parkinson's disease. The potential effects of altered mitochondrial transport or mitophagy on neuronal health and cell death reflect the particular dependence of neurons on mitochondrial metabolism.

INTRODUCTION

Mitochondria support many important neuronal functions including the mobilization of synaptic vesicles, assembly of the cytoskeleton within synaptic boutons, and reversal of ion influxes across their electrochemical gradients which is necessary for action potentials and maintenance of synaptic transmission. In addition to aerobic ATP production, mitochondria also serve to buffer Ca^{2+} at nerve terminals and may be involved in certain forms of short-term synaptic plasticity.[1,2] The loss of mitochondria from axon terminals impairs synaptic transmission.[3–6] Defective transport of axonal mitochondria is implicated in the pathogenesis of several major neurodegenerative diseases.[7]

It is thought that mitochondria are generated within the soma and delivered to their final destination via long neuronal processes. Mitochondria are transported down axons and stop at areas where metabolic demand is high, including active growth cones, axonal branches, synapses, nodes of Ranvier, myelination boundaries, and regions of demyelination.[8] Axonal distribution of mitochondria is characterized by dynamic and bidirectional movements along the microtubule (MT)-based cytoskeleton. Mitochondria frequently pause, change direction, or undergo persistent docking in certain regions. These complex mobility patterns are likely a result of mitochondrial coupling to the opposing anterograde and retrograde motors and docking/anchoring machinery.[9] Syntaphilin targets axonal mitochondria and mediates their stationary docking by anchoring to MTs.[2,10] Such a mechanism enables neurons to maintain proper mitochondrial density along axons and near synapses. While one-third of axonal mitochondria are mobile in mature cultured neurons, a large fraction remains stationary. The combination of dynamic saltatory and bidirectional movement results in a net velocity of 0.3–2.0 μm/sec.

Axonal and synaptic structures undergo spontaneous and activity-dependent remodeling. Mitochondrial mobility and distribution patterns continually change to support alterations in neuronal activity. Elevated cytosolic Ca^{2+} and increased synaptic activity result in recruitment of mobile mitochondria to the stationary pool near synapses.[11] In addition, mitochondrial trafficking and anchoring are tightly regulated during development. Thus, the mechanisms controlling the balance between the mobile and stationary phases of axonal mitochondria are targets of regulation. Recent identification of the mitochondrial KIF5 motor-adaptor complexes and Miro as a calcium sensor that controls mitochondrial movement may provide molecular targets for regulation.[12–15]

Conversely, the issue of how effete or damaged mitochondria are recognized and transported back to the soma for degradation represents an important emerging area for basic and disease-oriented research. Poorly functional mitochondria are not only less efficient at supporting energy demands at the synapses, but also may release potentially harmful mediators to include reactive oxygen species, dysregulated calcium and pro-apoptotic proteins. Macroautophagy (hereafter referred to as autophagy) is the major degradation system for organelles and long-lived proteins, playing important roles in mitochondrial quality control.[16–18] Sequestration of damaged mitochondria into autophagosomes may serve to limit potentially deleterious leakage of these mediators, prior to their eventual delivery to lysosomes for full degradation.

Microtubule-dependent processes are important for mammalian autophagosome delivery to lysosomes,[19] although the question of whether damaged mitochondria are directly transported back to the soma, or are transported as autophagic cargo, remains unresolved. Recent data implicate proteins involved in recessive forms of Parkinson's disease in regulating mitochondrial function, Miro-related trafficking and autophagy.[20–23] An improved understanding of regulatory mechanisms involved in anterograde and retrograde mitochondrial transport and mitophagy will thus advance our knowledge of fundamental neuronal processes that support axonal and synaptic homeostasis.

MOTOR PROTEINS DRIVING MITOCHONDRIAL TRANSPORT

The neuronal cytoskeleton is comprised of MTs, actin, and intermediate neurofilaments (NFs). The long-range neuronal transport of mitochondria predominantly occurs along MTs. While dendritic MTs exhibit mixed polarity, axonal MTs are organized into a polar array whose "plus" end is directed distally and "minus" end is directed toward the soma. The targeted transport of mitochondria to distal regions of axons and synaptic terminals depends upon MT polarity and organization.

Kinesin superfamily members (KIFs) and cytoplasmic dynein are the main MT-associated motor proteins that drive long-distance axonal transport via ATP hydrolysis.[24] While cytoplasmic dynein motors are responsible for retrograde movement toward the soma,[9] kinesin motors drive the anterograde transport of mitochondria to distal axonal regions and synaptic terminals (Figure 1).[3,25,26]

At least 45 different kinesin motor genes have been identified in humans and mice. Kinesin-1 (or KIF5), the first motor identified to drive plus end-directed transport *in vitro,* contains two heavy chains (KHCs) and two light chains (KLCs).[27] Its motor function arises from homo- or heterodimer formation in the coiled-coil regions of the heavy chains in the stalk domain (Figure 2A). Mammals have three KIF5 motor isoforms (A, B and C), while *Drosophila* have only one. KIF5B is present in most cell types. KIF5A and C are neuron-specific motors. Each KIF5 heavy chain contains an amino terminal motor domain that binds

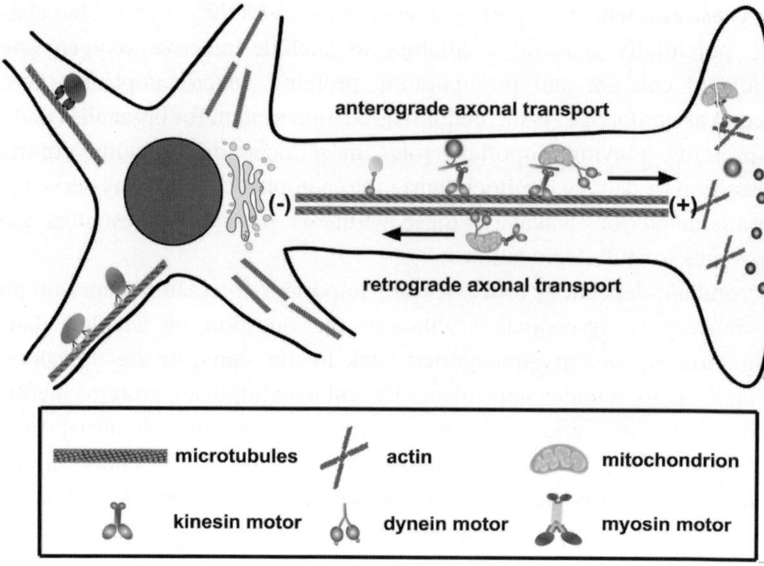

Figure 1. Axonal mitochondrial transport. Axonal microtubules are organized so that their plus (+) ends are directed toward axon terminals and their minus ends (-) toward the cell body. In dendrites, however, microtubules exhibit mixed orientation. Microtubule polarity and organization in axons allows motor proteins to facilitate the targeted transport of synaptic cargoes and organelles. Anterograde axonal mitochondrial transport is driven by plus-end directed kinesin motors, while minus-end directed dyneins mediate retrograde mitochondrial transport along axons. Short-range mitochondrial movement in nerve terminals involves transport by actin-based myosin motors. (Adapted with permission from Ref. 100.)

directly to MTs, whereas its carboxyl terminal domain, named the cargo-binding domain, mediates the association of KIF5 with KLC or directly interacts with mitochondrial adaptor proteins. Recent studies, however, indicate that the KIF5 cargo-binding domain attaches directly to mitochondrial adaptor proteins to facilitate mitochondrial transport.

KIF5 has a prominent role in the anterograde transport of axonal mitochondria.[3,26] Under normal conditions, mitochondria are present in the cytoplasm of undifferentiated extra-embryonic cells and are transported toward the cell periphery. Targeted disruption of *kif5b*, however, results in the perinuclear clustering of mitochondria.[25] *Drosophila kif5* mutants display impaired mitochondrial transport and reduced mitochondrial distribution in larval motor axons.[9] Biochemical and imaging analyses show that KIF5 motors are associated with brain mitochondria. Furthermore, hippocampal neurons expressing the KIF5 cargo-binding domain exhibit impaired axonal mitochondrial transport and reduced mitochondrial

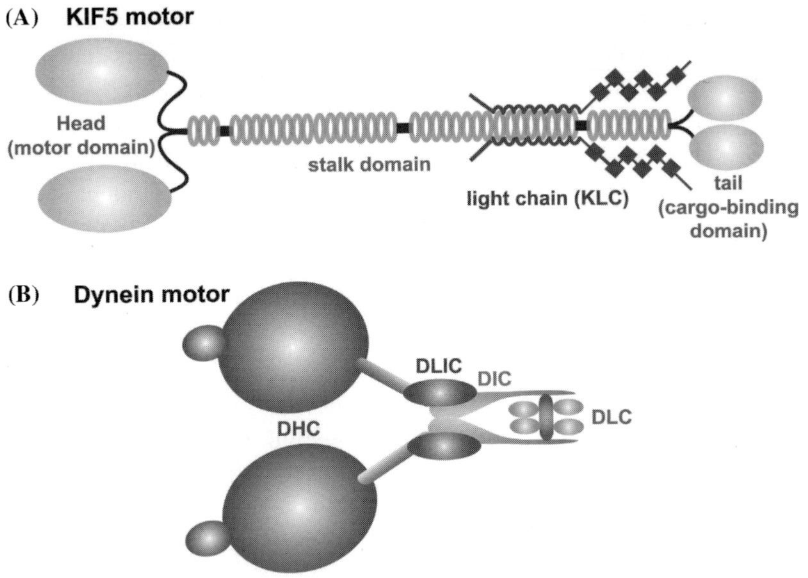

Figure 2. Structure of KIF5 and dynein motor proteins. (**A**) KIF5 motors form homodimers through coiled-coil regions in their stalk domains. KIF5 itself possesses motor function. It also binds to the light chain of kinesin-1 (KLC) through its domain located at the stalk and tail regions. Thus, there are two forms of KIF5 motor-cargo coupling. KIF5 can associate with organelles and transport cargoes directly though the cargo-binding region in its tail domain or indirectly via the C-terminal domain of KLC. (**B**) Cytoplasmic dyneins are composed of heavy chains (DHC), light intermediate chains (DLIH), intermediate chains (DIC) and light chains (DLC). To transport cargoes, cytoplasmic dynein needs to bind with the dynactin complex (not shown). (Adapted with permission from Ref. 101.)

density in distal axons due to disrupted linkage of KIF5 motor with mitochondrial adaptor.[26] These studies highlight KIF5's role in driving axonal anterograde transport of mitochondria.

Another brain-enriched kinesin, KIF1Bα of the kinesin–3 family, is reported to interact directly with mitochondria.[28] It has been shown to transport purified mitochondria at a rate of 0.5 μm/sec *in vitro*. Although *kif1B* mutation leads to peripheral neuropathies in mice, the proposed role of KIF1Bα in anterograde mitochondrial transport requires further characterization.

Retrograde axonal transport is driven primarily by cytoplasmic dynein. Dynein is a multisubunit complex composed of two heavy chains (DHCs) and several intermediate (DICs), light intermediate (DLICs) and light chains (DLCs). Dynein heavy chains function as motors, while other polypeptides are thought to mediate the association of the dynein motor with cargoes or to regulate motility (Figure 2B). In addition, dynactin, a large 11-subunit complex, can bind directly to cytoplasmic dynein and MTs through its p150^{Glued} component. Dynactin, while not essential, can enhance the processivity of the dynein motor, mediate interactions with cargoes,[29] and coordinate bidirectional axonal transport.[30]

Compared to the kinesin superfamily, few dynein heavy chains have been identified. Thus, a large number of accessory proteins, such as dynactin and DLCs, likely mediate cargo-binding specificity. Interestingly, although there is a strong bias towards the microtubule minus end, it has been shown that the dynein-dynactin complex can move bidirectionally.[31] This ability may enable the dynein motor to bypass obstacles during transport within the cell. Due to their length, motor neurons are especially vulnerable to defects in the dynein-dynactin complex and dynein function.

Dynein and dynactin associate with mitochondria purified from the *Drosophila* nervous system and are essential for mediating mitochondrial retrograde transport.[9] Mutations in the dynein heavy chain and dynactin p150^{Glued} disrupt bidirectional fast organelle transport and result in axonal swellings composed of mitochondria and other retrograde and anterograde cargoes. In addition, expressing a mutant p150^{Glued} causes mitochondrial accumulation within the cell body.[32] Dynein has been observed co-localized with mitochondria moving in the anterograde and retrograde directions.[33] It appears likely that the direction in which an individual mitochondrion moves is determined by a tug-of-war model between opposing kinesin and dynein motors. If this is the case, each mitochondrion likely couples to two opposing kinesin and dynein motors that coordinate bidirectional movement of axonal mitochondria. Several lines of evidence support the functional coupling of two opposing motors on individual mitochondrion. First, a *Drosophila* homologue of the mammalian kinesin-interacting JIP-1/JNK scaffold protein, APLIP1, is implicated in retrograde mitochondrial transport.[34] Second, a *Drosophila* mitochondria adaptor protein, dMiro, helps regulate both anterograde and retrograde axonal mitochondrial transport.[35] Further studies are necessary to determine if motor adaptors facilitate the crosstalk between the two opposing motors and how the motor-adaptor complexes coordinate bidirectional mitochondrial movement along axons in response to changes in neuronal physiology and synaptic activity.

While long-range mitochondrial transport along neuronal processes is carried out by kinesins and dyneins, local and short-range transport of mitochondria within actin-rich nerve terminals and growth cones is likely driven by myosin motors. Although actin-based mitochondrial mobility is well characterized in plant cells and budding yeast, myosin-mediated mitochondrial transport has not yet been demonstrated in neurons. It has been shown, however, that mitochondria and myosin V-associated cargoes move along axons at a similar rate that is markedly slower than other axonal organelles. The movement of mitochondria along actin filaments supports the idea that myosin motors may also drive short-range mitochondrial transport within certain regions of axons, particularly at nerve terminals (Figure 1).

It was recently demonstrated in cultured *Drosophila* neurons that myosin V and VI, but not myosin II, modulate axonal mitochondrial transport.[36] Myosin V depletion increases mitochondrial mean velocity in both anterograde and retrograde directions while myosin VI loss-of-function selectively increases mitochondrial retrograde transport in axons. As myosin V can form hetero-motor complexes by interacting with kinesin or the 8-kDa dynein light chain, a coupled "dual transport" model has been proposed in which actin-based myosin motors permit local short-range movement, while long-range axonal transport is carried out by MT-based motors. It is also possible that myosins may promote mitochondrial docking on the actin-based cytoskeleton by facilitating their dissociation from MT tracks. In this scenario, myosins may increase the frequency of short pauses during long-range bidirectional axonal transport. While these two models are attractive, further investigation is needed to determine how myosin motors help regulate MT-based motor activity and if they are required for synapse-directed mitochondrial transport.

ADAPTORS LINKING MOTORS TO MITOCHONDRIA

Cargo-containing vesicles attach to transport motors with a high degree of specificity to preserve cargo identity and enable targeted trafficking. Kinesin-driven anterograde neuronal transport is achieved by associating with their cargoes via adaptors (or receptors). To understand the mechanisms regulating neuronal mitochondrial transport and distribution, identification and characterization of mitochondrial receptors and motor-adaptor complexes is necessary.

The *Drosophila* protein Milton is a well-characterized KIF5 motor-adaptor component essential for mitochondrial transport in neurons. It attaches to mitochondria indirectly through an interaction with Miro, a member of the Rho-GTPase family present in the mitochondrial outer membrane. The *milton* mutation results in the depletion of mitochondria at synaptic terminals.[3] Miro contains two EF hand Ca^{2+}-binding motifs and is required for connecting Milton and KIF5 to mitochondria (Figure 3A). Mutation of *miro* impairs neurotransmitter release during prolonged stimulation of *Drosophila* neurons by disrupting anterograde mitochondrial transport and preventing mitochondria from entering the axon and synapses.[4]

Two mammalian Milton orthologues (TRAK1 and TRAK2) and two mammalian orthologues of dMiro (Miro1 and Miro2) have been identified. Several

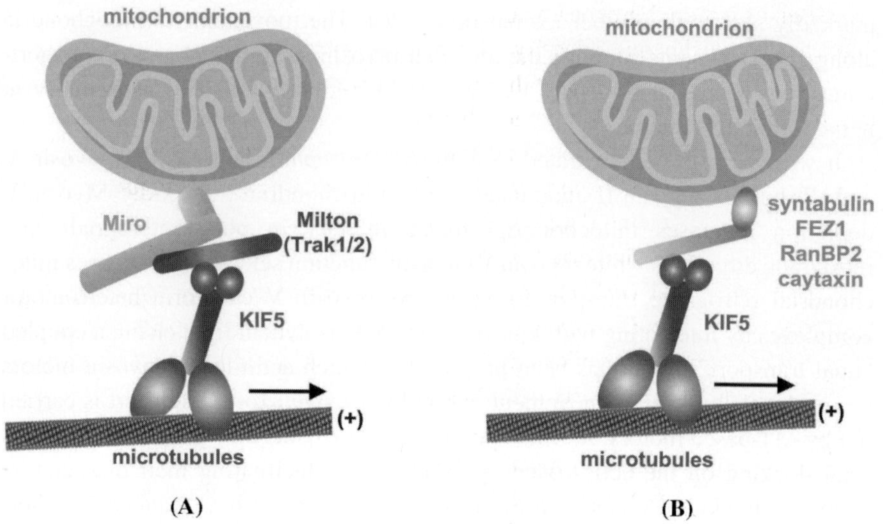

Figure 3. KIF5 motor adaptors for mitochondrial transport. (**A**) Milton-Miro adaptor complex. Miro is a mitochondrial outer membrane protein of the Rho-GTPase family. Independent of KLC, Milton recruits KIF5 to mitochondria by binding Miro. TRAK 1 and TRAK2 (mammalian Milton orthologues) can form complexes with Miro1 and Miro2 (mammalian orthologues of dMiro). (**B**) KIF5 associates with mitochondria and mediates mitochondrial anterograde transport via Syntabulin, a KIF5 adaptor that targets to mitochondria via its C-terminal transmembrane domain. FEZ1, RanBP2 and caytaxin have also been identified as additional candidate kinesin motor adaptors that may contribute to mitochondrial transport. (Adapted with permission from Ref. 101.)

functional similarities and differences exist between Milton and its TRAK orthologues. While Milton functions as a mitochondrial adaptor that links Miro to KIF5 in *Drosophila* (Figure 3A), TRAKs, each exhibiting ~34% homology with Milton, have transport roles of other intracellular cargos and organelles in addition to mitochondria. Although TRAKs are important for mitochondrial trafficking, it is unclear if they function just as a linker of Miro and KIF5 or as regulatory elements of motor activity.

Another KIF5 motor adaptor, syntabulin, is involved in the axonal transport of presynaptic components and is required for proper presynaptic assembly and activity-dependent plasticity during development.[37,38] In addition, syntabulin serves to link the KIF5 motor and mitochondria by inserting its carboxyl-terminal transmembrane domain into the outer mitochondrial membrane. This coupling plays an important role in the anterograde transport of neuronal mitochondria.[26] Knocking down syntabulin with siRNA or blocking the syntabulin–KIF5 interaction with binding-domain transgenes impairs mitochondrial transport to the distal processes of cultured hippocampal neurons. The resultant mitochondrial distribution, characterized by high-density clusters in the soma and low-density distribution along processes, is consistent with time-lapse imaging in live neurons. Syntabulin loss-of-function reduces anterograde, but not retrograde, transport of mitochondria along axonal processes. This phenotype suggests that syntabulin mediates mitochondrial anterograde transport by linking KIF5 to mitochondria (Figure 3B).

Several other proteins have been proposed as candidate adaptors that link the KIF5 motor and mitochondria (Figure 3B). One of these, fasciculation and elongation protein zeta-1 (FEZ1), is a brain-specific coiled-coil protein involved in axonal outgrowth. Its *Drosophila* orthologue, UNC-76, has been shown to enable axonal transport of organelles and vesicles via an association with the kinesin motor. Additionally, FEZ1 is essential for the anterograde axonal transport of mitochondria in the hippocampus. Caytaxin, a hereditary ataxia protein, associates with kinesin-1 through a direct interaction with KLC. It co-localizes with mitochondria in both axons and dendrites and is required for the proper distribution of mitochondria in distal neurites. It has been proposed that caytaxin serves as a protein scaffold in phosphatidylinositol-calcium-signaling pathways and may link cargoes to kinesin-1 by binding KLC. RanBP2, which co-localizes and interacts directly with KIF5B and KIF5C, is another candidate adaptor. The association of its KIF5-binding domain with KIF5B and KIF5C has been shown to regulate mitochondrial transport in non-neuronal cells. It has not been demonstrated, however, that this protein plays a role in axonal mitochondrial transport.

The mechanisms that link the dynein motor and mitochondria are not well characterized. For example, dynaction, traditionally thought to link dyneins and cargo membranes, was recently proposed to play a role in the coordination of opposing molecular motors.[30] Another model suggests that dyneins may link to membrane-associated proteins via their DLCs and DICs, independent of dynactin. Supporting this theory, a dynein light chain (Tctex1) is thought to associate with the mitochondrial outer membrane via an interaction with the voltage-dependent anion-selective channel 1.[39] Miro, a KIF5 adaptor for mitochondria, is required for dynein-mediated retrograde transport of mitochondria in addition to kinesin-driven anterograde transport in *Drosophila* axons.[35] Thus, although an association of Miro and dynein has not been demonstrated, it appears that Miro may serve as a general mitochondrial motor adaptor for both KIF5 and dynein.

MITOCHONDRIAL DOCKING IN AXONS

In mature axons, approximately two-thirds of the axonal mitochondrial population remains stationary at any given time. The presence of stationary mitochondria may reflect their dissociation from MTs or transport motors, or their association with static anchors or docking receptors that stably attach mitochondria to the axonal cytoskeleton. Neurons utilize special mechanisms to recruit mobile mitochondria to the stationary pool in response to changes in neuronal and synaptic activity to achieve the desired balance between mobile and stationary axonal mitochondria. These mitochondrial-docking mechanisms may serve to maintain the proper density of stationary mitochondria along axons and near synapses where energy production and calcium buffering capacity are especially important. As ATP has a limited diffusion capacity in the intracellular environment, it is likely that stationary mitochondria serve as local energy stations that provide the ATP necessary to maintain the activity of Na^+–K^+ ATPase, fast spike propagation, and synaptic transmission. Identifying axonal mitochondrial-docking receptors will provide new molecular targets that may allow regulation of axonal mitochondrial distribution and transport.

A recent study utilizing genetic mouse models and time-lapse imaging identified syntaphilin (SNPH) as a "static anchor" for axonal mitochondria.[2] It is neuron-specific and targets to axonal mitochondria via its C-terminal tail (Figure 4A). SNPH immobilizes mitochondria by anchoring to the MT-based cytoskeleton. Deleting the murine *snph* dramatically increases mitochondrial mobility and decreases the density of stationary mitochondria within axons (Figure 4B). Conversely, overexpressing SNPH in neurons arrests axonal mitochondrial transport. The study proposes that SNPH is an anchoring receptor that facilitates axonal mitochondrial docking, thereby allowing neurons to maintain proper stationary mitochondrial density in axons and synapses (Figure 4C). Additionally, the dynein light chain, LC8, can regulate SNPH-mediated docking, suggesting that the mitochondrial mobility can be regulated through the dynamic interactions between mitochondrial-docking receptors and the axonal cytoskeleton.[10] SNPH may be a useful molecular target to further investigation into how mobile mitochondria are recruited to the stationary pool in response to changes in axonal activity and synaptic modification. Studies concerning the physiological impact of mitochondrial docking on synaptic transmission, axonal homeostasis, and axonal degeneration will benefit from the *snph* mutant mouse model. There is mounting evidence that defective mitochondrial trafficking is involved in the pathology of several neurodegenerative diseases. It is expected that defective mitochondrial docking/anchoring could also affect neuronal function, particularly in motor neurons due to their long axonal processes. Future studies which cross *snph* mice with these disease mouse models may increase the transport of dysfunctional mitochondria to the soma where they may be repaired or degraded via mitophagy.

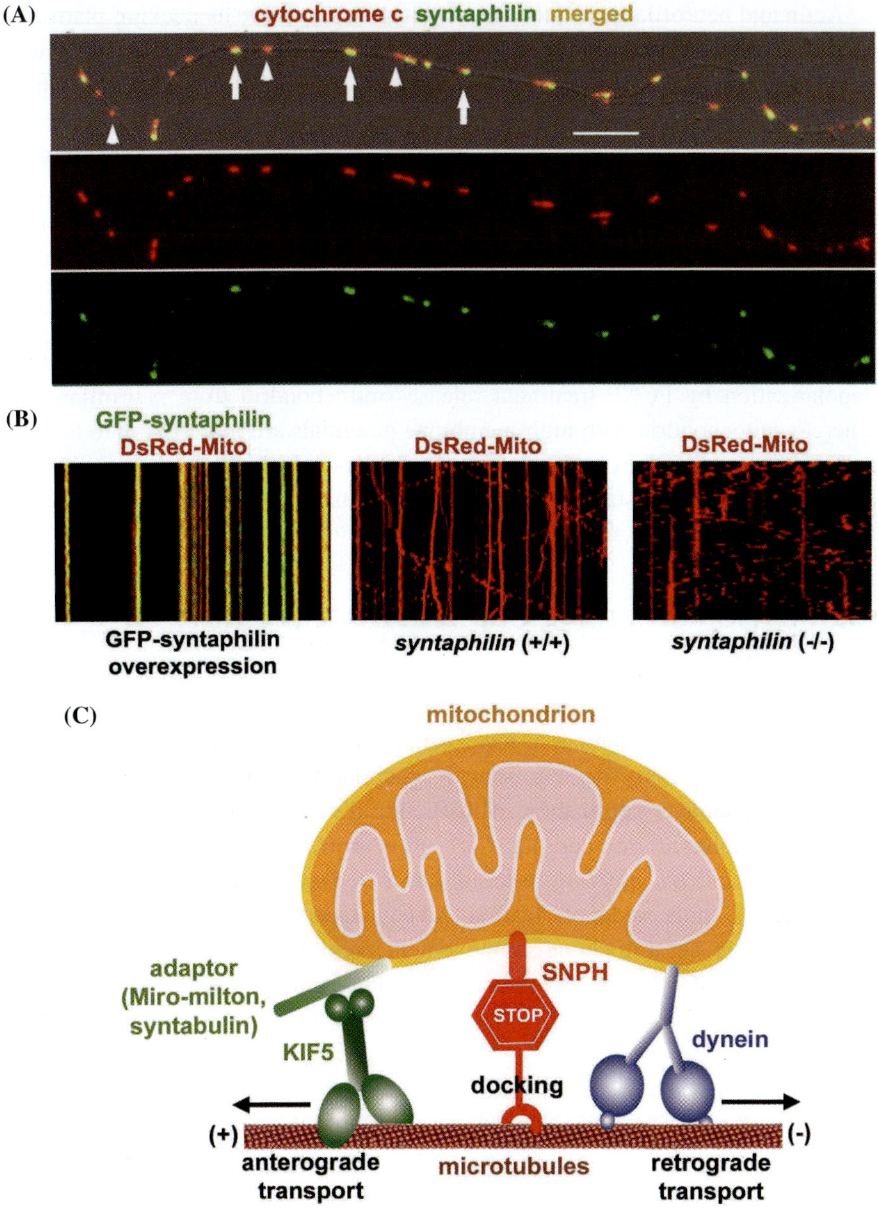

Figure 4. SNPH acts as a receptor for docking axonal mitochondria. (**A**) Axonal mitochondrial targeting of SNPH in cultured hippocampal neurons. Hippocampal neurons were co-immunostained for SNPH and mitochondrial marker cytochrome c. Arrows indicate axonal SNPH-associated mitochondria, and arrowheads point to axonal SNPH-negative mitochondria. Scale bars, 10 μm. (**B**) SNPH

Actin and neurofilaments have also been shown to serve as docking platforms for both axonal and dendritic mitochondria, although their docking receptors have yet to be identified. Actin is relatively enriched in neuronal subdomains where ATP production is in high demand, such as growth cones, presynaptic terminals, and dendritic spines. Neuronal activity induces actin-dependent mitochondrial retention within dendritic spines.[13] In axons, disrupting the actin cytoskeleton and actin-based myosin motors results in increased mitochondrial movement.[36] Additionally, actin is required for NGF-dependent local mitochondrial recruitment and tethering in axons.[40] In non-neuronal cells or *in vitro* systems, mitochondria tether to neurofilaments and neuron-specific intermediate filaments in a process likely depending upon their membrane potential. Inducing mitochondrial depolarization by FCCP treatment releases mitochondria from neurofilaments whereas mitochondria with high membrane potentials are more likely to bind to neurofilaments.[41] These observations support a model in which healthy mitochondria remain anchored stably along neurofilaments while dysfunctional ones are transported back to the soma for recycling.

REGULATION OF MITOCHONDRIAL TRANSPORT BY NEURONAL ACTIVITY

Axonal mitochondria undergo saltatory and bidirectional movements. Metabolic demand determines the relative levels of mobile and stationary mitochondria. Mitochondrial velocity decreases near synapses when local ATP is depleted following glutamate application. Mitochondrial mobility is also modulated in response to altered intracellular Ca^{2+} or neurotransmitter release. Blocking action potentials by applying tetrodotoxin increases mitochondrial movement, whereas KCL-induced depolarization and enhanced Ca^{2+} influx through NMDA receptors decreases mitochondrial movement in cultured neurons.[11] The application of N-type Ca^{2+} and Na^{2+} channel blockers inhibit synaptic activity and

Figure 4. (***Continued***) immobilizes axonal mitochondria and deletion of *snph* gene in mice increases the mobility of axonal mitochondria. The mobility of axonal mitochondria was assessed in live neurons one week after transfection. Motion data are represented in kymographs. Vertical lines represent stationary mitochondria, while slanted or cured lines indicate mobile ones. Left panel: wild-type neurons co-transfected at DIV6 with DsRed-mito (red) and GFP-SNPH (green); middle panel: wild-type neurons transfected with DsRed-Mito alone; right panel: *snph* (-/-) neurons transfected with DsRed-Mito alone. **(C)** Proposed role of SNPH. SNPH acts as a docking protein for anchoring mitochondria in axons and is required for maintaining a large number of axonal mitochondria in a stationary state by interacting with the microtubule-based cytoskeleton. (Images in A are adapted with permission from Ref. 2. Images in Figure 4B and the schematic diagram in Figure 4C are adapted with permission from Ref. 100.)

increase mitochondrial transport. Furthermore, the duration of intracellular Ca^{2+} elevation corresponds to decreased mitochondrial movements. Mitochondria are often retained at presynaptic terminals and postsynaptic dendritic spines where they help maintain Ca^{2+} homeostasis via Ca^{2+} sequestration and providing ATP to pump Ca^{2+} across the plasma membrane. During sustained synaptic activity neurons utilize special mechanisms to achieve activity-dependent retention of mitochondria.

Thus, an important question is: how are mobile mitochondria recruited to the stationary pool in response to elevated cytosolic Ca^{2+} and sustained synaptic activity? An important clue regarding Ca^{2+}-dependent regulation of mitochondrial mobility comes from our evolving understanding that the KIF5 motor-adaptor Miro can bind Ca^{2+} via EF hand domains. Synaptic recruitment of mitochondria in response to increased action potential firing rates and activation of glutamate receptors depends upon a Miro-mediated Ca^{2+}-sensing pathway. Mitochondria are immobilized when Miro binds Ca^{2+}. This Ca^{2+}-induced cessation of mitochondrial mobility was effectively blocked in neurons expressing a mutant Miro EF hand incapable of binding Ca^{2+}. Two mechanisms have been proposed to explain how Ca^{2} mediates suppression of mitochondrial mobility (Figure 5).

Wang and Schwarz[14] propose a motor-adaptor switch model. They have shown that the C-terminal tail of KIF5 binds mitochondria in the absence of Ca^{2+} through an interaction with the Milton-Miro complex, which allows its N-terminal motor domain to facilitate transport by engaging microtubules. Ca^{2+} binding prevents the KIF5-microtubule interaction by inducing a direct interaction between the motor domain and Miro. Thus, in this model, a Ca^{2+}-binding switch mechanism that turns motor-microtubule engagement "on" or "off" controls mitochondrial mobility (Figure 5A). Alternatively, Macaskill *et al.*[13] propose that Miro mediates mitochondrial transport by linking mitochondria to KIF5. In this model, Ca^{2+} binding to the EF hands releases KIF5 from Miro-bound mitochondria. Thus, activity-triggered Ca^{2+} influx could recruit mobile mitochondria to activated synapses (Figure 5B).

Although these models are attractive, several questions remain. Both anterograde and retrograde mitochondrial transport are affected by Miro-mediated and Ca^{2+}-dependent mitochondrial immobilization. It is unclear, however, how KIF5 motor-adaptor Miro affects the retrograde dynein transport complex. It has been suggested that bidirectional mobility patterns of axonal mitochondria arise from the competition between opposing KIF5 and dynein motors that are simultaneously bound to individual mitochondria. However, when kinesin-driven movement is disturbed, dynein does not simply take over. Thus, it is likely that motor-adaptors coordinate the activity of the opposing motors. The recent

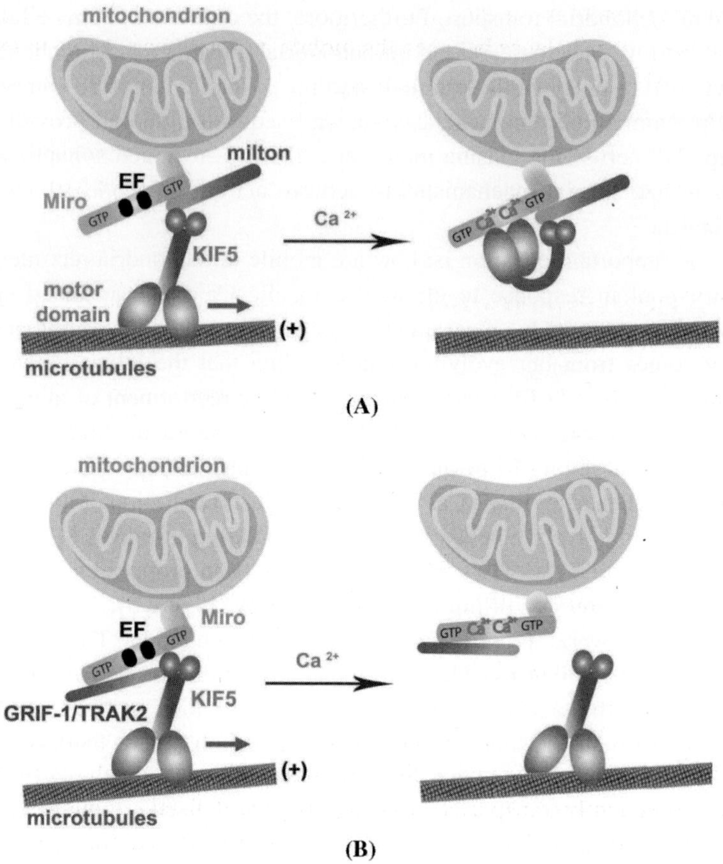

Figure 5. Two models of Miro as a Ca^{2+} sensor that regulates mitochondrial mobility. Miro contains two GTPase domains and calcium-binding EF hand motifs. Thus, it may regulate axonal mitochondrial mobility through GTP hydrolysis or in response to calcium binding induced by synaptic activity and calcium signaling pathways. **(A)** Model 1: Ca^{2+}-binding discourages KIF5 engagement with MTs. The KIF5 motor domain is free to engage MTs while its tail links to Miro via Milton in a Ca^{2+}-independent manner. Ca^{2+}-binding to the EF hands triggers the direct interaction of the motor domain with Miro, preventing the motor from engaging with MTs.[14] **(B)** Model 2: Ca^{2+}-binding detaches KIF5 from mitochondria. Ca^{2+}-binding to the EF hands dissociates Miro from KIF5 while KIF5-binding protein GRIF-1/TRAK2 (a mammalian homologue of Milton) remains bound to Miro1.[13] This diagram is adapted with permission from Ref. 15.)

identification of syntaphilin as a "static anchor" for axonal mitochondria presents another promising mechanism (Figure 4C).[2] Another question is whether the motor-adaptor complex and docking receptors share a single regulatory system. Additionally, do docking interactions drive the physical displacement of the motor-adaptor complex (or vice versa)? And what are the neuronal signals that maintain the proper balance between the mobile and stationary pools in response to changes in neuronal activity and regulate the interplay between motor-docking machineries? Future studies using genetic mouse models will provide insight into the molecular mechanisms underlying the complex regulation of mitochondrial bidirectional movement and their temporal-spatial localization in axons and near synapses.

REGULATION OF MITOCHONDRIAL MOBILITY BY NEURONAL SIGNALING PATHWAYS

Mitochondria are efficiently transported along nascent axons and recruited to the stationary pool within active growth cones, branches of developing neurons, nodes of Ranvier, myelination boundaries, sites of axonal protein synthesis, and synaptic terminals. Developing neurons exhibit high levels of anterograde mitochondrial movement and increased mitochondrial density along elongating axons.[9,42] The significant reduction in mitochondrial density that occurs when axons cease growing is likely due to an increase in retrograde mitochondrial movement.

The mitochondrial membrane potential influences the direction of their transport. Mitochondria exhibiting high membrane potentials move anterogradely, while those with low membrane potentials tend to move toward the cell body.[43] These findings support the notion that the soma generates healthy mitochondria that are subsequently transported along axons toward distal regions or synapses, while aged or damaged mitochondria are delivered to the soma where they undergo reparative fusion with healthy mitochondria or are degraded through mitophagy.[44] It must be noted, however, that early autophagosomes containing mitochondria have been reported in distal neuritic processes, indicating that such retrograde transport is not obligatory for mitophagy. Within growth cones of chick sensory neurons, receptor-mediated increases in mitochondrial membrane potential can be elicited by NGF and semaphorin signaling. This suggests that important local cell signaling pathways may regulate neuronal mitochondrial membrane potential, thus helping determine mitochondrial transport direction and distribution. This attractive model, however, remains to be further tested as a recent study could not find a significant difference in membrane potentials between stationary, anterogradely or retrogradely moving mitochondria.[45]

In dorsal root ganglion (DRG) neuron cultures, NGF serves as a docking signal that immobilizes axonal mitochondria. The process depends upon the PI(3) kinase phosphorylation pathway and docking probably occurs on F-actin filaments.[40] Axonal mitochondrial transport can be stimulated by activation of the 5-HT1A receptor and Akt pathway by serotonin.[46] In hippocampal neurons, dopamine inhibits mitochondrial mobility via the same signaling cascade. Class I PI(3) kinase regulates Akt activity, and mitochondrial transport is likely altered by NGF and serotonin/dopamine signaling pathways through a conserved regulatory mechanism controlled by Class I PI(3) kinase. It has been reported that nitric oxide (NO) can modulate mitochondrial mobility. Reduced mitochondrial membrane potentials, which result in rapid termination of mitochondrial movement, are associated with elevated NO levels.[47]

As organelle size affects mitochondrial mobility, alterations of mitochondrial fusion/fission events can also regulate mitochondrial transport. Fusion requires the apposition of two adjacent organelles. The coordination of neuronal mitochondrial transport and membrane fusion/fission events likely involves similar transport mechanisms. For example, mutation of a mitochondrial dynamin-like GTPase, Drp1, interferes with the fission process, resulting in impaired mitochondrial axonal transport and synaptic targeting.[5] Deletion or mutation of an outer mitochondrial membrane protein involved in mitochondrial fusion, mitofusin (Mfn2), impairs anterograde and retrograde neuronal mitochondrial transport. Expressing mutated Mfn2 in neurons significantly impairs mitochondrial transport in axons.[48] Therefore, there is probably crosstalk between membrane fusion/fission proteins and transport machinery that involves an interaction between Mfn2 and the Miro/Milton complex.

DEFECTIVE MITOCHONDRIAL TRANSPORT AND NEURODEGENERATION

Proper axonal mitochondrial transport and distribution are critical for neuronal function. Mitochondria in neurons must travel long distances — up to one meter, in some cases — from the cell body to their destinations. Defective trafficking of axonal mitochondria has been implicated in the pathologic processes of neurodegenerative diseases such as Alzheimer's, Huntington's, and amyotrophic lateral sclerosis (ALS).[7] Thus, further research into mitochondrial transport will benefit our understanding of neurodegeneration.

Recent studies demonstrate that Huntingtin-linked *Htt72Q* and *Htt150Q* transgenic mice display abnormal neuronal mitochondrial mobility characterized by slow movement, frequent stops and decreased range that occurs before mitochondrial dysfunction.[49,50] ALS is a late onset neurodegenerative disease that specifically affects motor neurons. The ALS-linked *SOD1*^{G93A} transgenic mouse line and NSC34 motor neuron cells or cultured neurons expressing the h*SOD1*^{G93A} mutant all show altered axonal mitochondrial transport.[51–53] Furthermore, mutant hSOD1 in ALS models has been shown to interact with both the anterograde motor protein kinesin-2 complex, via kinesin-associated protein 3 (KAP3), and the dynein-dynactin retrograde motor protein complex providing insight into the molecular mechanism by which SOD1 may interfere with axonal transport.

Whether altered mitochondrial transport plays a critical role in axonal degeneration or is a side effect of general axonal transport defects has been a subject for debate. Recent advances describe the mechanistic role of mutant SOD1 in altering axonal mitochondrial dynamics and in its association with impaired mitochondrial maintenance and recycling. These studies, however, do not directly address whether defective mitochondrial transport plays a causative role in axonal degeneration or simply reflects sequelae of general defects in axonal transport. No studies definitively implicate axonal transport of mitochondria as a causative factor in the pathogenesis of ALS-linked *SOD1*^{G93A} mice. Thus, it is important to assess the impact of mitochondrial mobility on axonal degeneration, in particular, to address whether enhanced axonal mitochondrial mobility may have a positive impact on the pathogenesis of the disease mouse model by contributing to the removal of dysfunctional mitochondria from synapses and distal axons.

MACROAUTOPHAGY: THE END OF THE MITOCHONDRIAL LIFE CYCLE

Autophagy comprises a set of lysosome-dependent degradation pathways, which play important roles in cellular homeostasis and pathophysiologic responses to nutritive, oxidative and other stressors. There are several mechanisms by which degradative cargo is delivered to the lysosome, including chaperone-mediated passage through a lysosomal translocation pore and direct invagination of the lysosomal membrane. However, cargo size is typically limiting for these mechanisms, and it is believed that mitochondrial autophagy in mammalian cells occurs predominantly through the process of macroautophagy (hereafter referred to as autophagy).

In brief, autophagic recycling is initiated by **integration** of a growing number of autophagy-promoting signaling pathways with pro-growth autophagy-inhibiting pathways such as the mammalian target of rapamycin (mTOR). This regulates the process of **nucleation** (Figure 6A) whereby the membrane site from which the nascent autophagosome will grow is specified. The process is regulated in part through alteration of the local phosphatidylinositol 3-phosphate content by the Class III PI(3) kinase (yeast Vps34; human *PIK3C3*) in complex with Beclin 1 and Atg14L. **Formation** of the autophagosome depends upon covalent conjugation events involving ubiquitin-like proteins Atg12 and microtubule associated protein 1 light chain 3 (LC3/Atg8). The Atg12-5 complex is targeted to the membrane through interaction with Atg16L to promote covalent conjugation of LC3 to phosphotidylethanolamine, which is necessary for extension and closure of autophagic membranes. As discussed below, LC3 is pivotal not only in autophagosome formation, but also in aspects of **cargo recognition/enrichment** (Figures 6B and 6C). Autophagosome **maturation** involves vesicular transport and fusion with components of the endolysosomal system, resulting in acidification and delivery of lysosomal hydrolases to complete the process of **lysosomal degradation**, followed by release and **biosynthetic re-utilization** of cellular building blocks.

Due to the continual dynamics of mitochondrial fission, fusion, transport and piecemeal delivery of newly synthesized nuclear encoded proteins for integration with mitochondrially translated respiratory chain subunits, it is important to realize that the concept of a mitochondrial life cycle represents a vast oversimplification. It is difficult to pinpoint mitochondrial "birth," as newly synthesized components are continuously added to pre-existing mitochondria, and fission causes more individual mitochondria, but without an increase in functional mass. It is important to recognize that individual mitochondrial proteins or complexes may undergo a more rapid degradative turnover than the whole organelle through several other mechanisms.[18] Nevertheless, under certain circumstances, entire segments of mitochondria can be triggered to undergo selective mitochondrial autophagy, or mitophagy.

SELECTIVE MITOPHAGY: A NEW PARADIGM

For many years, autophagy was considered to be a non-selective bulk degradation system, with some exceptions such as peroxisome-turnover in yeast. The observation of mitochondria within autophagosomes was thought to reflect bystander entrapment or observer bias, as this easily recognized cargo facilitates identification of an autophagosome. The first report demonstrating complete and selective autophagic degradation of mitochondria occurred in the context of neuronal cell death under conditions where rapid apoptotic execution was inhibited.[54] Apoptosis inhibitors serve only to delay, not prevent cell death, as mitochondrial removal then commits the neurons to death.

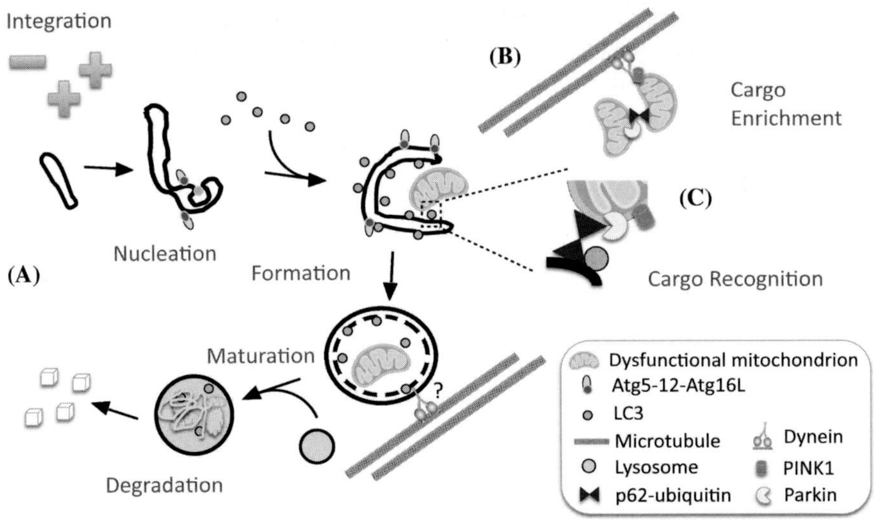

Figure 6. Schematic diagram illustrating major steps in mitochondrial autophagy. **(A)** Cellular conditions trigger-signaling pathways that promote and suppress autophagy. If integration of these signals results in autophagy induction, localized increases in endoplasmic reticulum phosphatidylinositol 3-phosphate content mediate the nucleation of autophagy protein deposition, to include the Atg5-12-16L complex, which serves to promote covalent conjugation of LC3 to phosphatidylethanolamine. LC3 lipidation is necessary for growth and complete formation of the autophagosome. Selective sequestration of cargo occurs during LC3-dependent extension of autophagic membrane, potentially through two distinct mechanisms that are not mutually exclusive **(B, C)**. The transiently double membrane autophagosome undergoes maturation by fusing with acidic compartments such as endosomes and lysosomes, a process that is often microtubule dependent. The cargo is degraded inside lysosomes, with release of cellular building blocks. **(B)** In the first model of cargo selection, enrichment for damaged mitochondria is mediated by selective transport to perinuclear microtubule organizing centers. Concentrating damaged cargo to be cleared in aggresomes, a process dependent upon p62 binding to ubiquitinated cargo, would enhance the relative efficiency of autophagic clearance even in the absence of more specific means of cargo enrichment. PINK1 interactions with Miro, or VDAC1 interactions with dynein may be involved in this process, as full length PINK1 is stabilized and VDAC1 is ubiquitinated by parkin on chemically depolarized mitochondria. **(C)** There is also data to indicate that autophagosomes can form and be transported within neurites of primary neurons and differentiated neuroblastoma cells, although the adapters remain to be identified. Selective mitophagy in this context would depend upon direct interactions involving the altered mitochondrial surface, or adaptor proteins that recognize damaged mitochondria, and LC3 itself. In a manner analogous to recognition of protein aggregates, PINK1 recruitment of parkin could result in protein ubiquitination at the surface of mitochondria, with recruitment of p62, an adapter protein that binds both ubiquitin and LC3. This would function in analogy to the Atg32-Atg11 system in yeast. Both of these models have precedent in non-neuronal and non-mitochondrial contexts for autophagic cargo selection, and further studies in primary neurons are needed to clarify mechanisms that function to ensure mitochondrial quality control in such a highly polarized, mitochondrially-dependent cell type.

Additional studies in mammalian and yeast cells have moved the concept of selective mitochondrial autophagy, or mitophagy, into the mainstream of autophagy research. During lens and erythrocyte maturation, mitochondria are selectively removed through mechanisms that include autophagy.[55] Clearance of presumably undamaged mitochondria during nitrogen starvation in yeast depends upon the presence of the outer mitochondrial membrane protein Uth1p[56] and Atg32,[57] the latter of which links mitochondria to the Atg11 adapter shared with other forms of selective yeast autophagy. These data indicate that entrapment of mitochondria does not occur simply on a stochastic basis during bulk phase starvation-induced cytoplasmic turnover, but involves recognition of outer membrane receptors.

Primary fibroblasts from patients with Mitochondrial Encephalopathy, Lactic Acidosis, and Stroke (MELAS) syndrome exhibit a higher degree of starvation-induced mitophagy than control fibroblasts,[58] interpreted as supporting the selective removal of mitochondria with damaged DNA. It remains to be determined, however, whether the degree of mutant mtDNA burden within a given cell is correlated with the degree of mitophagy activation, or even whether mitophagy can result in decreased heteroplasmy. Mitophagy preferentially involves small mitochondria as shown in aged cardiomyocytes,[59] and mitophagy that is induced in a Parkinsonian culture model is downregulated by inhibiting fission.[22,60] As fusion-fission events are often coupled, the failure of depolarized mitochondrial segments to re-fuse with the mitochondrial reticulum has been proposed as a mechanism for culling dysfunctional mitochondria in pancreatic beta cells.[61]

RECOGNITION OF MITOCHONDRIAL CARGO FOR MITOPHAGY

The regulation of mitophagy in response to mitochondrial injury involves at least two distinct components, which may or may not be interrelated. The first involves induction of autophagy through engagement of the core autophagy machinery. The second involves mechanisms to recognize and enrich for damaged mitochondria within the forming autophagosome. Reactive oxygen species produced by damaged mitochondria represents one set of signals that can both induce autophagy and potentially modify mitochondria for cargo recognition. ROS produced by PINK1-deficient mitochondrial elicits mitochondrial fission, which is necessary for effective mitophagy,[22] and may potentially modify phospholipids or other mitochondrial surface constituents.[62] The protein Nix is also involved in both autophagy induction and mitochondrial cargo enrichment.[23,55]

Mitochondrial depolarization is one of the major proposed signals for mitophagy. In liver cells, the mitochondrial permeability transition occurs prior to delivery of the mitochondrion into Lysotracker-labeled compartments.[63] Proton ionophores that uncouple mitochondria (FCCP/CCCP) induce mitochondrial translocation of overexpressed parkin, which functions to promote clearance of the chemically depolarized mitochondria.[20] Depolarization impairs the ability of mitochondria to import proteins and process the mitochondrial targeting sequence, resulting in accumulation of full-length PINK1 at the mitochondrial surface.[64–66] Wild type, but not mutant, PINK1 recruits overexpressed parkin to the mitochondrial surface,[67,68] where it ubiquitinates mitofusin in Drosophila[69] or SH-SY5Y cells.[70]

It was subsequently shown that parkin functions only in the cargo enrichment step, and is not necessary for FCCP-engagement of the general autophagy machinery.[23] Interestingly, parkin can also suppress general autophagy through mono-ubiquitination of Bcl2,[71] while overexpression of PINK1 downregulates general autophagy induced by 6-hydroxydopamine.[22] These differences could relate to subcellular compartmentalization, as autophagy suppression by PINK1 or by PKA is independent of their mitochondrial localization,[22,60] while mitochondrial targeting of PINK1 is essential for promoting parkin-related mitochondrial clearance.[65,72]

Unresolved questions include whether cargo enrichment by parkin involves direct mitochondrial interactions with the autophagy machinery, or if it reflects microtubule-related transport to perinuclear aggresome-organizing areas. The p62/sequestosome adaptor protein apposes ubiquitinated protein aggregates to LC3,[73] and p62 has been implicated in parkin-induced mitophagy.[74] However, others indicate that parkin controls perinuclear aggregation of FCCP-injured mitochondria[19,75] into "mito-aggresomes," rather than showing direct effects on mitophagy. Interestingly, parkin has been reported to ubiquitinate VDAC1,[74] which can interact with a dynein light chain.[39] Further support for the trafficking hypothesis is provided by recent studies showing that overexpressed PINK1-parkin mediates proteasomal degradation of outer mitochondrial membrane proteins, including Miro, the adapter that links mitochondria to kinesin.[18,76,77] Considering the great distance between synaptic structures and the perinuclear organizing region, these considerations may be cell-type specific, and additional studies employing primary neurons and *in vivo* systems will be necessary to elucidate mechanisms most relevant to neuronal homeostasis and neurodegenerative diseases.

While direct insults that lead to mitochondrial depolarization can trigger mitophagy, depolarization is not an obligate signal for mitophagy. During starvation, live cell imaging studies show loss of Tetramethyl Rhodamine Methyl Ester (TMRM) signal following, rather than preceding, completion of a GFP-LC3 ring around the mitochondrial segment,[78] implicating other recognition signals.

Mitochondrial depolarization also occurs downstream of autophagosome formation in reticulocytes.[79] In both cases, loss of membrane potential still precedes delivery to the lysosome, as originally described.[63] While alternative signals may include post-translational modifications of mitochondrial outer membrane proteins, our work implicates redistribution of specific mitochondrial phospholipids, which interact with the N-terminal domain of LC3, in cargo recognition for mitophagy (C. T. Chu, H. Bayir, V. K. Kagan, unpublished data).

Likewise, while accumulation of full length PINK1 at the surface of FCCP/CCCP-depolarized mitochondria recruits parkin, this is not the only mechanism for parkin-facilitated mitophagy. Parkin overexpression complements PINK1 deficiency in many model systems,[22,80] indicating that it does not depend upon PINK1 for all of its neuroprotective functions. Parkin compensates for loss of PINK1 by promoting mitophagy,[22] and the cell survival effects of parkin overexpression in PINK1-deficient cells is dependent upon intact autophagy function.[81] Interestingly, parkin recruitment to mitochondria in human fibroblasts injured by hydrogen peroxide does not require mitochondrial PINK1,[82] and increased expression of the DnaJ/Hsp40 chaperone is sufficient to restore mitochondrial localization of mutant parkin,[83] indicating alternative mechanisms of recruitment. Mitochondrial trafficking of extracellular signal regulated protein kinase ½ is observed in Parkinson's disease neurons,[84] and mitochondrially-localized ERK induces mitophagy even in the absence of added injury.[85] The regulatory mechanisms underlying multiple pathways to mitophagy may reflect differences in the initiating stimulus, such as oxidants or depolarization, or cell type.

Additional studies utilizing primary neurons and *in vivo* systems will be necessary, as neurons are less tolerant of losing mitochondrial mass than other cell types. Metabolic profiling of oxygen consumption in living neuron cultures reveal striking differences from HeLa cervical cancer cells; ATP levels are rapidly diminished in neurons treated with the mitochondrial uncoupler FCCP or with oligomycin to inhibit mitochondrial ATP synthase, while HeLa cell ATP levels are unaffected in the same time frame.[86] Interestingly, FCCP did not induce mitophagy in HeLa cells grown under conditions that promote mitochondrial dependence, just as it failed to increase PINK1-parkin recruitment to mitochondria in a variety of primary neuron types.[87] In contrast, parkin-recruitment to depolarized mitochondria is observed in neuronally-differentiated induced pluripotent stem cells.[88] Studies using metabolically characterized primary neurons and differentiated neuron-like cells will be necessary to investigate whether or not there are different regulatory mechanisms or thresholds for mitophagy in cells dependent upon mitochondria for the bulk of ATP generation.

MITOPHAGY IN NEURONAL SURVIVAL AND CELL DEATH

There is growing consensus that regulation of mitochondrial autophagy is likely to be important for neuronal health. In addition to regulating cell death, autophagy also regulates fatty acid metabolism and insulin responsiveness by affecting adipocyte mitochondrial mass.[89] The majority of mitophagy studies, however, have been conducted in non-neuronal cells, and thus, the significance of mitophagy in neuronal mitochondrial quality control is not well characterized. As discussed above, given the particular dependence of neurons upon mitochondrial metabolism, in contrast to readily transfected transformed cell lines, it is likely that additional regulatory mechanisms to prevent excessive mitochondrial clearance operate in neurons.[90] Indeed, complete and selective mitochondrial clearance, as observed in FCCP/CCCP-treated cells co-expressing parkin and PINK1, was initially described as a neuronal cell death process.[54] Beclin1-independent mitophagy also contributes to MPP^+ toxicity.[91] On the other hand, mitophagy plays a compensatory, neuroprotective role in PINK1-deficient SH-SY5Y neuroblastoma cells.[22] As in yeast, mitophagy may represent a quality control response at low levels, and a program of cell death initiated when cells are unable to recover.[92]

Another factor of particular importance to neurons is that protection of the soma from cell death may come at an expense to the support of distant neuritic-synaptic structures. Autophagy causes neurite degeneration in sympathetic neurons,[93] and in cortical neurons expressing mutant leucine rich repeat kinase 2 (LRRK2) as a model of autosomal dominant Parkinson's disease.[94] Increased mitophagy in retinal ganglion cells with mutations in Opa1 may also contribute to optic atrophy in this autosomal dominant disease,[95] although it is still unclear what stimulates autophagy or mitophagy in these contexts.

While both mitochondrial transport and mitophagy are regulated by membrane potential, whether and how mitochondrial transport or altered mobility impacts mitophagy is not clear. Additional unresolved questions include whether or not dysfunctional axonal mitochondria must undergo retrograde transport to perinuclear organizing areas for turnover. While the endoplasmic reticulum is an important source of autophagic membranes in the soma,[96] there is evidence in yeast and mammalian cells that mitochondria can donate membranes for autophagy,[97,98] potentially explaining the presence of early autophagosomes in neurites.[93,99] If so, mechanisms that regulate transport of early autophagosomes may be just as pertinent as mechanisms that regulate mitochondrial transport in determining effective cellular recycling of mitochondria. Ultimately, it will be interesting to determine whether enhancing mitochondrial mobility facilitates the turnover of dysfunctional mitochondria. Future studies using genetic mouse models or live cell tracking of mitochondrial mobility *in vivo*, will help elucidate the molecular and cellular mechanisms regulating mitochondrial mobility, distribution and factors impacting their turnover in axons, dendrites and near synapses. These studies will shed light on fundamental neuronal processes that may advance our understanding of human neurodegenerative disorders.

ACKNOWLEDGMENTS

Z. Sheng is supported by the Intramural Research Program of NINDS, NIH. C. T. Chu is supported in part by the National Institutes of Health (AG026389, NS065789), and she is recipient of an AFAR/Ellison Medical Foundation Julie Martin Mid-Career Award in Aging Research.

REFERENCES

1. Y. Tang and R. S. Zucker, Mitochondrial involvement in post-tetanic potentiation of synaptic transmission, *Neuron* **18**: 483–491 (1997).
2. J. S. Kang, J. H. Tian, P. Y. Pan, P. Zaid, C. Li, C. Deng and Z. H. Sheng, Docking of axonal mitochondria by syntaphilin controls their mobility and affects short-term facilitation, *Cell* **132**: 137–148 (2008).
3. R. S. Stowers, L. J. Megeath, J. Górska-Andrzejak, I. A. Meinertzhagen and T. L. Schwarz, Axonal transport of mitochondria to synapses depends on milton, a novel Drosophila protein, *Neuron* **36**: 1063–1077 (2002).
4. X. Guo, G. T. Macleod, A. Wellington, F. Hu, S. Panchumarthi, M. Schoenfield, L. Marin, M. P. Charlton, H. L. Atwood and K. E. Zinsmaier, The GTPase dMiro is required for axonal transport of mitochondria to Drosophila synapses, *Neuron* **47**: 379–393 (2005).
5. P. Verstreken, C. V. Ly, K. J. Venken, T. W. Koh, Y. Zhou and H. J. Bellen, Synaptic mitochondria are critical for mobilization of reserve pool vesicles at Drosophila neuromuscular junctions, *Neuron* **47**: 365–378 (2005).
6. H. Ma, Q. Cai, W. Lu, Z. H. Sheng and S. Mochida. KIF5B motor adaptor syntabulin maintains synaptic transmission in sympathetic neurons, *J Neurosci* **29**: 13019–13029 (2009).
7. E. A. Schon and S. Przedborski, Mitochondria: the next (neurode) generation, *Neuron* **70**: 1033–1053 (2011).
8. C. L. Zhang, P. L. Ho, D. B. Kintner, D. Sun and S. Y. Chiu, *J Neurosci* **30**: 3555–3566 (2010).
9. A. D. Pilling, D. Horiuchi, C. M. Lively and W. M. Saxton, Kinesin-1 and dynein are the primary motors for fast transport of mitochondria in Drosophila motor axons, *Mol Biol Cell* **17**: 2057–2068 (2006).

10. Y. M. Chen, C. Gerwin and Z. H. Sheng, Dynein light chain LC8 regulates syntaphilin-mediated mitochondrial docking in axons, *J Neurosci* **29**: 9429–9438 (2009).
11. G. L. Rintoul, A. J. Filiano, J. B. Brocard, G. J. Kress and I. J. Reynolds, Glutamate decreases mitochondrial size and movement in primary forebrain neurons, *J Neurosci* **23**: 7881–7888 (2003).
12. M. Saotome, D. Safiulina, G. Szabadkai, S. Das, A. Fransson, P. Aspenstrom, R. Rizzuto and G. Hajnóczky, Bidirectional Ca^{2+}-dependent control of mitochondrial dynamics by the Miro GTPase, *Proc Natl Acad Sci USA* **105**: 20728–20733 (2008).
13. A. F. Macaskill, J. E. Rinholm, A. E. Twelvetrees, I. L. Arancibia-Carcamo, J. Muir, A. Fransson, P. Aspenstrom, D. Attwell and J. T. Kittler, Miro1 is a calcium sensor for glutamate receptor-dependent localization of mitochondria at synapses, *Neuron* **61**: 541–555 (2009).
14. X. Wang and T. L. Schwarz, The mechanism of Ca^{2+}-dependent regulation of kinesin-mediated mitochondrial motility, *Cell* **136**: 163–174 (2009).
15. Q. Cai and Z. H. Sheng, Moving or stopping mitochondria: Miro as a traffic cop by sensing calcium, *Neuron* **61**: 493–496 (2009).
16. J. J. Jennings, Jr., J. H. Zhu, Y. Rbaibi, X. Luo, C. T. Chu and K. Kiselyov, Mitochondrial aberrations in mucolipidosis type IV, *J Biol Chem* **281**: 39041–39050 (2006).
17. Y. Zhang, H. Qi, R. Taylor, W. Xu, L. F. Liu and S. Jin, The role of autophagy in mitochondria maintenance: characterization of mitochondrial functions in autophagy-deficient *S. cerevisiae* strains, *Autophagy* **3**: 337–346 (2007).
18. S. R. Yoshii, C. Kishi, N. Ishihara and N. Mizushima, Parkin mediates proteasome-dependent protein degradation and rupture of the outer mitochondrial membrane, *J Biol Chem* **286**: 19630–19640 (2011).
19. J. Y. Lee, Y. Nagano, J. P. Taylor, K. L. Lim and T. P. Yao, Disease-causing mutations in parkin impair mitochondrial ubiquitination, aggregation, and HDAC6-dependent mitophagy, *J Cell Biol* **189**: 671–679 (2010).
20. D. Narendra, A. Tanaka, D. F. Suen and R. J. Youle, Parkin is recruited selectively to impaired mitochondria and promotes their autophagy, *J Cell Biol* **183**: 795–803 (2008).
21. A. Weihofen, K. J. Thomas, B. L. Ostaszewski, M. R. Cookson and D. J. Selkoe, Pink1 forms a multiprotein complex with Miro and Milton, linking Pink1 function to mitochondrial trafficking, *Biochemistry* **48**: 2045–2052 (2009).
22. R. K. Dagda, S. J. Cherra, 3rd, S. M. Kulich, A. Tandon, D. Park and C. T. Chu, Loss of pink1 function promotes mitophagy through effects on oxidative stress and mitochondrial fission, *J Biol Chem* **284**: 13843–13855 (2009).
23. W. X. Ding, H. M. Ni, M. Li, Y. Liao, X. Chen, D. B. Stolz, G. W. Dorn, 2nd and X. M. Yin, Nix is critical to two distinct phases of mitophagy, reactive oxygen species-mediated autophagy induction and Parkin-ubiquitin-p62-mediated mitochondrial priming, *J Biol Chem* **285**: 27879–27890 (2010).

24. P. J. Hollenbeck and W. M. Saxton, The axonal transport of mitochondria, *J Cell Sci* **118**: 5411–5419 (2005).
25. Y. Tanaka, Y. Kanai, Y. Okada, S. Nonaka, S. Takeda, A. Harada and N. Hirokawa, Targeted disruption of mouse conventional kinesin heavy chain, kif5B, results in abnormal perinuclear clustering of mitochondria, *Cell* **93**: 1147–1158 (1998).
26. Q. Cai, C. Gerwin and Z. H. Sheng, Syntabulin-mediated anterograde transport of mitochondria along neuronal processes, *J Cell Biol* **170**: 959–969 (2005).
27. N. Hirokawa and R. Takemura, Molecular motors and mechanisms of directional transport in neurons, *Nat Rev Neurosci* **6**: 201–214 (2005).
28. M. Nangaku and R. Sato-Yoshitake, Y. Okada, Y. Noda, R. Takemura, H. Yamazaki and N. Hirokawa, KIF1B, a novel microtubule plus end-directed monomeric motor protein for transport of mitochondria, *Cell* **79**: 1209–1220 (1994).
29. S. J. King and T. A. Schroer, Dynactin increases the processivity of the cytoplasmic dynein motor, *Nat Cell Biol* **2**: 20–24 (2000).
30. M. Haghnia, V. Cavalli, S. B. Shah, K. Schimmelpfeng, R. Brusch, G. Yang, C. Herrera, A. Pilling and L. S. Goldstein, Dynactin is required for coordinated bidirectional motility, but not for dynein membrane attachment, *Mol Biol Cell* **18**: 2081–2089 (2007).
31. R. Mallik, D. Petrov, S. A. Lex, S. J. King and S. P. Gross, Building complexity: an *in vitro* study of cytoplasmic dynein with *in vivo* implications, *Curr Biol* **15**: 2075–2085 (2005).
32. J. R. Levy, C. J. Sumner, J. P. Caviston, M. K. Tokito, S. Ranganathan, L. A. Ligon, K. E. Wallace, B. H. LaMonte, G. G. Harmison, I. Puls, K. H. Fischbeck, E. L. Holzbaur, A motor neuron disease-associated mutation in p150Glued perturbs dynactin function and induces protein aggregation, *J Cell Biol* **172**: 733–745 (2006).
33. N. Hirokawa, R. Sato-Yoshitake, T. Yoshida and T. Kawashima, Brain dynein (MAP1C) localizes on both anterogradely and retrogradely transported membranous organelles *in vivo*, *J Cell Biol* **111**: 1027–1037 (1990).
34. D. Horiuchi, R. V. Barkus, A. D. Pilling, A. Gassman and W. M. Saxton, APLIP1, a kinesin binding JIP-1/JNK scaffold protein, influences the axonal transport of both vesicles and mitochondria in Drosophila, *Curr Biol* **15**: 2137–2141 (2005).
35. G. J. Russo, K. Louie, A. Wellington, G. T. Macleod, F. Hu, S. Panchumarthi and K. E. Zinsmaier, Drosophila Miro is required for both anterograde and retrograde axonal mitochondrial transport, *J Neurosci* **29**: 5443–5455 (2009).
36. D. Pathak, K. J. Sepp and P. J. Hollenbeck, Evidence that myosin activity opposes microtubule-based axonal transport of mitochondria, *J Neurosci* **30**: 8984–8992 (2010).
37. Q. Su, Q. Cai, C. Gerwin, C. L. Smith and Z. H. Sheng, Syntabulin is a microtubule-associated protein implicated in syntaxin transport in neurons, *Nat Cell Biol* **6**: 941–953 (2004).

38. Q. Cai, P. Y. Pan and Z. H. Sheng, Syntabulin-kinesin-1 family member 5B-mediated axonal transport contributes to activity-dependent presynaptic assembly, *J Neurosci* **27**: 7284–7296 (2007).
39. C. Schwarzer, S. Barnikol-Watanabe, F. P. Thinnes and N. Hilschmann, Voltage-dependent anion-selective channel (VDAC) interacts with the dynein light chain Tctex1 and the heat-shock protein PBP74, *Int J Biochem Cell Biol* **34**: 1059–1070 (2002).
40. S. R. Chada and P. J. Hollenbeck, Nerve growth factor signaling regulates motility and docking of axonal mitochondria, *Curr Biol* **14**: 1272–1276 (2004).
41. O. I. Wagner, J. Lifshitz, P. A. Janmey, M. Linden, T. K. McIntosh and J. F. Leterrier, Mechanisms of mitochondria-neurofilament interactions, *J Neurosci* **23**: 9046–9058 (2003).
42. M. O'Toole, R. Latham, R. M. Baqri and M. E. Miller, Modeling mitochondrial dynamics during *in vivo* axonal elongation, *J Theor Biol* **255**: 369–377 (2008).
43. K. E. Miller and M. P. Sheetz, Axonal mitochondrial transport and potential are correlated, *J Cell Sci* **117**: 2791–2804 (2004).
44. D. T. Chang and I. J. Reynolds, Mitochondrial trafficking and morphology in healthy and injured neurons, *Prog Neurobiol* **80**: 241–268 (2006).
45. J. Verburg and P. J. Hollenbeck, Mitochondrial membrane potential in axons increases with local nerve growth factor or semaphorin signaling, *J Neurosci* **28**: 8306–8315 (2008).
46. S. Chen, G. C. Owens, K. L. Crossin and D. B. Edelman, Serotonin stimulates mitochondrial transport in hippocampal neurons, *Mol Cell Neurosci* **36**: 472–483 (2007).
47. G. L. Rintoul, V. J. Bennett, N. A. Papaconstandinou and I. J. Reynolds, Nitric oxide inhibits mitochondrial movement in forebrain neurons associated with disruption of mitochondrial membrane potential, *J Neurochem* **97**: 800–806 (2006).
48. R. H. Baloh, R. E. Schmidt, A. Pestronk and J. Milbrandt, Altered axonal mitochondrial transport in the pathogenesis of Charcot-Marie-Tooth disease from mitofusin 2 mutations, *J Neurosci* **27**: 422–430 (2007).
49. E. Trushina, *et al.* Mutant huntingtin impairs axonal trafficking in mammalian neurons *in vivo* and *in vitro*, *Mol Cell Biol* **24**: 8195–8209 (2004).
50. A. L. Orr, S. Li, C. E. Wang, H. Li, J. Wang, J. Rong, X. Xu, P. G. Mastroberardino, J. T. Greenamyre and X. J. Li, N-terminal mutant huntingtin associates with mitochondria and impairs mitochondrial trafficking, *J Neurosci* **28**: 2783–2792 (2008).
51. K. J. De Vos, A. L. Chapman, M. E. Tennant, C. Manser, E. L. Tudor, K. F. Lau, J. Brownlees, S. Ackerley, P. J. Shaw, D. M. McLoughlin, C. E. Shaw, P. N. Leigh, C. C. Miller and A. J. Grierson, Familial amyotrophic lateral sclerosis-linked SOD1 mutants perturb fast axonal transport to reduce axonal mitochondria content, *Hum Mol Genet* **16**: 2720–2728 (2007).

52. J. Magrané, I. Hervias, M. S. Henning, M. Damiano, H. Kawamata and G. Manfredi, Mutant SOD1 in neuronal mitochondria causes toxicity and mitochondrial dynamics abnormalities, *Hum Mol Genet* **18**: 4552–4564 (2009).
53. E. Perlson, G. B. Jeong, J. L. Ross, R. Dixit, K. E. Wallace, R. G. Kalb and E. L. Holzbaur, A switch in retrograde signaling from survival to stress in rapid-onset neurodegeneration, *J Neurosci* **29**: 9903–9917 (2009).
54. L. Xue, G. C. Fletcher and A. M. Tolkovsky, Mitochondria are selectively eliminated from eukaryotic cells after blockade of caspases during apoptosis, *Curr Biol* **11**: 361–365 (2001).
55. I. Novak, V. Kirkin, D. G. McEwan, J. Zhang, P. Wild, A. Rozenknop, V. Rogov, F. Lohr, D. Popovic, A. Occhipinti, A. S. Reichert, J. Terzic, V. Dotsch, P. A. Ney and I. Dikic, Nix is a selective autophagy receptor for mitochondrial clearance, *EMBO Rep* **11**: 45–51 (2010).
56. I. Kissova, M. Deffieu, S. Manon and N. Camougrand, Uth1p is involved in the autophagic degradation of mitochondria, *J Biol Chem* **279**: 39068–39074 (2004).
57. T. Kanki, K. Wang, Y. Cao, M. Baba and D. J. Klionsky, Atg32 is a mitochondrial protein that confers selectivity during mitophagy, *Dev Cell* **17**: 98–109 (2009).
58. Y. Gu, C. Wang and A. Cohen, Effect of IGF-1 on the balance between autophagy of dysfunctional mitochondria and apoptosis, *FEBS Lett* **577**: 357–360 (2004).
59. A. Terman, H. Dalen, J. W. Eaton, J. Neuzil and U. T. Brunk, Mitochondrial recycling and aging of cardiac myocytes: the role of autophagocytosis, *Exp Gerontol* **38**: 863–876 (2003).
60. R. K. Dagda, A. M. Gusdon, I. Pien, S. Strack, S. Green, B. Van Houten, S. J. Cherra, 3rd and C. T. Chu, Mitochondrially localized PKA reverses mitochondrial pathology and dysfunction in a cellular model of Parkinson's disease, *Cell Death Differ* **18**: 1–10 (2011).
61. G. Twig, A. Elorza, A. J. Molina, H. Mohamed, J. D. Wikstrom, G. Walzer, L. Stiles, S. E. Haigh, S. Katz, G. Las, J. Alroy, M. Wu, B. F. Py, J. Yuan, J. T. Deeney, B. E. Corkey and O. S. Shirihai, Fission and selective fusion govern mitochondrial segregation and elimination by autophagy, *EMBO J* **27**: 433–446 (2008).
62. R. A. Kirkland, R. M. Adibhatla, J. F. Hatcher and J. L. Franklin, Loss of cardiolipin and mitochondria during programmed neuronal death: evidence of a role for lipid peroxidation and autophagy, *Neuroscience* **115**: 587–602 (2002).
63. S. P. Elmore, T. Qian, S. F. Grissom and J. J. Lemasters, The mitochondrial permeability transition initiates autophagy in rat hepatocytes, *FASEB J* **15**: 2286–2287 (2001).
64. W. Lin and U. J. Kang, Characterization of PINK1 processing, stability, and subcellular localization, *J Neurochem* **106**: 464–474 (2008).
65. D. P. Narendra, S. M. Jin, A. Tanaka, D. F. Suen, C. A. Gautier, J. Shen, M. R. Cookson and R. J. Youle, PINK1 is selectively stabilized on impaired mitochondria to activate parkin, *PLoS Biol* **8**: e1000298 (2010).

66. N. Matsuda, S. Sato, K. Shiba, K. Okatsu, K. Saisho, C. A. Gautier, Y. S. Sou, S. Saiki, S. Kawajiri, F. Sato, M. Kimura, M. Komatsu, N. Hattori and K. Tanaka, PINK1 stabilized by mitochondrial depolarization recruits Parkin to damaged mitochondria and activates latent Parkin for mitophagy, *J Cell Biol* **189**: 211–221 (2010).
67. S. Kawajiri, S. Saiki, S. Sato, F. Sato, T. Hatano, H. Eguchi and N. Hattori, PINK1 is recruited to mitochondria with parkin and associates with LC3 in mitophagy, *FEBS Lett* **584**: 1073–1079 (2010).
68. S. Geisler, K. M. Holmstrom, A. Treis, D. Skujat, S. S. Weber, F. C. Fiesel, P. J. Kahle and W. Springer, The PINK1/Parkin-mediated mitophagy is compromised by PD-associated mutations, *Autophagy* **6**: 871–878 (2010).
69. A. C. Poole, R. E. Thomas, S. Yu, E. S. Vincow and L. Pallanck, The mitochondrial fusion-promoting factor mitofusin is a substrate of the PINK1/parkin pathway, *PLoS One* **5**: e10054 (2010).
70. M. E. Gegg, J. M. Cooper, K. Y. Chau, M. Rojo, A. H. Schapira and J. W. Taanman, Mitofusin 1 and mitofusin 2 are ubiquitinated in a PINK1/parkin-dependent manner upon induction of mitophagy, *Hum Mol Genet* **19**: 4861–4870 (2010).
71. D. Chen, F. Gao, B. Li, H. Wang, Y. Xu, C. Zhu and G. Wang, Parkin monoubiquitinates BCL-2 and regulates autophagy, *J Biol Chem* **285**: 38214–38223 (2010).
72. S. Michiorri, V. Gelmetti, E. Giarda, F. Lombardi, F. Romano, R. Marongiu, S. Nerini-Molteni, P. Sale, R. Vago, G. Arena, L. Torosantucci, L. Cassina, M. A. Russo, B. Dallapiccola, E. M. Valente and G. Casari, The Parkinson-associated protein PINK1 interacts with Beclin1 and promotes autophagy, *Cell Death Differ* **17**: 962–974 (2010).
73. S. Pankiv, T. H. Clausen, T. Lamark, A. Brech, J. A. Bruun, H. Outzen, A. Øvervatn, G. Bjørkøy and T. Johansen, p62/SQSTM1 binds directly to Atg8/LC3 to facilitate degradation of ubiquitinated protein aggregates by autophagy, *J Biol Chem* **282**: 24131–24145 (2007).
74. S. Geisler, K. M. Holmstrom, D. Skujat, F. C. Fiesel, O. C. Rothfuss, P. J. Kahle and W. Springer, PINK1/Parkin-mediated mitophagy is dependent on VDAC1 and p62/SQSTM1, *Nat Cell Biol* **12**: 119–131 (2010).
75. C. Vives-Bauza, C. Zhou, Y. Huang, M. Cui, R. L. de Vries, J. Kim, J. May, M. A. Tocilescu, W. Liu, H. S. Ko, J. Magrane, D. J. Moore, V. L. Dawson, R. Grailhe, T. M. Dawson, C. Li, K. Tieu and S. Przedborski, PINK1-dependent recruitment of Parkin to mitochondria in mitophagy, *Proc Natl Acad Sci USA* **107**: 378–383 (2010).
76. X. Wang, D. Winter, G. Ashrafi, J. Schlehe, Y. L. Wong, D. Selkoe, S. Rice, J. Steen, M. J. Lavoie and T. L. Schwarz, PINK1 and Parkin target Miro for phosphorylation and degradation to arrest mitochondrial motility, *Cell* **147**: 893–906 (2011).
77. N. C. Chan, A. M. Salazar, A. H. Pham, M. J. Sweredoski, N. J. Kolawa, R. L. Graham, S. Hess and D. C. Chan, Broad activation of the ubiquitin-proteasome system by Parkin is critical for mitophagy, *Hum Mol Genet* **20**: 1726–1737 (2011).

78. I. Kim, S. Rodriguez-Enriquez and J. J. Lemasters, Selective degradation of mitochondria by mitophagy, *Arch Biochem Biophys* **462**: 245–253 (2007).
79. J. Zhang and P. A. Ney, Autophagy-dependent and -independent mechanisms of mitochondrial clearance during reticulocyte maturation, *Autophagy* **5**: 1064–1065 (2009).
80. I. E. Clark, M. W. Dodson, C. Jiang, J. H. Cao, J. R. Huh, J. H. Seol, S. J. Yoo, B. A. Hay and M. Guo, Drosophila pink1 is required for mitochondrial function and interacts genetically with parkin, *Nature* **441**: 1162–1166 (2006).
81. R. K. Dagda and C. T. Chu, Mitochondrial quality control: insights on how Parkinson's disease related genes PINK1, parkin, and Omi/HtrA2 interact to maintain mitochondrial homeostasis, *J Bioenerg Biomembr* **41**: 473–479 (2009).
82. A. Rakovic, A. Grunewald, P. Seibler, A. Ramirez, N. Kock, S. Orolicki, K. Lohmann and C. Klein, Effect of endogenous mutant and wild-type PINK1 on Parkin in fibroblasts from Parkinson disease patients, *Hum Mol Genet* **19**: 3124–3137 (2010).
83. J. M. Rose, S. S. Novoselov, P. A. Robinson and M. E. Cheetham, Molecular chaperone-mediated rescue of mitophagy by a Parkin RING1 domain mutant, *Hum Mol Genet* **20**: 16–27 (2010).
84. J. H. Zhu, F. Guo, J. Shelburne, S. Watkins and C. T. Chu, Localization of phosphorylated ERK/MAP kinases to mitochondria and autophagosomes in Lewy body diseases, *Brain Pathol* **13**: 473–481 (2003).
85. R. K. Dagda, J. Zhu, S. M. Kulich and C. T. Chu, Mitochondrially localized ERK2 regulates mitophagy and autophagic cell stress: implications for Parkinson's disease, *Autophagy* **4**: 770–782 (2008).
86. A. M. Gusdon and C. T. Chu, To eat or not to eat: neuronal metabolism, mitophagy, and Parkinson's disease, *Antioxid Redox Signal* **14**: 1979–1987 (2011).
87. V. S. Van Laar, B. Arnold, S. J. Cassady, C. T. Chu, E. A. Burton and S. B. Berman, Bioenergetics of neurons inhibit the translocation response of Parkin following rapid mitochondrial depolarization, *Hum Mol Genet* **20**: 927–940 (2011).
88. P. Seibler, J. Graziotto, H. Jeong, F. Simunovic, C. Klein and D. Krainc, Mitochondrial Parkin recruitment is impaired in neurons derived from mutant PINK1 induced pluripotent stem cells, *J Neurosci* **31**: 5970–5976 (2011).
89. Y. Zhang, S. Goldman, R. Baerga, Y. Zhao, M. Komatsu and S. Jin, Adipose-specific deletion of *autophagy-related gene 7 (atg7)* in mice reveals a role in adipogenesis, *Proc Natl Acad Sci USA* **106**: 19860–19865 (2009)
90. V. S. Van Laar, B. Arnold, S. J. Cassady, C. T. Chu, E. A. Burton and S. B. Berman, Bioenergetics of neurons inhibit the translocation response of Parkin following rapid mitochondrial depolarization, *Hum Mol Genet* **20**: 927–940 (2011).
91. J. H. Zhu, C. Horbinski, F. Guo, S. Watkins, Y. Uchiyama and C. T. Chu, Regulation of autophagy by extracellular signal-regulated protein kinases during 1-methyl-4-phenylpyridinium-induced cell death, *Am J Pathol* **170**: 75–86 (2007).

92. M. Priault, B. Salin, J. Schaeffer, F. M. Vallette, J. P. di Rago and J. C. Martinou, Impairing the bioenergetic status and the biogenesis of mitochondria triggers mitophagy in yeast, *Cell Death Differ* **12**: 1613–1621 (2005).
93. Y. Yang, K. Fukui, T. Koike and X. Zheng, Induction of autophagy in neurite degeneration of mouse superior cervical ganglion neurons, *Eur J Neurosci* **26**: 2979–2988 (2007).
94. S. J. Cherra III, S. M. Kulich, G. Uechi, M. Balasubramani, J. Mountzouris, B. W. Day and C. T. Chu, Regulation of the autophagy protein LC3 by phosphorylation. *J Cell Biol* **190**: 533–539 (2010).
95. K. E. White, V. J. Davies, V. E. Hogan, M. J. Piechota, P. P. Nichols, D. M. Turnbull and M. Votruba, OPA1 deficiency associated with increased autophagy in retinal ganglion cells in a murine model of dominant optic atrophy, *Invest Ophthalmol Vis Sci* **50**: 2567–2571 (2009).
96. W. A. Dunn, Jr., Studies on the mechanisms of autophagy: formation of the autophagic vacuole, *J Cell Biol* **110**: 1923–1933 (1990).
97. F. Reggiori, T. Shintani, U. Nair and D. J. Klionsky, Atg9 cycles between mitochondria and the pre-autophagosomal structure in yeasts, *Autophagy* **1**: 101–109 (2005).
98. D. W. Hailey, A. S. Rambold, P. Satpute-Krishnan, K. Mitra, R. Sougrat, P. K. Kim and J. Lippincott-Schwartz, Mitochondria supply membranes for autophagosome biogenesis during starvation, *Cell* **141**: 656–667 (2010).
99. Q. J. Wang, Y. Ding, D. S. Kohtz, N. Mizushima, I. M. Cristea, M. P. Rout, B. T. Chait, Y. Zhong, N. Heintz and Z. Yue, Induction of autophagy in axonal dystrophy and degeneration, *J Neurosci* **26**: 8057–8068 (2006).
100. Q. Cai and Z.-H. Sheng, Mitochondrial transport and docking in axons, *Exp Neurol* **218**: 257–267 (2009).
101. Q. Cai, M. L. Davis and Z.-H. Sheng, Regulation of axonal mitochondrial transport and its impact on synaptic transmission, *Neurosci Res* **70**: 9–15 (2011).

CHAPTER 17

ROLE OF AUTOPHAGY IN NEURITE DEGENERATION *IN VITRO*

Yi Yang, Xiaoxiang Zheng and Tatsuro Koike

ABSTRACT

Accumulating evidence shows that axons and dendrites undergo degeneration, independently of cell soma, often associated with decreased levels of cellular ATP caused by deficiency of trophic factor support, which eventually activate autophagy. This review will cover recent experimental evidence on autophagy induction, autophagosome dynamics, and certain aspects of deregulation of autophagy during neuritic degeneration *in vitro*.

INTRODUCTION

There are multiple cellular compartments in a single neuron, including cell body, axon, dendrites and synapses. The question as to how each compartment coordinates together and executes its own functions independently has drawn much attention. Various anterograde and retrograde trophic signals as well as death signals play important roles in determining the fate of each compartment.[1,2] The term "trophic" was first introduced by Waller to describe the influence of the cell body on its axon and target tissue based on the finding that degeneration initially occurs in the distal stump of transected axons.[3]

Self-destructive programs are activated without damaging the cell body during development when exuberant axon branches and synapses are eliminated or pruned.[4] In particular, axons and dendrites (collectively called neurites) occupy much larger surface areas than the neuronal cell body where most of the structural and functional proteins are synthesized, rendering the neurites particularly vulnerable to energy crisis. In a mutant mouse strain carrying the *wld*^{s} (Wallerian degeneration slow) gene, the degeneration of transected axons is markedly delayed.[5,6] The key gene encoding nicotinamide mononucleotide adenylytransferase (NMNAT) 1 is highly expressed in the mutant mouse. It is an indispensable enzyme for the biosynthesis of NAD^+ whose levels are in equilibrium with ATP in the stationary state. This enzyme is normally localized in the nucleus but is also found in axons in the mutant mouse, suggesting that elevated levels of NAD^+ provide axonal protection by preventing local energy failure and/or by activating cofactor-requiring enzymes. Possible candidates that utilize NAD^+ and suppress axon degeneration in the mutant mouse include the sirtuins, the mammalian homologs of the yeast life span-extending molecule.[7–9]

Neurite degeneration involves caspase-independent degeneration associated with decreased levels of ATP or NAD^+.[10–12] In contrast, caspase-dependent (apoptotic) pathways play a pivotal role in somal death where cellular ATP levels are well maintained. Accumulating evidence suggests that this caspase-independent pathway of neurite degeneration involves autophagy as a major mechanism. This review will cover recent experimental evidence on the role of autophagy in neurite degeneration under physiological and pathological conditions.[13,14] Particular emphasis is placed on neuritic autophagy induction, autophagosome dynamics, and certain aspects of the regulatory mechanism of neuritic autophagy using neuronal cell culture models.

THE INDUCTION OF AUTOPHAGY DURING NEURITE DEGENERATION

Autophagy is a well-characterized catabolic mechanism whereby cells degrade proteins and recycle cytoplasmic components and intracellular organelles in response to nutrient starvation. It involves the formation of double-membrane structures, termed autophagosomes or autophagic vacuoles (AVs), which fuse with the lysosomal membrane to form the autolysosome, where the contents are degraded.[15] Coordinated increases in autophagosomes and autolysosomes have been widely accepted as indicators of autophagy induction, although acute lysosomal failure may elicit similar increases. The conversion of cytosolic microtubule-associated protein light chain 3-I (LC3-I) to LC3-II (lipidated form) is a reliable marker for autophagic vacuoles.[16]

Emerging lines of evidence show that autophagy is associated with various human diseases, including acute and chronic neurological disorders, such as stroke, brain trauma, neurotoxin injury, Alzheimer's disease (AD), Parkinson's disease (PD), and Huntington's disease.[17] Pathological accumulation of autophagosomes or autolysosomes in neuronal cell soma and projections are observed in patients with neurological diseases as well as in *in vitro* and *in vivo* experimental paradigms.

Two ubiquitin-like molecules, Atg12 and LC3/Atg8, are involved in autophagosome biogenesis. Enhanced immunoreactivity of Atg12 was observed in tau-immunopositive dystrophic neurites and some neurofibrillary tangles in brain sections of patients with AD.[18] In addition, the accumulation of autophagosomes was found in neurites in a transgenic mouse model of AD, and in neurites following stroke and brain trauma.[19–21] Electron microscopic analysis of human brain biopsies from patients with prion diseases revealed the presence of AVs in axon terminals, which was consistent with axonal degeneration.[22] Furthermore, the induction of autophagy was closely correlated with axonal dystrophy and degeneration in Purkinje cells during excitotoxic neurodegeneration.[23] In our previous study, the accumulation of autophagosomes and autolysosomes was detected in degenerating neurites of cultured mouse superior cervical ganglion (SCG) neurons initiated by three different experimental paradigms, including *in vitro* Wallerian degeneration, nerve growth factor (NGF) deprivation and microtubule disruption.[24] Notably, these autophagosomes/autolysosomes that contained collapsed cytoskeletal proteins were associated with neuritic beadings during degeneration. This implies impaired neuritic transport may induce the focal accumulation of organelles, resulting in the morphology of beading. Furthermore, expression of the autophagy related protein LC3-II was dramatically upregulated during the early stage of neurite degeneration.

Methamphetamine (METH) has been proposed to selectively destroy the neurites of dopaminergic neurons, leaving cell soma intact. Larsen and colleagues found that the autophagic process was induced in response to METH, and appeared to correlate with the neurite loss. These observations are in contrast to a previous hypothesis that neurite loss occurs non-specifically, by structural collapse after attenuation of cellular energy stores.[25] Therefore, this research group suggested that METH-induced overproduction of oxyradicals promoted protein damage and/or dysfunction, resulting in the upregulation of the autophagic process and selective degeneration of dopaminergic neurites.

Neuritic retraction has been accepted as a prominent feature of neuronal degeneration. Mutation of leucine rich repeat kinase 2 (LRRK2), which is implicated in both familial Parkinsonism and sporadic PD, can lead to progressive neurite shortening.[26] Interestingly, LRRK2-mediated neurite retraction was accompanied with autophagy activation in SH-SY5Y cells and in primary cortical neurons.[27] Moreover, the mitogen activated protein kinase/extracellular signal regulated protein kinase (MAPK/ERK) kinase (MEK) signal pathway contributes to autophagy-related neurite shortening, as treatment with the MEK inhibitor U0126, reduced mutant LRRK2-induced neurite autophagy and shortening.[26]

OPA1 is a key mitochondrial protein, which is essential for maintaining the integrity of the mitochondrial network. Mutations in the *OPA1* gene caused structural abnormalities of the optic nerve axons and autosomal dominant optic atrophy (ADOA).[28] Meanwhile, enhanced autophagy activation was observed, which is likely to result from an increase in abnormal mitochondria and could contribute to retinal ganglion cell loss and subsequent optic atrophy seen in ADOA.[28] Even though many studies reveal that the autophagic process plays an active role in axonal and dendritic degeneration. Park and colleagues illustrated that lysosomes were required for myelin ovoid formation after sciatic nerve injury, while the autophagy inhibitor 3-methyladenine could not suppress myelin sheath degradation.[29]

DYNAMICS OF AUTOPHAGOSOMES IN NEURITES UNDERGOING DEGENERATION

Owing to the unique architecture of neurons, compartments in distal neurites can be far away from the cell soma. Therefore, efficient delivery of autophagosomes to the proper location for local needs seems to be crucial to maintain the functions of neurons.

A variety of reporters have been developed to trace the distribution and dynamics of autophagosomes in live cells. By using the fluorescent dye monodansylcadaverine (MDC) and a green fluorescent protein (GFP)-AtATG8e fusion protein, the highly dynamic movement of labeled-autophagosomes was observed in Arabidopsis.[30] These autophagosomes moved across the cell with an average speed of approximately 2 μm/s. Rapid directional movement of GFP-LC3-labeled autophagosomes was also detected in Chinese hamster ovary (CHO) cells, in which a typical autophagosome traveled for about 3 μm during a period of ten seconds, followed by short, random movements or pauses.[31] In cultured HeLa cells, the maximum speed of GFP-LC3-labeled autophagosomes was 5.3 ± 2.6 μm/s.[32]

The dynamic transport of autophagosomes was also observed in highly specialized neuronal compartments, such as axons. It is suggested that autophagosomes depend on microtubule and dynein for their retrograde transport. For example, in cultured cerebellar granule neurons, LC3-labeled autophagosomes traveled along axons towards cell soma with a mean velocity of 0.25 ± 0.02 μm/s.[33,34] The movement appeared to be mediated by the motor protein dynein, as application of the dynein inhibitor EHNA (erythro-9-[3-(2-hydroxynonyl)] adenine) largely blocked such vectorial movement. These results suggest that autophagosomes play important roles in axons and are dynamically regulated. In addition to the involvement of dynein, recent studies showed that autophagosomes could move in a microtubule- and dynein-dynactin motor complex-dependent manner in HeLa cells.[32] After formation, autophagosomes traveled along microtubules towards the centrosome, where lysosomes are usually concentrated.

REGULATORY MECHANISMS OF NEURITE DEGENERATION USING TISSUE CULTURE MODELS

Cell death pathways have been well delineated in cultured rodent sympathetic neurons, which undergo apoptosis when they are deprived of NGF, the target-derived neurotrophic factor.[35] Caspase-dependent (apoptotic) pathways of cell death play a pivotal role on cell soma degeneration where cellular ATP levels are well maintained. In contrast, axons and dendrites undergo caspase-independent degeneration associated with altered levels of ATP[10–12] that seems to involve autophagy as a major degeneration mechanism.[14,a] Morphologically, the initial events include growth cone collapse and defasciculation, which are followed by appearance of membrane blebs and neuritic beading. The cytoskeletal fragmentation does not take place until much later points of time. Two-dimensional gel electrophoresis analysis of neuritic proteins isolated from cultured sympathetic neurons revealed that both collapsin response mediator protein-2 (crmp-2) and collapsin response mediator protein-4 (CRMP-4) are cleaved in parallel with beading formation after NGF deprivation.[38] CRMP-2 is known to play a critical role in neuronal polarity, axonal guidance and axon regeneration.[39,40] This is a globular protein with the C-terminus region attached on the surface, suggesting that this region is particularly vulnerable to proteolysis.[41] Native CRMP-2 is capable of binding to both tubulin dimers and actin filaments, thus acting as a transporter of these molecules along microtubules in association with the anterograde motor protein kinesin-1.[42] Removal of the C-terminus region will render the CRMP-2 incapable of binding to actin filaments and kinesin-1, which results in impaired transport toward the axon terminal, although the truncated form still holds the binding capacity to microtubules.

Using different protease inhibitors and small interfering (si) RNAs, we have demonstrated that a truncated form (58 kDa) of CRMP-2 is produced by calpain1-mediated cleavage of native CRMP-2 (65 kDa). The same truncation is

[a]We have here limited our discussion to the case of sympathetic neurons. It has been shown that caspase activation occurs during neurite degeneration in other neuronal cell types, depending on experimental paradigms employed.[36,37]

observed during *in vitro* Wallerian degeneration or following treatment with colchicine. More diverse forms of cleaved CRMP-2 are known to be produced under acute traumatic and neurotoxic injury,[43] epilepsy,[44] and AD.[45] Therefore, CRMP-2 (as well as CRMP-4) truncation is associated with various forms of neurodegeneration. Indeed, overexpression of the truncated form of CRMP-2 in cultured cerebellar granule neurons leads to beading formation and neurite degeneration (Koike *et al.*, unpublished). Moreover, a proteomics study shows that the native form of CRMP-2 is highly upregulated in the *wld*^{s} mutant mouse, suggesting its involvement in a pro-survival signaling,[46] whereas the truncated forms are associated with neuronal degeneration.[38]

Cathepsins are involved in the mechanism of autophagy.[15] Cathepsin inhibitors, but not calpain inhibitors, suppress neurite degeneration caused by NGF deprivation, suggesting that autophagy plays a major role in this type of neurite degeneration.[36] Neurite degeneration is delayed when the expression of autophagy genes including *beclin1*, *atg7*, and *atg4* (autophagin) is suppressed by using RNA interference (RNAi) technique.[24] Moreover, BAPTA-AM, an intracellular Ca^{2+} chelator, suppresses neurite degeneration (Kato and Koike, unpublished), but not somal apoptosis.[47] Thapsigargin, an inhibitor of the endoplasmic reticulum (ER) ATPase that increases the level of intracellular Ca^{2+}, suppresses cell body apoptosis, but destabilizes neurites in SCG cells and cerebellar granule neurons.[48–50] However, mobilization of intracellular Ca^{2+} is also known to activate autophagy[51] or cause neuronal death.[52] Thus, an alteration of intracellular Ca^{2+} level that is beneficial to neuronal somal survival can be detrimental to neuritic stability by inducing autophagy.

The involvement of intracellular Ca^{2+} levels in the regulation of autophagy is still in an early phase of investigation. However, emerging evidence suggests a direct link between massive Ca^{2+} influx and autophagy activation both *in vivo* and *in vitro,*[53,54] which will be more fully discussed in the next section. As a possible mechanism underlying this link, the following molecular steps will be considered here: AMP-activated protein kinase (AMPK) cascade activation and the initial steps for autophagosome formation (Figure 1).

First, AMPK is a cellular switch that activates catabolic pathways and turns off anabolic processes in response to perturbed cellular energy levels.[55] Under physiological situations, AMPK is activated by an increased AMP/ATP ratio upon energy deficiency, as well as by metabolic insults including glucose or oxygen deprivation. In addition, AMPK activation occurs *via* Ca^{2+}-mediated activation of calcium/calmodulin-dependent protein kinase kinase (CaMKK)-alpha/beta. Activated AMPK is known to inhibit mTOR (mammalian target of rapamycin), a negative regulator of autophagy.[55] Consequently, it is likely that excitotoxicity-induced increase in the level of cytoplasmic Ca^{2+} can activate AMPK *via* CaMKK-alpha/beta and lead to

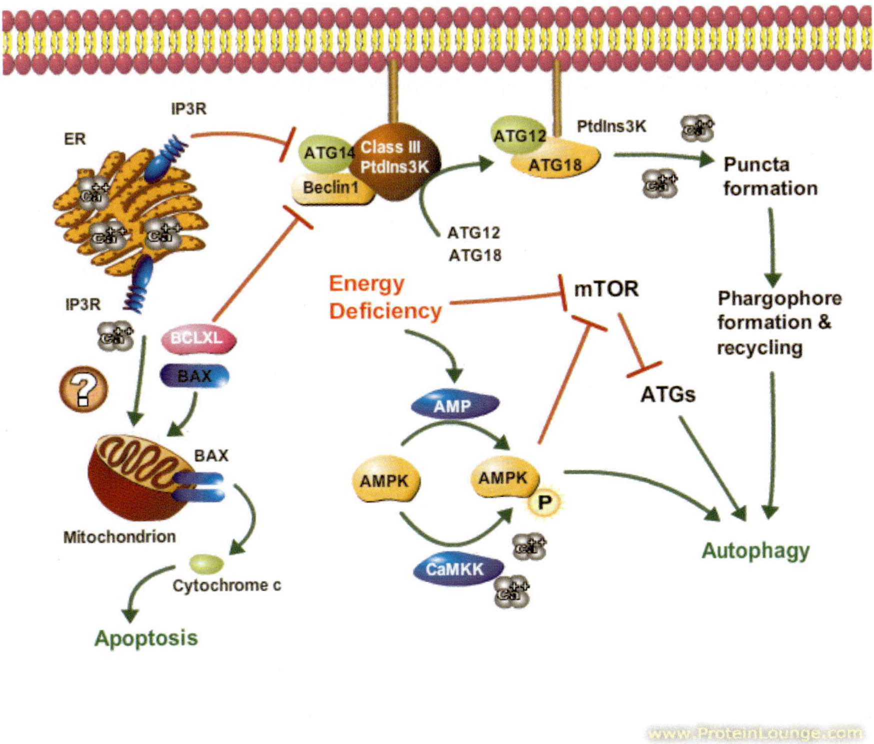

Figure 1. Involvement of cytoplasmic Ca^{2+} in regulating neuronal apoptosis and autophagy. Both neuronal activity and neurotrophic factors are known to support neuronal survival.[34,48] For example, the survival of sympathetic neurons is dependent on nerve growth factor (NGF), the target-derived trophic factor, both *in vivo* and *in vitro*. After withdrawal of NGF, these neurons die *via* the mitochondrial pathway of apoptosis in which Bax is transferred from cytosol to mitochondria and causes subsequent release of cytochrome c, which leads to caspase activation[14,34] (the lower left pathway). This pathway is regulated by constitutive Ca^{2+} release from inositol 1,4,5-trisphosphate receptors (PtdIns(1,4,5)P3R) localized in the endoplasmic reticulum (ER).[56] If this step is altered, the apoptotic pathway may be compromised. The ER also regulates autophagy *via* the direct interaction between PtdIns(1,4,5)P3R and Beclin1, a BH3-only-protein essential for autophagy functions.[57] Autophagy signaling pathways crucially involve the activity of class III phosphatidylinositol 3-kinase (PtdIns3K). This enzyme can generate phosphatidylinositol-3 phosphate (PtdIns(3)P) which forms a complex with Beclin1 and p16 and activates subsequent autophagic processes (from the middle to right pathway).[58] Normally, the mammalian target of rapamycin (mTOR) acts to inactivate the autophagic mechanism.[59] Thus, there are two potential mechanisms for the involvement of elevated Ca^{2+} levels in regulating the autophagic mechanism. First, autophagy is induced by inhibition of mTOR. This is accomplished by calcium/calmodulin-dependent protein kinase kinase (CaMKK)-alpha/beta-mediated activation of AMP-dependent protein kinase (AMPK).[60] Second, the puncta formation of the Atg18-Atg2 complex in the autophagosomal membranes is stimulated by elevated intracellular Ca^{2+} (the right pathway).

autophagy activation. However, it should be noted that Ca^{2+} influx does not always lead to autophagy activation. In fact, membrane depolarization-induced Ca^{2+} influx suppresses both somal apoptosis and neuritic autophagy.[48] Only aberrant and persistent Ca^{2+} influx can lead to autophagy induction.[53,54]

Second, autophagy signaling pathways crucially involve the activity of class III phosphatidylinositol 3-kinase (PtdIns3K). This enzyme can generate phosphatidylinositol-3-phosphate (PtdIns3P), which then recruits the Atg18-Atg2 complex to autophagosomal membranes through a specific binding of Atg18 to PtdIns3P.[61] As the complex is localized to autophagosomal membranes, it forms dot-like structures called puncta. Recent study using an AMPK-deficient mouse model shows that this puncta formation is promoted by an increase in the level of cytoplasmic Ca^{2+} and suppressed by Ca^{2+} chelating agents.[60] Thus, it is likely that thapsigargin can enhance autophagy independently of AMPK activation in a Ca^{2+}-dependent manner.

Ca^{2+} present in the ER lumen is released through the ryanodine receptors or inositol 1,4,5-trisphosphate receptors (PtdIns(1,4,5)P3R). Diverse physiological processes, including the mitochondrial pathway of apoptosis, are controlled by constitutive Ca^{2+} fluxes from the ER to mitochondria.[56] PtdIns(1,4,5)P3R also regulates autophagy by its binding to Beclin1 through a direct protein-protein interaction.[57] Thus, PtdIns(1,4,5)P3R has a critical role in the cross-talk between apoptosis and autophagy, which is illustrated in Figure 1. Aberrant Ca^{2+} efflux from the ER to the cytoplasm *via* PtdIns(1,4,5)P3R might also stimulate autophagy through the above-mentioned calcium-dependent mechanism. Ca^{2+} influx from extracellular milieu may play a similar role in this cross-talk during degeneration of fully differentiated axons.

The reason as to why neurite degeneration occurs *via* autophagy rather than apoptosis is unclear. There is evidence suggesting that both apoptosis and autophagy proceeds in parallel in NGF-deprived cultured SCG neurons.[62] However, caspases are activated in the cell body, but not in neurites.[10,11] Interestingly, it has been shown that mitochondria in the neurites of SCG are incapable of buffering an increase in cytoplasmic Ca^{2+} levels mediated by caffeine-induced Ca^{2+} release from the ER.[63] Thus, the weaker Ca^{2+} buffering by

Figure 1. (***Continued***) The reason why autophagy is predominant over apoptosis in sympathetic neurites, but not in somata, in the degeneration caused by NGF deprivation is still not clear, but it is likely that the functional cross-talk between mitochondria and the ER may be compromised, as illustrated by the "?" on the left side of the figure. Alternatively, there may be scavenging mechanisms to prevent propagation of apoptotic signals as suggested.[64] As a result, the apoptotic mechanism begins to work, but is terminated halfway, whereas autophagy is continuously activated in the neurites under augmented Ca^{2+} release from the ER.

mitochondria in the neurites suggests that the cross-talk between mitochondria and the ER in the neurites might be compromised (the " ? " mark in Figure1). This notion is consistent with the finding that cultured SCG neurites are dependent on glycolysis rather than aerobic respiration.[58] As discussed earlier regarding the role of calcium in autophagy activation, an increased release of Ca^{2+} from the ER or from extracellular milieu in the neurites is likely to activate autophagy. Thus, autophagy may become the major pathway of degeneration, whereas the mitochondrial pathway of apoptosis begins to work at the same time, but is terminated halfway in the neurites (Figure1). Alternatively, there may be scavenging mechanisms to prevent propagation of apoptotic signals as suggested.[64] Further study will be required to reveal the mechanism by which different levels of intracellular Ca^{2+} may have opposite effects on the maintenance and activation of autophagy.[65]

AUTOPHAGY IN NEURITE DEGENERATION: A DOUBLE-EDGED SWORD WITH THERAPEUTIC POTENTIAL

Accumulating evidence shows that autophagy has been linked to various human neurodegenerative diseases. However, whether autophagy has a pro-death or pro-survival function in these diseases is controversial. Besides, the mechanism by which autophagy causes neurites to die is still unclear.

Studies with nematode *Caenorhabditis elegans* revealed an involvement of autophagy in GABAA receptor degradative trafficking, which is critical for synaptogenesis.[66] The mTOR signaling pathway, a negative regulator of autophagy as discussed previously, has been proposed to regulate dendritic protein synthesis in cultured hippocampal neurons.[67] Moreover, decreased activity of mTOR was found during neuroblastoma N2a cell differentiation. Thus, higher autophagic activity is related to differentiation. Importantly, inhibiting the autophagic process abrogated cell differentiation.[68] These observations strongly suggest the important role of autophagy in neuritic development and synaptogenesis.

The autophagy inhibitor 3-methyladenine efficiently suppressed neurite degeneration by protecting neurites from the loss of viability and mitochondrial function.[24] Furthermore, knocking down the key autophagy-related gene *atg7* or *beclin1* by RNA interference significantly delayed axonal and dendritic degeneration after NGF deprivation in cultured SCG neurons. Reduced expression of Atg7 also suppressed neurite fragmentation after transaction. Prolonged autophagy in cell soma and neurites has been demonstrated to contribute to cell death in N-methyl-D-aspartate (NMDA)-mediated excitotoxicity in cultured cerebellar granule neurons.[53] The autophagy inhibitor 3-methyladenine significantly reduced

NMDA-induced LC3 lipidation as well as the accumulation of autophagosomes. Application of autophagy inhibitor or knocking down the key autophagy gene *atg7* protected cell bodies and neurites against NMDA-induced excitotoxicity.[53] It has been shown that dendritic degeneration in cortical neurons expressing mutant LRRK2 is suppressed by inhibiting autophagy *via* LC3 phosphorylation.[27]

In vivo studies further provided evidence showing that autophagy inhibition plays an active role in preventing axonal degeneration. By utilizing a unique epifluorescence imaging approach, axonal degeneration in the optic nerve of the living rat was visualized *in vivo*.[54] Crush lesion of the optic nerve induced a rapid Ca^{2+} influx through calcium channels, which resulted in a secondary induction of autophagy that participates actively in axonal degradation, suggesting the close correlation between autophagy and intraaxonal Ca^{2+} level. Furthermore, in this experimental paradigm, autophagy inhibition by 3-methyladenine delayed axonal degeneration.

Given the fact that autophagy inhibition can protect neurites against degeneration initiated by different stress signals, it is possible that induction of autophagy has a pro-death function. However, the interpretation of activated autophagy in neurite degeneration is controversial. This is because autophagy may act as a cellular protection mechanism during axonal and dendritic degeneration.

Axonal dystrophy, a hallmark of axonopathy, has been described in the central and, less frequently, peripheral nervous systems in a variety of genetic and metabolic settings and with aging. It is characterized by focal dilations (swellings, varicosities) that interrupt the continuity of axons or terminal end bulbs (retraction or degeneration balls). This can occur in dysfunctional or degenerating neurons. Autophagy has important roles in the well-being of axons, as further discussed below.

Yue and his research group showed the accumulation of autophagosomes in dystrophic axons of Purkinje cells in *Lurcher* mice, a genetic model for excitotoxicity of neurons.[23] In these animals, a mutant glutamate receptor δ2 (GluR δ2Lc) is constitutively activated and triggers Purkinje cell soma death and axonal degeneration.[23] Moreover, phosphorylated microtubule-associated protein 1B (MAP1B) interacted with LC3 and accumulated in axonal dystrophic swellings of degenerating Purkinje cells. In addition, specific ablation of the essential autophagy gene, *atg7*, caused cell-autonomous, progressive dystrophy and degeneration of the axon terminals in Purkinje cells. This suggests the importance of neuronal autophagy for the maintenance of local homeostasis of axon terminals and protection against axonal degeneration.[69] Interestingly, axonal dystrophy of mutant Purkinje cells preceded the dendritic or spine atrophy, indicating that axon terminals are much more vulnerable to autophagy impairment than dendrites in this experimental paradigm. Consistent with these observations, another autophagy-related gene, *atg5*, also shows its critical role in maintaining the axon morphology and membrane structures. Loss of *atg5* leads to axonal swelling, followed by progressive neurodegeneration in Purkinje cells.[70]

A recent report characterized a functional role of FIP200 (FAK family-interacting protein of 200 kDa) in regulating neuronal cell homeostasis. This is a conserved protein currently identified as a potential mammalian counterpart of the yeast autophagy protein Atg17. Neural-specific deletion of FIP200 resulted in cell death and neurite degeneration in the cerebellum, providing novel insight into the essential role of autophagy in the prevention of neurodegenerative diseases.[71] These findings emphasize the indispensable role of neuronal autophagy in the maintenance of axonal and dendritic homeostasis, particularly in local membrane trafficking and turnover. Therefore, dysfunctional autophagy may contribute to axonopathy and subsequent neurodegeneration.

CONCLUSION

Autophagy contributes to the homeostasis and maintenance of the neurites. Future work is needed to elucidate its precise role in regulating axonal and dendritic degeneration. In addition, since local protein synthesis and degradation plays critical roles axonal regeneration,[72] it has been speculated that the autophagic process may also be involved in regulating neurite regeneration.

ACKNOWLEDGMENT

We would like to thank Dr. J. Chang, Department of Neurobiology and Developmental Science, University of Arkansas Medical System, for his invaluable comments.

REFERENCES

1. C. M. McCann, Q. T. Nguyen, H. Santo Neto and J. W. Lichtman, Rapid synapse elimination after postsynaptic protein synthesis inhibition *in vivo*, *J Neurosci* **27**: 6064–6067 (2007).
2. C. D. Deppmann, S. Mihalas, N. Sharma, B. E. Lonze, E. Niebur and D. D. Ginty, A model for neuronal competition during development, *Science* **320**: 369–373 (2008).
3. A. V. Waller, Experiments on the section of the glossopharyngeal and hypoglossal nerves of the frog, and observations on the alterations produced thereby in the structure of their primitive fibres, *Philos Trans Roy Soc Lond* **140**: 423–429 (1850).

4. J. W. Song, T. Misgeld, H. Kang, S. Knecht, J. Lu, Y. Cao, S. L. Cotman, D. L. Bishop and J. W. Lichtman, Lysosomal activity associated with developmental axon pruning, *J Neurosci* **28**: 8993–9001 (2008).
5. J. Wang and Z. He, NAD and axon degeneration: from the *Wlds* gene to neurochemistry, *Cell Adh Migr* **3**: 77–87 (2009).
6. M. P. Coleman and M. R. Freeman, Wallerian degeneration, Wld^s and Nmnat, *Annu Rev Neurosci* **33**: 245–267 (2010).
7. T. Araki, Y. Sasaki and J. Milbrandt, Increased nuclear NAD biosynthesis and SIRT1 activation prevent axonal degeneration, *Science* **305**: 1010–1013 (2004).
8. K. Suzuki and T. Koike, Mammalian Sir2-related protein (SIRT) 2-mediated modulation of resistance to axonal degeneration in slow Wallerian degeneration mice: a crucial role of tubulin deacetylation, *Neuroscience* **147**: 599–612 (2007).
9. K. Suzuki and T. Koike, Resveratrol abolishes resistance to axonal degeneration in slow Wallerian degeneration (Wlds) mice: activation of SIRT2, an NAD-dependent tubulin deacetylase, *Biochem Biophys Res Commun* **359**: 665–671 (2007).
10. J. T. Finn, M. Weil, F. Archer, R. Siman, A. Srinivasan and M. C. Raff, Evidence that Wallerian degeneration and localized axon degeneration induced by local neurotrophin deprivation do not involve caspases, *J Neurosci* **20**: 1333–1341 (2000).
11. K. Ikegami and T. Koike, Non-apoptotic neurite degeneration in apoptotic neuronal death: pivotal role of mitochondrial function in neurites, *Neuroscience* **122**: 617–626 (2003).
12. K. Ikegami, S. Kato and T. Koike, N-alpha-p-tosyl-L-lysine chloromethyl ketone (TLCK) suppresses neuritic degeneration caused by different experimental paradigms including *in vitro* Wallerian degeneration, *Brain Res* **1030**: 81–93 (2004).
13. M. C. Raff, A. V. Whitmore and J. T. Finn, Axonal self-destruction and neurodegeneration, *Science* **296**: 868–781 (2002).
14. T. Koike, Y. Yang, K. Suzuki and X. Zheng, Axon and dendrite degeneration: its mechanisms and protective experimental paradigms, *Neurochem Int* **52**: 751–760 (2008).
15. D. J. Klionsky and S. D. Emr, Autophagy as a regulated pathway of cellular degradation, *Science* **290**: 1717–1721 (2000).
16. Y. Kabeya, N. Mizushima, T. Ueno, A. Yamamoto, T. Kirisako, T. Noda, E. Kominami, Y. Ohsumi and T. Yoshimoro, LC3, a mammalian homologue of yeast Apg8p, is localized in autophagosome membranes after processing, *EMBO J* **19**: 5720–5728 (2000).
17. D. C. Rubinsztein, M. DiFiglia, N. Heintz, R. A. Nixon, Z. H. Qin, B. Ravikumar, L. Stefanis and A. Tolkovsky, Autophagy and its possible roles in nervous system diseases, damage and repair, *Autophagy* **1**: 11–22 (2005).
18. J. F. Ma, Y. Huang, S. D. Chen and G. Halliday, Immunohistochemical evidence for macroautophagy in neurons and endothelial cells in Alzheimer's disease, *Neuropathol Appl Neurobiol* **36**: 312–319 (2010).
19. W. H. Yu, A. M. Cuervo, A. Kumar, C. M. Peterhoff, S. D. Schmidt, J. H. Lee, P. S. Mohan, M. Mercken, M. R. Farmery, L. O. Tjernberg, Y. Jiang, K. Duff, Y. Uchiyama, J. Näslund, P. M. Mathews, A. M. Cataldo and R. A. Nixon,

Macroautophagy — a novel beta-amyloid peptide-generating pathway activated in Alzheimer's disease, *J Cell Biol* **171**: 87–98 (2005).

20. F. Adhami, G. Liao, Y. M. Morozov, A. Schloemer, V. J. Schmithorst, J. N. Lorenz, R. S. Dunn, C. V. Vorhees, M. Wills-Karp, J. L. Degen, R. J. Davis, N. Mizushima, P. Rakic, B. J. Dardzinski, S. K. Holland, F. R. Sharp and C. Y. Kuan, Cerebral ischemia-hypoxia induces intravascular coagulation and autophagy, *Am J Pathol* **169**: 566–583 (2006).
21. Y. Lai, R. W. Hickey, Y. Chen, H. Bayir, M.L. Sullivan, C. T. Chu, P. M. Kochanek, C. E. Dixon, L. W. Jenkins, S. H. Graham, S. C. Watkins and R. S. Clark, Autophagy is increased after traumatic brain injury in mice and is partially inhibited by the antioxidant gamma-glutamylcysteinyl ethyl ester, *J Cereb Blood Flow Metab* **28**: 540–550 (2008).
22. B. Sikorska, P. P. Liberski, P. Giraud, N. Kopp and P. Brown, Autophagy is a part of ultrastructural synaptic pathology in Greutzfeldt-Jakob disease: a brain biopsy study, *Int J Biochem Cell Biol* **36**: 2563–2573 (2004).
23. Q. J. Wang, Y. Ding, D. S. Kohtz, N. Mizushima, I. M. Cristea, M. P. Rout, B. T. Chait, Y. Zhong, N. Heintz and Z. Yue Z, Induction of autophagy in axonal dystrophy and degeneration, *J Neurosci* **26**: 8057–8068 (2006).
24. Y. Yang, K. Fukui, T. Koike and X. X. Zheng, Induction of autophagy in neurite degeneration of mouse superior cervical ganglion (SCG) neurons, *Eur J Neurosci* **26**: 2979–2988 (2007).
25. K. E. Larsen, E. A. Fon, T. G. Hastings, R. H. Edwards and D. Sulzer, Methamphetamine-induced degeneration of dopaminergic neurons involves autophagy and upregulation of dopamine synthesis, *J Neurosci* **22**: 8951–8960 (2002).
26. E. D. Plowey, S. J. Cherra 3rd, Y. J. Liu and C. T. Chu, Role of autophagy in G2019S-LRRK2-associated neurite shortening in differentiated SH-SY5Y cells, *J Neurochem* **105**: 1048–1056 (2008).
27. S. J. Cherra 3rd, S. M. Kulich, G. Uechi, M. Balasubramani, J. Mountzouris, B. W. Day and C. Chu, Regulation of the autophagy protein LC3 by phosphorylation, *J Cell Biol* **190**: 533–539 (2010).
28. K. E. White, V. J. Davies, V. E. Hogan, M. J. Piechota, P. P. Nichols, D. M. Turnbull and M. Votruba, OPA1 deficiency associated with increased autophagy in retinal ganglion cells in a murine model of dominant optic atrophy, *Invest Ophthalmol Vis Sci* **50**: 2567–2571 (2009).
29. H. K. Lee, Y. K. Shin, J. Jung, S. Y. Seo, S. Y. Baek and H. T. Park, Proteasome inhibition suppresses Schwann cell dedifferentiation *in vitro* and *in vivo*, *Glia* **57**: 1825–1834 (2009).
30. A. L. Contento, Y. Xiong and D. C. Bassham, Visualization of autophagy in Arabidopsis using the fluorescent dye monodansylcadaverine and a GFP-AtATG8e fusion protein, *Plant J* **42**: 598–608 (2005).
31. E. Fass, E. Shvets, I. Degani, K. Hirschberg and Z. Elazar, Microtubules support production of starvation-induced autophagosomes but not their targeting and fusion with lysosomes, *J Biol Chem* **281**: 36303–36316 (2006).

32. S. Kimura, T. Noda and T. Yoshimori, Dynein-dependent movement of autophagosomes mediates efficient encounters with lysosomes, *Cell Struct Funct* **33**: 109–122 (2008).
33. Z. Y. Yue, Regulation of neuronal autophagy in axon: implication of autophagy in axonal function and dysfunction/degeneration, *Autophagy* **3**: 139–141 (2007).
34. K. Katsumata, J. Nishiyama, T. Inoue, N. Mizushima, J. Takeda and M. Yuzaki, Dynein- and activity-dependent retrograde transport of autophagosomes in neuronal axons, *Autophagy* **6**: 1–8 (2010).
35. G. V. Putcha and E. M. Johnson Jr., Men are but worms: neuronal cell death in *C. elegans* and vertebrates, *Cell Death Differ* **11**: 38–48 (2004).
36. K. J. Ivins, E. T. Bui and C. W. Cotman, Beta-amyloid induces local neurite degeneration in cultured hippocampal neurons: evidence for neuritic apoptosis, *Neurobiol Dis* **5**: 365–378 (1998).
37. B. F. El-Khodor and R. E. Burke, Medial forebrain bundle axotomy during development induces apoptosis in dopamine neurons of the substantia nigra and activation of caspases in their degenerating axons, *J Comp Neurol* **452**: 65–79 (2002).
38. E. Touma, S. Kato, K. Fukui and T. Koike, Calpain-mediated cleavage of collapsin response mediator protein (CRMP)-2 during neurite degeneration in mice, *Eur J Neurosci* **26**: 3368–3381 (2007).
39. N. Arimura, C. Menager, Y. Fukata and K. Kaibuchi, Role of CRMP-2 in neuronal polarity, *J Neurobiol* **58**: 34–47 (2004).
40. T. Yoshimura, N. Arimura and K. Kaibuchi, Molecular mechanisms of axon specification and neuronal disorders, *Ann N Y Acad Sci* **1086**: 116–125 (2006).
41. V. Majava, N. Löytynoja, W. O. Chen, G. Lubec and P. Kursula, Crystal and solution structure, stability and post-translational modifications of collapsin response mediator protein 2, *FEBS J* **275**: 4583–4596 (2008).
42. Y. Kawano, T. Yoshimura, D. Tsuboi, S. Kawabata, T. Kaneko-Kawano, H. Shirataki, T. Takenawa and K. Kaibuchi, CRMP-2 is involved in kinesin-1-dependent transport of the Sra-1/WAVE1 complex and axon formation, *Mol Cell Biol* **25**: 9920–9935 (2005).
43. Z. Zhang, A. K. Ottens, S. Sadasivan, F. H. Kobeissy, T. Fang, R. L. Hayes and K. K. Wang, Calpain-mediated collapsin response mediator protein-1, -2, and -4 proteolysis after neurotoxic and traumatic brain injury, *J Neurotrauma* **24**: 460–472 (2007).
44. T. Czech, J. W. Yang, E. Csaszar, J. Kappler, C. Baumgartner and G. Lubec, Reduction of hippocampal collapsin response mediated protein-2 in patients with mesial temporal lobe epilepsy, *Neurochem Res* **29**: 2189–2196 (2004).
45. Y. Gu, N. Hamajima and Y. Ihara, Neurofibrillary tangle-associated collapsin response mediator protein-2 (CRMP-2) is highly phosphorylated on Thr-509, Ser-518, and Ser-522, *Biochemistry* **39**: 4267–4275 (2000).
46. T. M. Wishart, J. M. Paterson, D. M. Short, S. Meredith, K. A. Robertson, C. Sutherland, A. M. Cousin, M. B. Dutia and T. H. Gillingwater, Differential proteomics analysis of synaptic proteins identifies potential cellular targets and protein mediators of synaptic neuroprotection conferred by the slow Wallerian degeneration (Wlds) gene, *Mol Cell Proteom* **6**: 1318–1330 (2007).

47. D. P. Martin, A. Ito, K. Horigome, P. A. Lampe and E. M. Johnson Jr., Biochemical characterization of programmed cell death in NGF-deprived sympathetic neurons, *J Neurobiol* **23**: 1205–1220 (1992).
48. T. Koike and S. Tanaka, Evidence that nerve growth factor dependence of sympathetic neurons for survival *in vitro* may be determined by levels of cytoplasmic free Ca^{2+}, *Proc Natl Acad Sci USA* **88**: 3892–3896 (1991).
49. P. A. Lampe, E. B. Cornbrooks, A. Juhasz, E. M. Johnson Jr. and J. L. Franklin, Suppression of programmed neuronal death by a thapsigargin-induced Ca^{2+} influx, *J Neurobiol* **26**: 205–212 (1995).
50. V. Levick, H. Coffey and S. R. D'Mello, Opposing effects of thapsigargin on the survival of developing cerebellar granule neurons in culture, *Brain Res* **676**: 325–335 (1995).
51. M. Høyer-Hansen, L. Bastholm, P. Szyniarowski, M. Campanella, G. Szabadkai, T. Farkas, K. Bianchi, N. Fehrenbacher, F. Elling, R. Rizzuto, I. S. Mathiasen and M. Jäättelä, Control of macroautophagy by calcium, calmodulin-dependent kinase kinase-beta, and Bcl-2, *Mol Cell* **25**: 193–205 (2007).
52. K. Hernández-Fonseca and L. Massieu, Disruption of endoplasmic reticulum calcium stores is involved in neuronal death induced by glycolysis inhibition in cultured hippocampal neurons, *J Neurosci Res* **82**: 196–205 (2005).
53. S. Sadasivan, Z. Zhang, S. F. Larner, M. C. Liu, W. Zheng, F. H. Kobeissy, R. L. Hayes and K. K. Wang, Acute NMDA toxicity in cultured rat cerebellar granule neurons is accompanied by autophagy induction and late onset autophagic cell death phenotype, *BMC Neurosci* **18**: 11–21 (2010).
54. J. Knöferle, J. C. Koch, T. Ostendorf, U. Michel, V. Planchamp, P. Vutova, L. Tönges, C. Stadelmann, W. Brück, M. Bähr and P. Lingor, Mechanisms of acute axonal degeneration in the optic nerve *in vivo*, *Proc Natl Acad Sci USA* **107**: 6064–6069 (2010).
55. G. V. Ronnett, S. Ramamurthy, A. M. Kleman, L. E. Landree and S. Aja, AMPK in the brain: its roles in energy balance and neuroprotection, *J Neurochem* **109** (Suppl. 1): 17–23 (2009).
56. C. Cárdenas, R. A. Miller, I. Smith, T. Bui, J. Molgó, M. Müller, H. Vais, K. H. Cheung, J. Yang, I. Parker, C. B. Thompson, M. J. Birnbaum, K. R. Hallows and J. K. Foskett, Essential regulation of cell bioenergetics by constitutive InsP3 receptor Ca^{2+} transfer to mitochondria, *Cell* **142**: 270–283 (2010).
57. J. M. Vicencio, C. Ortiz, A. Criollo, A. W. Jones, O. Kepp, L. Galluzzi, N. Joza, I. Vitale, E. Morselli, M. Tailler, M. Castedo, M. C. Maiuri, J. Molgó, G. Szabadkai, S. Lavandero and G. Kroemer, The inositol 1,4,5-trisphosphate receptor regulates autophagy through its interaction with Beclin 1, *Cell Death Differ* **16**: 1006–1017 (2009).
58. T. Kawataki, K. Osafune, M. Suzuki and T. Koike, Neuronal maturation-associated resistance of neurite degeneration caused by trophic factor deprivation or microtubule-disrupting agents, *Brain Res* **1230**: 37–49 (2008).
59. H. C. Cheng, S. R. Kim, T. F. Oo, T. Kareva, O. Yarygina, M. Rzhetskaya, C. Wang, Z. Talloczy, K. Tanaka, M. Komatsu, K. Kobayashi, H. Okano, N. Kholodilov and R. E. Burke, Akt suppresses retrograde degeneration of dopaminergic axons by inhibition of macroautophagy, *J Neurosci* **31**: 2125–2135 (2011).

60. A. Grotemeier, S. Alers, S. G. Pfisterer, F. Paasch, M. Daubrawa, A. Dieterle, B. Viollet, S. Wesselborg, T. Proikas-Cezanne and B. Stork, AMPK-independent induction of autophagy by cytosolic Ca^{2+} increase, *Cell Signal* **22**: 914–925 (2010).
61. K. Obara, T. Sekito, K. Niimi and Y. Ohsumi, The Atg18-Atg2 complex is recruited to autophagic membranes *via* phosphatidylinositol 3-phosphate and exerts an essential function, *J Biol Chem* **283**: 2397–2380 (2008).
62. L. Xue, G. C. Fletcher and T. M. Tolkovsky, Autophagy is activated by apoptotic signalling in sympathetic neurons: an alternative mechanism of death execution, *Mol Cell Neurosci* **14**: 180–198 (1999).
63. L. Núñez, L. Senovilla, S. Sanz-Blasco, P. Chamero, M. T. Alonso, C. Villalobos and J. García-Sancho, Bioluminescence imaging of mitochondrial Ca^{2+} dynamics in soma and neurites of individual adult mouse sympathetic neurons, *J Physiol* **580**(Pt. 2): 385–395 (2007).
64. H. Bayir, A. A. Kapralov, J. Jiang, Z. Huang, Y. Y. Tyurina, V. A. Tyurin, Q. Zhao, N. A. Belikova, I. I. Vlasova, A. Maeda, J. Zhu, H. M. Na, P. G. Mastroberardino, L. J. Sparvero, A. A. Amoscato, C. T. Chu, J. T. Greenamyre and V. E. Kagan, Peroxidase mechanism of lipid-dependent cross-linking of synuclein with cytochrome C: protection against apoptosis versus delayed oxidative stress in Parkinson disease, *J Biol Chem.* **284**: 15951–15969 (2009).
65. H. G. Xia, L. Zhang, G. Chen, T. Zhang, J. Liu, M. Jin, X. Ma, D. Ma and J. Yuan, Control of basal autophagy by calpain1 mediated cleavage of ATG5, *Autophagy* **6**: 61–66 (2010).
66. A. M. Rowland, J. E. Richmond, J. G. Olsen, D. H. Hall and B. A. Bamber, Presynaptic terminals independently regulate synaptic clustering and autophagy of GABAA receptors in *Caenorhabditis elegans*, *J Neurosci* **26**: 1711–1720 (2006).
67. R. Gong, C. S. Park, N. R. Abbassi and S. J. Tang, Roles of glutamate receptors and the mammalian target of rapamycin (mTOR) signaling pathway in activity-dependent dendritic protein synthesis in hippocampal neurons, *J Biol Chem* **281**: 18802–18815 (2006).
68. M. Zeng and J. N. Zhou, Roles of autophagy and mTOR signaling in neuronal differentiation of mouse neuroblastoma cells, *Cell Signal* **20**: 659–665 (2008).
69. M. Komatsu, Q. J. Wang, G. R. Holstein, V. L. Friedrich Jr., J. Iwata, E. Kominami, B. T. Chait, K. Tanaka and Z. Yue, Essential role for autophagy protein Atg7 in the maintenance of axonal homeostasis and the prevention of axonal degeneration, *Proc Natl Acad Sci USA* **104**: 14489–14494 (2007).
70. J. Nishiyama, E. Miura, N. Mizushima, M. Watanabe and M. Yazaki, Aberrant membranes and double-membrane structures accumulate in the axons of Atg5-null Purkinje cells before neuronal death, *Autophagy* **3**: 591–596 (2007).
71. C. C. Liang, C. Wang, X. Peng, B. Gan and J. L. Guan, Neural-specific deletion of FIP200 leads to cerebellar degeneration caused by increased neuronal death and axon degeneration, *J Biol Chem* **285**: 3499–3509 (2010).
72. L. F. Gumy, C. L. Tan and J. W. Fawcett, The role of local protein synthesis and degradation in axon regeneration, *Exp Neurol* **223**: 28–37 (2010).

INDEX